How to use this book

Welcome to your OCR AS/A level Physics A student book. In this book you will find a number of features designed to support your learning.

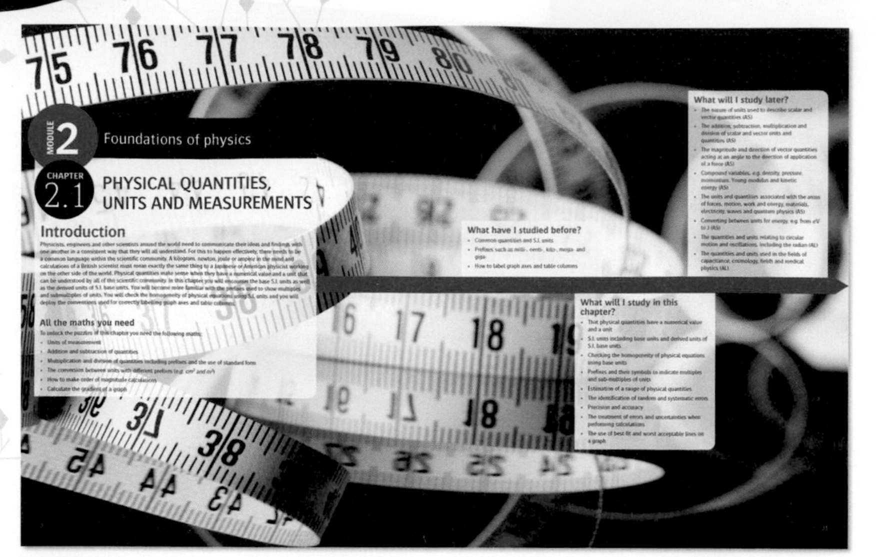

Chapter openers

Each chapter starts by setting the context for that chapter's learning:

- Links to other areas of Physics are shown, including previous knowledge that is built on in the chapter and future learning that you will cover later in your course.
- The **All the maths I need** checklist helps you to know what maths skills will be required.

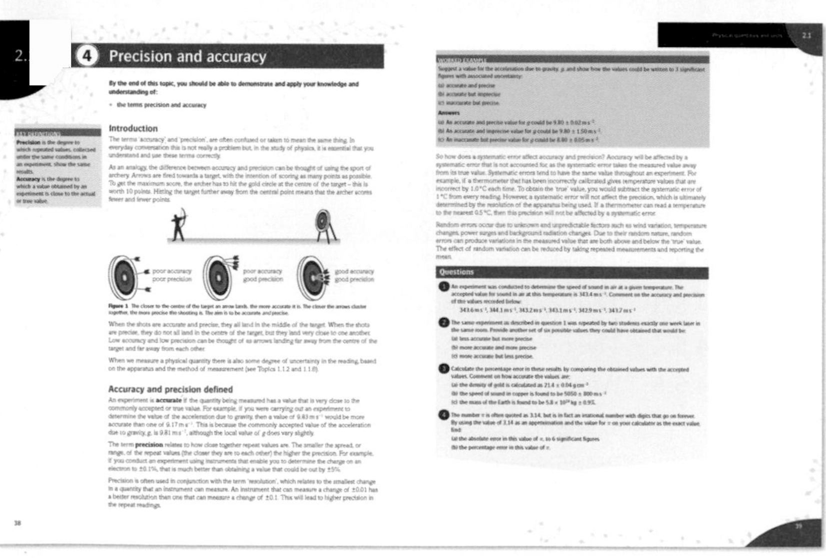

Main content

The main part of the chapter covers all of the points from the specification you need to learn. The text is supported by diagrams and photos that will help you understand the concepts.

Within each topic, you will find the following features:

- **Learning objectives** at the beginning of each topic highlight what you need to know and understand.
- **Key terms** are shown in bold and defined within the relevant topic for easy reference.
- **Worked examples** show you how to work through questions, and how your calculations should be set out.
- **Investigations** provide a summary of practical experiments that explore key concepts.
- **Learning tips** help you focus your learning and avoid common errors.
- **Did you know?** boxes feature interesting facts to help you remember the key concepts.

At the end of each topic, you will find **questions** that cover what you have just learned. You can use these questions to help you check whether you have understood what you have just read, and to identify anything that you need to look at again.

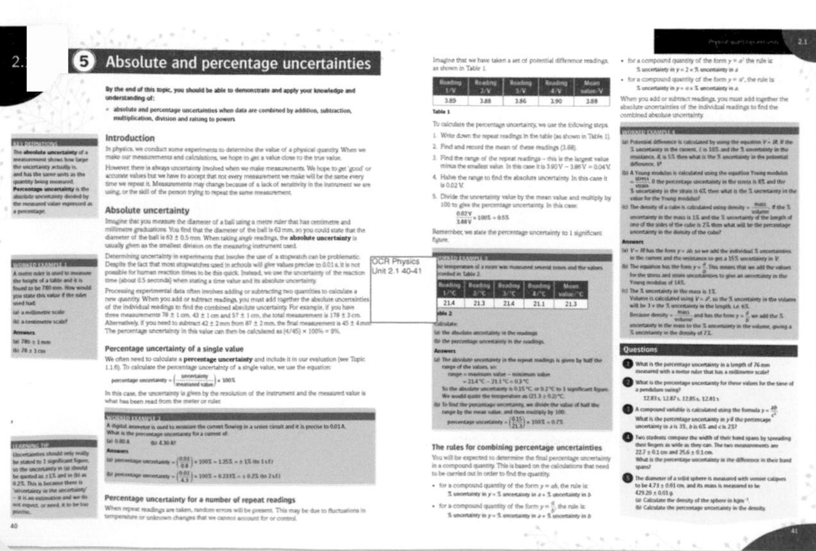

Thinking Bigger

At the end of each chapter there is an opportunity to read and work with real-life research and writing about science. These sections will help you to expand your knowledge and develop your own research and writing techniques. The questions and tasks will help you to apply your knowledge to new contexts and to bring together different aspects of your learning from across the whole course. The timeline at the bottom of the spread highlights which other chapters of your book the material relates to.

These spreads will give you opportunities to:

- read real-life material that's relevant to your course
- analyse how scientists write
- think critically and consider relevant issues
- develop your own writing
- understand how different aspects of your learning piece together.

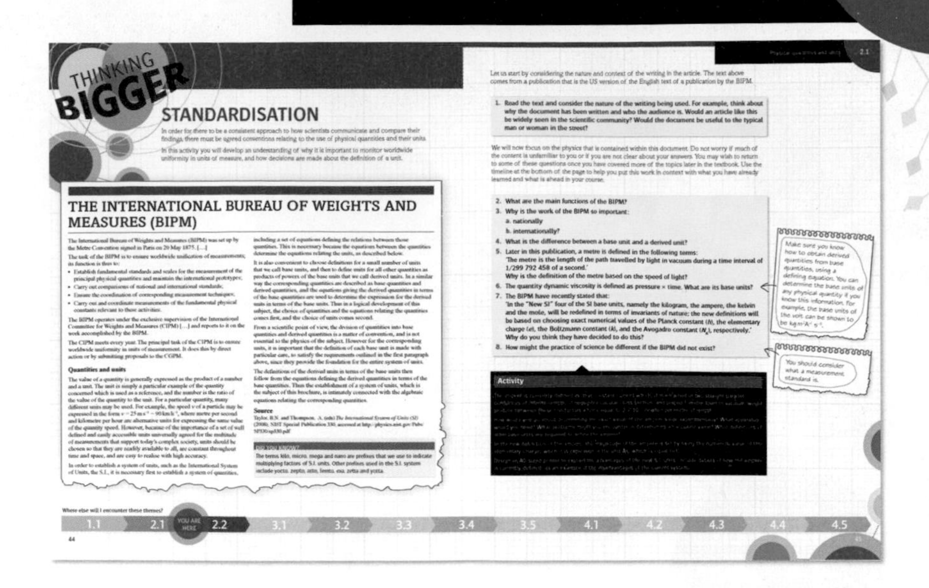

Practice questions

At the end of each chapter, there are **practice questions** to test how fully you have understood the learning.

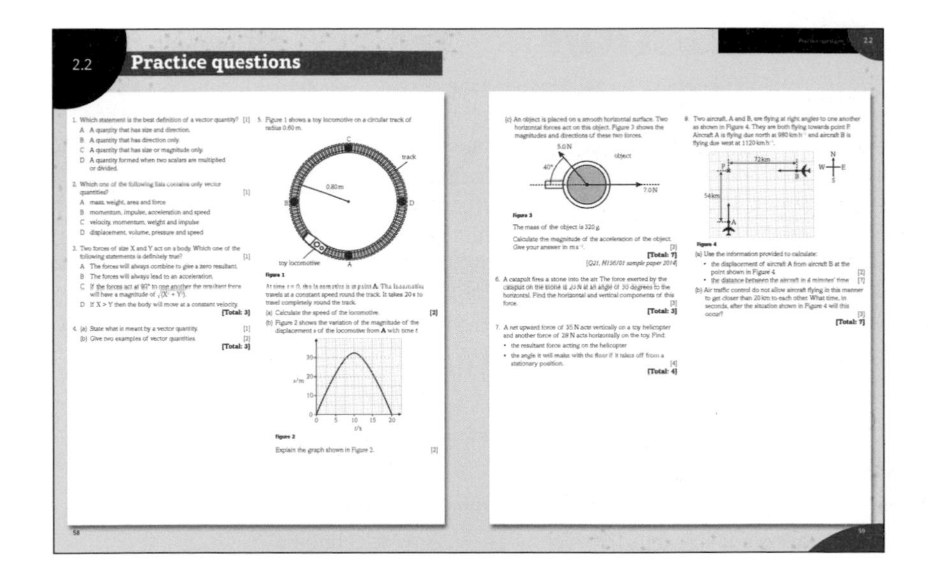

Getting the most from your ActiveBook

Your ActiveBook is the perfect way to personalise your learning as you progress through your OCR AS/A level Physics A course. You can:

- access your content online, anytime, anywhere
- use the inbuilt highlighting and annotation tools to personalise the content and make it really relevant to you
- search the content quickly.

Highlight tool

Use this to pick out key terms or topics so you are ready and prepared for revision.

Annotations tool

Use this to add your own notes, for example, links to your wider reading, such as websites or other files. Or make a note to remind yourself about work that you need to do.

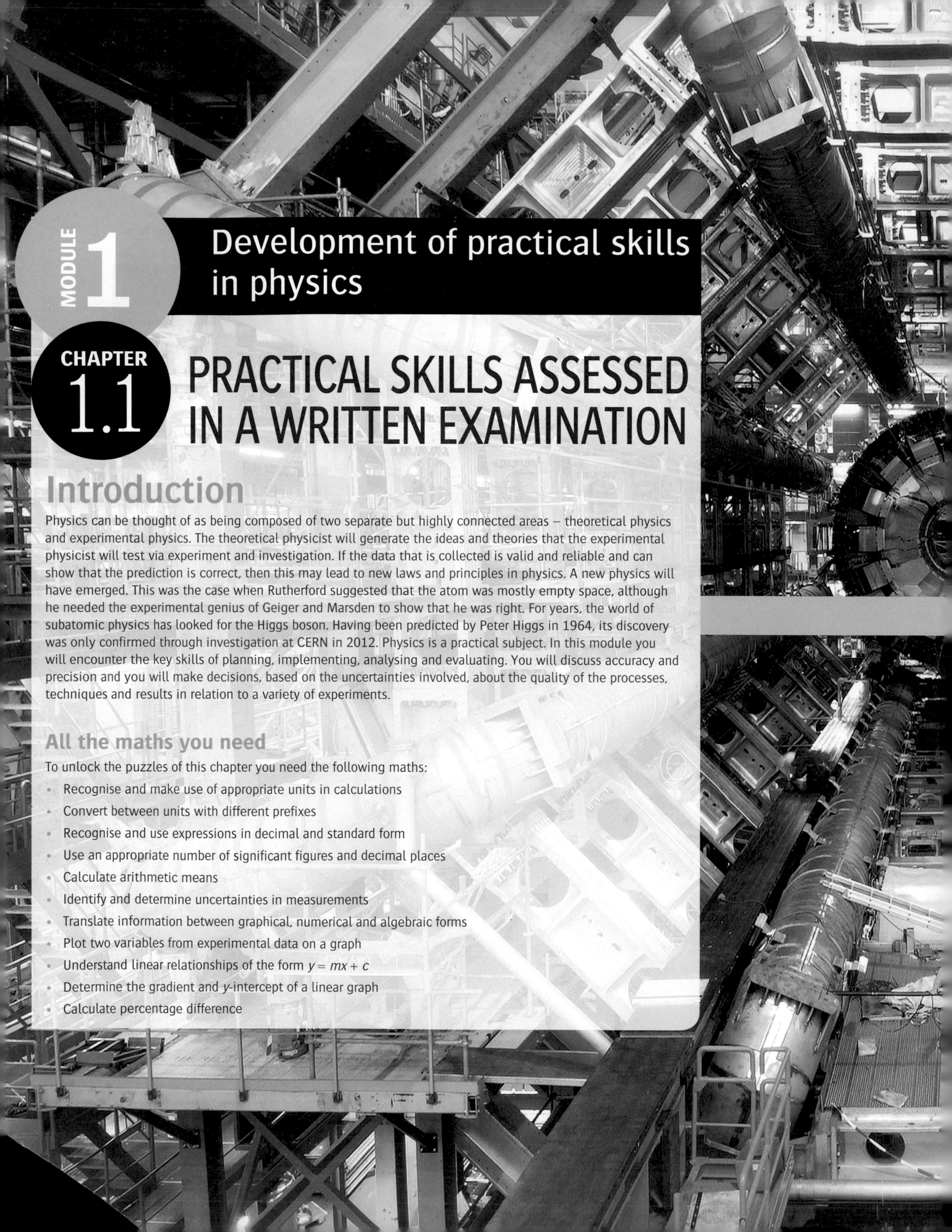

MODULE 1

Development of practical skills in physics

CHAPTER 1.1

PRACTICAL SKILLS ASSESSED IN A WRITTEN EXAMINATION

Introduction

Physics can be thought of as being composed of two separate but highly connected areas – theoretical physics and experimental physics. The theoretical physicist will generate the ideas and theories that the experimental physicist will test via experiment and investigation. If the data that is collected is valid and reliable and can show that the prediction is correct, then this may lead to new laws and principles in physics. A new physics will have emerged. This was the case when Rutherford suggested that the atom was mostly empty space, although he needed the experimental genius of Geiger and Marsden to show that he was right. For years, the world of subatomic physics has looked for the Higgs boson. Having been predicted by Peter Higgs in 1964, its discovery was only confirmed through investigation at CERN in 2012. Physics is a practical subject. In this module you will encounter the key skills of planning, implementing, analysing and evaluating. You will discuss accuracy and precision and you will make decisions, based on the uncertainties involved, about the quality of the processes, techniques and results in relation to a variety of experiments.

All the maths you need

To unlock the puzzles of this chapter you need the following maths:

- Recognise and make use of appropriate units in calculations
- Convert between units with different prefixes
- Recognise and use expressions in decimal and standard form
- Use an appropriate number of significant figures and decimal places
- Calculate arithmetic means
- Identify and determine uncertainties in measurements
- Translate information between graphical, numerical and algebraic forms
- Plot two variables from experimental data on a graph
- Understand linear relationships of the form $y = mx + c$
- Determine the gradient and y-intercept of a linear graph
- Calculate percentage difference

OCR

Oxford Cambridge and RSA

This is an OCR endorsed resource

OCR AS/A level Physics A

Second Edition

1

Mike O'Neill

ALWAYS LEARNING

PEARS

Published by Pearson Education Limited, 80 Strand, London, WC2R 0RL.

www.pearsonschoolsandfecolleges.co.uk

Text © Pearson Education Limited 2015
Edited by Sue Gardner, Tony Clappison and Gillian Lindsey
Designed by Elizabeth Arnoux for Pearson Education Limited
Typeset by Tech-Set Ltd, Gateshead
Original illustrations © Pearson Education Limited 2015
Illustrated by Tech-Set Ltd, Gateshead
Cover design by Juice Creative
Picture research by Alison Prior

The right of Mike O'Neill to be identified as author of this work has been asserted by him in accordance with the Copyright, Designs and Patents Act 1988.

First edition published 2008

This edition published 2015

18 17 16 15
10 9 8 7 6 5

British Library Cataloguing in Publication Data
A catalogue record for this book is available from the British Library

ISBN 978 1 447 99082 6

Printed in Slovakia by Neografia

This resource is endorsed by OCR for use with specification OCR Level 3 Advanced Subsidiary GCE in Physics A (H156) and OCR Level 3 Advanced GCE in Physics A (H556).

In order to gain OCR endorsement this resource has undergone an independent quality check. OCR has not paid for the production of this resource, nor does OCR receive any royalties from its sale. For more information about the endorsement process please visit the OCR website www.ocr.org.uk

Acknowledgements
The publisher would like to thank the following for their kind permission to reproduce their photographs:

(Key: b-bottom; c-centre; l-left; r-right; t-top)

Alamy Images: David J Green 154 (Resistors), Eileen Langsley Gymnastics 88, Image Broker 123, Jeff J Daly 122 (Metals), pbpgalleries 177 (F1 Car), Pjr Studio 95, Science Photos 219tc, 219tr; **Brian Otis & Babak Parviz:** 204; **Corbis:** Toronto Star / Zuma / ICON SMI 177 (Usain Bolt); **Fotolia.com:** algre 86 (Fig 1), Chris Brignell 177 (Hairdryer), Georgejmclittle 114, Joggie Botma 86 (Fig 2), Kalafoto 115, Kochtopf 86 (Fig 3), Kosziru 98tl, magik555 130tl, Pressmaster 98bl, SeanPavonePhoto 186-187, The Photos 177 (Aircraft), Youcanmore 153cl; **FotoLibra:** Marie-Laure Stone 153br; **Getty Images:** AFP / Johannes Eisele 112, Amos Chapple 223, Charlie Shuck / Digital Vision 122 (Chewing gum), Christopher Furlong 210 (Wave), Claudia Dewald 210 (Earthquake), Cormac McCreesh 111bl, Frank Robichon / AFP 109tl, IMAGEMORE Co, Ltd 104-105, John P Kelly 60-61, Martin Ruegner 130, Maxbmx 122 (Springs), Nick Veasey 210 (X-ray), Oxford Science Archive / Print Collector 109tr, Pierre Bourrier 122 (Skin), Popperfoto 222bl, Raymond Boyd 113, The Life Picture Collection 83, Universal Images Group 256t, 256c; **Imagemore Co., Ltd:** 177 (Power station); **Martyn F. Chillmaid:** 98cl, 173, 228, 241b; **Pearson Education Ltd:** Jules Selmes 154 (Thermistor), Studio 8 153cr; **PhotoDisc:** Jules Frazier 122 (Bands);

Science Photo Library Ltd: 222cl, Andrew Lambert Photography 219cr, 219br, 235, 241t, 244, 245, 268, Colin Cuthbert 130tc, Dr Jeremy Burgess 86 (Fig 4), Edward Kinsman 142cl, Erich Schrempp 208-209, GIPhotostock 122 (3 Springs), 154 (Meters), 238, Keith Kent 111cl, Library of Congress 243, Martyn F Chillmaid 71cl (Lightgate), 154 (LDR), 196, Maximilien Brice, CERN 8-9, NASA / MSFC 271, Patrice Loiez, CERN 254-255, Peter Aprahamaian / Sharples Stress Engineers Ltd 226cl, Ria Novosti 80-81, Royal Institute of Great Britain 235cr, Russell Knightley 220, Science Source 154 (LEDs), Sheila Terry 256b, Trevor Clifford Photography 154 (Variable resistor), TRL Ltd 142cr; **Shutterstock.com:** 2009fotofriends 246c, Andrea Danti 136-137, Andrey Pavlov 177 (Ant), ArtisticPhoto 164-165, Becky Stares 48-49, berna namoglu 30-31, Dudarev Mikhail 56, Epic Stock Media 211, James BO Insogna 150-151, Jan van der Hoeven 218, Michelangelus 120-121, Michelle D Millman 246b, Sapsiwai 147; **TopFoto:** ImageWorks / Peter Hvizdak 71cl (Accelerameter); **Veer / Corbis:** alexraths 210 (Ultrasound), chaoss 226tl, Perkmeup Imagery 227

Cover images: *Front:* **Shutterstock.com:** yuri4u80

We are grateful to the following for permission to reproduce copyright material:

Figures
Figure on page 100 adapted from *Beckham as a physicist?*, Physics Education doi:10.1088/0031-9120/36/1/301, 36 10 (Gren Ireson), © IOP Publishing. Reproduced with permission. All rights reserved and with permission from the author Professor Gren Ireson; Figure on page 160 adapted from *Physical principles of defibrillators*, Anaesthesia & Intensive Care Medicine (Michael T. Weisz 2009), Copyright © 2009 Elsevier Ltd. All rights reserved, Reprinted from Anaesthesia & Intensive Care Medicine, Vol 10, Michael T. Weisz, Physical principles of defibrillators, pp 367,369, Copyright (2009), with permission from Elsevier

Text
Article on page 44 adapted from *The International System of Units* Special Publication 330, National Institute of Standards and Technology (US Department of Commerce 2008), reproduced with permission of the BIPM, which retains full internationally protected copyright; Article on page 56 adapted from Discover Magazine (Maia Weinstock), Kalmbach Publishing Co, from Discover, February 05, 2004 © 2004 Discover Media. All rights reserved. Used by permission and protected by the Copyright Laws of the United States. The printing, copying, redistribution, or retransmission of this Content without express written permission is prohibited; Article on page 76 adapted from *Terminal velocity of skydivers*, Physics Education doi:10.1088/0031-9120/35/4/11, 35 281 (D C Agrawal), © IOP Publishing. Reproduced with permission. All rights reserved and with permission from the author Dr. Dulli Chandra Agrawal Professor of Physics; Article on page 100 adapted from *Beckham as a physicist?*, Physics Education doi:10.1088/0031-9120/36/1/301, 36 10 (Gren Ireson), © IOP Publishing. Reproduced with permission. All rights reserved and with permission from the author Professor Gren Ireson; Article on page 132 adapted from *Fire resistance of framed buildings* doi:10.1088/0031-9120/37/5/305, Physics Education 37 390 Vol 37 No 5 (Ian Burgess), © IOP Publishing. Reproduced with permission. All rights reserved and with permission from the author Professor Ian Burgess, Professor of Structural Engineering; Article on page 146 adapted from *Car safety features that avoid accidents*, http://www.cbsnews.com/news/car-safety-features-that-avoid-accidents/, Jerry Edgerton, Moneywatch 2/10/2013, CBS News; Article on page 160 adapted from *Physical principles of defibrillators*, Anaesthesia & Intensive Care Medicine (Michael T. Weisz 2009), Copyright © 2009 Elsevier Ltd. All rights reserved, Reprinted from Anaesthesia & Intensive Care Medicine, Vol 10, Michael T. Weisz, Physical principles of defibrillators, pp 367,369, Copyright (2009), with permission from Elsevier; Article on page 182 adapted from *An introduction to electrical resistivity in geophysics*, American Journal of Physics, 69, 943 -952. Reproduced with permission from the author Dr Rhett Herman; Article on page 204 from *Trees could be the ultimate in green power* (Colin Barras), http://www.newscientist.com/article/dn17767-trees-could-be-the-ultimate-in-green-power.html#.VOMFe-asXkg; Article on page 250 adapted from *What Is Seismology?* http://www.geo.mtu.edu/UPSeis/waves.html, ©2007 Michigan Technological University, 2007, UPSeis, an educational site for budding seismologists, Houghton, MI USA, viewed 20th January 2015, http://www.geo.mtu.edu/UPSeis/waves.html; Article on page 270 adapted from *My Way: Doppler shift and energy transfer to a solar sail* doi:10.1088/0031-9120/37/5/401, Physics Education, 37 No 5 (Warren Hirsch and Mark Kobrak 2002), © IOP Publishing. Reproduced with permission. All rights reserved and with permission of the authors Mark Kobrak and Warren Hirsch; Article on page 116 from *This is the next generation of renewable energy technologies*, The Telegraph, 13/12/2014 (Rebecca Burn-Callander), copyright © Telegraph Media Group Limited

Every effort has been made to contact copyright holders of material reproduced in this book. Any omissions will be rectified in subsequent printings if notice is given to the publishers.

Contents

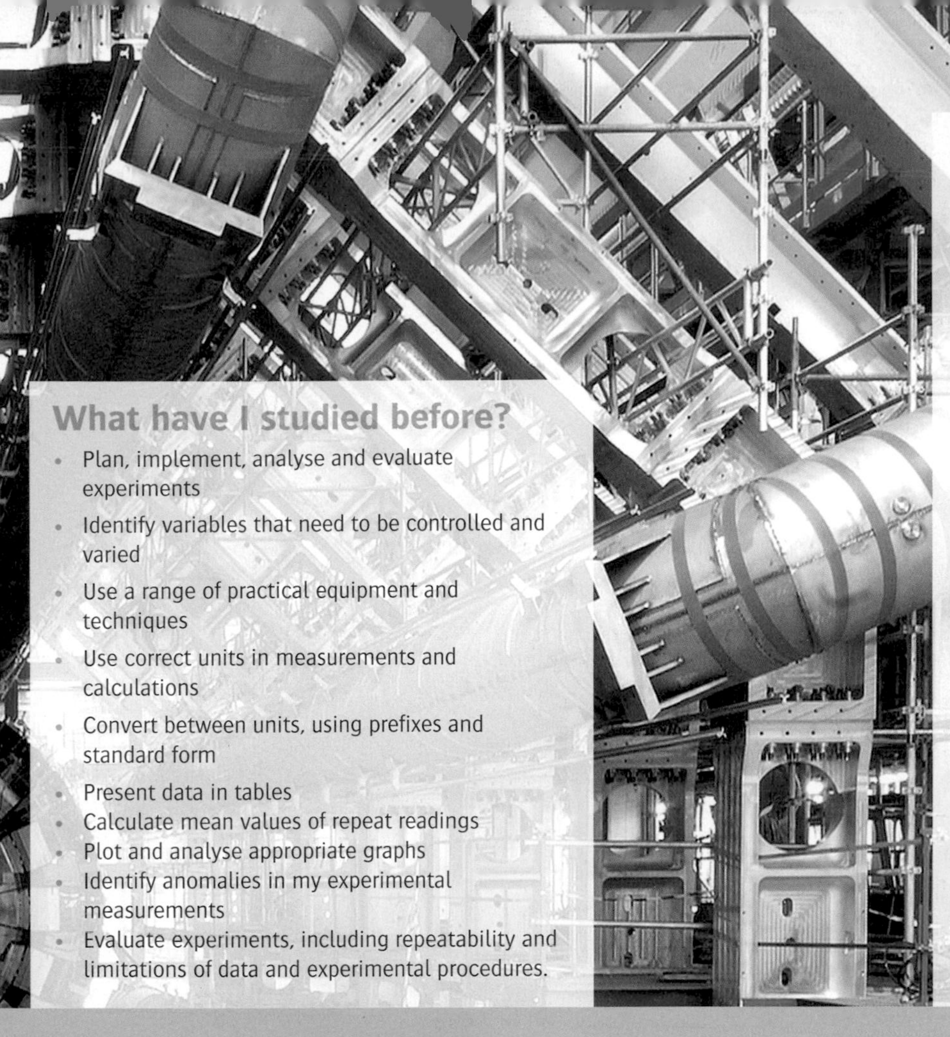

What have I studied before?

- Plan, implement, analyse and evaluate experiments
- Identify variables that need to be controlled and varied
- Use a range of practical equipment and techniques
- Use correct units in measurements and calculations
- Convert between units, using prefixes and standard form
- Present data in tables
- Calculate mean values of repeat readings
- Plot and analyse appropriate graphs
- Identify anomalies in my experimental measurements
- Evaluate experiments, including repeatability and limitations of data and experimental procedures.

What will I study later?

- Precision, accuracy and uncertainties in measurements and data (AS)
- The identification of systematic errors and random uncertainties and methods used to deal with them (AS)
- Rules for combining absolute and percentage uncertainties when performing a variety of calculations (AS)
- Graphical treatment of errors and uncertainties, including line of best fit and worst line (AS)
- Use stopwatch and light gates for accurate timing during investigations (AS)
- Use calipers and micrometers for the accurate measurement of small distances (AS)
- Use a signal generator and oscilloscope to investigate waves (AS)
- Use a laser or light source to investigate characteristics of light, including interference and diffraction (AS)
- Use ICT such as computer modelling, or a data logger with a variety of sensors to collect data, or use software to process data (AS)
- Use ionising radiation, including detectors (AL)
- Use error bars effectively in the plotting of graphs to help determine the sizes of errors (AL)

What will I study in this chapter?

- Experimental design, including how to solve problems in a practical context
- Evaluate an experimental method to determine that it is appropriate to meet expected outcomes
- Use a wide range of practical apparatus and techniques correctly
- Present and use observations, units and data in an appropriate format
- The conventions used for labelling graph axes and table columns
- Process, analyse and interpret qualitative and quantitative experimental results
- Use appropriate mathematical techniques to analyse quantitative data
- Use an appropriate number of significant figures and decimal places
- Plot and interpret suitable graphs from experimental results
- The limitations of experimental procedures, including anomalies and uncertainties in measurements

Planning and experimental design

By the end of this topic, you should be able to demonstrate and apply your knowledge and understanding of:

* **experimental design, including how to solve problems set in a practical context**
* **identification of variables that must be controlled, where appropriate**
* **evaluation that an experimental method is appropriate to meet the expected outcomes**

The importance of good design

The Nobel prize-winning physicist Richard Feynman once stated that, 'It doesn't matter how beautiful your theory is, it doesn't matter how smart you are. If it doesn't agree with experiment, it's wrong.'

What might seem to be a quite direct and straight comment is, essentially, one of the most important truths in physics. If we cannot obtain qualitative and quantitative findings showing that our theory or hypothesis is correct, then we cannot be sure the theory is correct.

This means that when planning an experiment, you need to focus on a number of important ingredients that will give your experiment the best chance of obtaining meaningful data from which you can draw meaningful, accurate and honest conclusions.

These ingredients include:

- selecting suitable apparatus and equipment
- planning the use of suitable techniques to collect **valid** data
- using appropriate scientific knowledge from the specification and in context with the practical work being undertaken to form a hypothesis that can be tested
- identifying variables that must be controlled
- evaluating if experimental method is appropriate for reaching the expected outcomes.

LEARNING TIP

Data is **valid** if the experimental design allows the question to be answered (consider, for example, what you think your measurements tell you about the experimental task) and the procedures and instruments are suitable (for example, the control variables are kept constant and in your judgement the measurement uncertainties are within acceptable limits). A valid conclusion must be supported by valid data, collected under the specified conditions.

Use of appropriate apparatus and techniques

Evidently, to give ourselves the best chance of obtaining valid data for subsequent analysis and for drawing conclusions, we need good techniques and suitable apparatus. Table 1 shows some of the experiments you might carry out during the first year of your A level Physics course. The table shows the title of the experiment, the key theory behind it, the apparatus you might consider using and the reasons for why this might be the case. It is always a good idea to think of these things before embarking on an investigation – failing to plan is planning to fail.

A student is trying to determine the relationship between the time period of a pendulum (how long it takes to make one complete oscillation) and the length of the pendulum. He uses the apparatus shown in Figure 1.

The student has realised a number of important issues in relation to the experiment:

- the need for the length of pendulum to be determined accurately, using a metre rule, from the point of suspension to the centre of the pendulum bob
- the need for a stopwatch to be used with times precise to 0.01 s
- the need to determine his own reaction time so that he can evaluate how much this affects the readings he obtains

The student also needs to consider the following points:

- A **fiducial mark** will be used as a reference point. This makes it easier to see when the bob passes a clear, sharply defined reference point.
- He will time for 20 oscillations because the time for just 1 oscillation could lead to major errors in his stopwatch readings and high percentage uncertainties in his data.

KEY DEFINITION

A **fiducial mark**, in this example, is an object placed in the field of view for the observer to use as a point of reference.

Investigation	Key theory and objectives	Apparatus	Reasoning
Determining the resistivity of a sample of metal wire	The resistivity of a wire is independent of the wire's geometry and is a fixed quantity for a sample, determined from the equation $\rho = RA/l$. Obtaining values for the resistance of wires of different lengths will allow us to plot a graph of R against l. If we also know the cross-sectional area of the wire then we can determine the value for the resistivity, ρ.	• power pack • ammeter • voltmeter • switch • metre rule • micrometer • crocodile clips • rheostat	Potential difference and current readings are needed so that the resistance can be calculated for each length. The wire will have a very small diameter, so a micrometer will be needed to measure it so that the wire's cross-sectional area can be calculated. A switch will be used so that heating effects are minimised. Crocodile clips will give good contact at a well-defined point on the wire so that the length recorded on the tight wire will be accurate. The rheostat will ensure that the current flowing is of an appropriate size.
Determining the Young modulus of a sample of metal wire	The Young modulus of a material is defined as $\frac{\text{stress}}{\text{strain}}$ and has a constant value for a given material. By determining the value of the Young modulus for a number of wires made of the same material of different dimensions, we can determine whether this is true.	• Searle's apparatus • micrometer • masses • electronic balance • long ruler • eye protection	Searle's apparatus allows the test wire to be taut and weights to be added to extend a wire. The Vernier scale on the apparatus allows the extension to be measured precisely using the spirit level arrangement. The micrometer is needed to measure the diameter of the wire precisely because its cross-sectional area has to be determined for the stress to be calculated. A long ruler has to be used to measure the original length of the wire precisely, which needs to be in excess of 2 m.
Determining the terminal velocity of a body falling in a viscous liquid	Based on a knowledge of drag and the resistance forces encountered in fluids, a falling body should reach its terminal velocity in a viscous liquid relatively quickly. Furthermore, the viscosity of the liquid could also be determined based on the measurements made relating to the ball's velocity, its dimensions and the acceleration due to gravity.	• measuring cylinder • beaker containing a viscous liquid • access to a balance and micrometre screw gauge • tube filled with the viscous liquid • elastic bands or other method of marking distances along tube • steel ball bearings • magnet • metre rule • stopwatch • paper towels	Steel ball bearings are dropped into the long tube containing the viscous liquid and allowed to fall some distance through it. The elastic bands should be wrapped around the tube at equal distance intervals. When these equal distance intervals correspond to equal spacing between the elastic bands then we know that the ball must be moving at its terminal velocity. The rule enables us to measure distance precisely to the nearest millimetre and the stopwatch will allow us to measure to the nearest 0.01 s, although the reaction time of the experimenter will determine the uncertainty in the readings, not the resolution of the stopwatch.

Table 1 Examples of things to consider when planning an effective investigation.

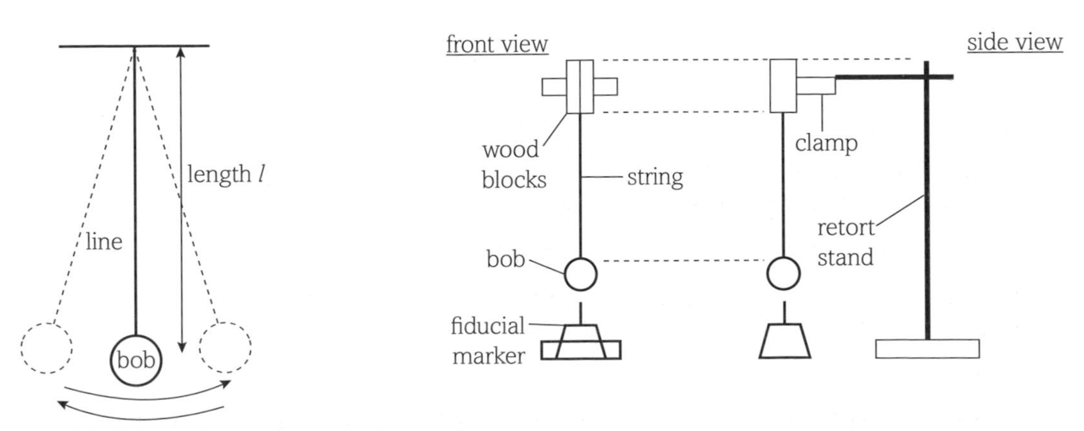

Figure 1 A typical set up for an investigation into how the time period of a simple pendulum is affected by its length.

Range of measurements

You should select a range of at least six values for the independent variable, and test each value three times. For example, the student could take a set of 9 readings for the length of the pendulum, equally spaced between 0.100 m and 1.000 m. This number of readings should allow him to identify a trend, although he may decide to take more measurements.

Identifying the variables that must be controlled

When investigating the relationship between any two variables, all others need to be controlled because otherwise it would not be possible to state that they did not affect the results.

- If he is investigating the relationship between time period and length, then everything else that can change must be controlled and kept the same for each reading taken. The following need to be controlled:
  - the mass of the pendulum bob – regardless of whether or not the mass has an effect on the time period, it must be kept constant in this experiment
  - the angle from which the pendulum bob is released has to be the same, preferably small too, based on his initial research
  - the pendulum should also only be allowed to swing in a two-dimensional plane, backwards and forwards only, without changing into a conical swing. This may also have an impact on the results, so if it happens, he will need to stop the run and start that trial again.

You cannot really control the room temperature and the absence of any draughts but could mention these as possible factors.

LEARNING TIP

In order to evaluate your method you must understand that there is always an uncertainty in any measurement and that the equipment and the way it is used can result in both random errors and systematic errors (see Topics 2.1.3 and 2.1.4). Every instrument also has a limit to its resolution (see Topic 1.1.2).

Evaluating the experimental method

Having planned the procedure, the student may decide that the equipment, and the techniques he plans to use, are not as good as they could be and that this might have an adverse effect on the quality of the data that he obtains. By carefully evaluating the proposed method, he might decide to change the investigation set up to enable the collection of more accurate, precise and valid data.

Reasons for this new arrangement might include:

- a more precise measurement of the time period using a laser beam and light-dependent resistor so that the light beam is interrupted periodically by the pendulum bob.
- a higher, but constant, mass to avoid any conical motion of the pendulum.

He may modify what he does after carrying out a trial experiment, for example to take more readings in a certain range if there is a large change between two values.

Questions

1. A student conducts an investigation to determine the resistance of a length of wire. She sets up a series circuit with a cell, ammeter, voltmeter and length of nichrome wire.
   (a) She measures the length of the wire with a ruler.
   (b) She takes one reading for the current in the wire and one reading for the potential difference across the wire for lengths of 15 cm, 25.5 cm, 50 cm and 80 cm.
   (c) She records the data from an analogue voltmeter which measures from 1 to 20 V in 1 V intervals. She records the data from a digital ammeter which records the current to the nearest 0.01 A.
   (d) She calculates the resistance and gives her answer to 3 significant figures.
   Explain the ways in which this investigation could be improved.

2. A student wants to conduct an experiment to determine which liquids are best to use in a domestic central heating system. What steps should she take to ensure that the variables that must be controlled did not affect the outcome?

3. How would you plan an experiment to determine the average speed of cars in your street? Consider the following:
   (a) equations and terms you need to consider and data you would collect
   (b) equipment you would use, with an explanation as to why
   (c) variables that would need to be controlled
   (d) how you would ensure that your experimental method is appropriate to meet the expected outcomes.

4. A student carried out an experiment to measure the acceleration of free fall. He used an analogue clock with a second hand to measure the time taken for a card to fall a measured distance from the roof of a house to the ground. Give two reasons why this is not a valid procedure to find the acceleration of free fall.

Implementation

By the end of this topic, you should be able to demonstrate and apply your knowledge and understanding of:

- how to use a wide range of practical apparatus and techniques correctly
- appropriate units for measurements
- presenting observations and data in an appropriate format
- the conventions used for labelling table columns

Using a wide range of practical apparatus correctly

Having planned your experiment, the quality of the data that you obtain will be largely dependent on how you implement your plan, collect your data and use your apparatus. Well-chosen instruments, a systematic method and a great degree of care will ultimately result in more accurate, precise and valid results.

During your course you will use a variety of practical equipment in a number of different contexts. You will probably already be familiar with much of this apparatus. A list of the apparatus, and key information relating to its use, is shown in Table 1.

Name of apparatus	Function of apparatus	When it is to be used	Typical range and resolution
ammeter	To measure the electric current flowing through a component	To measure the current through a bulb, diode or resistor when investigating I–V characteristics	0.00 A to 10.00 A ±0.01 A
voltmeter	To measure the potential difference across a component	To calculate the potential difference across a component when determining its electrical resistance	0.00 V to 20.00 V ±0.01 V
metre ruler	To measure the length of an object of moderate length	To measure the length and extension of a spring when trying to find a value for its spring constant, k	0.000 m to 1.000 m ±0.001 m
vernier calipers	To measure short lengths	To measure internal and external lengths such as the internal diameter of a tube	0.00 cm to 12.00 cm ±0.01 cm
micrometer screw gauge	To measure small values of width, thickness or diameter	To measure the diameter of a metal wire to determine its cross-sectional area in m^2. This is done when investigating electric circuits or stress–strain curves	0.00 mm to 25.00 mm ±0.01 mm
stopwatch	To measure periods of time	To measure the time taken for a car to travel a distance along a road to determine its average speed or velocity	0.00 s to 9999.99 s ±0.01 s
oscilloscope	To display waves and measure their frequencies	To determine the amplitude or time period of a sound wave from a signal generator or tuning fork. From this, frequencies, wave speeds and wavelengths can be determined	0 to 50 kHz ±1 Hz
laser	To provide a monochromatic source of light	To provide a light source for investigating diffraction using Young's double slits or a diffraction grating	Wavelengths range from 450 nm to 650 nm

Table 1 Apparatus you will meet in the first year of your A level.

Selecting suitable equipment

When planning and implementing a practical investigation, it is very important to decide which apparatus is most suitable and fit for purpose. For example:

- You would not use a metre rule to determine the diameter of a thin sample of nichrome wire because the resolution is not fine enough.
- You would not use a voltmeter with a range of 0 V to 5 V to measure the potential difference across a component connected to a 12 V cell.
- You may decide that digital ammeters and voltmeters are a more suitable, accurate and precise way of determining values, as opposed to their analogue counterparts. Analogue instruments can lead to 'parallax errors' that distort the true value (Figure 1).

Figure 1 Parallax error with an analogue meter occurs when a scale reading is not read from directly in front of the needle. The image on the right shows how the same meter would appear with the experimenter positioned to the right of the scale. This is remedied in some analogue meters by having a mirror behind the needle – you know the reading is correct when the needle and its image in the mirror coincide.

- You may decide that using a switch in an electric circuit would be a good choice so that temperature increases caused by a constantly present current can be minimised and controlled.
- A faulty stopwatch that doesn't respond to being pressed may be replaced with another.
- When performing an investigation into standing waves in a tube open at one end, the frequency used to determine the wavelength of sound in air might be chosen to have a low value to reduce the percentage error in any subsequent analysis and calculations.

Using a vernier scale

A metre rule can only measure to ± 0.5 mm, by estimating to half of the smallest scale division. Vernier calipers enable lengths to be measured to one tenth of a division on a main scale (so 0.1 mm for a scale marked in millimetres). The vernier scale has 10 scale divisions for every 9 main scale divisions. As the vernier moves along the main scale, you find the mark on the sliding scale which lines up exactly with one of the marks on the fixed scale. If the diameter on the main scale is between 7 and 8 mm and the line on the vernier scale lines up with the third line along the main scale, the measurement is 7.3 mm.

Vernier scales are also used in some other instruments, such as the micrometer screw gauge (Figure 2). The barrel of a micrometer screw gauge has a circular scale with 50 equally spaced divisions. For each complete rotation of the circular scale, it travels 0.5 mm along the millimetre scale. So the resolution is 1/50th of 0.5 mm, or 0.01 mm.

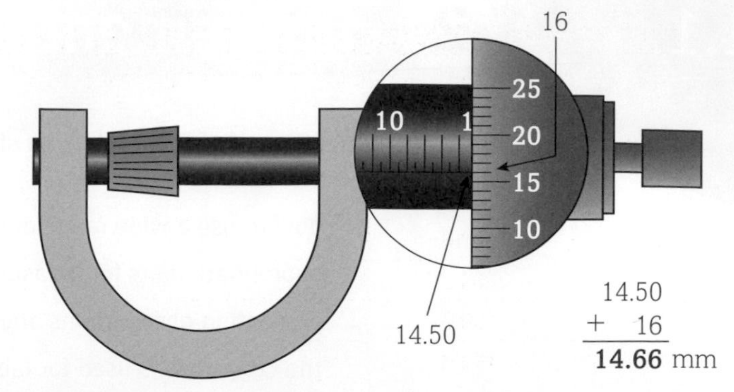

Figure 2 A micrometer screw gauge is used to measure very small widths, thicknesses or diameters, especially of thin wires. The thickness of the bung is found by adding the value of 14.50 mm on the barrel to the value of 0.16 mm on the sleeve, giving a value of 14.66 mm, which is precise to ±0.01 mm.

WORKED EXAMPLE 1

What apparatus would you use to measure the following quantities precisely?

(a) the length of a long piece (>80 cm) of metal wire
(b) the thickness of a piece of A4 card
(c) the current flowing through an electrical insulator
(d) the time taken for a car to travel 20 m.

Answers

(a) A metre ruler or a tape measure with m, cm and mm markings.
(b) A micrometer – the paper would be a fraction of a millimetre thick.
(c) A milliammeter or a microammeter – best to start with one that reads from 0 to 10 A just in case a short circuit caused the ammeter chosen to be damaged.
(d) A stopwatch, precise to 0.01 s.

Careful measurement

When making measurements in physics practical work, it is important to consider that every measurement has an uncertainty, called the **absolute uncertainty**, related to the resolution of the method of measurement. For example, if you are using a micrometer to measure the diameter of a wire, the uncertainty in the measurement is 0.01 mm. There may also be random or systematic errors associated with the instrument or how it is used (see Topic 2.1.3). For example, you should always check the reading is exactly zero with the jaws closed. If the zero reading is –0.02 cm due to wear or misuse, then all measurements will be systematically too small and a zero correction will be needed.

To minimise parallax error when reading scales you should always make sure your line of sight is at right angles to the scale (Figure 1).

Presenting observations and data

A sensible range of readings should be taken (see Topic 1.1.1), and the values should all be stated to an appropriate number of significant figures or decimal places. The columns in any table should have both a quantity and a unit in their heading.

All the readings you take should be plotted including those you suspect may be anomalous. Of course if you take repeat readings then it would be the mean value that you would plot.

WORKED EXAMPLE 2

Explain why the collection and presentation of data in Table 2 is better than that in Table 3.

Distance/m	Time/s
20.0	
40.0	
60.0	
80.0	
100.0	
120.0	
140.0	

Table 2

Metres	*T*
22	
28.5	
60	
72.22	
134.6	

Table 3

Answers

Possible answers include:

- There are enough (6 or more) readings in Table 2, but too few (5 readings) in Table 3.
- The data in Table 2 is given to a consistent and sensible number of decimal places, whereas in Table 3 there is no consistency.
- The data in Table 2 is equally spaced over the range. The data in Table 3 has a similar range, but the data is not equally spaced leading to results that are too close together in places, and widely spaced in others.
- The columns in Table 2 have a quantity and a unit clearly stated, whereas it is unclear what has been recorded in Table 3 and there is inconsistent use of quantities and units.

LEARNING TIP

When labelling table columns, you should separate the name of the quantity from its unit by a solidus (/), as shown in Table 2.

Questions

1 You are investigating the change in temperature of a liquid that is being heated.

(a) What apparatus would you use to investigate the change in temperature of a liquid?

(b) What readings would you take and how often would you take them?

2 The temperature of a hot drink that was cooling down on a laboratory bench was taken every minute using a mercury-in-glass thermometer and the data recorded in a table. What are the problems with the data that has been provided?

Temperature	Time
87.00	60 s
74.50	120 s
72.0	3 minutes
68.25	240 s
44.00	300

Table 4 Temperature and times for a cooling drink.

3 What is wrong with the following statements in relation to making measurements?

(a) 'I measured the thickness of the string with a 30 cm ruler because it has a millimetre scale.'

(b) 'The micrometer reading told me that the wire has a thickness of 1.5 mm.'

(c) 'My meter ruler has a millimetre scale and the desk's length was 34 cm.'

(d) 'I took four readings of the voltage and current and plotted them on a graph.'

(e) 'The potential difference values I collected with my voltmeter were 3.45 V, 3.4 V, 3 V and 3.6 V.'

4 What would be suitable units to use to measure the following quantities?

(a) the length of a room

(b) the volume of a cylinder

(c) the thickness of a metal wire

(d) the mass of a metal block

(e) the density of a metal block

(f) the wavelength of a visible light wave

(g) the frequency of a sound wave

(h) the amplitude of a water wave in a ripple tank.

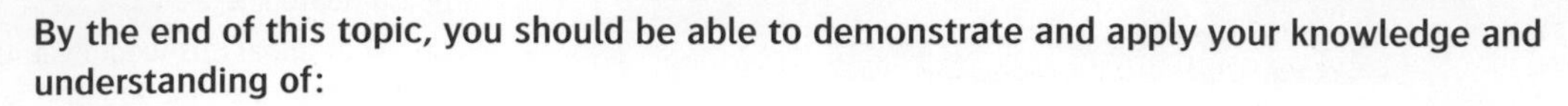

3 Analysing data

By the end of this topic, you should be able to demonstrate and apply your knowledge and understanding of:

- processing, analysing and interpreting qualitative and quantitative experimental results
- the identification of anomalies in experimental measurements
- use of appropriate mathematical skills for analysis of quantitative data

Having planned and implemented your experiment, you will now be at the stage where you have collected a range of data, with repeats, to be analysed and from which you can, hopefully, draw some clear and definite conclusions.

Dealing with anomalies or unexpected results

When collecting data you should continually check whether or not that data is suitable. Imagine that you collected the following readings for the thickness of a wire, using a micrometer screw gauge at different points along the length of the wire:

0.12 mm, 0.13 mm, 0.12 mm, 0.12 mm, 0.13 mm, 0.06 mm.

The first five readings are very close to one another. However, the sixth appears to be inconsistent with the others. On closer inspection of the wire, it might be that you notice that the wire is thinner at that point, so you would need to use another wire that has a constant diameter throughout. This method of working allows you to make the changes needed as you carry out the experiment.

KEY DEFINITION

Anomalous values in a set of results do not fit the overall trend in the data and so are judged not to be part of the inherent variation.

Anomalies are data points that do not fit the overall trend in the data. Anomalous readings can be easily identified in a data table showing repeated measurements of the same quantity, or on a graph or chart. They should be ignored when calculating a mean of repeated measurements, when drawing the line of best fit and also not included in the data set that is used to draw conclusions.

Using mathematical skills to analyse quantitative data

As part of your physics course you will need to become competent in the use of a variety of mathematical skills, both when applying the physics content to answer exam questions and also when dealing with data from experiments. The analytical skills that you need to consider include:

- Using physical constants in standard form, along with their correct units – e.g. writing the speed of light in a vacuum as $3.00 \times 10^{8}\,\mathrm{m\,s^{-1}}$ or the resistivity of nichrome as $1.1 \times 10^{-6}\,\Omega\,\mathrm{m}$.
- Reporting calculations to an appropriate number of significant figures when you have been given raw data quoted to a varying number of significant figures (see Topic 1.1.4).
- Calculating a mean value for repeated experimental readings.
- Translating information between graphical, numerical and algebraic forms, for example plotting a graph from tabulated data of displacement and time, and calculating a rate of change (instantaneous velocity) from the tangent to the curve at any point.
- Understanding that if a relationship obeys the straight-line equation $y = mx + c$ then the gradient and the y-intercept will provide values that can be analysed to draw conclusions.
- Finding the area under a graph, including estimating the area under graphs that are not linear.

WORKED EXAMPLE

The following results were obtained as part of an experiment to determine the resistivity of an unknown metal. The planning and implementation stages enabled 'good' results to be gathered, using ammeter and voltmeter readings taken to 2 decimal places, and five readings for the diameter of the wire were obtained to a precision of ±0.01 mm.

The five readings obtained for the diameter of the wire, at different places along its length, were:

0.14 mm, 0.15 mm, 0.14 mm, 0.14 mm and 0.14 mm.

Length of wire/m	Potential difference/V		Current/A		Average resistance/Ω
0.100	0.52	0.53	0.28	0.27	
0.200	0.65	0.66	0.20	0.21	
0.300	0.71	0.71	0.18	0.18	
0.400	0.75	0.75	0.15	0.16	
0.500	0.77	0.78	0.14	0.14	
0.600	0.81	0.81	0.11	0.11	
0.700	0.87	0.88	0.10	0.10	

Table 1 Data obtained for a resistivity investigation.

Determine:

(a) the mean values for the resistance in Ω for each of the lengths
(b) the mean value for the diameter of the wire in mm
(c) the cross-sectional area of the wire in m^2
(d) the equation needed to work out the resistivity of the metal wire
(e) the graph that should be drawn
(f) an explanation of how to use the graph and the equation to find a value for the resistivity of the metal wire.

Answers

(a) The values for the average resistance, going from 0.100 m to 0.700 m are obtained by adding the two potential difference values and dividing by two, followed by doing the same for the current values at each length. Having found the mean values for V and I, work out the average resistance at each length using

$$\text{average resistance} = \frac{\text{average potential difference}}{\text{average current}}$$

(b) Use the five values of the wire diameter to work out the mean diameter:

$$\frac{(0.14 + 0.15 + 0.14 + 0.14 + 0.14)}{5} = 0.14\text{ mm}$$

(c) So the mean radius will be 0.07 mm. You convert from mm to m by dividing by 1000 and this gives a radius of 7.0×10^{-5} m. You then need to use $A = \pi r^2$ to work out the cross-sectional area of the wire:

$$A = \pi \times (7.0 \times 10^{-5})^2$$
$$= 1.5 \times 10^{-8}\text{ m}^2$$

(d) Resistivity $= \left(\frac{\text{resistance}}{\text{length}}\right) \times$ cross-sectional area $\left(\rho = \frac{RA}{l}\right)$

(e) Plotting a graph of resistance against length gives a straight-line graph that goes through the origin.

(f) The gradient of the graph will be $\frac{\Delta R}{\Delta l}$. Looking at the equation, if you multiply this value by the cross-sectional area of the wire, you will be able to determine a value for the resistivity of the wire from the equation $\rho = \left(\frac{\Delta R}{\Delta l}\right) \times A$, or resistivity = gradient of graph × cross-sectional area of wire.

LEARNING TIP

Remember, you always plot the independent variable (the one that is varied) on the x-axis. The dependent variable (in this case the resistance) goes on the y-axis.

Questions

1 Look at Table 1.

(a) Draw a graph of average resistance against length for the values shown in Table 1.
(b) Obtain a value for the gradient of the graph.
(c) Find a value for the resistivity of the wire.

2 What graphs would need to be plotted and what subsequent analysis would be required for you to determine:

(a) a value for the Young modulus of a material from a graph of stress against strain?
(b) a value for the Young modulus of a material from a graph of force against extension?

3 Explain the data, graph and analysis that would be needed to determine a value for the speed of sound based on a series of echoes detected from a wall.

4 Alex collects seven values for the diameter of a wire of constant cross-sectional area. She records these values as 0.38 mm, 0.39 mm, 0.37 mm, 0.67 mm, 0.38 mm, 0.39 mm and 0.38 mm. Which of these values is likely to be anomalous and what will be the size of the average mean value if it is:

(a) included?
(b) not included in the data set?

4 Significant figures

By the end of this topic, you should be able to demonstrate and apply your knowledge and understanding of:

- appropriate use of significant figures

Introduction

When recording data and calculating results, it is important to be consistent about the number of digits in a value that are known with certainty. This will depend on the resolution of the measuring instrument. One of the areas that causes confusion is the use of **significant figures** and decimal places. Many students use an appropriate number of significant figures when recording measured quantities or when writing the value of a calculated quantity, but many do not understand why this is done.

Many students believe, incorrectly, that the use of a higher number of decimal places or significant figures will increase the accuracy of the experiment. So we need to consider the meanings of these terms and then look at the mathematics behind why the correct use of significant figures is so important.

KEY DEFINITION

The number of **significant figures** in a measured or calculated quantity indicates the number of digits that have a meaning and about which we can be certain.

Significant figures and decimal places

Let us start by looking at numbers that contain no zero digits: in these cases it is quite simple to determine the number of significant figures and decimal places. The number of significant figures is just the total number of digits present in the value and the number of decimal places is the number of digits after the decimal point.

LEARNING TIP

The first rule to remember in this situation is: 'Start from the first *non-zero* digit and count from the left to the right. However, many digits there are, is the number of significant figures.' Zeros to the left of the first non-zero digit are not significant.

WORKED EXAMPLE 1

To how many significant figures and decimal places are the following numbers stated?

(a) 234.73
(b) 54 564.9
(c) 12.3426
(d) 4.762×10^{20}

Answers

(a) 5 significant figures and 2 decimal places
(b) 6 significant figures and 1 decimal place
(c) 6 significant figures and 4 decimal places
(d) 4 significant figures and 3 decimal places – ignore the power term

Significant figures involving zeros

The first problem we encounter when dealing with significant figures occurs when the number has a zero digit.

WORKED EXAMPLE 2

To how many significant figures are the following numbers stated?

(a) 0.000438
(b) 70 457.4
(c) 3.23×10^{-6}
(d) 5.064 06
(e) 0.0760

Answers

(a) 3 significant figures – you start counting when you get to the '4'
(b) 6 significant figures – the zero after the '7' is significant because it is between other significant digits
(c) 3 significant figures – ignore the power term
(d) 6 significant figures
(e) 3 significant figures

Significant figures involving zeros at the end of a number *without* a decimal point

When numbers are written with one or more zeros at the end of a number that has no decimal point, the situation can become ambiguous. In other words it is not possible to state with certainty how many significant figures a number has.

LEARNING TIP

The second key rule is: 'Numbers that end in one or more zeros, without a decimal point have an ambiguous number of significant figures.'

Table 1 shows some examples.

Number	Number of significant figures	Explanation
20 000	Any number from 1 to 5	The zero digits after the '2' could all be actual zero values or the number could have been rounded
850	2 or 3	The zero after the '5' could be an actual zero or the number could have been rounded to 2 significant figures
65 000	2, 3, 4 or 5	The zero digits after the '5' could be real or there could be rounding to the nearest 1000, 100, 10 or 1
9670	3 or 4	The actual value could be 9670 to 4 significant figures or it could be rounded to 3 significant figures

Table 1 Significant figures.

Significant figures involving zeros, but with a non-zero digit at the end

When a non-zero digit appears at the end of a value, this makes all the zero digits between the non-zero digits significant.

LEARNING TIP

The third key rule is 'Zero values between non-zero numbers are significant.'

Compared to the numbers in Table 1:

- 20 006 has 5 significant figures
- 851 has 3 significant figures
- 65 008 has 5 significant figures
- 9673 has 4 significant figures.

Significant figures involving zeros at the end of a number *with* a decimal point

In a number with a decimal point all the numbers to the left of the decimal point, regardless of whether they are zero or non-zero, are significant. For example:

- 30 000 Ω is a resistance value that could have 1, 2, 3, 4 or 5 significant figures.
- 30 000.0 Ω is a resistance value that is written to 6 significant figures. The decimal point makes all the zero values to the left of it significant. The zero after the decimal point shows that we are confident of the value to six significant figures.

Choosing how to best write numbers to show the number of significant figures

As we have already seen, we do not know for certain how many significant figures the quantity 30 000 kΩ has. However, the way we write numbers can help:

- $R = 30\,000\ \Omega$ could have 1, 2, 3, 4 or 5 significant figures
- $R = 3 \times 10^4\ \Omega$ is written to 1 significant figure
- $R = 3.0 \times 10^4\ \Omega$ is written to 2 significant figures
- $R = 3.00 \times 10^4\ \Omega$ is written to 3 significant figures
- $R = 30\ \text{k}\Omega$ is written to 1 significant figure
- $R = 30.0\ \text{k}\Omega$ is written to 3 significant figures

Using scientific notation makes it clear how many significant figures the values have been stated to.

The correct use of significant figures in calculated values

You may need to use the raw data you have collected to make a calculation. For example, when trying to determine the volume of a cube in an experiment relating to density calculations, you would be expected to measure the length of the side of the cube and then raise it to the power 3 to work out the volume.

Side of cube, l/cm	Volume of cube, l^3/cm^3
8.43	599
8.44	601
8.45	603

Table 2 Values for length and length^3 to 3 significant figures.

In Table 2 note that both the measurements and calculated results are stated to the same number of significant figures.

Questions

1. To how many significant figures and decimal places are these numbers stated?
   (a) 56 002
   (b) 65.78
   (c) 45 000
   (d) 45 000.0
   (e) 0.000 000 87
   (f) 2.4×10^9
   (g) 5×10^5

2. Copy and complete Table 3 using numbers written to the correct number of significant figures.

Length of box, x/cm	Area of face of box, x^2/cm^2	Volume of box, l^3/cm^3
25.8		
25.9		
26.0		

Table 3

3. Round the following values to 3 significant figures:
   60.02, 60.00, 60.07, 5421, 5428, 542.3, 542.8, 53 320 000, 43 480 000, 4.696×10^9, 1.6056×10^{-4}

4. A sphere has a radius of 2.0×10^{-3} m. Calculate its volume to an appropriate number of significant figures.

5 Plotting and interpreting graphs

By the end of this topic, you should be able to demonstrate and apply your knowledge and understanding of:

- plotting and interpreting suitable graphs from experimental results
- selecting and labelling of axes with appropriate scales, quantities and units
- measuring of gradients and intercepts

Introduction

Having collected and tabulated your data, the next step often involves plotting a graph so that trends and patterns can be identified and important values can be obtained from analysis of the gradient and the *y*-intercept.

Good habits to adopt when plotting graphs

It is conventional to plot the independent variable (the one you change) on the *x*-axis and the dependent variable on the *y*-axis.

When plotting graphs, it is important to consider the importance of the following factors.

- **Choice of scale** – it needs to be big enough to accommodate all the collected values in as much of the graph paper as possible. At least half of the graph grid should be occupied in both the *x* and *y* directions. Scales should be clearly indicated and have suitable, sensible ranges that are easy to work with (for example, avoid using scales with multiples of 3). The scales should increase outwards and upwards from the origin. Each axis should be labelled with the quantity that is being plotted, along with the correct unit.
- **Labelling the axes** – label each axis with the name of the quantity and its unit. For example I/A means current in amperes Note that the solidus (/) is used to separate the quantity and the unit.
- **Plotting of points** – points should be plotted so that they all fit on the graph grid and not outside it. All values should be plotted and the points must be precise to within half a small square. Points must be clear, and not obscured by the line of best fit, and they need to be plotted so that they are thin. There should be at least six 'good' points plotted on the graph, with major outliers identified.
- **Line or curve of best fit** – there should be equal numbers of points above and below the line of best fit. A clear plastic ruler will help you do this. The line should not be forced to go through the origin, and the points plotted should not be joined up with a line that is too thick or joined up 'dot-to-dot' like a frequency polygon.

  Outliers (anomalous values) that have not been subject to checking during the implementation stage, should be ignored if they are obviously wildly incorrect as they will have an unjustifiably large effect on the gradient of the line of best fit.
- **Calculating the gradient** – the calculation needs to be shown, including the correct substitution of identified, accurately plotted from the axes into the equation that will be of the form $m = \frac{\Delta y}{\Delta x}$.

  The triangle used to calculate the gradient should be drawn on the graph and it needs to be as large as possible – small triangles are not acceptable for working out a gradient. When using the results from a table of values, the triangle that is used to obtain the gradient should have points that lie on the line of best fit.
- **Determining the *y*-intercept** – the *y*-intercept is the *y* value obtained where the line crosses the *y*-axis – on the line $x = 0$. You can apply your knowledge of the equation $y = mx + c$ if the best fit line does not cross the *y*-axis along the line $x = 0$. Values should be read accurately from the graph, with the scale on the *y*-axis being interpreted correctly.

WORKED EXAMPLE 1

Two graphs are shown in Figures 1 and 2. One of the graphs has been reasonably well drawn, the other has a number of significant errors. Identify:

(a) which of the two graphs is the better and explain why
(b) improvements you would make to graph A
(c) improvements you would make to graph B.

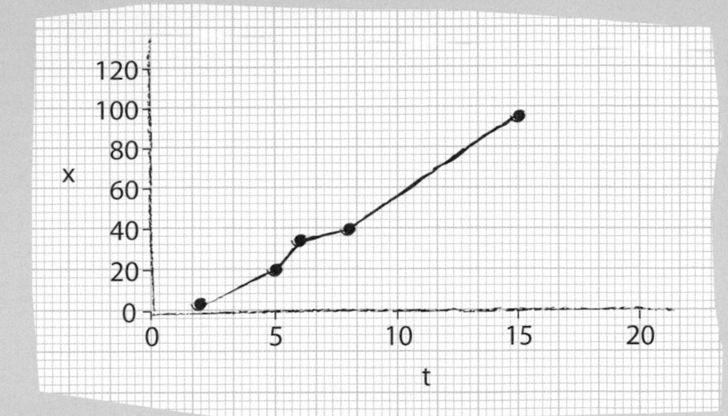

Figure 1 Graph A.

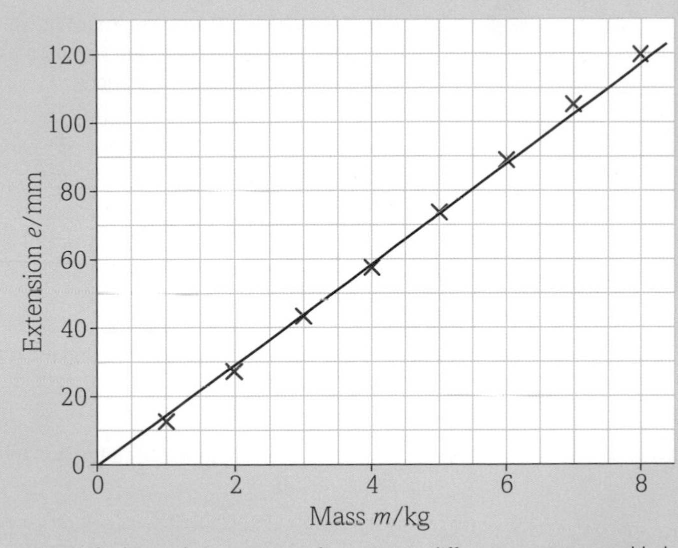

Figure 2 Graph B – this graph shows the extension of a spring as different masses are added.

Answers

(a) Graph B, despite not being perfect, is the better graph:

- it has a line of best fit with points properly distributed above, below and on the line of best fit
- it has more than enough 'good' points plotted
- the axes are labelled and the spacing of the axes and graph size are appropriate.

Graph A is deficient in the following:

- there is no title – it is not possible to determine what the graph is showing, or even the context
- there are no units or quantities are stated on the axes
- there is no line of best fit – points are joined up 'dot-to-dot'
- the points plotted are too thick
- there are not enough points plotted – there should be at least six
- the points plotted are not spaced equally
- no gradient shown and no gradient calculations are given.

(b) Graph B, despite being better than Graph A would benefit from:

- a title – e.g. 'Graph to show how the extension of a spring varies as the mass added to it increases'
- values on the x-axis could be shown to a higher number of significant figures, such as '2.000' instead of '2' for the first kilogram label.
- a gradient triangle shown that uses the maximum values e.g. gradient $= \frac{(120 - 0)\text{mm}}{(8 - 0)\text{kg}}$ which gives a value of 15 mm kg^{-1}. The experimenter could then argue that each kilogram causes an extension of 15 mm in their conclusion.

(c) You could argue that force (N) would be a more suitable x-axis variable to plot, although this is open to debate. Error bars could be included if necessary and if you are familiar with their use.

WORKED EXAMPLE 2

The equation $V = E - Ir$ is of the form $y = mx + c$ and so values can be plotted to give a straight line. A typical example of such a graph is shown in Figure 3.

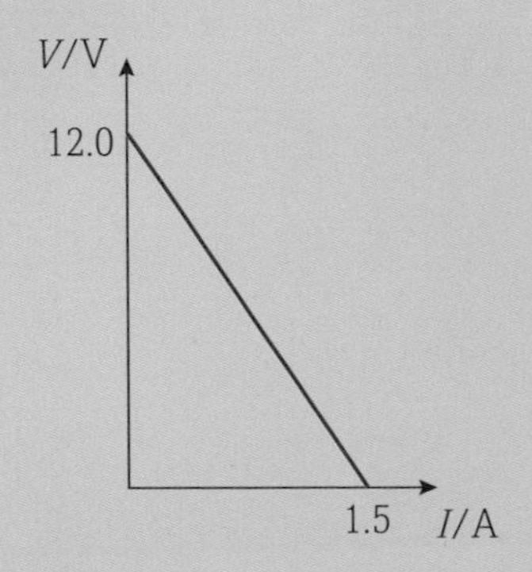

Figure 3 Graph of V against I for the equation $V = E - Ir$.

From the shape of the graph and the equations provided determine values for E and r.

Answers

E is the y-intercept, so is equal to 12.0 V; r is the gradient of the graph and corresponds to $\frac{\Delta y}{\Delta x}$ or $\frac{12}{1.5}$, which gives a value of 8. The units are VA^{-1}, which is equivalent to ohms (Ω).

Questions

1 Plot a graph of average resistance (Ω) against length of wire (m) for the results in Table 1. Your graph should show every feature needed for maximum marks, based on what has been discussed in this topic.

Length of wire/m	Volts/V			Current/A			Resistance/Ω		
	1	2	3	1	2	3	1	2	3
0.100	0.47	0.47	0.47	0.24	0.23	0.23	1.96	2.04	2.04
0.200	0.60	0.59	0.58	0.16	0.17	0.17	3.75	3.47	3.41
0.300	0.65	0.64	0.64	0.14	0.13	0.13	4.64	4.92	4.92
0.400	0.69	0.69	0.68	0.11	0.11	0.11	6.27	6.27	6.18
0.500	0.72	0.72	0.72	0.10	0.09	0.08	7.20	8.00	9.13
0.600	0.76	0.76	0.76	0.07	0.07	0.07	10.90	10.90	10.90
0.700	0.82	0.82	0.82	0.06	0.06	0.06	13.67	13.67	13.67

Table 1 Results from an investigation on how the resistance of a wire changes with length.

2 Table 2 shows the results of an experiment in which a datalogger is used to measure the time taken for a small metal sphere to fall from rest for a range of different distances.
$s = \frac{1}{2}at^2$

Height fallen, h/m	0.30	0.40	0.50	0.60	0.70
Mean time to fall this distance, t/s	0.246	0.282	0.324	0.341	0.375
t^2					

Table 2 Results from an investigation to find the acceleration of free fall, using the trap door and electromagnet method.

(a) Fill in the values for t^2.

(b) Plot a graph of t^2 against height fallen and identify any anomalous results.

(c) From your graph, determine the acceleration of free fall, g.

(d) Comment on the number of data points obtained.

3 For each of the equations below, what graph should be plotted so as to give a straight line from which a value for the unknown quantity can be determined? State the horizontal and vertical axes.

(a) $v = \frac{1}{2L}\sqrt{\frac{T}{\mu}}$, to find a value for μ, where v is the wave speed of progressive waves that make up a standing wave on a vibrating guitar string, T is the tension, L is the length and μ is the mass per unit length.

(b) $v^2 = 2as$, to find a value for a.

6 Evaluating experiments

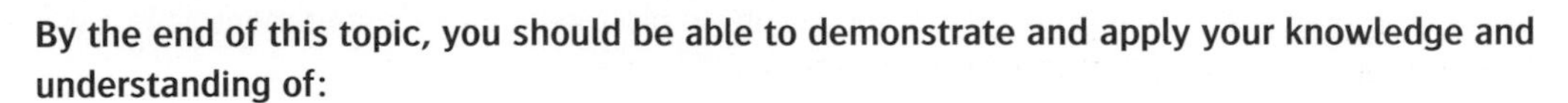

By the end of this topic, you should be able to demonstrate and apply your knowledge and understanding of:

- how to evaluate results and draw conclusions
- the limitations of experimental procedures
- precision and accuracy of measurements and data, including margins of error, percentage errors and uncertainties in apparatus
- the refining of experimental design by suggesting improvements to the procedures and apparatus

What is evaluation?

When drawing a conclusion you should be able to evaluate the experimental procedure, and also evaluate the quality of the data. How much confidence do you have that the method used to collect the data has produced good evidence – for example, how much uncertainty is there in the final value? What simple improvements can you suggest that would increase the accuracy of the experiment? This is the part that many students find the most difficult and it often results in low marks being awarded in questions.

We shall look at the key issues involved in the evaluation process.

Calculating percentage uncertainty of the apparatus used

The uncertainty in a measurement (see Topic 2.1.5) is related to the resolution or smallest scale division of the measuring instrument. For example:

- when using a protractor, angles are measured to the nearest degree
- when using a ruler, lengths are measured to the nearest millimetre
- when using a stopwatch, time is measured to the nearest 0.01 seconds, but your reaction time is 0.1 to 0.5 seconds so a degree of precision of 0.1 to 0.5 seconds is much more reasonable.

The percentage of uncertainty in any single reading taken using the equipment is found using:

$$\frac{\text{resolution or measurement error}}{\text{measured volume}} \times 100\%$$

For example, if a voltmeter records a reading of 3.24 V and its resolution is 0.01 V, then the percentage uncertainty in the reading is $\frac{0.01}{3.24} \times 100\% = 0.3\%$

The final value would then be quoted as 3.24 ± 0.3%.

Notice that for a particular piece of apparatus, the larger the value measured, the smaller the percentage error. For example, when measuring the length of a piece of wire there will be a greater percentage error in measuring a length of 6.7 cm than in measuring a length of 29.8 cm using the same ruler.

LEARNING TIP

When measuring with analogue instruments you must round up or down to the nearest scale division. The uncertainty in the measurement is therefore *half* the smallest scale division.

Comparing an experimental value and an accepted value

There are a number of physical quantities whose values have been precisely measured and verified. One such example is the value *g*, for the acceleration due to gravity, which we commonly take as 9.81 m s^{-2}.

The value for g can be determined in a number of ways – most commonly through the use of a pendulum or a falling ball bearing that hits a trap door (see Topic 3.1.5). In both cases a graph can be plotted and the value of g calculated from the gradient of the best fit line.

Imagine that the graph we plotted after one experiment gave a value for g of 9.74 m s^{-2}. The **percentage error** would be calculated as:

$$\text{percentage error} = \frac{\text{difference between accepted value and experimental value}}{\text{accepted value}} \times 100\%$$

In this case, it would be:

$$\frac{(9.81 - 9.74)}{9.81} \times 100\% = 0.7\%$$

A result is accurate if its percentage uncertainty is greater than its percentage difference from the 'true' result. For example if the measured value is 9.74 m s^{-2} with an uncertainty of 1%, the range of possible values (the **margin of error**) is from 9.64 to 9.84 m s^{-2}, which includes the true value.

KEY DEFINITION

A **margin of error** shows the range within which a value lies.

Commenting on the scatter of points in a graph

It is good practice to refer to the scatter of the points about the line of best fit and to compare how close they lie to the line. When you comment on the scatter of graph points, you are making reference to how confident you are that you have identified a relationship between the variables. The smaller the scatter and the closer the points are to the line of best fit, the stronger the correlation, and therefore the greater the confidence in the relationship between the variables. You could discuss the scatter of the data in a qualitative and quantitative way, for example, 'The values that I have plotted on my graph of Y against X are generally very close to the line of best fit. In fact all the points plotted lie well within 10% of the value given by the line of best fit. For example, the biggest anomaly occurred when X was 2, giving Y a value of 38.6, compared with the line of best fit value, which is 37.4. This is a percentage difference of only 2%, which shows that my experimental values have an acceptable degreee of accuracy'.

Commenting on precision of measurements

Precision refers to how close repeat readings are to one another. When using a micrometer to measure the diameter of a wire, you would take several readings at different points along the length of the wire to check that its diameter was constant and that the cross-section area along the wire is constant through the whole length. You may find that repeat readings give accurate values of 0.86 mm, 0.86 mm, 0.87 mm, 0.86 mm and 0.87 mm. The precision of ±0.01 mm allows you to state with some confidence that these values are accurate to within one-hundredth of a millimetre. Conversely, if you were making a rough *estimate* that the resistance of a wire was 'about 20 Ω' then you would not need to worry about high levels of precision.

Commenting on anomalies

Identify each point that is anomalous (see Topic 1.1.3) and which you have ignored when plotting your line or curve of best fit. For example, circle these outliers and mention why you have rejected them.

Limitations and improvements

Even if the experimental result is close to the true value, there are always opportunities to identify limitations of the procedure and suggest improvements to the apparatus or measurement techniques. The items to include here could involve:

- avoiding parallax error when reading scales (see Topic 1.1.2)
- taking steps to reduce random errors and systematic errors (see Topic 2.1.3)
- equipment that may not have been working properly
- variables that were difficult to control
- using better equipment with greater precision and resolution (see Topic 2.1.4)
- using a fiducial mark to act as a clear reference point – e.g. when measuring time period of a pendulum using a stopwatch (see Topic 1.1.1)

- reducing heating effects that could lead to erroneous or distorted results – e.g. use wires of larger cross-sectional area or use a smaller current and a milliammeter
- sources of error – e.g. the sample of wire may not have a uniform cross-section so it would be better to measure the diameter at several different places and in two mutually perpendicular directions at every place.

You can refer explicitly to the impact that changing a variable will have on the outcome, but you need to be specific. For example, you might mention that 'Resistance will increase if the time that the current is flowing for increases.' Similarly, you may state that 'the intensity of light will decrease as the distance from the source to the detector increases'.

Putting it all together to draw a conclusion

Imagine that you are conducting an investigation into the behaviour of elastic materials under the influence of tensile forces. This involves the calculation of a quantity called the Young modulus (see Topic 3.4.3), which is defined as $\frac{\text{stress}}{\text{strain}}$ or $\frac{F \times l}{A \times e}$, where:

F is the tensile force in N, l is the original length of the wire in m, e is the extension of the wire in m and A is the cross-sectional area of the cylindrical wire in m^2.

You may determine the following percentage uncertainties:

- uncertainty in F is 1% – based on the uncertainty in mass of the slotted masses used as loads
- uncertainty in l is 0.1% – 1 mm in a 1 m length of wire
- uncertainty in e is 1% – based on the vernier scale on Searle's apparatus, which measures to 0.05 mm
- uncertainty in the radius of wire is 1% – meaning that the percentage uncertainty in the calculated area is 2% (when you square a value, you double the percentage uncertainty associated with it).

Because the Young modulus = $\frac{F \times l}{A \times e}$, and because you add uncertainties irrespective of whether you are multiplying or dividing, we end up with a total uncertainty of 1% + 0.1% + 1% + 2% = 4.1%, or 4% to 1 significant figure.

This is the uncertainty from calculation, using the uncertainty values in the measurements. You can also determine the uncertainty from the difference between the gradients of line of best fit and the worst acceptable line (see Topic 2.1.6). This might turn out to be a percentage uncertainty of 8%. If this is the case, then always use the largest uncertainty value when stating your answer and use this to decide how accurate your final answer is when comparing it with the accepted value.

LEARNING TIP

When data are combined by addition, subtraction, multiplication, division and raising to powers, there are simple rules for finding the percentage uncertainty in the calculated result. These rules are explained in Topic 2.1.5.

Questions

1. State five key things you should include when writing the evaluation of an experiment.
2. Explain what is meant by the following terms and say how you would determine their size:
   (a) percentage uncertainty from a single reading
   (b) percentage uncertainty from a gradient
   (c) percentage difference.
3. What limitations could most easily be improved in experiments?
4. A balance records a mass of 17.61 g and the uncertainty is 0.01 g. Calculate the percentage uncertainty in this reading.
5. Calculate the percentage uncertainty in a measured length of 6.9 ± 0.1 cm.
6. A dog has an accurate mass of 3.65 kg. When weighed on a defective scale, its mass is measured as 3.80 kg.
   (a) What is the percentage error in measurement to the nearest $\frac{1}{10}$th of 1%?
   (b) If a cat has a mass of 1.40 kg on the same scale, what is its actual mass to the nearest 0.01 kg?

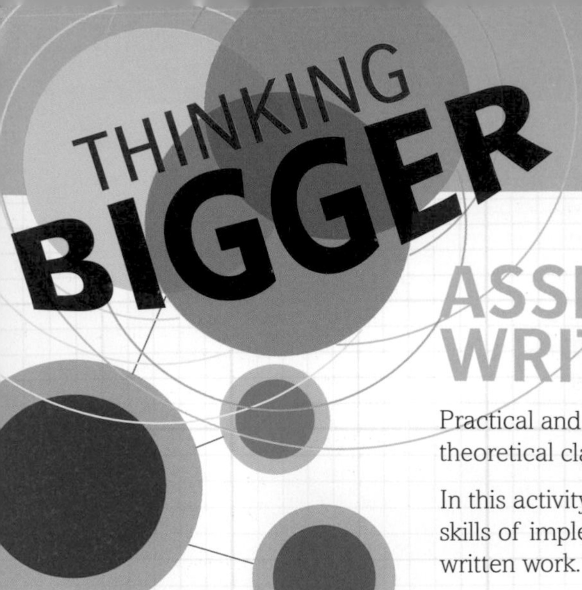

ASSESSING A PRACTICAL WRITE-UP

Practical and investigative work is the lifeblood of physics. Without evidence to back up our theoretical claims, we can have no physical laws.

In this activity you will develop an understanding of how to improve practical techniques and the skills of implementing, planning, analysing and evaluating via the analysis of a sample of practical written work.

A student has been given the following task to plan, implement, analyse and evaluate.

Task: Find the relationship between force and extension for a single spring, springs in series and springs in parallel.

General introduction

When a force of tension is applied to a spring, the extension will increase as the force increases in accordance with Hooke's law. The law states that the extension will be directly proportional to the load provided that the elastic limit is not exceeded.

Design and carry out a procedure to determine the spring constant for a single spring, springs in series and springs in parallel. Having done this, provide an explanation for how the value of the spring constant is different for the single spring, the series arrangement and the parallel arrangement. Your procedure should include:

1. A plan to include suitable apparatus and identification of variables to control.
2. A method which focuses on the practical techniques, suitable units and an appropriate format for the presentation of your data.
3. An analysis of the data, using correct equations, significant figures and values from graphs.
4. An evaluation of your procedure which includes a conclusion, a mention of accuracy and precision, anomalies in your data, percentage errors and any limitations in experimental procedures. You should also explain how to improve your procedures for future investigations.

Student's write-up

Aim

To find out about Hooke's law.

Method

1. I set up the apparatus as shown in my diagram for a single spring.
2. I measured the length of the spring using a metre ruler that had cm rulings on it. I recorded this value.
3. I added masses to the spring and recorded the extension each time. I repeated this procedure to get two sets of data.
4. I did the same thing for two springs arranged in parallel, then I did it again for two springs in series.
5. I worked out the spring constants for each arrangement and compared them. My data is shown in Table 1.

Calculations

The single spring had a value for the spring constant of 4 N cm. The two springs in parallel had a spring constant of 8.0 N cm and the two springs in series had an effective spring constant of 2 N cm. I plotted graphs of extension against force and found the gradient of each graph which gave me these values.

Evaluation

My values are accurate as they agree with Hooke's law. I could make my results more accurate by stating them to more decimal places and by taking more repeats. There are no anomalies in my data and they are all precise. I would guess that my values have a percentage error of about 10%.

Where else will I encounter these themes?

1. Identify as many mistakes as you can in the planning and implementation stages of this investigation.
2. Comment on the data in Table 1. How is it presented well? How is it presented poorly?

Single spring			Two springs in parallel			Two springs in series		
Force/N	Extension/cm		Force/N	Extension/cm		Force/N	Extension/cm	
1	4	4.0	1.0	8	8.0	2	4	4.0
3	12	12.2	2.0	16.5	17	4	8	9
4	15.5	16	3.0	25.5	26.0	6	12.0	12
7	28.8	29	4.0	31	41	8	17.5	18
11	44	43.5	5.0	40.00	40	10	20	23

Table 1 The student's experimental results.

3. How accurate are the values stated in the conclusion?
4. Evaluate the student's evaluation of their own investigation.

Look at the specification to check what is expected in order to produce a high-quality investigation write up. Familiarise yourself with the key elements of planning, designing, carrying out, analysing and evaluating a practical task.

DID YOU KNOW?

If you measure the time period of an oscillating spring, including two springs in series and two springs arranged in parallel, you can get an accurate value for the spring constant of the arrangement. This gives you another method for calculating, and verifying, the value obtained from the force–extension method.

Activity

Re-write the student's write-up, so that the issues identified and addressed in your responses to the questions above are incorporated. Pay particular attention to the following key areas:

- Experimental design
- Identification of variables that need to be controlled and changed
- Use of a wide range of apparatus
- Appropriate units of measurement
- Appropriate presentation of data
- Use of appropriate mathematical skills
- Correct use of significant figures
- Plotting of graphs
- Measurement of gradient
- Accuracy of conclusions
- Percentage errors
- Identification of anomalies
- Suggesting improvements to the procedures, techniques and apparatus

Practice questions

1. The diameter of a cylindrical wire is found to have a mean value of 0.34 mm. Which one of the following is true? [1]

   A The diameter will be 0.34 mm at every point along the wire.

   B The diameter will have been obtained accurately with a metre ruler.

   C The values will have been obtained accurately with a micrometer.

   D The cross-sectional area of the wire will be 0.34 mm^2.

2. The relationship between two variables p and q is given by $p = \frac{1}{2}a + q^2$ where a is a constant value.

   Which of the following will produce a straight line of constant gradient? [1]

   A p plotted against q

   B p plotted against a

   C p plotted against root q

   D p plotted against q^2

3. An experiment is being conducted to determine how the increase in temperature of a room depends on the time for which it is being heated by a heater.

   Which of the following need to be controlled? [1]

   (i) the volume of the room
   (ii) the power rating of the heater
   (iii) the nature of the insulation and number of windows

   A (i), (ii) and (iii)

   B only (i) and (ii)

   C only (ii) and (iii)

   D only (i)

[Total: 3]

4. The diagram shows an arrangement used to investigate how the kinetic energy of a toy car varies with its distance d from the top of the ramp.

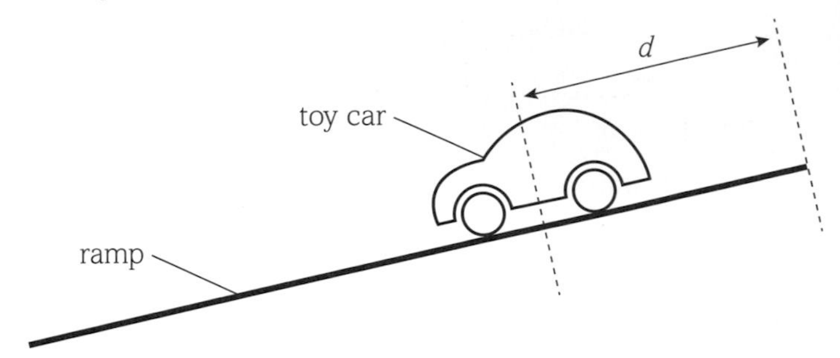

Design a laboratory experiment to determine the kinetic energy of the car at one particular distance d from the top of the ramp.

In your description pay particular attention to:

- how the apparatus is used
- what measurements are taken
- how the data is analysed. [3]

[Total: 3]

[*Q22(b), H156/01 sample paper 2014*]

5. The micrometer screw gauge used to determine the diameter of a wire had a zero error. The diameter recorded by a student was larger than it should have been.

   Discuss how this error would affect the calculation of the Young modulus. [3]

[Total: 3]

6. The resistivity of a metal, calculated from an experiment, is larger than the value shown in a data book.

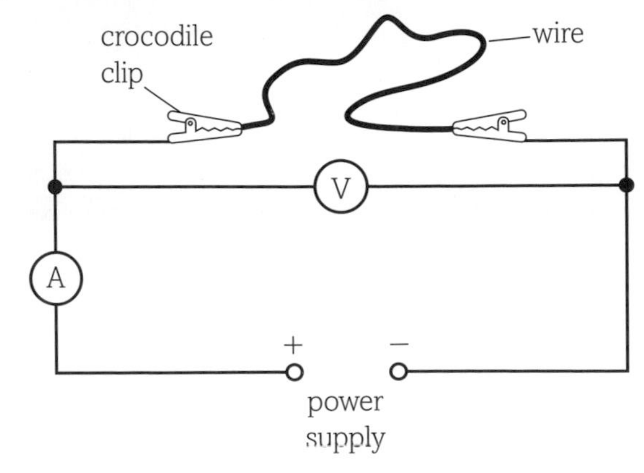

Explain two possible limitations of the experiment. [3]

[Total: 3]

[*Q24(c), H156/01 sample paper 2014*]

7. Look at the set up of apparatus below to determine the electrical resistivity of a sample of nichrome wire.

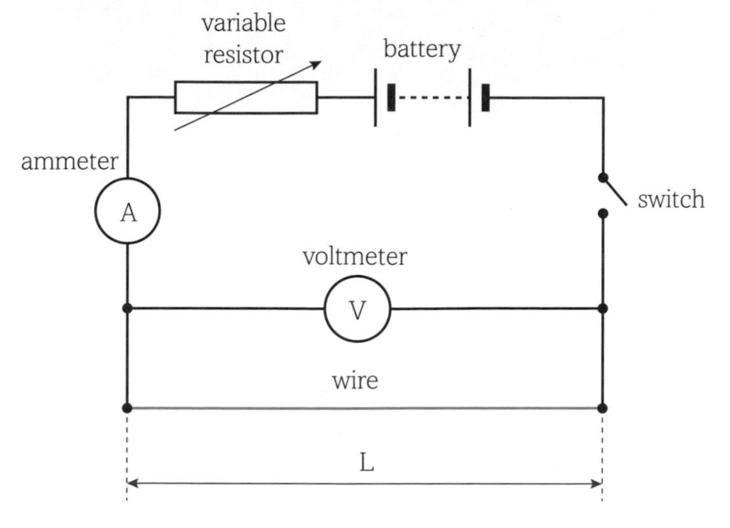

Explain:

- what other equipment you would need in order to determine the resistivity of the wire [1]
- how the resistivity can be determined from the apparatus and results [3]
- what random and systematic errors are and how these errors may occur in the experiment [4]
- how these errors may or may not be removed or reduced. [2]

[Total: 10]

MODULE 2

Foundations of physics

CHAPTER 2.1

PHYSICAL QUANTITIES, UNITS AND MEASUREMENTS

Introduction

Physicists, engineers and other scientists around the world need to communicate their ideas and findings with one another in a consistent way that they will all understand. For this to happen effectively, there needs to be a common language within the scientific community. A kilogram, newton, joule or ampere in the mind and calculations of a British scientist must mean exactly the same thing to a Japanese or American physicist working on the other side of the world. Physical quantities make sense when they have a numerical value and a unit that can be understood by all of the scientific community. In this chapter you will encounter the base S.I. units as well as the derived units of S.I. base units. You will become more familiar with the prefixes used to show multiples and submultiples of units. You will check the homogeneity of physical equations using S.I. units and you will deploy the conventions used for correctly labelling graph axes and table columns.

All the maths you need

To unlock the puzzles of this chapter you need the following maths:

- Units of measurement
- Addition and subtraction of quantities
- Multiplication and division of quantities including prefixes and the use of standard form
- The conversion between units with different prefixes (*e.g.* cm^3 *and* m^3)
- How to make order of magnitude calculations
- Calculate the gradient of a graph

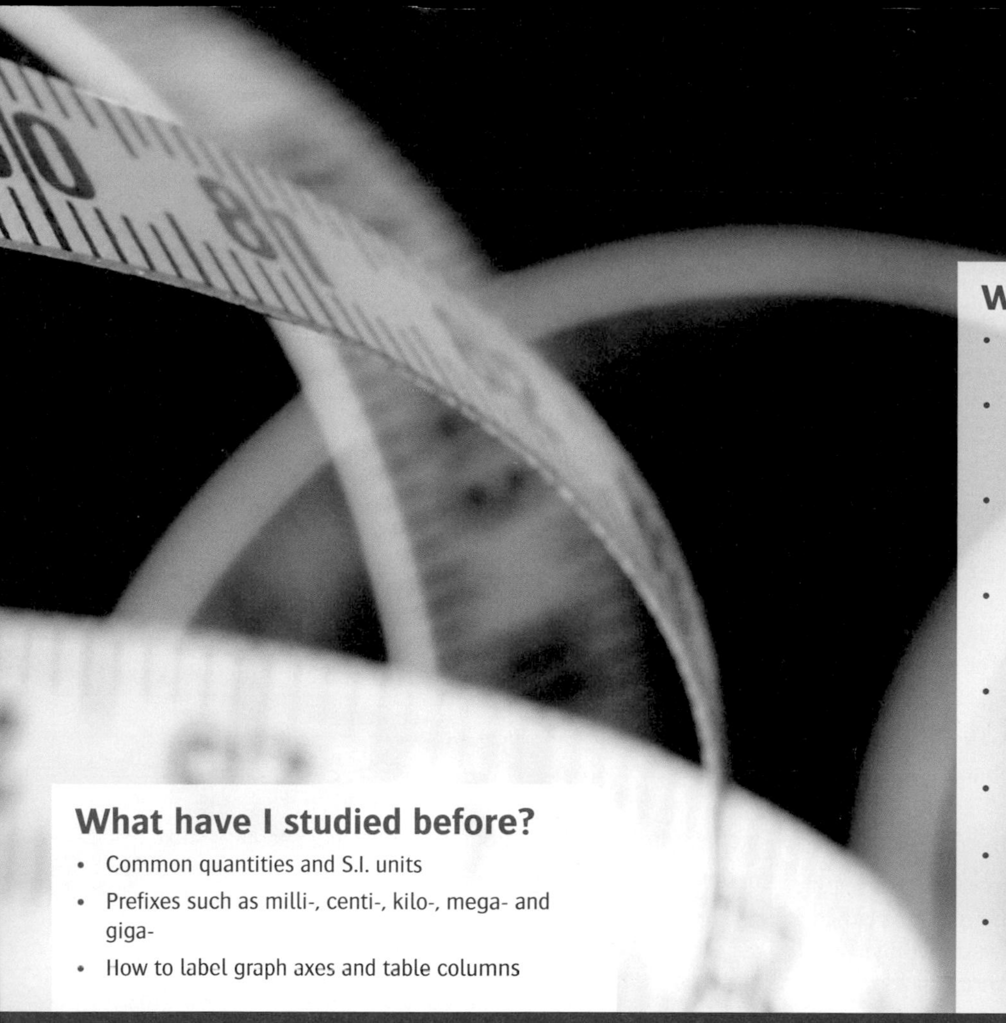

What have I studied before?

- Common quantities and S.I. units
- Prefixes such as milli-, centi-, kilo-, mega- and giga-
- How to label graph axes and table columns

What will I study in this chapter?

- That physical quantities have a numerical value and a unit
- S.I. units including base units and derived units of S.I. base units
- Checking the homogeneity of physical equations using base units
- Prefixes and their symbols to indicate multiples and sub-multiples of units
- Estimation of a range of physical quantities
- The identification of random and systematic errors
- Precision and accuracy
- The treatment of errors and uncertainties when performing calculations
- The use of best-fit and worst acceptable lines on a graph

What will I study later?

- The nature of units used to describe scalar and vector quantities (AS)
- The addition, subtraction, multiplication and division of scalar and vector units and quantities (AS)
- The magnitude and direction of vector quantities acting at an angle to the direction of application of a force (AS)
- Compound variables, e.g. density, pressure, momentum, Young modulus and kinetic energy (AS)
- The units and quantities associated with the areas of forces, motion, work and energy, materials, electricity, waves and quantum physics (AS)
- Converting between units for energy, e.g. from eV to J (AS)
- The quantities and units relating to circular motion and oscillations, including the radian (AL)
- The quantities and units used in the fields of capacitance, cosmology, fields and medical physics (AL)

Physical quantities and units

By the end of this topic, you should be able to demonstrate and apply your knowledge and understanding of:

* physical quantities have a numerical value and a unit
* Système International (S.I.) base quantities and their units
* derived units of S.I. base units
* prefixes and their symbols to indicate decimal submultiples or multiples of units
* checking the homogeneity of physical equations using S.I. base units

Physical quantities

In physics, a *quantity* is a measurement of something. Any quantity will have a numerical value indicating its size and an appropriate unit.

For example, there are many units that can be used to measure length. If we wanted to state the distance from Newcastle to Dubai, we would use kilometres, for example:

distance from Newcastle to Dubai = 5642 km.

If we wanted to state this distance in cm, then we would convert it from km to cm by multiplying by 100 000, so the distance would be stated as 564 200 000 cm. However, we usually state longer distances in km and shorter distances in m, cm, mm or even smaller units that are fit for purpose.

LEARNING TIP

You may be able to work out what formula to use by looking at the units of the quantity you want to calculate. For example, momentum is measured in $kg\,m\,s^{-1}$, so to find the momentum of an object you need to multiply mass (kg) by velocity ($m\,s^{-1}$).

Units

We can express quantities in many different units. For example, we could refer to the length of a field in metres, yards, inches, furlongs, miles, kilometres or even cubits. However, this can lead to confusion when we need to convert from one unit to another. In 1960, the international scientific community agreed to adopt a single unit for each quantity. These are known as Système International (S.I.) units. The seven base units are shown in
Table 1. The units for every other scientific quantity can be derived from these base units, and some of these are shown in Table 2.

Quantity	Unit	Symbol
mass	kilogram	kg
length	metre	m
time	second	s
temperature	kelvin	K
electrical current	ampere	A
amount of substance	mole	mol
luminous intensity	candela	cd

Table 1 The seven S.I. base quantities and their associated units.

Quantity	Unit	Symbol
acceleration	metres per second squared	$m\,s^{-2}$
density	kilograms per cubic metre	$kg\,m^{-3}$
electric charge	coulomb	C
energy	joule	J
force	newton	N
momentum	kilogram metres per second	$kg\,m\,s^{-1}$
potential difference	volt	V
power	watt	W
pressure	pascal	Pa
velocity	metres per second	$m\,s^{-1}$

Table 2 Some derived quantities and their S.I. units.

Unit prefixes

There is huge variation in the size of things. For example, the diameter of the Universe is about 10^{24} m, and the diameter of an atomic nucleus about 10^{-15} m. We often use *standard form* to help us write down very big and very small numbers in a quick and efficient way, rather than typing very long numbers with lots of digits into calculators or writing them out in full when performing written calculations. We can also use *prefixes* to show the size of a quantity in comparison to the S.I. unit at a power of ten expressed in standard form. These prefixes are shown in Table 3 and you must learn them.

Prefix	Abbreviation	Standard form
pico-	p	10^{-12}
nano-	n	10^{-9}
micro-	μ	10^{-6}
milli-	m	10^{-3}
centi-	c	10^{-2}
deci-	d	10^{-1}
kilo-	k	10^{3}
mega-	M	10^{6}
giga-	G	10^{9}
tera-	T	10^{12}

Table 3 S.I. prefixes and their abbreviations.

LEARNING TIP

These prefixes are the cause of many lost exam marks! Every time you convert into standard form, just do a quick check. If you are converting mm into m and have calculated that 6.4 mm = 6400 m, checking will show that you have *multiplied* by 1000 instead of *dividing* by 1000 (i.e. 6.4 mm = 0.0064 m). Check your work at every stage in a calculation, not just at the end. Areas and volumes can be even trickier – for example, the area of a rectangle 3 cm × 5 cm is not 0.15 m^2. You need to change the lengths into metres *before* doing the multiplication:
area = 3 cm × 5 cm = 0.03 m × 0.05 m = 0.0015 m^2.

Checking the homogeneity of equations

You can use base units to check whether an equation representing physical quantities could be correct. If the equation is correct, the units must be of the same type (homogeneous) for each quantity in the equation. For example, it would make no sense to add quantities such as 32 kg + 4.3 N, or equate 3 N = 1 kg × 3 s. However, having homogeneous units does not prove that the equation is correct.

WORKED EXAMPLE

Use units to show that the equation $v = u + at$ is homogeneous.

Answer

Writing down the units for each side we have:
v and u are both velocities, and have base units of m s^{-1}
at is acceleration × time and so has base units of m s^{-2} × s = m s^{-1}
The units of all three terms in the equation are m s^{-1}, so the equation is homogeneous.

Questions

1 Write the following in terms of fundamental units, using standard form.
For example: the wavelength of red light = 600 nm; 600 nm = 6.0×10^{-7} m.
(a) A raindrop has a diameter of 0.1 mm.
(b) The distance from Land's End to John O'Groats is 1000 km.
(c) The mass of a person is 60 000 g. (Note that kilogram is the only fundamental unit that itself has a prefix.)
(d) An X-ray has a wavelength of 0.46 nm.
(e) One day.
(f) One year.
(g) The area of a sheet of A4 paper is 29.7 cm × 21.0 cm = 624 cm^2.
(h) The volume of a sphere is $\frac{4\pi r^3}{3}$. What is the volume of a ball with a diameter of 4 mm?
(i) An electric current of 400 μA.

2 What are the prefixes that are:
(a) smaller than pico
(b) bigger than tera?

3 A component in a microprocessor is 60 nm long. Write down this length in μm, mm, m and km using the powers of ten notation.

4 Copy Table 4 and complete it using the correct prefix and S.I. unit.

Quantity	Calculation	Correct value, prefix and S.I. unit
Area of table = 85 cm × 120 cm	Area = 0.85 m × 1.20 m	Area = 1.02 m^2
Length of a car = 4200 mm		
Volume of a room = 1000 cm × 3400 cm × 85 000 mm		
Resistance = $\frac{420\text{ kV}}{105\text{ mA}}$		
A speed of 20 km h^{-1}		

Table 4

5 Do the following calculations and conversions without using a calculator:
(a) Convert 4.6×10^{-3} m to mm.
(b) Convert 780 nm to m.
(c) Express the area of a 50 cm by 80 cm table in m^2.
(d) Find the prefix that is equal to $\frac{\text{kilo}}{\text{nano}} \times \text{micro} \times \frac{\text{mega}}{\text{tera}}$
(e) A storage facility has the dimensions 10 m × 5 m × 4 m. How many biscuit tins measuring 50 cm × 50 cm × 10 cm could be fitted into the building?

6 Show that the unit of charge, C, can also be written as As.

7 Show that the unit of force, N, can also be written in base units as kg m s^{-2}.

8 Derive the units of power, work and pressure in S.I. base units, using their defining equation.

9 A physical quantity is defined as mass per unit length. Determine the units of this quantity.

10 Use units to verify that the equation $p = \rho g h$ for the pressure p in a fluid of depth h is homogeneous.

2 Estimating physical quantities

By the end of this topic, you should be able to demonstrate and apply your knowledge and understanding of:

* making estimates of physical quantities

What is estimation?

When we estimate a value, we use sensible and simple substitutes in order to perform a calculation – the sort that we can perform mentally if required. An estimation is an educated guess resulting in an answer that is close, but not exactly equal, to the true answer.

Suppose you were given the following numbers to add:

97, 105, 48, 980, 312, 8, 53, 202

You might try to add them up mentally but it would probably lead to an arithmetical error because there are too many digits to memorise. Instead, you could round the numbers to 1 significant figure, leading to:

100, 100, 50, 1000, 300, 10, 50, 200

This would give an 'estimation value' of 1810 – the true value is 1805. This means that the estimation is better than 99% accurate!

WORKED EXAMPLE 1

To calculate the resistance of an electrical component, we use the equation:

$$\text{resistance} = \frac{\text{potential difference}}{\text{current}}$$

It is found that a potential difference of 24.5 V will produce a current of 0.95 A in a component (see Figure 1).

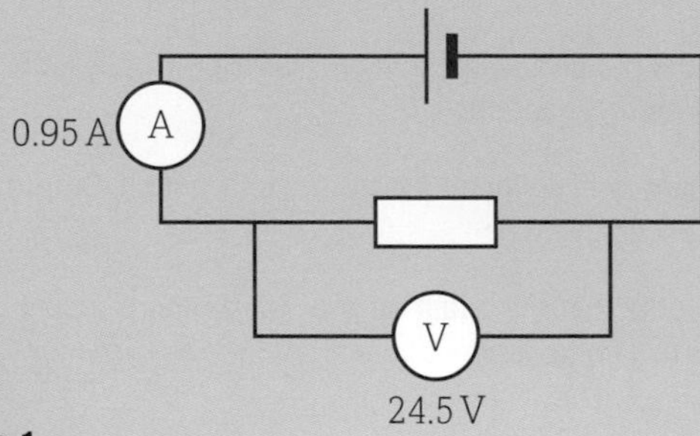

Figure 1

(a) What is the value of the potential difference to 2 significant figures?

(b) What is the value of the current to 1 significant figure?

(c) Use these values to estimate the resistance of the component.

Answers

(a) 24.5 V is stated to 3 significant figures. To round to 2 significant figures we look at the third digit and see that it is equal to 5, so we round up from 24.5 to 25 V.

(b) 0.95 A becomes 1 A to 1 significant figure.

(c) Resistance $= \frac{25\text{ V}}{1\text{ A}} = 25\,\Omega$

LEARNING TIP

When you are doing a calculation you should use one or two more significant figures than are in the data given. Then round your final answer to the correct number of significant figures.

WORKED EXAMPLE 2

The dimensions of a rectangular cuboid are 48.9 cm, 103.2 cm and 12.7 cm, as shown in Figure 2.

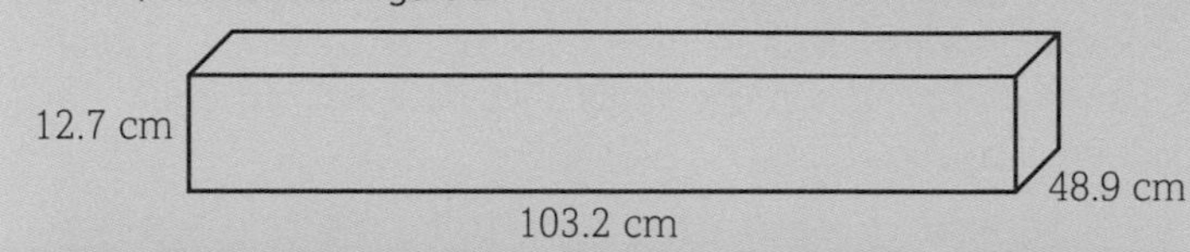

Figure 2 The dimensions of the rectangular cuboid.

(a) Calculate the exact volume of the cuboid in cm^3 using $V = b \times l \times h$.

(b) Make an estimate for the volume of the cuboid in cm^3 by rounding each value to 1 significant figure.

(c) What is the percentage accuracy of the estimate compared with the true value?

Answers

(a) $V = 48.9\text{ cm} \times 103.2\text{ cm} \times 12.7\text{ cm} = 64\,090.296\text{ cm}^3$

(b) $V = 50\text{ cm} \times 100\text{ cm} \times 10\text{ cm} = 50\,000\text{ cm}^3$

(c) $\%\text{ accuracy} = \left(\frac{50\,000}{64\,090.296}\right) \times 100 = 78\%$

It is useful to be able to estimate quantities that are difficult or impossible to measure directly. In some cases this will involve an approximation, for example describing an apple as a sphere, or rounding the value of π to 3.

LEARNING TIP

Many students muddle units when estimating and end up with unrealistic answers such as:

- very fast speeds for cars in kinematics questions, e.g. the car that travels at 5000 m s^{-1}
- the mass of an apple being 100 kg instead of 100 g.

Always double check your calculations or use a different method and see if you get a similar answer.

Estimates and orders of magnitude

When estimating a value, round your estimate to the nearest power of ten. Your answer should be accurate to the correct 'order of magnitude' – that is, your answer should be within one power of ten of the correct answer.

WORKED EXAMPLE 3

How many times heavier than a proton is a DNA molecule?

Answer

The mass of a DNA molecule is 3.02×10^{-18} kg and the mass of a proton is 1.67×10^{-27} kg.

Using standard form, an estimation of the ratio of mass of DNA : mass of a proton is:

$10^{-18} : 10^{-27}$, so the DNA is 10^{9} times heavier.

The answer from the full calculation is 1.80×10^{9}, which is the same order of magnitude as the estimation.

WORKED EXAMPLE 4

Estimate the volume of a ten-pound note.

Answer

Estimate the length and width, for example about 7 cm × 15 cm.

Estimate the thickness by comparison with the pages in a book. If a book of width of 1 cm has about 200 pages, the thickness of a sheet of paper is about 0.005 cm, or 5×10^{-3} cm.

The volume of the note is therefore about $7\,\text{cm} \times 15\,\text{cm} \times 5 \times 10^{-3}\,\text{cm} = 0.525\,\text{cm}^3$

To the nearest power of ten this is about $1\,\text{cm}^3$.

Your estimate is probably within an order of magnitude of the true value, that is the true value is probably between 10^{-1} and $10\,\text{cm}^3$.

	Mass/kg		Distance/m		Power/W
galaxy	1.8×10^{41}	Earth to Sun	1.5×10^{11}	power station	1.0×10^{9}
Sun	2.0×10^{30}	Sun's radius Earth's radius	7×10^{8} 6.4×10^{6}	family car	5.3×10^{5}
uranium-238 atom	4×10^{-25}	London–Paris	3.4×10^{5}		
proton	1.6×10^{-27}	light wavelength	6.2×10^{-7}		
electron	9×10^{-31}				

Table 1 Data and orders of magnitude relating to mass, distance, time and power.

Questions

1. Use the data in Table 1 to estimate:
   (a) how long it will take light to reach the Earth from the Sun
   (b) how long it will take sound to travel from London to Paris
   (c) how many stars there are in a galaxy
   (d) how many protons you could fit into a uranium nucleus
   (e) how many cars you could run from a power station.
2. A car travels across Africa taking a total time of 3 weeks, during which it covers a distance of 5468 km. Estimate the car's speed in $\text{m}\,\text{s}^{-1}$.
3. We calculate the density of a material using the formula
   density = $\dfrac{\text{mass}}{\text{volume}}$.
   An object will float in water if it has a density lower than $1\,\text{g}\,\text{cm}^{-3}$.
   (a) Estimate the density of the cuboid shown in Figure 3.
   (b) Will it float or sink when placed in water?

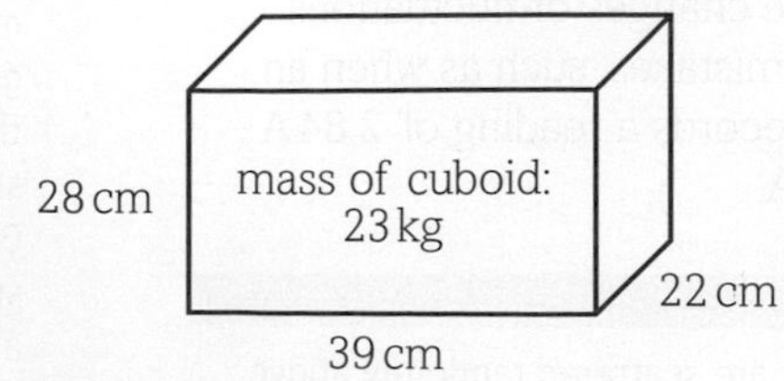

Figure 3

4. Which value is the best order of magnitude value for the mass of a mouse?
   10^{-1} kg, 10^{-2} kg, 10^{-3} kg
5. Estimate the mass in kg of a 1 mm diameter raindrop.
6. Estimate the mass of a bathtub filled with 1 pence pieces.

Systematic errors and random errors

By the end of this topic, you should be able to demonstrate and apply your knowledge and understanding of:

* systematic errors and random errors in measurement

What is an error?

No matter how careful we try to be when planning and implementing an experimental or investigative procedure, there will always be occasions where errors occur in the collection of data and in the subsequent calculation process.

We come across the term 'error' frequently in everyday language, and some common definitions are:

- the condition of having incorrect or false knowledge
- an incorrect belief or judgement
- a mistake or wrongdoing
- the difference between a computed value or a measured value and a true or theoretically correct one.

The last of these definitions is possibly the best one to use when considering errors in experimental science.

When conducting experiments it is important to minimise errors because they will lead to an incorrect conclusion, value or theory being produced.

Random errors

Random errors in experimental measurements are caused by unknown and unpredictable changes during the experiment. These may occur because of changes in the instruments or in the environmental conditions.

Humans can also make errors when reading or recording a value. A human error is often a 'one-off', caused by a simple mistake. This means that results will be inconsistent when repeated, so it is often easy to spot a human error because it is wildly different from the corresponding values from other repeats.

Random errors and human errors are not really the same.Random errors will always occur due to small changes in environmental conditions, such as small temperature changes or fluctuations in light levels. Human errors are just mistakes, such as when an experimenter misreads a value and records a reading of 2.84 A instead of the correct value of 2.48 A.

KEY DEFINITION

Random errors give measurements that are scattered randomly above and below the true value when the measurement is repeated. A better result can be obtained by finding the mean value of the results of several readings.

WORKED EXAMPLE 1

A student was collecting data in order to determine the relationship between the length of a metal wire and its electrical resistance. He had planned the experiment well and used suitable and appropriate apparatus to collect the data he needed.

- He measured the length of the wire to the nearest millimetre, the current to 0.01 A and the voltage to 0.01 V.
- He recorded his average resistance values to 1 decimal place.

Table 1 shows his collected results.

(a) Can you identify possible random errors and human errors in his results table?

(b) What suggestions would you make to help the student to recognise these errors quickly?

Answers

(a) The current values for 0.200 m (20.0 cm) appear to have been written incorrectly (as 0.80 A instead of 0.08 A) on both occasions. The student may have misread the value, or their partner (if they had one) may have read or stated the value incorrectly. Either way, the figure is incorrect, leading to a value for the resistance that is ten times too small.

You could also argue that there is variation in some repeat readings for the voltage and current, which may have been caused by heating or by a power surge. For example, the potential difference readings for the 0.400 m length are 2.91 V and 2.87 V – close but not identical. This constitutes a random error, caused by unknown and unpredictable changes in the circuit conditions. Random errors tend to even out under statistical analysis.

(b) When the graph of resistance against length is plotted, it will be obvious that the 0.200 m reading is anomalous. However, it is better to spot such errors early, so that new readings can be taken using the same equipment, in the same conditions and at the same time as the other readings. Minor variations and fluctuations in the equipment and conditions can provide different values, even though this should not be the case theoretically.

You could suggest that the student looks for a pattern in the results that would be expected from the planning and hypothesis generation. For this investigation, the hypothesis may have been that the resistance is directly proportional to the length of the wire, so reductions in resistance or increases in current from 0.100 m to 0.200 m should ring alarm bells and be checked. In addition, the student could take a third set of results (a second set of repeats), although this is not always possible, or even necessary if the results are close.

Length/cm	Potential difference 1/V	Current 1/A	Potential difference 2/V	Current 2/A	Average resistance/Ω
10.0	1.10	0.09	1.00	0.09	11.7
20.0	1.86	0.80	1.90	0.80	2.3
30.0	2.74	0.07	2.74	0.08	34.2
40.0	2.91	0.06	2.87	0.06	48.2
50.0	2.92	0.05	2.92	0.05	58.3
60.0	3.52	0.05	3.52	0.05	70.3
70.0	3.83	0.04	3.83	0.04	95.8

Table 1 A student's results.

Examples of random errors

- Variable heating in circuits causing variations in the current being measured.
- An unexpectedly large extension of a metal wire due to a fault in its structure.

KEY DEFINITIONS

A **systematic error** is an error that does not happen by chance but instead is introduced by an inaccuracy in the apparatus or its use by the person conducting the investigation. This type of error tends to shift all measurements in the same direction.
A **zero error** is a type of systematic error caused when an instrument is not properly calibrated or adjusted, and so gives a non-zero value when the true value is zero.

Systematic errors

Systematic errors in experimental observations usually involve the measuring instruments. They may occur because:

- there is something wrong with the instrument or its data-handling system
- the instrument is used wrongly by the student conducting the experiment.

Systematic errors are often present in all the readings taken and can be removed once identified. For example, if the needle on a set of scales is pointing to 25 g when nothing is on the pan, all values recorded will be bigger than the true mass by 25 g. To obtain the true value, simply subtract 25 g from each of the readings collected.

Examples of systematic errors

- The scale printed on a metre rule is incorrect and the ruler scale is only 99.0 cm long.
- The needle on an ammeter points to 0.1 A when no current is flowing. Each value recorded will, therefore, be bigger than the true value by 0.1 A. This is an example of a **zero error** – the apparatus shows a non-zero value when it should be registering a value of exactly zero.
- A thermometer has been incorrectly calibrated, so it constantly gives temperature readings that are 2 °C lower than the true temperature.
- A parallax error is caused by reading a scale at the wrong angle, for example when your eye is not parallel with the meniscus when using a measuring cylinder.

WORKED EXAMPLE 2

What changes would you make to the results that you collected if you were using an ammeter that looked like Figure 1 before it was connected?

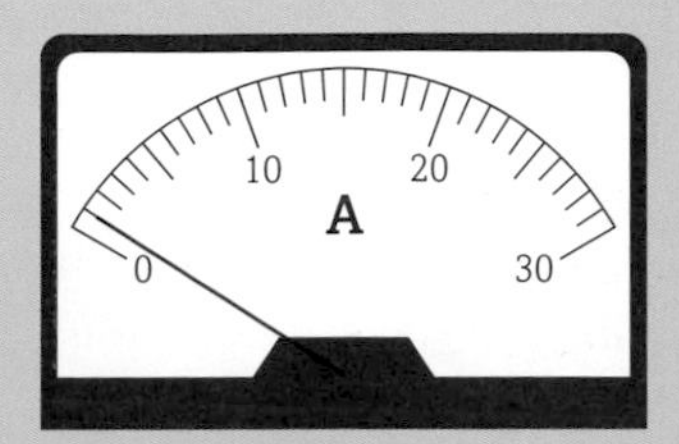

Figure 1 An ammeter scale.

Answers

The ammeter has a zero error – it reads 1 A even when it is not in use. You need to reset the needle so that it points to 0 A when not in use, or subtract 1 A from each value taken during the experiment.

Questions

1. Suggest a reason why some metre rulers are more accurate than others.
2. List one systematic and one random error possible when using a top pan balance to measure mass.
3. List possible systematic errors when measuring the density of a fluid.
4. Read the statements below and explain how each situation could be caused by a random error and/or a systematic error.
   (a) 'This voltmeter is registering a value of 0.1 V even though it is not connected in the circuit!'
   (b) 'This micrometer points to 0.001 mm even when it is supposed to read exactly 0.'
   (c) 'My current readings for the length of 40 cm were 1.56 A, 1.54 A and 1.58 A.'
   (d) 'The wind was blowing and causing the temperature of my water bath to decrease.'
   (e) 'The stopwatch readings are all different when I time the period of oscillation of a pendulum.'

2.1

Precision and accuracy

By the end of this topic, you should be able to demonstrate and apply your knowledge and understanding of:

* the terms precision and accuracy

KEY DEFINITIONS

Precision is the degree to which repeated values, collected under the same conditions in an experiment, show the same results.

Accuracy is the degree to which a value obtained by an experiment is close to the actual or true value.

Introduction

The terms 'accuracy' and 'precision', are often confused or taken to mean the same thing. In everyday conversation this is not really a problem but, in the study of physics, it is essential that you understand and use these terms correctly.

As an analogy, the difference between accuracy and precision can be thought of using the sport of archery. Arrows are fired towards a target, with the intention of scoring as many points as possible. To get the maximum score, the archer has to hit the gold circle at the centre of the target – this is worth 10 points. Hitting the target further away from the central point means that the archer scores fewer and fewer points.

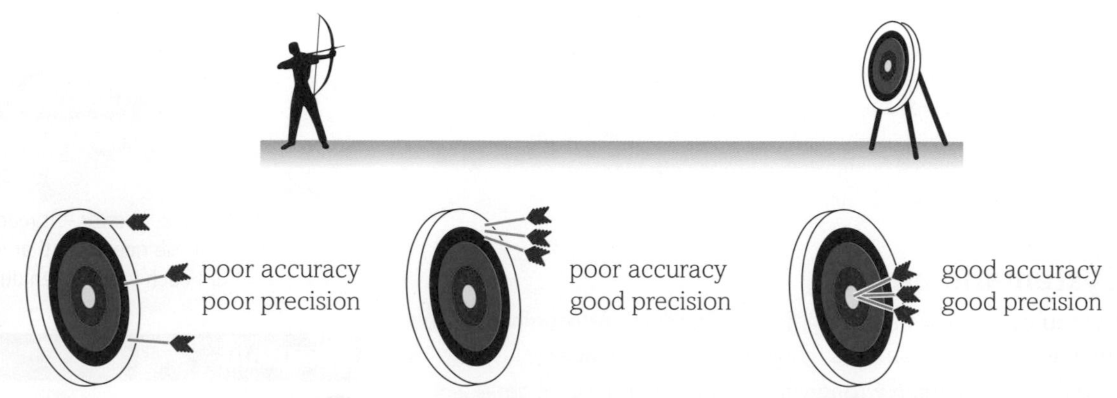

Figure 1 The closer to the centre of the target an arrow lands, the more accurate it is. The closer the arrows cluster together, the more precise the shooting is. The aim is to be accurate *and* precise.

When the shots are accurate and precise, they all land in the middle of the target. When the shots are precise, they do not all land in the centre of the target, but they land very close to one another. Low accuracy and low precision can be thought of as arrows landing far away from the centre of the target and far away from each other.

When we measure a physical quantity there is also some degree of uncertainty in the reading, based on the apparatus and the method of measurement (see Topics 1.1.2 and 1.1.6).

Accuracy and precision defined

An experiment is **accurate** if the quantity being measured has a value that is very close to the commonly accepted or true value. For example, if you were carrying out an experiment to determine the value of the acceleration due to gravity, then a value of $9.83\,\mathrm{m\,s^{-2}}$ would be more accurate than one of $9.17\,\mathrm{m\,s^{-2}}$. This is because the commonly accepted value of the acceleration due to gravity, g, is $9.81\,\mathrm{m\,s^{-2}}$, although the local value of g does vary slightly.

The term **precision** relates to how close together repeat values are. The smaller the spread, or range, of the repeat values (the closer they are to each other) the higher the precision. For example, if you conduct an experiment using instruments that enable you to determine the charge on an electron to ±0.1%, that is much better than obtaining a value that could be out by ±5%.

Precision is often used in conjunction with the term 'resolution', which relates to the smallest change in a quantity that an instrument can measure. An instrument that can measure a change of ±0.01 has a better resolution than one that can measure a change of ±0.1. This will lead to higher precision in the repeat readings.

WORKED EXAMPLE

Suggest a value for the acceleration due to gravity, *g*, and show how the values could be written to 3 significant figures with associated uncertainty:

(a) accurate and precise

(b) accurate but imprecise

(c) inaccurate but precise.

Answers

(a) An accurate and precise value for *g* could be $9.80 \pm 0.02\ \text{m s}^{-2}$.

(b) An accurate and imprecise value for *g* could be $9.80 \pm 1.50\ \text{m s}^{-2}$.

(c) An inaccurate but precise value for *g* could be $8.80 \pm 0.05\ \text{m s}^{-2}$.

So how does a systematic error affect accuracy and precision? Accuracy will be affected by a systematic error that is not accounted for, as the systematic error takes the measured value away from its true value. Systematic errors tend to have the same value throughout an experiment. For example, if a thermometer that has been incorrectly calibrated gives temperature values that are incorrect by 1.0 °C each time. To obtain the 'true' value, you would subtract the systematic error of 1 °C from every reading. However, a systematic error will not affect the precision, which is ultimately determined by the resolution of the apparatus being used. If a thermometer can read a temperature to the nearest 0.5 °C, then this precision will not be affected by a systematic error.

Random errors occur due to unknown and unpredictable factors such as wind variation, temperature changes, power surges and background radiation changes. Due to their random nature, random errors can produce variations in the measured value that are both above and below the 'true' value. The effect of random variation can be reduced by taking repeated measurements and reporting the mean.

Questions

1 An experiment was conducted to determine the speed of sound in air at a given temperature. The accepted value for sound in air at this temperature is $343.4\ \text{m s}^{-1}$. Comment on the accuracy and precision of the values recorded below:

$343.6\ \text{m s}^{-1}$, $344.1\ \text{m s}^{-1}$, $343.2\ \text{m s}^{-1}$, $343.1\ \text{m s}^{-1}$, $342.9\ \text{m s}^{-1}$, $343.7\ \text{m s}^{-1}$

2 The same experiment as described in question 1 was repeated by two students exactly one week later in the same room. Provide another set of six possible values they could have obtained that would be:

(a) less accurate but more precise

(b) more accurate and more precise

(c) more accurate but less precise.

3 Calculate the percentage error in these results by comparing the obtained values with the accepted values. Comment on how accurate the values are:

(a) the density of gold is calculated as $21.4 \pm 0.04\ \text{g cm}^{-3}$

(b) the speed of sound in copper is found to be $5050 \pm 800\ \text{m s}^{-1}$

(c) the mass of the Earth is found to be $5.8 \times 10^{24}\ \text{kg} \pm 0.9\%$.

4 The number π is often quoted as 3.14, but is in fact an irrational number with digits that go on forever. By using the value of 3.14 as an approximation and the value for π on your calculator as the exact value, find:

(a) the absolute error in this value of π, to 6 significant figures

(b) the percentage error in this value of π.

Absolute and percentage uncertainties

By the end of this topic, you should be able to demonstrate and apply your knowledge and understanding of:

* absolute and percentage uncertainties when data are combined by addition, subtraction, multiplication, division and raising to powers

Introduction

In physics, we conduct some experiments to determine the value of a physical quantity. When we make our measurements and calculations, we hope to get a value close to the true value.

However, there is always uncertainty involved when we make measurements. We hope to get 'good' or accurate values but we have to accept that not every measurement we make will be the same every time we repeat it. Measurements may change because of a lack of sensitivity in the instrument we are using, or the skill of the person trying to repeat the same measurement.

KEY DEFINITIONS

The **absolute uncertainty** of a measurement shows how large the uncertainty actually is, and has the same units as the quantity being measured.

Percentage uncertainty is the absolute uncertainty divided by the measured value expressed as a percentage.

Absolute uncertainty

Imagine that you measure the diameter of a ball using a metre ruler that has centimetre and millimetre graduations. You find that the diameter of the ball is 63 mm, so you could state that the diameter of the ball is 63 ± 0.5 mm. When taking *single* readings, the **absolute uncertainty** is usually given as the smallest division on the measuring instrument used.

Determining uncertainty in experiments that involve the use of a stopwatch can be problematic. Despite the fact that most stopwatches used in schools will give values precise to 0.01 s, it is not possible for human reaction times to be this quick. Instead, we use the uncertainty of the reaction time (about 0.5 seconds) when stating a time value and its absolute uncertainty.

Processing experimental data often involves adding or subtracting two quantities to calculate a new quantity. When you add or subtract readings, you must add together the absolute uncertainties of the individual readings to find the combined absolute uncertainty. For example, if you have three measurements 78 ± 1 cm, 43 ± 1 cm and 57 ± 1 cm, the total measurement is 178 ± 3 cm. Alternatively, if you need to subtract 42 ± 2 mm from 87 ± 2 mm, the final measurement is 45 ± 4 mm. The percentage uncertainty in this value can then be calculated as (4/45) × 100% = 9%.

WORKED EXAMPLE 1

A metre ruler is used to measure the height of a table and it is found to be 780 mm. How would you state this value if the ruler used had:

(a) a millimetre scale

(b) a centimetre scale?

Answers

(a) 780 ± 1 mm

(b) 78 ± 1 cm

Percentage uncertainty of a single value

We often need to calculate a **percentage uncertainty** and include it in our evaluation (see Topic 1.1.6). To calculate the percentage uncertainty of a single value, we use the equation:

$$\text{percentage uncertainty} = \left(\frac{\text{uncertainty}}{\text{measured value}}\right) \times 100\%$$

In this case, the uncertainty is given by the resolution of the instrument and the measured value is what has been read from the meter or ruler.

WORKED EXAMPLE 2

A digital ammeter is used to measure the current flowing in a series circuit and it is precise to 0.01 A. What is the percentage uncertainty for a current of:

(a) 0.80 A (b) 4.30 A?

Answers

(a) $\text{percentage uncertainty} = \left(\frac{0.01}{0.8}\right) \times 100\% = 1.25\% = \pm 1\%$ (to 1 s.f.)

(b) $\text{percentage uncertainty} = \left(\frac{0.01}{4.3}\right) \times 100\% = 0.233\% = \pm 0.2\%$ (to 2 s.f.)

LEARNING TIP

Uncertainties should only really be stated to 1 significant figure, so the uncertainty in (a) should be quoted as ±1% and in (b) as 0.2%. This is because there is 'uncertainty in the uncertainty' – it is an estimation and we do not expect, or need, it to be too precise.

Percentage uncertainty for a number of repeat readings

When repeat readings are taken, random errors will be present. This may be due to fluctuations in temperature or unknown changes that we cannot account for or control.

Imagine that we have taken a set of potential difference readings, as shown in Table 1.

Reading 1/V	Reading 2/V	Reading 3/V	Reading 4/V	Mean value/V
3.89	3.88	3.86	3.90	3.88

Table 1

To calculate the percentage uncertainty, we use the following steps.

1. Write down the repeat readings in the table (as shown in Table 1).
2. Find and record the mean of these readings (3.88).
3. Find the range of the repeat readings – this is the largest value minus the smallest value. In this case it is 3.90 V − 3.86 V = 0.04 V.
4. Halve the range to find the absolute uncertainty. In this case it is 0.02 V.
5. Divide the uncertainty value by the mean value and multiply by 100 to give the percentage uncertainty. In this case:

$$\frac{0.02\text{ V}}{3.88\text{ V}} \times 100\% = 0.5\%$$

Remember, we state the percentage uncertainty to 1 significant figure.

WORKED EXAMPLE 3

The temperature of a room was measured several times and the values recorded in Table 2.

Reading 1/°C	Reading 2/°C	Reading 3/°C	Reading 4/°C	Mean value/°C
21.4	21.3	21.4	21.1	21.3

Table 2

Calculate:

(a) the absolute uncertainty in the readings

(b) the percentage uncertainty in the readings.

Answers

(a) The absolute uncertainty in the repeat readings is given by half the range of the values, so:

range = maximum value − minimum value

= 21.4 °C − 21.1 °C = 0.3 °C

So the absolute uncertainty is 0.15 °C, or 0.2 °C to 1 significant figure. We would quote the temperature as (21.3 ± 0.2) °C.

(b) To find the percentage uncertainty, we divide the value of half the range by the mean value, and then multiply by 100:

$$\text{percentage uncertainty} = \left(\frac{0.15}{21.3}\right) \times 100\% = 0.7\%$$

The rules for combining percentage uncertainties

You will be expected to determine the final percentage uncertainty in a compound quantity. This is based on the calculations that need to be carried out in order to find the quantity.

- for a compound quantity of the form $y = ab$, the rule is:
  % uncertainty in y = % uncertainty in a + % uncertainty in b
- for a compound quantity of the form $y = \frac{a}{b}$, the rule is:
  % uncertainty in y = % uncertainty in a + % uncertainty in b
- for a compound quantity of the form $y = a^2$ the rule is:
  % uncertainty in $y = 2 \times$ % uncertainty in a
- for a compound quantity of the form $y = a^n$, the rule is:
  % uncertainty in $y = n \times$ % uncertainty in a.

When you add or subtract readings, you must add together the absolute uncertainties of the individual readings to find the combined absolute uncertainty.

WORKED EXAMPLE 4

(a) Potential difference is calculated by using the equation $V = IR$. If the % uncertainty in the current, I, is 10% and the % uncertainty in the resistance, R, is 5% then what is the % uncertainty in the potential difference, V?

(b) A Young modulus is calculated using the equation Young modulus $= \frac{\text{stress}}{\text{strain}}$. If the percentage uncertainty in the stress is 8% and the % uncertainty in the strain is 6% then what is the % uncertainty in the value for the Young modulus?

(c) The density of a cube is calculated using density $= \frac{\text{mass}}{\text{volume}}$. If the % uncertainty in the mass is 1% and the % uncertainty of the length of one of the sides of the cube is 2% then what will be the percentage uncertainty in the density of the cube?

Answers

(a) $V = IR$ has the form $y = ab$, so we add the individual % uncertainties in the current and the resistance to get a 15% uncertainty in V.

(b) The equation has the form $y = \frac{a}{b}$. This means that we add the values for the stress and strain uncertainties to give an uncertainty in the Young modulus of 14%.

(c) The % uncertainty in the mass is 1%.
Volume is calculated using $V = a^3$, so the % uncertainty in the volume will be 3 × the % uncertainty in the length, i.e. 6%.
Because density $= \frac{\text{mass}}{\text{volume}}$ and has the form $y = \frac{a}{b}$, we add the % uncertainty in the mass to the % uncertainty in the volume, giving a % uncertainty in the density of 7%.

Questions

1. What is the percentage uncertainty in a length of 76 mm measured with a metre ruler that has a millimetre scale?
2. What is the percentage uncertainty for these values for the time of a pendulum swing?
   12.83 s, 12.87 s, 12.85 s, 12.81 s
3. A compound variable is calculated using the formula $y = \frac{ab}{c^3}$.
   What is the percentage uncertainty in y if the percentage uncertainty in a is 3%, b is 6% and c is 2%?
4. Two students compare the width of their hand spans by spreading their fingers as wide as they can. The two measurements are 22.7 ± 0.1 cm and 25.6 ± 0.1 cm.
   What is the percentage uncertainty in the difference in their hand spans?
5. The diameter of a solid sphere is measured with vernier calipers to be 4.73 ± 0.01 cm, and its mass is measured to be 429.20 ± 0.01 g.
   (a) Calculate the density of the sphere in kgm^{-3}.
   (b) Calculate the percentage uncertainty in the density.

2.1

6 Graphical treatment of errors and uncertainties

By the end of this topic, you should be able to demonstrate and apply your knowledge and understanding of:

* **graphical treatment of errors and uncertainties; line of best fit; worst acceptable line; absolute and percentage uncertainties; percentage difference**

LEARNING TIP

Using a large triangle to calculate the gradient reduces the percentage error when reading from the axes.

Using graphs

We make full use of graphs in physics to show relationships between pairs of dependent variables and independent variables. Graphs are a useful, highly visual way of demonstrating the relationship between two variables, showing patterns and trends and allowing us to determine values from measurements of the gradient and the *y*-intercept.

Graphs are most effective when:

- the scale of the graph has been chosen so that the plotted points cover as much of the graph paper as possible in both directions
- the points are plotted clearly
- the lines of best fit and worst fit are drawn clearly
- the gradient can be calculated easily using two points on the line that are as far apart as possible, but within the measured range.
- the *y*-intercept can be read clearly and accurately using the scale on the *y*-axis.

KEY DEFINITIONS

Percentage difference is the difference between two values, divided by the average and shown as a percentage.

Error bars on a graph represent the absolute uncertainty in measurements and can be plotted in the x and y directions.

A **worst fit** line is the worst acceptable line, still passing through all the error bars. This will be either the steepest possible line of fit, or the least steep line of fit.

Determining the uncertainty in the gradient from the maximum and minimum gradients

It is possible to determine the uncertainty in a gradient by drawing lines of maximum and minimum gradient through the appropriate points on the graph. If there is only a small amount of scatter then **error bars** can be incorporated into the graph to help this to happen.

The uncertainty in a gradient can be determined as follows:

1. Add error bars to each point. The size of the error bars are usually the same for each measurement.
2. Draw a line of best fit through the scattered points and within the error bars. The line of best fit should go through as many points as possible, with equal numbers of points above and below the line. Discard any major outliers.
3. Calculate the gradient of the line of best fit.
4. Do the same for the **worst fit line**, which may be more steep or less steep than the line of best fit.
5. To find the uncertainty from the graph, work out the difference between the gradients of the line of best fit and the line of worst fit. This should be expressed as a positive value (the modulus). The equation you use is:

   uncertainty = (gradient of best fit line) − (gradient of worst fit line)
6. Calculate the percentage uncertainty in the gradient using the following equation:

$$\text{percentage uncertainty} = \left(\frac{\text{uncertainty}}{\text{gradient of best fit line}}\right) \times 100\%$$

For example, if the gradient of the line of best fit has a value of 3.7 and the gradient of the worst acceptable line was 4.3, the percentage uncertainty would be:

$$\frac{(4.3 - 3.7)}{3.7} \times 100\% = 16\%$$

Alternatively, you can draw a graph that has a line of best fit, a maximum gradient line and a minimum gradient line. In this case, the uncertainty is half the difference between the maximum and minimum gradients, as shown in Worked example 1.

WORKED EXAMPLE 1

An experiment is performed to determine the relationship between the current flowing through an electrical component and the potential difference across it. The current is measured to ±2.5 mA and the potential difference to ±0.1 V; these are the values used to plot the error bars on the *y*- and *x*-axes respectively.

Calculate:

(a) the gradient of the line of best fit

(b) the gradient of the line of minimum gradient

(c) the gradient of the line of maximum gradient

(d) the uncertainty in the gradient.

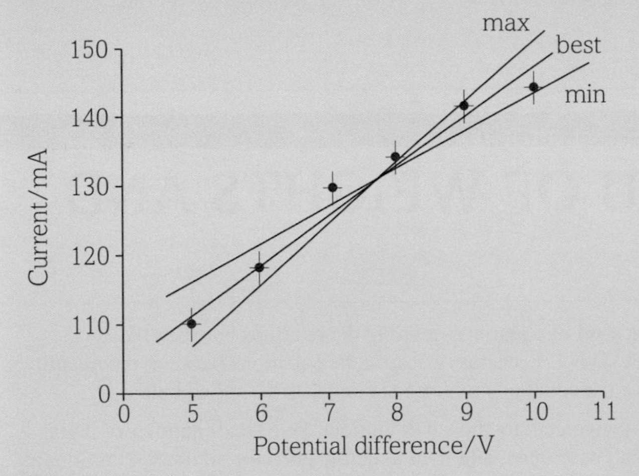

Figure 1 Lines of best fit.

Answers

(a) Gradient $= \frac{147\text{ mA} - 107\text{ mA}}{10\text{ V} - 4.5\text{ V}} = 7.27\text{ mA V}^{-1}$

(b) Gradient $= \frac{145\text{ mA} - 115\text{ mA}}{10.5\text{ V} - 5.0\text{ V}} = 5.45\text{ mA V}^{-1}$

(c) Gradient $= \frac{152\text{ mA} - 106\text{ mA}}{10\text{ V} - 5.0\text{ V}} = 9.20\text{ mA V}^{-1}$

(d) The uncertainty in the gradient is half the difference between the minimum line gradient and the maximum line gradient.

$$\text{uncertainty} = \tfrac{1}{2} \times (9.20 - 5.45) = 1.875\text{ mA V}^{-1}$$

The gradient can now be written as $(7.3 \pm 2)\text{ mA V}^{-1}$ with the gradient given to 2 significant figures and the uncertainty quoted to 1 significant figure.

Determining the uncertainty in the *y*-intercept from the maximum and minimum gradients

You may be asked to determine the uncertainty in the *y*-intercept by using the lines of maximum and minimum gradient and the line of best fit. As before, you can draw through points to obtain these lines or you can draw the lines within the error bars. Once the graph is plotted, with or without error bars, you need to do the following:

1. Draw the line of best fit and extrapolate it to find the 'best' value of the *y*-intercept.
2. Draw the lines of maximum and minimum gradient and extrapolate them back to get the maximum and minimum values for the *y*-intercept.
3. Determine the difference between the maximum and minimum values then halve it to find the uncertainty in the *y*-intercept value.
4. To find the percentage uncertainty in the *y*-intercept value, use the equation:

$$\text{percentage uncertainty} = \left(\frac{\text{uncertainty}}{\text{'best' } y\text{-intercept value}}\right) \times 100\%$$

LEARNING TIP

If the *y*-intercept does not fit on your graph when the time of best fit is extrapolated, you can use the gradient and the equation $y = mx + c$ to estimate the *y*-intercept *c*.

WORKED EXAMPLE 2

In an experiment to find the terminal p.d. of a cell, a graph of potential difference against current gives a best fit line *y*-intercept value of 12.8 V, and maximum and minimum line intercept values of 11.4 V and 14.7 V respectively. Use this information to find:

(a) the uncertainty

(b) the % uncertainty in the *y*-intercept value

(c) the answer that should be stated.

Answers

(a) Uncertainty $= \tfrac{1}{2}(14.7 - 11.4)$

$= \tfrac{1}{2} \times 3.3 = \pm 1.65\text{ V}$

(b) Percentage uncertainty $= \left(\frac{1.65}{12.8}\right) \times 100\%$

$= \pm 12.9\%$

(c) 12.8 ± 1.7 V or 12.8 ± 2 V

Questions

1 By extrapolating the line of best fit, the minimum line and the maximum line to the *y*-axis in Figure 1, work out the % uncertainty in the *y*-intercept.

2 Copy the graph shown in Figure 2 and work out the % uncertainty in the gradient and in the *y* intercept. The line of best fit has been drawn for you.

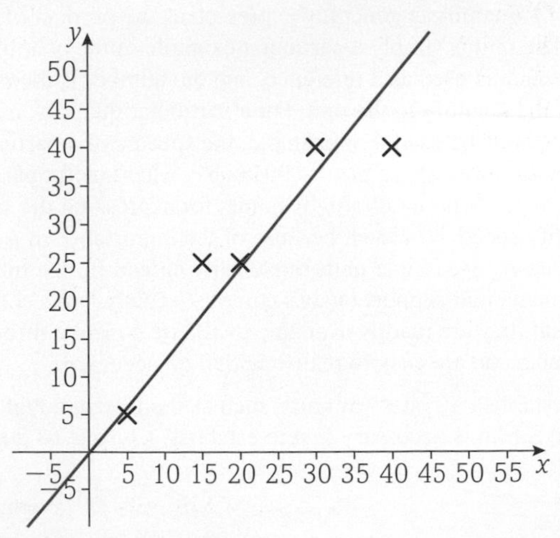

Figure 2

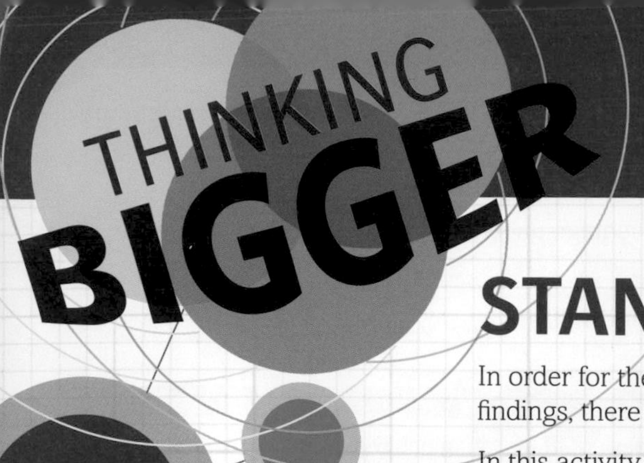

STANDARDISATION

In order for there to be a consistent approach to how scientists communicate and compare their findings, there must be agreed conventions relating to the use of physical quantities and their units.

In this activity you will develop an understanding of why it is important to monitor worldwide uniformity in units of measure, and how decisions are made about the definition of a unit.

THE INTERNATIONAL BUREAU OF WEIGHTS AND MEASURES (BIPM)

The International Bureau of Weights and Measures (BIPM) was set up by the Metre Convention signed in Paris on 20 May 1875. [...]

The task of the BIPM is to ensure worldwide unification of measurements; its function is thus to:

- Establish fundamental standards and scales for the measurement of the principal physical quantities and maintain the international prototypes;
- Carry out comparisons of national and international standards;
- Ensure the coordination of corresponding measurement techniques;
- Carry out and coordinate measurements of the fundamental physical constants relevant to these activities.

The BIPM operates under the exclusive supervision of the International Committee for Weights and Measures (CIPM) [...] and reports to it on the work accomplished by the BIPM.

The CIPM meets every year. The principal task of the CIPM is to ensure worldwide uniformity in units of measurement. It does this by direct action or by submitting proposals to the CGPM.

Quantities and units

The value of a quantity is generally expressed as the product of a number and a unit. The unit is simply a particular example of the quantity concerned which is used as a reference, and the number is the ratio of the value of the quantity to the unit. For a particular quantity, many different units may be used. For example, the speed v of a particle may be expressed in the form $v = 25\ \text{m s}^{-1} = 90\ \text{km h}^{-1}$, where metre per second and kilometre per hour are alternative units for expressing the same value of the quantity speed. However, because of the importance of a set of well defined and easily accessible units universally agreed for the multitude of measurements that support today's complex society, units should be chosen so that they are readily available to all, are constant throughout time and space, and are easy to realise with high accuracy.

In order to establish a system of units, such as the International System of Units, the S.I., it is necessary first to establish a system of quantities, including a set of equations defining the relations between those quantities. This is necessary because the equations between the quantities determine the equations relating the units, as described below.

It is also convenient to choose definitions for a small number of units that we call base units, and then to define units for all other quantities as products of powers of the base units that we call derived units. In a similar way the corresponding quantities are described as base quantities and derived quantities, and the equations giving the derived quantities in terms of the base quantities are used to determine the expression for the derived units in terms of the base units. Thus in a logical development of this subject, the choice of quantities and the equations relating the quantities comes first, and the choice of units comes second.

From a scientific point of view, the division of quantities into base quantities and derived quantities is a matter of convention, and is not essential to the physics of the subject. However for the corresponding units, it is important that the definition of each base unit is made with particular care, to satisfy the requirements outlined in the first paragraph above, since they provide the foundation for the entire system of units.

The definitions of the derived units in terms of the base units then follow from the equations defining the derived quantities in terms of the base quantities. Thus the establishment of a system of units, which is the subject of this brochure, is intimately connected with the algebraic equations relating the corresponding quantities.

Source

Taylor, B.N. and Thompson, A. (eds) *The International System of Units (SI)* (2008), NIST Special Publication 330, accessed at http://physics.nist.gov/Pubs/SP330/sp330.pdf

DID YOU KNOW?

The terms kilo, micro, mega and nano are prefixes that we use to indicate multiplying factors of S.I. units. Other prefixes used in the S.I. system include yocto, zepto, atto, femto, exa, zetta and yotta.

Where else will I encounter these themes?

1.1 | 2.1 | YOU ARE HERE 2.2 | 3.1 | 3.2 | 3.3

Let us start by considering the nature and context of the writing in the article. The text above comes from a publication that is the US version of the English text of a publication by the BIPM.

1. **Read the text and consider the nature of the writing being used. For example, think about why the document has been written and who the audience is. Would an article like this be widely seen in the scientific community? Would the document be useful to the typical man or woman in the street?**

We will now focus on the physics that is contained within this document. Do not worry if much of the content is unfamiliar to you or if you are not clear about your answers. You may wish to return to some of these questions once you have covered more of the topics later in the textbook. Use the timeline at the bottom of the page to help you put this work in context with what you have already learned and what is ahead in your course.

2. What are the main functions of the BIPM?
3. Why is the work of the BIPM so important:
   a. nationally
   b. internationally?
4. What is the difference between a base unit and a derived unit?
5. Later in this publication, a metre is defined in the following terms:
   'The metre is the length of the path travelled by light in vacuum during a time interval of 1/299 792 458 of a second.'
   Why is the definition of the metre based on the speed of light?
6. The quantity dynamic viscosity is defined as pressure × time. What are its base units?
7. The BIPM have recently stated that:
   'In the "New SI" four of the SI base units, namely the kilogram, the ampere, the kelvin and the mole, will be redefined in terms of invariants of nature; the new definitions will be based on choosing exact numerical values of the Planck constant (h), the elementary charge (e), the Boltzmann constant (k), and the Avogadro constant (N_A), respectively.'
   Why do you think they have decided to do this?
8. How might the practice of science be different if the BIPM did not exist?

Make sure you know how to obtain derived quantities from base quantities, using a defining equation. You can determine the base units of any physical quantity if you know this information. For example, the base units of the volt can be shown to be kg m^{2}A^{-1} s^{-3}.

You should consider what a measurement standard is.

Activity

The ampere is currently defined as 'that constant current which, if maintained in two straight parallel conductors of infinite length, of negligible circular cross section, and placed 1 metre apart in vacuum, would produce between these conductors a force equal to 2×10^{-7} newton per metre of length'.

How would you go about determining the exact value of the ampere value experimentally? What apparatus would you need? What problems might you encounter in determining an accurate value? What definitions of other base units are required to define the ampere?

In the new definition of the ampere, the magnitude of the ampere is set by fixing the numerical value of the elementary charge, when it is expressed in the unit As, which is equal to C.

Design an A0-sized poster to explain the advantages of the new S.I. units. Include details of how the ampere is currently defined, as an example of the disadvantages of the current system.

2.1 Practice questions

1. Which of the following is greatest in value? [1]
   - A mega × nano
   - B mega ÷ micro
   - C giga ÷ micro
   - D kilo ÷ (milli)^3

2. Which quantity below has units that could be expressed as kg m s^{-2}? [1]
   - A momentum
   - B energy
   - C impulse
   - D force

3. The radius of a metal sphere is 4.95 cm. What would be a good estimate of its volume? [1]
   - A 80 cm^3
   - B 500 cm^3
   - C 1500 cm^3
   - D 750 cm^3

4. Which of the following are the base units for the ohm? [1]
   - A kg m^3 s^{-1}
   - B kg m^2 s^{-3}A^{-2}
   - C kg m^{-1}A^{-1}
   - D V A^{-1}

5. An analogue ammeter registers a value of 0.50 A when no current is flowing through it. The experimenter then records three repeat values of 3.50 A, 3.52 A and 3.46 A.

   What is the nature of the error(s) associated with the current measurement? [1]
   - A There are random errors but no systematic errors.
   - B There are systematic errors but no random errors.
   - C There are no errors present.
   - D There are both systematic and random errors.

[Total: 5]

6. Draw a line from the unit on the left-hand side to the equivalent unit on the right-hand side. [2]

joule (J)	kgm s^{-2}
watt (W)	Nm
newton (N)	J s^{-1}

[Total: 2]

7. Match the prefix calculation on the left-hand side with the correct power on the right-hand side. [3]

kilo × mega	10^{-6}
kilo ÷ mega	10^9
nano ÷ milli	10^{-3}
micro × milli	10^{-9}

[Total: 3]

8. An experiment to determine the acceleration due to gravity, g, uses the equation $g = \frac{4\pi l^2}{T^2}$ where l is the length of the pendulum in metres and T is the time period of oscillation of the pendulum in seconds.

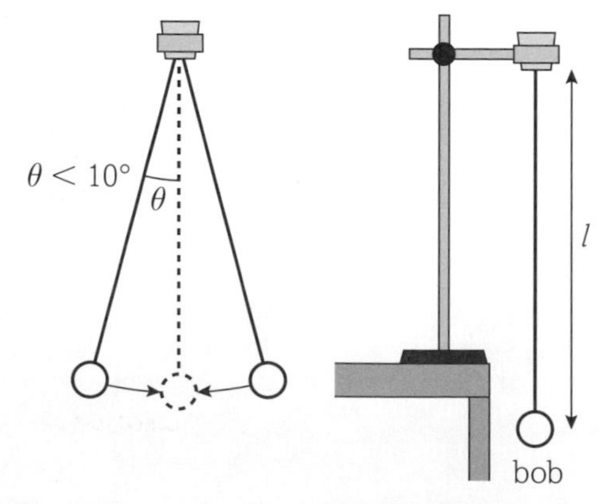

The readings taken were: $l = 1.250 \pm 0.001$ m and $T = 2.25 \pm 0.02$ s.

Calculate:

- the value of g from this data [2]
- the percentage uncertainty in g. [2]

[Total: 4]

9. The force F acting on a body is obtained by using the equation $F = \frac{mv^2}{6(x_2 - x_1)}$. The values obtained in the investigation were $m = 50.0 \pm 0.5$ kg, $v = 6.0 \pm 0.2$ m s^{-1}, $x_2 = 4.8 \pm 0.2$ m and $x_1 = 3.2 \pm 0.1$ m. Calculate:
   - the size of the force, F, acting on the body in N [2]
   - the percentage uncertainty in the value of F. [2]

   [Total: 4]

10. An experiment is conducted to determine the density of a rectangular metal block. The mass of the block and its dimensions are as stated below:

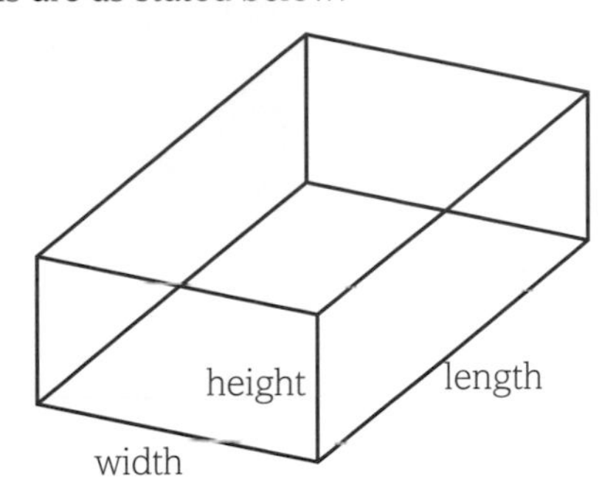

- Length = 96 ± 0.5 mm
- Height = 15 ± 0.5 mm
- Width = 42 ± 0.5 mm
- Mass = 532 ± 0.5 g

The density calculation is performed twice – once with the actual numbers, and secondly with the highest possible numbers on the top of the equation and the lowest possible numbers on the bottom of the equation.

As density $= \frac{\text{mass}}{\text{volume}}$, show that the answer can be stated as

density = 8800 ± 500 kg. [5]

[Total: 5]

11. A force, F, acts vertically downwards on a circular area, *A*, as shown in the diagram below. If the percentage uncertainty in *F* is 8% and the percentage uncertainty in the radius of the circle is 3% then what will be the percentage uncertainty in the value obtained of the pressure exerted by the force? [3]

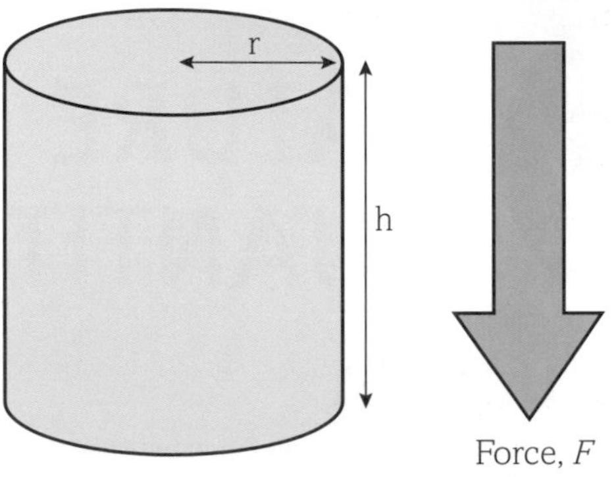

[Total: 3]

MODULE 2 Foundations of physics

CHAPTER 2.2

NATURE OF QUANTITIES

Introduction

Physics is a subject concerned with both magnitude and direction. From the very beginning of the Universe, some 14 billion years ago, to the most contemporary scientific discoveries, we have become fascinated with collisions and explosions, including their respective sizes and directions. The Universe we inhabit is expanding, accelerating as it does so, with the velocities of galaxies increasing as they get further and further away from us. The smallest of subatomic particles, discovered for the first time in particle accelerators such as CERN were found through high-energy head-on collisions. The space programmes that took man beyond the Earth's atmosphere were planned, designed and built by great minds – focusing on the equations of energy, momentum, force, power and distance that would propel craft upwards and beyond the sky, sometimes to return, sometimes not. In this chapter you will learn about scalar and vector quantities. Both of these have a size or a magnitude, but only vectors have a direction associated with them. You will learn the importance of these quantities, how they are used and how they describe so much of the physical world that we see and feel around us.

All the maths you need

To unlock the puzzles of this chapter you need the following maths:

- Units of measurement
- Ratios
- How to calculate volume
- How to use sine and cosine of an angle
- How to use Pythagoras' theorem
- How to construct a scale drawing using triangles

What will I study later?

- How horizontal and vertical components of vectors apply to work and power calculations (AS)
- How to apply vectors to the areas of kinematics and dynamics (AS)
- How to use the Kelvin scale to measure absolute temperature and its relationship to the Celsius scale (AL)
- The mole and Avogadro's number (AL)
- The mean square speed and the root mean square speed when describing the motion of particles in a gas (AL)
- The exponential laws that govern the charging and discharging of a capacitor (AL)
- How to apply scalar and vector calculations to gravitational and electrical fields (AL)
- The vector and scalar quantities involving electric, magnetic fields and gravitational (AL)
- The exponential and logarithmic relationships between half-life, decay constant and activity (AL)
- The attenuation coefficient of X-rays and the exponential link to the thickness of material through which the X-rays are travelling (AL)

What have I studied before?

- Units of measurement
- Addition and subtraction of scalar quantities
- Rearranging formulae and substitution of values
- Standard form

What will I study in this chapter?

- The nature of scalar quantities and examples of scalars
- The nature of vector quantities and examples of vectors
- Calculations involving only scalar quantities
- Calculations involving only vector quantities
- Using the vector triangle to determine the resultant of any two coplanar vectors
- The resolving of a vector into its perpendicular components, F_x and F_y

2.2 1 Scalar and vector quantities

By the end of this topic, you should be able to demonstrate and apply your knowledge and understanding of:

* scalar and vector quantities

KEY DEFINITIONS

A **scalar** quantity has magnitude but not direction.
A **vector** quantity has magnitude and direction.

Scalar and vector quantities

In physics, we often deal with scalar and vector quantities in problems and calculations. Scalar and vector quantities share something in common, but they are also different in a very definite and important way.

Tables 1 and 2 list examples of scalar and vector quantities with common units, along with an example of how each term could be used in physics.

Scalar quantity	Units	Definition
mass	kg	The matter content of a body
density	$kg\,m^{-3}$	Mass per unit volume
volume	m^3	Three-dimensional space occupied by a body
distance	m	Length from one point to another
speed	$m\,s^{-1}$	Distance travelled per unit time
energy	J	Work done
power	W or $J\,s^{-1}$	Energy converted or work done per unit time

Table 1 Scalar quantities.

Vector quantity	Units	Definition
weight	N	Force, acting downwards through a body, due to gravity
pressure	Pa or $N\,m^{-2}$	Force per unit area
impulse	$kg\,m\,s^{-1}$	force × time
displacement	m	Distance from a specific point in a particular direction
velocity	$m\,s^{-1}$	Displacement per unit time
acceleration	$m\,s^{-2}$	Change in velocity per unit time
momentum	$kg\,m\,s^{-1}$	mass × velocity

Table 2 Vector quantities.

Figure 1 This man has a mass of 78 kg. This tells us the amount of matter in his body. Mass is expressed as a size or quantity, but it has no direction associated with it – hence it is a scalar quantity.

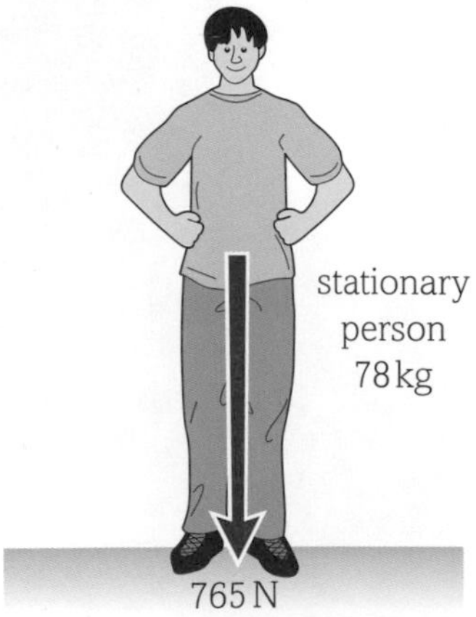

stationary person 78 kg

765 N

Figure 2 This man has a mass of 78 kg. His weight, calculated by using the formula $W = m \times g$, will be 78 kg × 9.81 $N\,kg^{-1}$ or approximately 765 N downwards. Notice how the weight has both a size and a direction because it is a vector quantity.

How are scalar quantities different from their vector equivalents?

We often refer to mass as 'weight'. People will say they 'weigh' 78 kg, when actually they are referring to their mass. Mass is a scalar quantity and has size only – in this case, a value of 78 kg. This is a measure of how much matter the person is composed of. It has no direction associated with it.

Conversely, weight is a force and has a size that we can calculate from the equation weight = mass × gravitational field strength, or $W = m \times g$. On Earth, the value for g is about 9.8 $N\,kg^{-1}$, so the weight of a person of mass 78 kg is about 7645 N, downwards. Weight always acts towards the centre of the Earth, so it has a direction associated with it.

Two very similar quantities that we meet in physics problems are speed and velocity. Speed is a scalar quantity; velocity is a vector quantity. Imagine a 400 m race in athletics. The average speed of the athlete is given by the equation

average speed $= \frac{\text{distance}}{\text{time}}$. So, if the athlete runs 400 m in 40 seconds, then the average speed is 10 $m\,s^{-1}$.

Distance is a scalar quantity. If we want to consider the direction of travel as well as the distance, then we need to use the vector quantity 'displacement'. The average total velocity of the athlete is given by

$$\text{average velocity} = \frac{\text{total displacement}}{\text{time}}$$

KEY DEFINITION

Displacement is the distance moved by an object in a particular direction. It is a vector quantity.

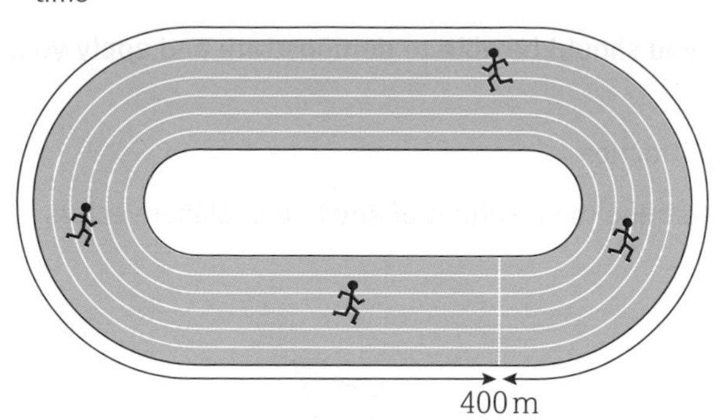

Figure 3 As the athlete runs round the track he covers, on average, a distance of 10 m each second, so his average speed is $10\,\text{m}\,\text{s}^{-1}$. He covers the distance of 400 m in exactly 40 s. Although this athlete has an average speed of $10\,\text{m}\,\text{s}^{-1}$ at the end of the race, his net displacement from the starting line is zero. This means that his average velocity, by definition, must also be zero. The vector nature of velocity means that we need to consider the direction of motion as well as the magnitude, which is his speed.

The athlete's displacement is how far he is from his starting point at the end of the race. Although he has run 400 m, his displacement is 0 m. Although his average speed is non-zero, his average velocity must be zero because his total displacement is zero. Another way to think about this is that his velocity is positive for the first half of the race but negative for the other half, because his direction has reversed. The velocities sum to zero. So, the athlete has an average speed of $10\,\text{m}\,\text{s}^{-1}$ but an average velocity of zero. No matter how fast he runs, his average velocity for the race will be zero!

Imagine a swimmer doing lengths in a swimming pool at a steady speed of $1\,\text{m}\,\text{s}^{-1}$. After two lengths the swimmer is back where she started and her displacement is zero. Her average velocity is also zero. Another way to think about it is that the velocities of $+1\,\text{m}\,\text{s}^{-1}$ and $-1\,\text{m}\,\text{s}^{-1}$ sum to zero.

Questions

1. Explain whether the following statements contain scalar or vector elements.
   (a) The car had an average speed of $34\,\text{m}\,\text{s}^{-1}$.
   (b) The car had an average speed of $34\,\text{m}\,\text{s}^{-1}$ in a southerly direction.
   (c) I can add these masses together to get a total mass of 45.7 kg.
   (d) I am a total of 45 m away from my starting point.
   (e) The temperature has increased from 280 K to 320 K.
   (f) The density of air is $1.290\,\text{kg}\,\text{m}^{-3}$.
   (g) The change in momentum was $11\,\text{kg}\,\text{m}\,\text{s}^{-1}$.

2. A woman goes into her local supermarket and walks straight ahead for 20 m before turning right to enter the aisles. She turns right again after 5 m and stops after another 8 m.
   (a) What is her total distance travelled? (b) What is her total displacement?

3. A disc of radius 0.20 m rotates once about its centre in 4.0 s. A point, P, is marked on its circumference as shown in Figure 4. P is the most southerly point of the disc at $t = 0$ s.

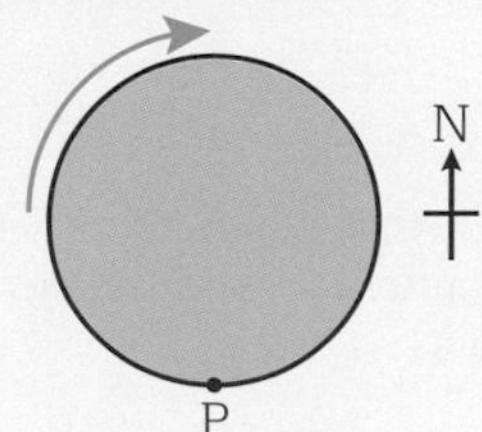

Figure 4

   (a) Calculate the magnitude and direction of the displacement of P from its initial position at $t = 1.0$ s, 2.0 s, 3.0 s and 4.0 s.
   (b) Calculate the magnitude and direction of the displacement of P from its original position at $t = 3\frac{1}{3}$ s. What distance has point P travelled in this time?

2.2  Scalar and vector calculations

By the end of this topic, you should be able to demonstrate and apply your knowledge and understanding of:

* vector addition and subtraction
* vector triangles to determine the resultant of any two coplanar vectors

Scalar arithmetic

Scalar addition and subtraction is very straightforward. We simply add or subtract the quantities, provided they are expressed in the same units.

For example, masses of 6.2 kg, 4.3 kg and 3.9 kg give a total mass of 14.4 kg. We have simply added the three quantities together because they are all expressed in kg. If the third mass above, 3.9 kg, was given as 3900 g we would have to convert it to kilograms by dividing by 1000 before adding it to the other values.

For scalar subtraction we subtract one value from another. For example, if a bathroom wall had a total area of 12.87 m^2 and 8.30 m^2 had been tiled, then 4.57 m^2 would be left untiled. We simply subtract 8.30 m^2 from 12.87 m^2. No conversion between units is required because both values are expressed in m^2. If one of the values is given in cm^2 we need to convert it to m^2 before we do the subtraction. We can do this by dividing the cm^2 value by 10 000 to convert it to m^2.

Also, we cannot add scalar quantities that have different units. For example, we cannot add 4 g to 25 s because mass and time are different physical quantities and adding them together is meaningless.

Vector arithmetic

Vectors are often represented by arrows. The length of an arrow represents its magnitude and the orientation and position of the arrow shows its direction. A scale can be used to determine the vector's magnitude (size) as shown in Figures 1, 2 and 3.

N

(scale: 1 cm = 5 m s^{-1})

Figure 1 This arrow represents the magnitude and direction of the velocity of a moving vehicle. It is 5 cm long and it is pointing north. Each 1 cm of length represents a velocity of 5 m s^{-1}. Therefore the vehicle must be travelling north at 25 m s^{-1}, which is how we state its velocity.

NE

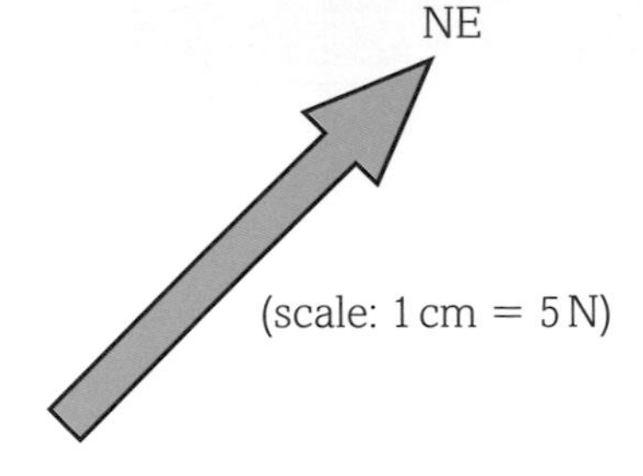

(scale: 1 cm = 5 N)

Figure 2 This arrow represents a force, which is also a vector because it has a size and a direction. Each 1 cm of the arrow's length represents a force of 5 N. Because the arrow is 4 cm long, the force must have a magnitude of 20 N. It is acting at 45° to the vertical, so we can state its direction as '45° to the vertical', '45° to the horizontal' or 'north east' (NE).

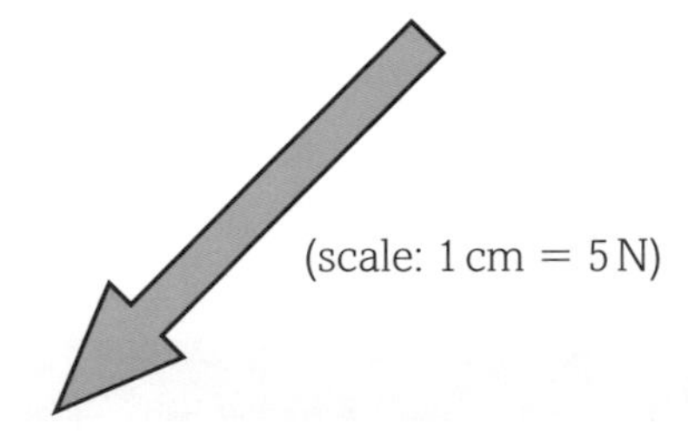

(scale: 1 cm = 5 N)

Figure 3 This vector has an equal magnitude but the opposite direction to the vector shown in Figure 2. These vectors have equal magnitudes and opposite directions so the resultant force would be zero if they acted on the same object.

When performing vector addition, both the magnitude and direction of the vectors are important and must be taken into consideration.

WORKED EXAMPLE 1

Two men push a car to get it moving (Figure 4). The driver then manages to get the engine started. What is the total force acting on the car?

Figure 4

Answer

The driving force of the car and the pushing forces from the two men are all acting in the same direction so we simply add them together. The resultant force is 390 N forwards.

WORKED EXAMPLE 2

Later on (Figure 5) the car is accelerating and experiencing drag. What is the overall net (or resultant) force acting on the car now?

These forces are acting in opposite directions (they are anti-parallel) so we perform the sum:

resultant force acting on the car = 3450 N + −1200 N = 3450 N − 1200 N = 2250 N to the right.

Notice how the resulting force has both a size (2250 N) and a direction (to the right).

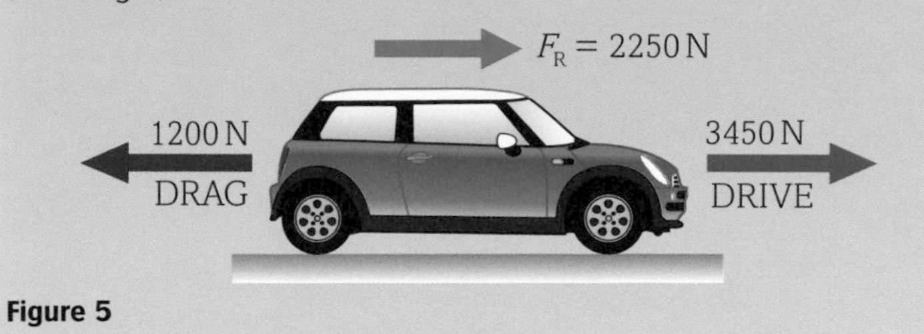

Figure 5

Vector triangles

Obviously, vectors do not always act parallel or anti-parallel to one another. Figure 6 shows an aircraft cruising at a speed of 360 m s^{-1} north with the wind blowing at a speed of 85 m s^{-1} from the east.

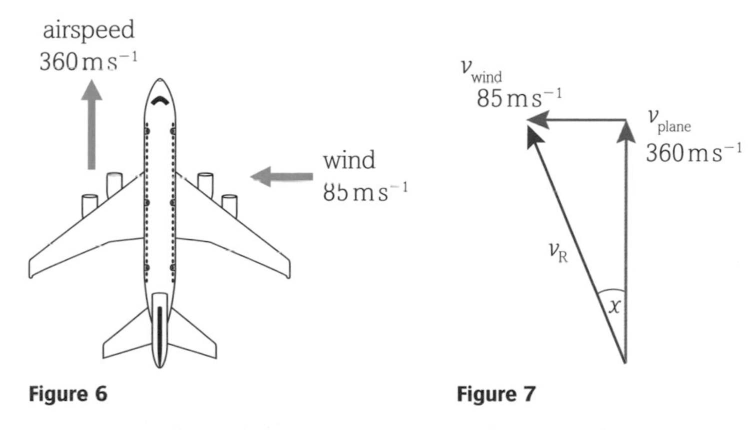

Figure 6 **Figure 7**

Figure 7 shows how this scenario can be illustrated using a vector.

KEY DEFINITIONS

A **vector triangle** is a type of scale diagram with two vectors, drawn tip-to-tail, to show how they can be added together.

The **resultant vector** or the sum of the two vectors forms the third side of the triangle.

We can work out the 'vector sum' of these velocities to find the actual speed and direction of the aircraft. Since the two vectors act as right angles to each other we use Pythagoras' theorem and trigonometry to calculate the magnitude of the **resultant vector**, v_R:

$$v_R = \sqrt{(v_{wind}^2 + v_{plane}^2)}$$
$$= \sqrt{(7225 + 129\,600)}$$
$$= 370 \text{ m s}^{-1}$$

We use trigonometry to determine the direction of v_R:

The angle, x, is given by $x = \tan^{-1}\left(\frac{85}{360}\right) = 13.3°$ (to 3 significant figures), or a bearing of 346.7

LEARNING TIP

Vector calculations can be performed on any vector quantities – velocity, force, momentum, acceleration, etc.

Adding vectors that are not at right angles

Vector triangles are also used to find the resultant of two vectors not at right angles. However, Pythagoras' theorem cannot be used in these cases. You must make a scale drawing of the vector triangle and then use a ruler to measure the size of the resultant and a protractor to find its direction. Always state the angle with reference to a specified direction, for example measured clockwise from north or anticlockwise from the positive x direction.

Questions

1 Calculate the resultant velocity and the angle shown in Figure 8.

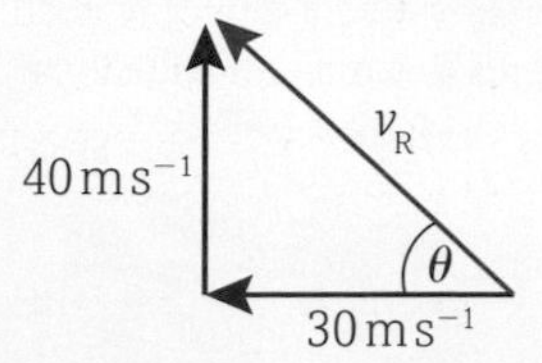

Figure 8

2 Calculate the resultant force and the angle shown in Figure 9.

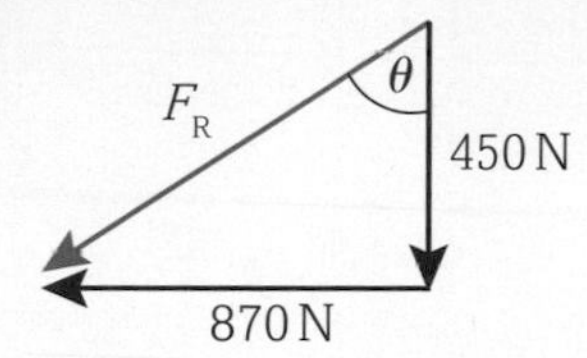

Figure 9

3 Calculate the resultant acceleration and the angle in Figure 10.

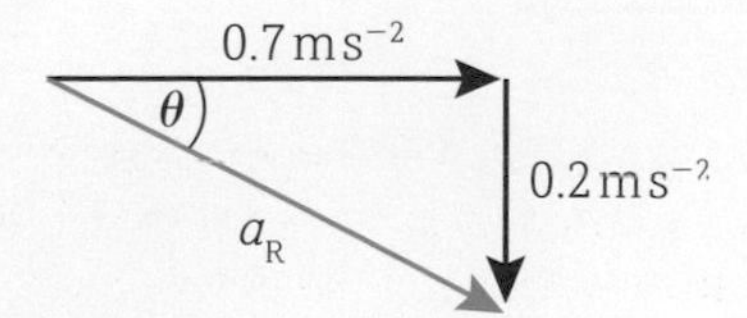

Figure 10

4 A plane has a velocity through the air of 240 m s^{-1} due north. The wind velocity is 90 m s^{-1} from the west. Calculate the velocity of the plane over the ground.

5 A plane completes a journey in two stages. The displacements at each stage are 4000 km at a compass bearing of 130° and 3000 km at a bearing of 340°. Use a scale diagram to find the total displacement and its bearing.

6 An aircraft sets a course northeast with a speed of 150 km h^{-1}. There is a wind of 40 km h^{-1} from the west. Find the resultant velocity of the aircraft (the actual motion of the aircraft over the ground), first by scale drawing and then by calculation. (Hint – use the cosine rule.)

Resolving vectors

By the end of this topic, you should be able to demonstrate and apply your knowledge and understanding of:

* resolving a vector into two perpendicular components; $F_x = F\cos\theta$ and $F_y = F\sin\theta$

Resolving vectors

As shown earlier, a vector triangle can be used to find a resultant vector when two vectors are at right angles to each other. We use Pythagoras' theorem to calculate the magnitude (size) of the resultant vector, and we use trigonometry to determine the direction of the resultant vector.

WORKED EXAMPLE 1

Figure 1 shows a speedboat travelling directly across a river of width 1200 m at a velocity of $12\,\text{m s}^{-1}$. The river is flowing at a velocity of $5\,\text{m s}^{-1}$ from west to east.

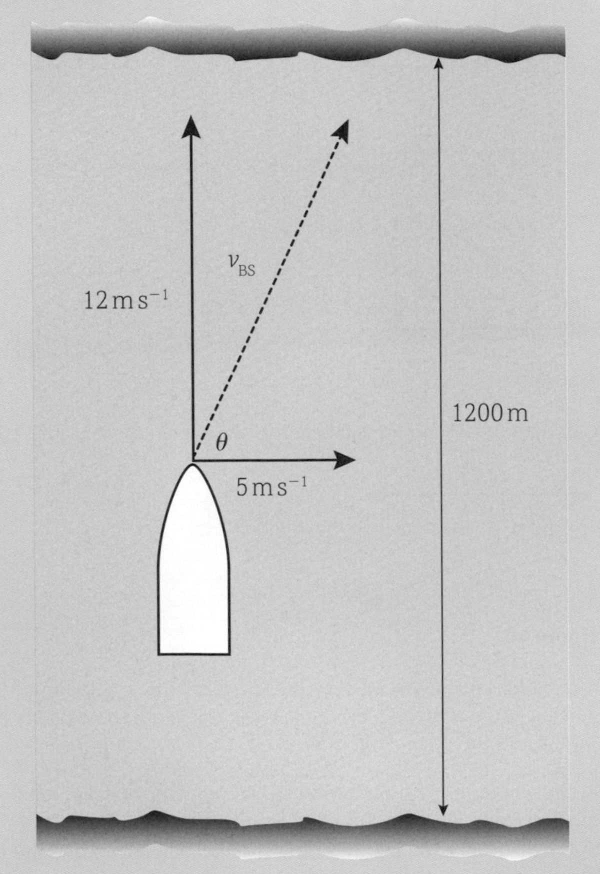

Figure 1

Calculate:

(a) the resultant velocity of the boat, v_{BS}

(b) the angle, θ, at which the boat travels with respect to the horizontal

(c) the time it takes the boat to cross the river

(d) the distance to the east that the boat arrives at the other side of the river compared with its starting position.

Answers

(a) $v_{BS} = \sqrt{(v_{boat}{}^2 + v_{current}{}^2)} = \sqrt{(5^2 + 12^2)} = 13\,\text{m s}^{-1}$

(b) $\theta = \tan^{-1}\left(\frac{12}{5}\right) = 67.4°$

(c) $\text{time} = \frac{\text{distance travelled}}{\text{speed}}$

$= \frac{1200\,\text{m}}{12\,\text{m s}^{-1}} = 100\,\text{s}$

(d) The distance east is found using trigonometry. In Figure 2 the distance travelled east, relative to the starting point is:
$x = 1200\,\text{m} \times \tan 22.6° = 500\,\text{m}$ east of the point directly north of the starting position.

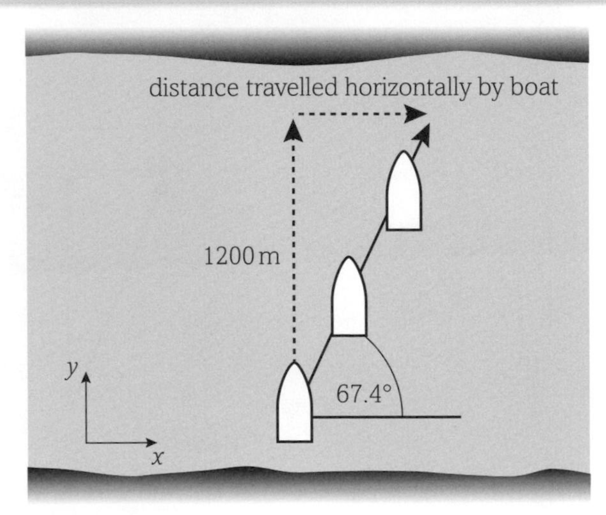

Figure 2

Resolving a vector involves the production of two vectors, at right angles to one another, the sum of which is equal to the original vector. These vectors are often shown as the horizontal component and the vertical component of the original vector.

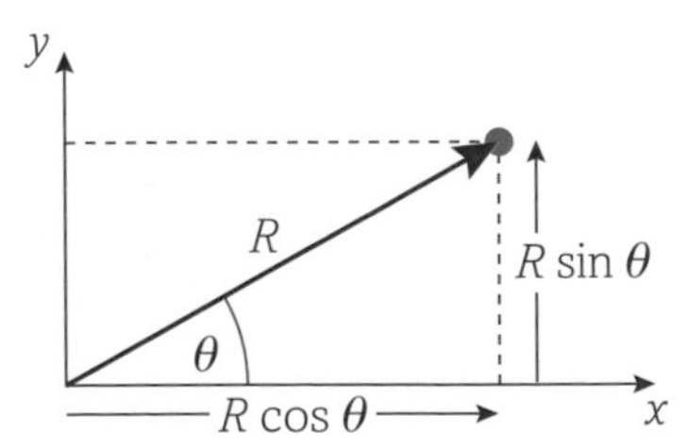

Figure 3

When you resolve a resultant vector into its horizontal and vertical **components**, you can make a problem much easier to solve. The reason behind this is that the vertical component determines the vertical behaviour of the object's motion only, and the horizontal component describes the object's horizontal motion only. In other words, the vertical and horizontal components of an object's motion are independent – they have no effect on each other at all. This is because the angle between the two components is 90° and cos 90° = 0, so the value of the horizontal component is zero when we consider the vertical component, and vice versa.

In part (c) of Worked example 1, we used a value of 12 m s^{-1}, rather than the resultant velocity of 13 m s^{-1}. This is because we used the direct distance across the river – 1200 m – and the component of the speedboat's velocity in this direction is 12 m s^{-1}. This gives a total time of 100 seconds. If we had used the resultant velocity in our calculation, then we would have needed to work out how far the boat had travelled in this direction. This is the hypotenuse of the vector triangle, a distance of 1300 m. However, using these figures still gives a time of 100 s.

In part (d) there is no need to use trigonometry. You could consider the horizontal component of motion for the boat – it travels 500 m at a speed of 5 m s^{-1}, again in a time of 100 seconds. This use of components is very important and is discussed further below.

LEARNING TIP

To calculate the distance travelled horizontally we use the horizontal component of its motion. To calculate the distance an object travels vertically we use the vertical component of its motion.

WORKED EXAMPLE 2

A javelin is thrown at a velocity of 12 m s^{-1} as shown in Figure 4. Calculate:

(a) the horizontal component of its velocity
(b) the vertical component of its velocity
(c) the horizontal distance it will travel if it is in the air for 3.6 seconds.

Answers

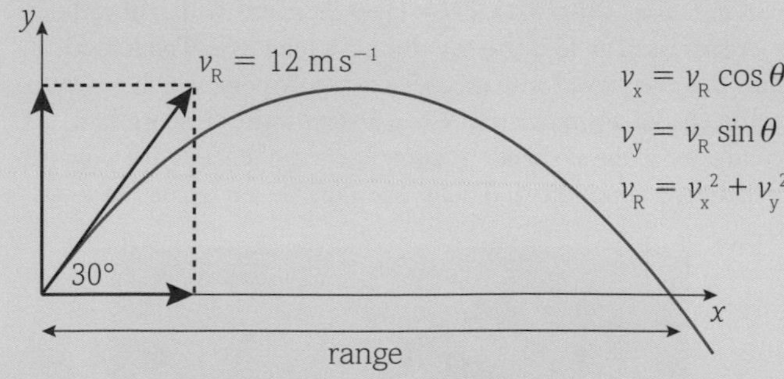

Figure 4

(a) The horizontal component is $v_R \times \cos 30° = 12 \times 0.866 = 10.4$ m s^{-1}
This is how fast the javelin is travelling in the horizontal direction – i.e. parallel to the ground.
(b) The vertical component of its velocity is $v_R \times \sin 30 = 6$ m s^{-1}. This is the component of the javelin's velocity in the upwards direction.
(c) This is found using horizontal distance = horizontal velocity × time
= 10.4 m s^{-1} × 3.6 s = 37.4 m

KEY DEFINITIONS

The **components** of a vector are the parts of a vector in two perpendicular directions. The process is called **resolving** the vector.

Questions

1. Resolve the vector shown in Figure 5 into its horizontal and vertical components.

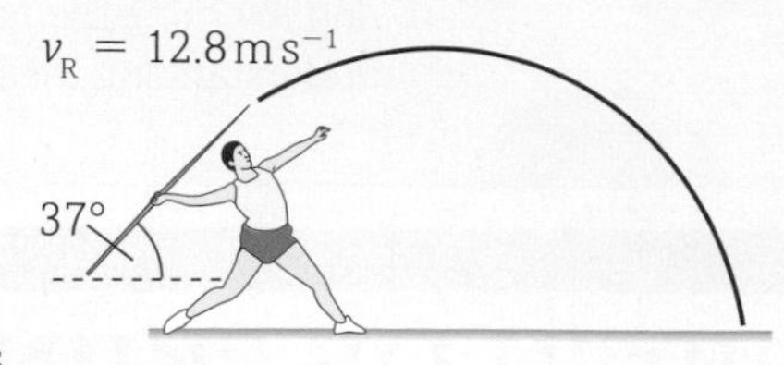

Figure 5

2. A car with a weight of 6100 N is at rest on a hill of gradient 9.0°. Calculate the component of its weight along the slope.

3. A boy is swimming across a river 65 m wide. He is moving through the water with a resultant velocity of 2.3 m s^{-1}. The current in the river has a velocity of 1.1 m s^{-1} parallel to the bank.
   (a) Sketch a vector triangle showing the two given vectors and a third vector for the velocity of the boy perpendicular to the bank.
   (b) Calculate the boy's velocity towards the opposite bank.
   (c) How long does it take him to cross the river?

4. A cannonball is fired at 42° to the horizontal at a speed of 18 m s^{-1}.
   Calculate values for:
   (a) the horizontal component of its velocity
   (b) the vertical component of its velocity
   (c) the horizontal distance it will travel if it is in the air for 22.5 seconds.

5. Explain what is happening in Figure 6 in terms of the horizontal and vertical components of motion.

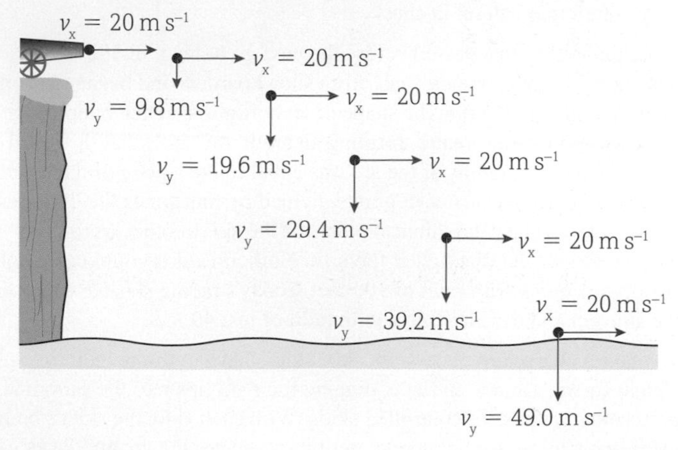

Figure 6

THINKING BIGGER

SCALARS AND VECTORS

The quantities that we come across in our day to day lives are either scalars or vectors – some have a size, others have a size and a direction.

In skiing, a number of scalar and vector quantities are involved. In this activity you will consider and examine the nature of these quantities and how optimising performance in skiing depends on understanding these quantities.

THE BIRTH OF SKIING

One brilliant winter morning in 1929, a young boy named David Lind flung himself down a snowy hill in rural Washington and promptly fell in love. His skis, which he had carved by hand from two hickory planks, weren't much different from the wooden runners and root bindings used by trappers more than 4,500 years ago, but they made for an exhilarating ride. 'They were 7 or 8 feet long and didn't have much flexibility', Lind says. They got him down the hill, but just barely in one piece.

Since that day, Lind has become one of the country's preeminent authorities on the science of skiing. [...] Science now governs every aspect of the sport, from fabricated snow to skis that are lighter, shorter, more flexible, and far easier to turn. The science behind the skis has triggered nothing short of a revolution in the sport, allowing almost anyone to carve up a mountain slope.

The father of modern skiing was a Norwegian potato farmer named Sondre Norheim. An imaginative freestyler with a habit of jumping off snowcapped rooftops, Norheim created the first heel-strap bindings in the 1860s. He also popularized the zigzag manoeuvre that became the basis of downhill skiing. A skier who heads straight down a 30-degree slope can reach 150 miles per hour, but traversing back and forth across the slope keeps one's momentum in check.

More than 100 years passed before the next great leap. In 1989, ski designers in Slovenia took a cue from snowboarders and began to shorten their skis and cut them in the shape of an hourglass. When put on edge, these skis bent at the centre, forming an arc in the snow that the board could follow. The radius of the arc was equal to the square of the length between the ski's two contact points divided by four times the difference between the widest and thinnest points of the ski. In other words, the deeper the side cut, the tighter the turn. Norheim's skis could carve only swooping curves with radii of 300 feet. Today's racing skis have four times the side cut and can carve arcs with radii of just 40 feet.

To make a sharp turn on straight skis, skiers have to throw their weight forward and to the side, digging their ski tips into the snow and performing a series of controlled skids. With each skid, the skier's body faces in a slightly new direction, until it completes the arc and faces forward again. On shorter skis with an hourglass shape, skiers can simply roll their ankles to one side and put their skis on edge. As the wide tip and tail dig into the snow, the ski bends in the middle and begins to turn, carving a path down the slope on its edge. Carving is less physically taxing than skidding. And though the new skis are slower on straightaways – their centres push deeper into the snow, increasing friction – they are faster overall because of their turning ability. They are also just as stable as regular skis.

Ski materials have received an equally extensive makeover. Engineers are continually working to make skis lighter and more flexible to absorb bumps in the snow, while keeping them rigid enough to hold their shape during turns. For that reason, metal skis were introduced in the 1950s. Today's skis, because of their broader tips and tails, have to endure torsional forces that skis of the past could not have withstood. Most skis are now made of sandwiches of fibreglass, wood, aluminium alloys, glue, and polymers.

The Volant ski company, for instance, uses a heat-treated stainless-steel top sheet, which doesn't twist much, helping ski edges dig into compact snow during turns. Other ski makers layer their skis with stiff carbon fibres, crisscrossed at 45 degrees to the ski's long axis. The K2 ski company has even toyed with piezoelectric polymers, which generate an electric charge when twisted. When a skier makes a sharp turn, a semiconductor in the ski sends a 'counter charge' back to the polymer, which dampens vibrations and helps keep the ski's edge on the snow.

Where else will I encounter these themes?
1.1 2.1 2.2 YOU ARE HERE 3.1 3.2 3.3

Let us start by considering the nature of the writing in the article. This article was taken from the magazine *Discover – Science for the curious* and the whole article was entitled 'The physics of … skiing'.

1. **Having read the article, comment on the type of writing that is being used. For example, who might have written the article? Is the audience the typical citizen or researchers in a scientific laboratory? What are the words or terms in the article that would influence you when determining who the article is intended for? How might you change the article to make it more suitable for a younger or less-informed audience?**

We will now look at the physics that is in the article. Do not worry if the physics content or the mathematics is challenging at this stage. You can always return to the article later in your course, once some of the related topics have been studied in more depth. Use the timeline at the bottom of the page to help you put this work in context with what you have already learned and what is ahead in your course.

2. By referring to the article, make a list of the areas of physics that are involved in the sport of skiing.
3. A skier has a mass of 76 kg and is skiing at 28 ° to the horizontal. Sketch a diagram showing the forces acting on the skier and determine how long he will take to travel a distance of 400 m along the slope. State any assumptions that you have made.
4. Is it better to be in the air or on the slope when skiing? Explain your answer
5. What would a skier need to do, in terms of their body position, when moving from a smooth piste into deeper snow? Explain this in terms of the forces acting on their body.
6. Perform calculations to show that 'A skier who heads straight down a 30-degree slope can reach 150 miles per hour, but traversing back and forth across the slope keeps one's momentum in check', as stated in the text.
7. Explain the following terms used in the article:
   a. torsional forces
   b. carving
   c. skidding
   d. 'dampens vibrations'.

DID YOU KNOW?

Harrison Schmitt, a crew member of the space shuttle Apollo 17, said that astronauts heading to the Moon should learn the art of cross-country skiing. He believed that the techniques involved in skiing helped walking on the Moon and even envisioned people setting off for 'lunar skiing holidays'.

Speed skiing is one of the fastest non-motorised sports on land. Italian skier Simone Origone currently holds the world record for the fastest downhill ski, clocking 156.2 miles per hour.

Activity

You have been appointed as Chief Skiing Physicist of the next Winter Olympic Games Team in order to help members of the team improve their skiing techniques. In particular, they wish to be able to ski more quickly and safely. In order to improve their skiing, they need to know how physics applies to them. In particular, they need to be aware of how the various scalars and vectors involved in the sport can be utilised to help them maximise performance.

The skiers will receive one half-hour session entitled 'The physics of skiing', where you will focus on the nature of scalars and vectors and how they apply to the sport. You must cover:

- A brief summary of the scalar and vector quantities that a skier will encounter.
- An explanation of how a variety of forces can increase or decrease their speed.
- A reference to the terms carving and skidding and the forces acting on the skier when these moves are being performed.
- Diagrams and simple calculations showing the size and direction of any vector quantities involved in downhill skiing.

Research key terms, such as skidding and carving, in more detail to determine how forces acting on the skier are different during a ski run.

2.2 Practice questions

1. Which statement is the best definition of a vector quantity? [1]
   - A A quantity that has size and direction.
   - B A quantity that has direction only.
   - C A quantity that has size or magnitude only.
   - D A quantity formed when two scalars are multiplied or divided.

2. Which one of the following lists contains only vector quantities? [1]
   - A mass, weight, area and force
   - B momentum, impulse, acceleration and speed
   - C velocity, momentum, weight and impulse
   - D displacement, volume, pressure and speed

3. Two forces of size X and Y act on a body. Which one of the following statements is definitely true? [1]
   - A The forces will always combine to give a zero resultant.
   - B The forces will always lead to an acceleration.
   - C If the forces act at 90° to one another the resultant force will have a magnitude of $\sqrt{(X^2 + Y^2)}$.
   - D If X > Y then the body will move at a constant velocity.

   [Total: 3]

4. (a) State what is meant by a vector quantity. [1]

   (b) Give two examples of vector quantities. [2]

   [Total: 3]

5. Figure 1 shows a toy locomotive on a circular track of radius 0.60 m.

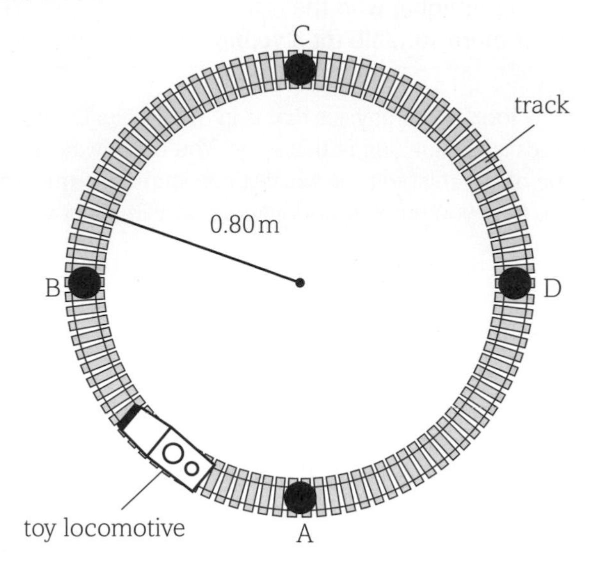

Figure 1

At time $t = 0$, the locomotive is at point **A**. The locomotive travels at a constant speed round the track. It takes 20 s to travel completely round the track.

(a) Calculate the speed of the locomotive. [2]

(b) Figure 2 shows the variation of the magnitude of the displacement s of the locomotive from **A** with time t.

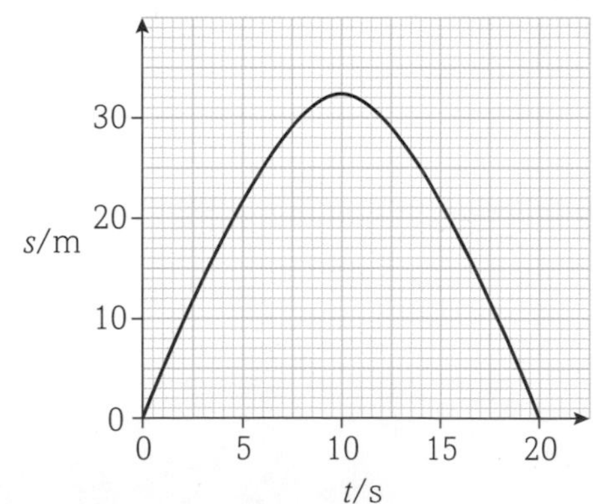

Figure 2

Explain the graph shown in Figure 2. [2]

(c) An object is placed on a smooth horizontal surface. Two horizontal forces act on this object. Figure 3 shows the magnitudes and directions of these two forces.

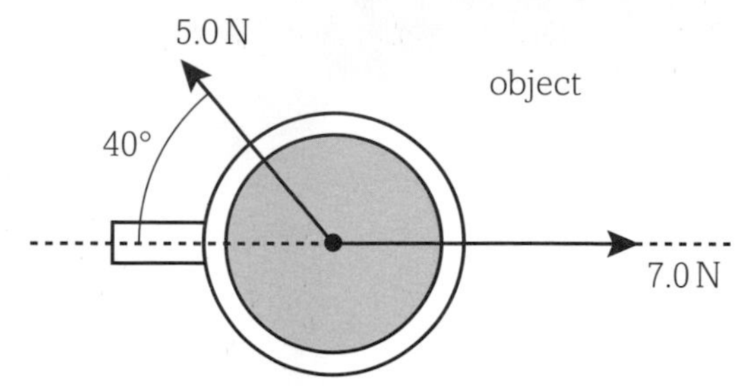

Figure 3

The mass of the object is 320 g.

Calculate the magnitude of the acceleration of the object. Give your answer in $m\,s^{-2}$. [3]

[Total: 7]

[Q21, H156/01 sample paper 2014]

6. A catapult fires a stone into the air. The force exerted by the catapult on the stone is 20 N at an angle of 30 degrees to the horizontal. Find the horizontal and vertical components of this force. [3]

[Total: 3]

7. A net upward force of 35 N acts vertically on a toy helicopter and another force of 28 N acts horizontally on the toy. Find:
   - the resultant force acting on the helicopter
   - the angle it will make with the floor if it takes off from a stationary position. [4]

[Total: 4]

8. Two aircraft, A and B, are flying at right angles to one another as shown in Figure 4. They are both flying towards point P. Aircraft A is flying due north at 980 $km\,h^{-1}$ and aircraft B is flying due west at 1120 $km\,h^{-1}$.

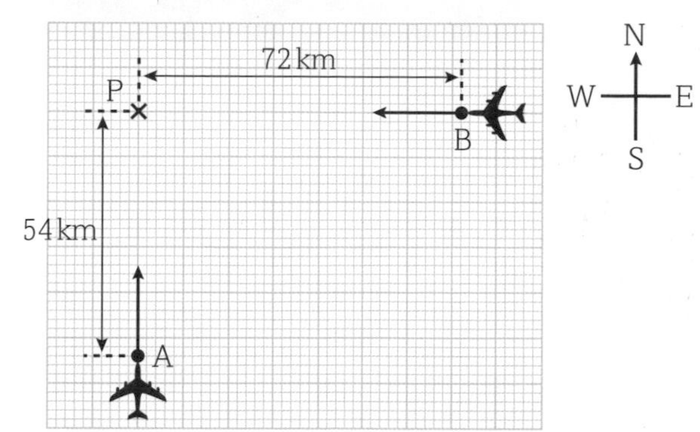

Figure 4

(a) Use the information provided to calculate:
   - the displacement of aircraft A from aircraft B at the point shown in Figure 4. [2]
   - the distance between the aircraft in 4 minutes' time. [2]

(b) Air traffic control do not allow aircraft flying in this manner to get closer than 20 km to each other. What time, in seconds, after the situation shown in Figure 4 will this occur? [3]

[Total: 7]

MODULE 3

Forces and motion

CHAPTER 3.1

MOTION

Introduction

Everything is moving – from the air molecules that surround us to the aircraft that carry us across the huge oceans and continental land masses. Physics is a subject that deals with the kinetic aspects of nature – from the obvious movement of the car in our highly visible and macroscopic everyday world, to the mysterious motion of the atoms, electrons and molecules that occupy the invisible and microscopic world – below and beyond what we can physically observe with light. The atoms that compose our skin, the water molecules that make up the hot or cold drink, the car, bus or bike that gets us to school and work are all moving. Furthermore, their motion and behaviour can all be explained by the same language and the same equations of motion. Regardless of whether we are describing the skydiver or the raindrop, the Olympic cyclist or the hummingbird, the same terms apply – speed, velocity, acceleration, distance and displacement. In addition, their movement can be accurately measured and predicted with the same laws of physics and the same equations of motion. In this chapter you will learn about the fundamental principles of motion and how things behave the way they do when they move and when they fall.

All the maths you need

To unlock the puzzles of this chapter you need the following maths:

- Units of measurement
- Interpreting the slope and intercept of a graph
- Calculating or estimating the area under a graph
- Calculating rate of change of velocity (acceleration) from the gradient of a linear velocity–time graph
- Drawing and using the slope of a tangent to a curve as a measure of rate of change
- Solving algebraic equations
- Using sine and cosine of an angle
- Rearranging formulae and substitution of values
- Direct proportionality

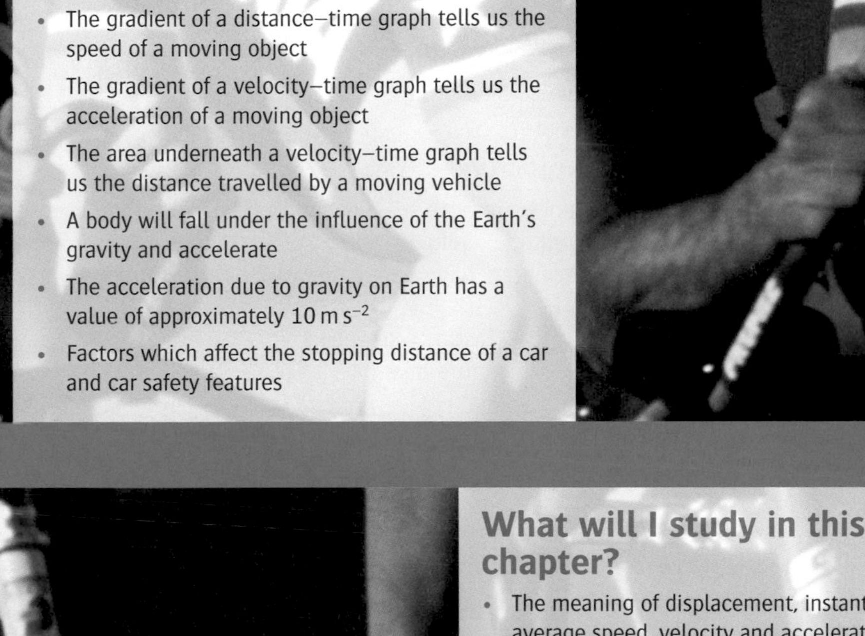

What have I studied before?

- Forces are measured in newtons and have a size and a direction
- Speed is the distance travelled per unit time
- Velocity is the distance travelled per unit time in a given direction
- Acceleration is the change in velocity per unit time
- The gradient of a distance–time graph tells us the speed of a moving object
- The gradient of a velocity–time graph tells us the acceleration of a moving object
- The area underneath a velocity–time graph tells us the distance travelled by a moving vehicle
- A body will fall under the influence of the Earth's gravity and accelerate
- The acceleration due to gravity on Earth has a value of approximately $10\ \mathrm{m\ s^{-2}}$
- Factors which affect the stopping distance of a car and car safety features

What will I study later?

- The behaviour of bodies travelling with non-uniform motion (AS)
- The ideas of drag and terminal velocity and their impact on the motion of moving objects in fluids (AS)
- Kinetic and potential energies – their definitions and how they are calculated (AS)
- The relationship between kinematics and dynamics (AS)
- Newton's laws of motion and their relation to momentum and impulse (AS)
- The momentum and energy considerations involved during collisions and explosions (AS)
- The forces and pressures exerted by solids, liquids and gases (AS)
- The nature of the centripetal, electrical, gravitational and strong nuclear force. (AL)
- Free and forced oscillations in a range of damped systems (AL)
- Force–distance graphs for bodies undergoing motion in electric and gravitational fields (AL)

What will I study in this chapter?

- The meaning of displacement, instantaneous speed, average speed, velocity and acceleration
- The graphical representation of displacement, speed, velocity and acceleration.
- Plotting and interpreting displacement–time graphs, with velocity being the gradient of the graph
- Plotting and interpreting velocity–time graphs, with the gradient being acceleration and the area beneath the graph representing displacement
- Estimation of the area beneath a non-linear velocity–time graph to provide an estimate of the displacement
- The four suvat equations – the equations for constant, linear acceleration, including the motion of bodies falling in a uniform gravitational field in the absence of air resistance
- The techniques and procedures used to investigate the motion of objects
- An understanding of the acceleration due to gravity on Earth, *g*, as well as techniques and procedures to determine its value
- An understanding of the terms reaction time, braking distance, thinking distance and stopping distance for a moving vehicle
- An appreciation of the factors that affect reaction time, thinking distance and braking distance

3.1 ① Definitions in kinematics

By the end of this topic, you should be able to demonstrate and apply your knowledge and understanding of:

* displacement, instantaneous speed, average speed, velocity and acceleration

Introduction

Kinematics is the study of the motion of objects and their spatial relationships, without any reference to either their masses or the forces that cause them to change direction or speed. When working in kinematics, you need to focus on the following quantities and definitions.

Quantity	S.I. unit	Definition
speed	$m\,s^{-1}$ or $km\,h^{-1}$	distance per unit time
displacement	m or km	distance moved in a particular direction from a reference point
velocity	$m\,s^{-1}$ or $km\,h^{-1}$	displacement per unit time
acceleration	$m\,s^{-2}$	change in velocity per unit time

Table 1 Quantities, units and definitions.

Displacement

If a ship sails a distance of 10 km in a southerly direction, we can say that the ship has a **displacement** of 10 km south. Displacement is a vector quantity because both its size (magnitude) and direction are stated.

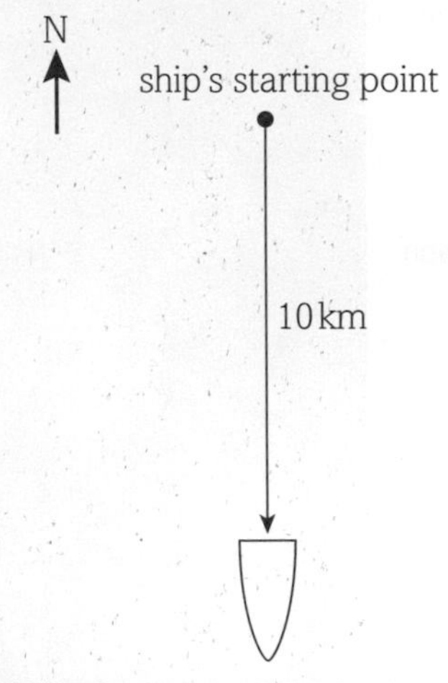

Figure 1 The ship is 10 km from its starting point, in a southerly direction – so its displacement is 10 km south.

Imagine the car journey in Figure 2. The car ends up 237 km away from its starting point A in a southwesterly direction. We could say that its displacement is 237 km south west or 237 km on a bearing of 225 degrees from north. During the journey, the car may have made many changes to its direction, taking an overall route that is actually 255 km in length, as shown. We say that the total distance travelled by the car is 255 km, but its displacement is 237 km south west.

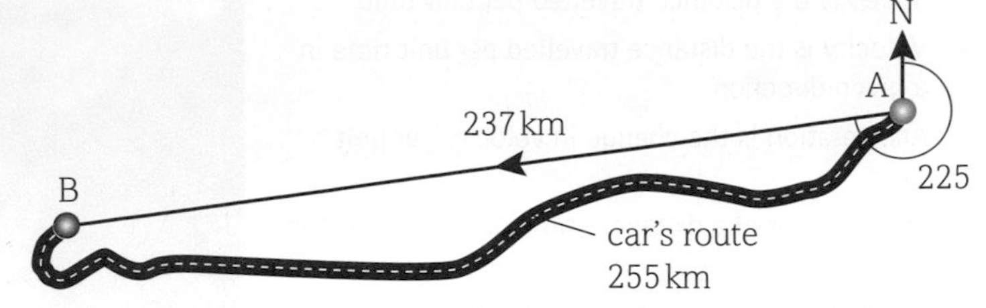

Figure 2 The displacement is less than the actual distance travelled.

Speed and velocity

Speed and distance are both scalar quantities because they have only a size (magnitude); velocity and displacement are vector quantities because they have both a size and a direction. In equation form:

$$\text{average speed} = \frac{\text{total distance travelled}}{\text{time taken}}$$

$$\text{average velocity} = \frac{\text{total displacement}}{\text{time taken}}$$

Using the car journey example shown in Figure 2, and assuming that the journey took a time of 3 hours, we get:

$$\text{average speed of car} = \frac{255\,\text{km}}{3\,\text{h}} = 85\,\text{km}\,\text{h}^{-1}$$

$$\text{average velocity of car} = \frac{237\,\text{km}}{3\,\text{h}} = 79\,\text{km}\,\text{h}^{-1}$$ in a southwesterly direction.

Note that the *average speed* is the total distance travelled/the total time taken for a journey. *Instantaneous speed* is the speed at an actual point in time during a journey. An average speed for a journey has a fixed value, whereas an instantaneous speed of a car changes during a journey as it speeds up, slows down, stops at traffic lights, etc.

Interpreting displacement–time graphs

Look at Figure 3, which represents the motion of a long-distance runner during a training run.

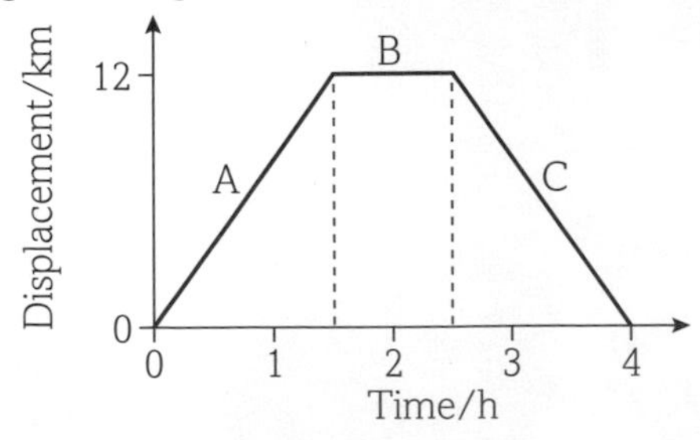

Figure 3

From this we can see that:

- The total distance covered was 24 km – that is, 12 km outwards and 12 km back home.
- The total displacement was zero – he is a distance of 0 km from his starting point at the end of the run.
- The average velocity of the runner was zero because the displacement is zero and

$$\text{average velocity} = \frac{\text{total displacement}}{\text{time}}.$$

- The average speed was $6\,\text{km}\,\text{h}^{-1}$ because 24 km was covered in a time of 4 hours.

Acceleration

Acceleration is the rate of change of velocity, given by

$$\text{acceleration} = \frac{\text{change in velocity}}{\text{time taken}}$$

A car may accelerate from $2.0\,\text{m}\,\text{s}^{-1}$ to $14.0\,\text{m}\,\text{s}^{-1}$ over a period of 4 seconds. The acceleration of the car is given by:

$$\text{acceleration} = \frac{(14.0\,\text{m}\,\text{s}^{-1} - 2.0\,\text{m}\,\text{s}^{-1})}{4.0\,\text{s}} = 3.0\,\text{m}\,\text{s}^{-2}$$

The value $3.0\,\text{m}\,\text{s}^{-2}$ means that over the 4.0 second time period, the car increased its speed by $3.0\,\text{m}\,\text{s}^{-1}$ each second.

Acceleration values can also be negative. If a car starts off with a velocity in the positive direction of $18.0\,\text{m}\,\text{s}^{-1}$ and slows down to $3.0\,\text{m}\,\text{s}^{-1}$ in 3.0 s, then its acceleration is given by:

$$\text{acceleration} = \frac{(3.0\,\text{m}\,\text{s}^{-1} - 18.0\,\text{m}\,\text{s}^{-1})}{3.0\,\text{s}} = -5.0\,\text{m}\,\text{s}^{-2}$$

This means that the car slowed down by $5.0\,\text{m}\,\text{s}^{-1}$ each second over the course of the three-second journey.

For an object moving in the negative direction, a negative acceleration indicates the object is accelerating in that direction.

LEARNING TIP

A deceleration of $9.81\,\text{m}\,\text{s}^{-2}$ is the same as an acceleration of $-9.81\,\text{m}\,\text{s}^{-2}$. When a cricket ball is thrown upwards it experiences a deceleration of $9.81\,\text{m}\,\text{s}^{-2}$, or an acceleration of $-9.81\,\text{m}\,\text{s}^{-2}$, because gravity is causing it to slow down as the force pulls it back to Earth in the opposite direction to the original motion.

We often use the equation $a = \frac{(v - u)}{t}$ to calculate the acceleration of an object. In this equation a is the acceleration, v is the final velocity, u is the initial velocity and t is the time taken. If v is higher than u, we have a positive acceleration and the object has speeded up; if v is lower than u, we have a negative acceleration and the object has slowed down.

Questions

1. Bob set out from home and ran a distance of 3 km east, before turning and running a distance of 4 km due south. This took him a total time of 63 minutes.
   Find the values of the total distance travelled, final displacement, average speed and average velocity in Bob's journey.
   Give your answers in units of metres and seconds.
2. Calculate the acceleration required for a car to reach a velocity of $30\,\text{m}\,\text{s}^{-1}$ in 5.5 s from rest.
3. What is the acceleration of a lorry that starts with a velocity of $30\,\text{m}\,\text{s}^{-1}$ and 4.8 s later has a velocity of $8\,\text{m}\,\text{s}^{-1}$?
4. A cyclist accelerates from a standing start at a uniform rate of $4\,\text{m}\,\text{s}^{-2}$ to a top speed of $20\,\text{m}\,\text{s}^{-1}$. For how long did the cyclist accelerate?
5. An athlete runs around a 400 m track shown in a time of 48 s.
   What is:
   (a) the average speed of the athlete
   (b) the average velocity of the athlete?

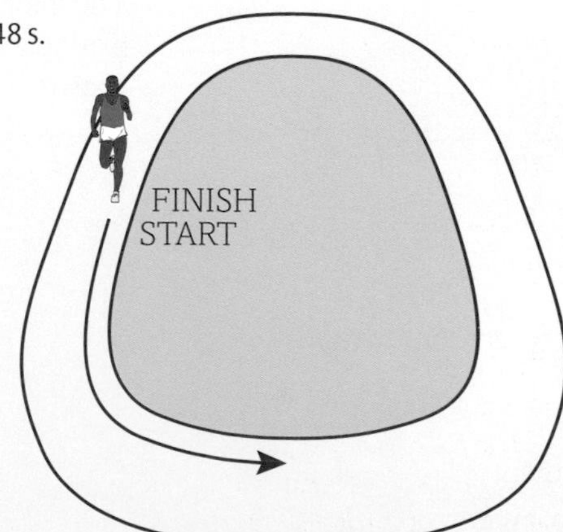

Figure 4

2 Graphs of motion

By the end of this topic, you should be able to demonstrate and apply your knowledge and understanding of:

* graphical representations of displacement, speed, velocity and acceleration
* velocity as the gradient of displacement–time graphs;
* acceleration as the gradient of velocity–time graphs; displacement as area under graph

Displacement–time graphs

A displacement–time graph shows how the displacement of a moving object varies with time. Displacement is plotted on the vertical axis and time along the horizontal axis. The gradient, or slope, of a displacement–time graph is equal to the instantaneous velocity of the moving object.

Interpreting displacement–time graphs

Consider this description of a journey taken by a cyclist.

For 30 seconds, the cyclist travelled at a constant velocity, reaching a displacement of 300 m from home. For the next 20 seconds, the cyclist increased her velocity and travelled a further 400 m. She then stopped at traffic lights for 1 minute, before moving a displacement of 100 m in the direction of home in a time of 20 s. She then travelled a further displacement of 200 m away from home for 20 seconds at a constant acceleration.

The graph in Figure 1 is a simplified displacement–time graph for this cyclist. As you can see, the total displacement of the cyclist is 800 m and the total time taken is 150 s. The journey can be split into five separate stages, A–E.

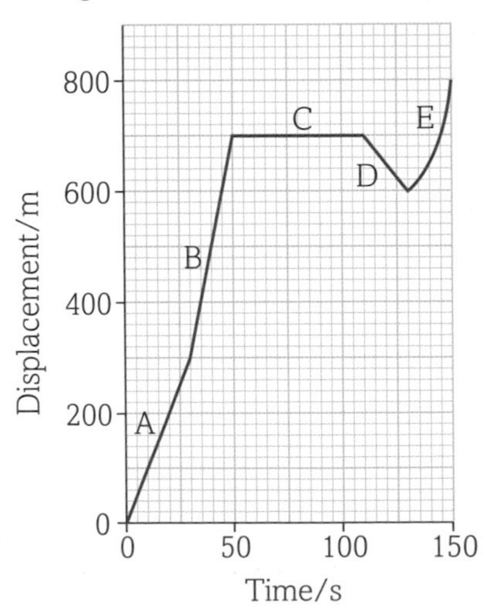

Figure 1 A displacement–time graph for a cyclist.

- A straight diagonal line on a displacement–time graph represents a constant velocity.
- A velocity is calculated using:

$$\text{velocity} = \frac{\text{change in displacement}}{\text{time taken for this change}}$$

- A flat or horizontal line on a displacement–time graph represents a stationary object – i.e. zero velocity.
- A curve indicates the velocity is not uniform. The gradient of the tangent to the curve at any point gives the instantaneous velocity (see Figure 1 in Topic 3.1.3).
- A curve of increasing positive gradient shows that the object is accelerating (speeding up).
- A curve of decreasing positive gradient shows that the object is decelerating (slowing down).

WORKED EXAMPLE 1

For each of the stages A–E in Figure 1, describe the motion of the cyclist and calculate any values for her velocity or acceleration.

Answers

A – The cyclist's displacement is 300 m after 30 s. The average velocity, therefore, is $\frac{300\text{ m}}{30\text{ s}} = 10\text{ m s}^{-1}$. This is a constant velocity because the gradient of the line is constant throughout this section of the journey.

B – The cyclist's average velocity for this part is $\frac{400\text{ m}}{20\text{ s}}$, which is 20 m s^{-1}. Again, the constant slope shows that this is a constant velocity.

C – At this point the gradient is zero and the velocity is zero. The cyclist does not move for this 60 s period of time.

D – The average velocity here is $\frac{-100\text{ m}}{20\text{ s}} = -5\text{ m s}^{-1}$. The negative value for the velocity comes from the fact that velocity is a vector and the cyclist has reversed her direction.

E – The upwards sloping curve means that the velocity is increasing.

Velocity–time graphs

The information from the graph in Figure 1 and the worked example can be used to plot a velocity–time graph, which is shown in Figure 2. Instead of plotting displacement on the

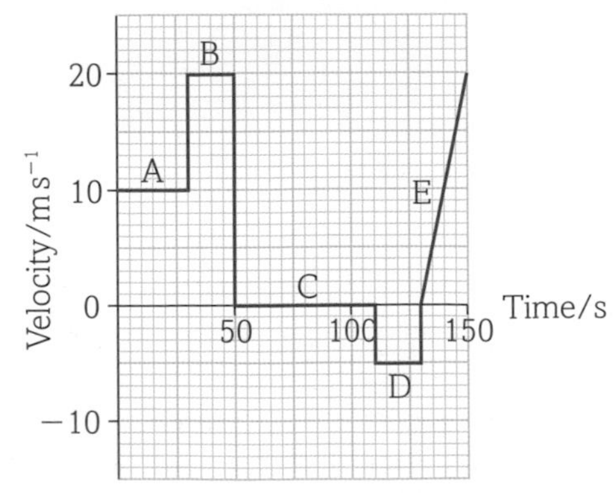

Figure 2 A velocity–time graph for the data in Figure 1.

vertical axis we are now plotting velocity. Hence the graph has a totally different shape from that of the displacement–time graph. Notice that velocity can have negative and positive values. For motion in a straight line it is usual to state one direction as positive and the opposite direction as negative.

The gradient of velocity–time graphs

The gradient of a velocity–time graph represents acceleration. A horizontal line represents a constant velocity. A line along the *x*-axis at a value of zero on the *y*-axis represents zero velocity.

An acceleration is calculated using:

$$\text{acceleration} = \frac{\text{change in velocity}}{\text{time taken for the change}}$$

- A diagonal line with a positive gradient represents a constant acceleration.
- A diagonal line with a negative gradient represents a constant negative acceleration.
- A curve indicates the velocity change is not uniform – the gradient of the tangent to the curve at any point gives the rate of change of velocity at that instant, i.e. the acceleration.

WORKED EXAMPLE 2

Use the velocity–time graph in Figure 2 to determine an accurate value for the acceleration of the cyclist in stage E.

Answer

$\text{Acceleration} = \frac{\text{change in velocity}}{\text{time taken}}$. So, $a = \frac{20\ \text{m s}^{-1}}{20\ \text{s}} = 1\ \text{m s}^{-2}$

The area beneath a velocity–time graph

The graph shown in Figure 3 is a velocity–time graph for a car travelling for 7.0 s with a constant velocity of 30 m s^{-1}. The shaded area of the graph has a height of 30 m s^{-1} and a width of 7.0 s. The area of the rectangle is 30 m s^{-1} × 7.0 s = 210 m. Note that the units of time seconds and seconds^{-1} cancel out.

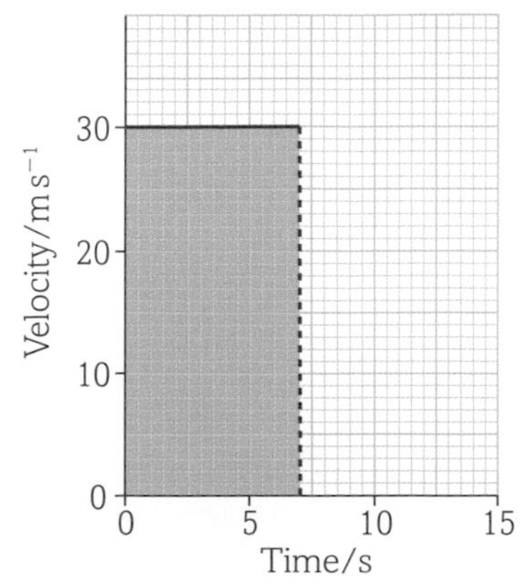

Figure 3 Area beneath a velocity–time graph.

The answer gives the displacement of the car in this period of time.

The area beneath a velocity–time graph represents the change of the displacement. The fact that the area beneath a graph like this represents displacement can be applied to any graph shape. However, the calculations involved are more difficult if the graph is not a horizontal straight line (non-uniform velocity).

WORKED EXAMPLE 3

Use the velocity–time graph in Figure 2 to determine by calculation that the total displacement for this journey was 800 m.

Answer

The displacement can be found by adding together the areas under the graph for each stage.

A – area = 30 s × 10 m s^{-1} = 300 m
B – area = 20 s × 20 m s^{-1} = 400 m
C – area = zero
D – area = 20 s × = –5 m s^{-1} = –100 m
E – area = $\frac{1}{2}$ × 20 s × 20 m s^{-1} = 200 m
Total area = total displacement = 300 m + 400 m – 100 m + 200 m = 800 m

If the velocity is not uniform and simple shapes such as triangles cannot be used, then the area can be found by counting squares on the graph paper. You will need to count the separate squares and then find the distance that corresponds to one square.

Questions

1 Use the three velocity–time graphs shown in Figure 4 to calculate the displacement in each case.

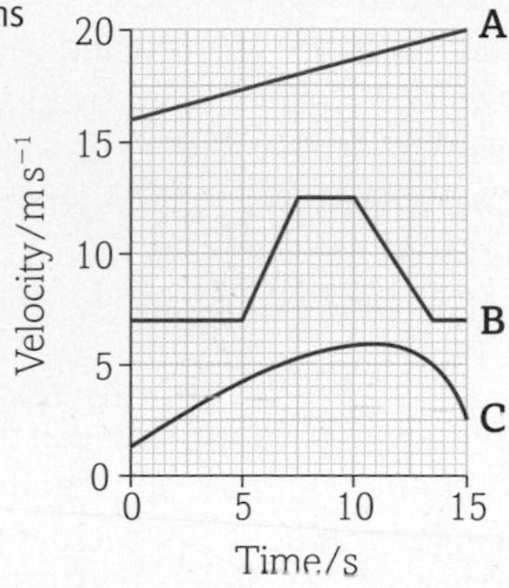

Figure 4

2 Figure 5 shows the distance–time graph for an underground train travelling between two stations.

(a) Sketch a graph to show how the speed of the train varies with time *t* during the journey. The train starts from rest at *t* = 0.

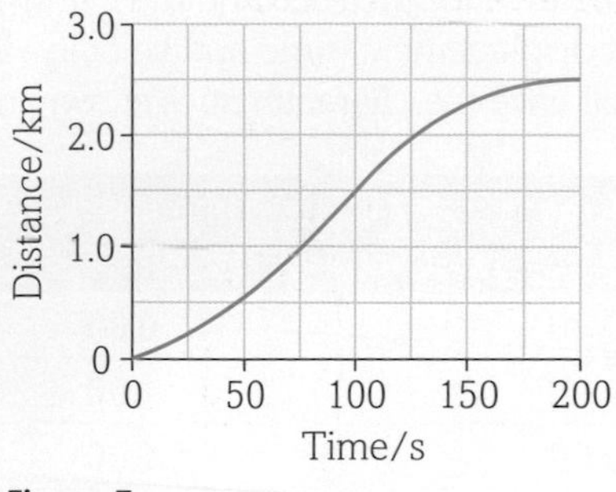

Figure 5

(b) Use the gradient of Figure 5 to find the maximum speed of the train in m s^{-1}.

(c) Calculate the average speed of the train.

3 The following speed–time data were collected during a car journey.

Time/s	0	10	20	30	40	50	60	70	80	90	100
Speed/m s^{-1}	0	15	30	30	30	30	0	0	10	10	0

(a) Draw a speed–time graph of the car journey.

(b) Describe the motion of the car during the journey.

(c) During which time period was the maximum acceleration of the car? Calculate its value.

(d) Calculate the total distance travelled in 100 s.

Constant acceleration equations

By the end of this topic, you should be able to demonstrate and apply your knowledge and understanding of:

* the equations of motion for constant acceleration in a straight line, including motion of bodies falling in a uniform gravitational field without air resistance
  - $v = u + at$
  - $s = \frac{1}{2}(u + v)t$
  - $s = ut + \frac{1}{2}at^2$
  - $v^2 = u^2 + 2as$

Introduction

When a resultant force acts on a body, the body will accelerate. The force may cause the body to speed up or to slow down. If the change in velocity of the object is the same for each second of its motion, then we say that the body has a constant acceleration. If drag forces are ignored and we drop a rock from the top of a cliff, its velocity will increase by the same amount each second as it falls. The force is due to its weight, and the typical acceleration is equal to $9.81\ \text{m s}^{-2}$ although this value does vary slightly over the Earth. At the start of a race a car could speed up by $3\ \text{m s}^{-1}$ for each second of its journey. Both of these scenarios are described in Tables 1 and 2.

As you can see from Table 2, the car starts from rest and experiences a linear acceleration of $3.00\ \text{m s}^{-2}$ for 5.00 s. The change in velocity per second is the acceleration of the vehicle – leading to an increase in the distance travelled each second.

The data for the rock and the car in Tables 1 and 2 can be plotted as displacement–time and velocity–time graphs, and these graphs will have the characteristics shown in Figures 1 and 2.

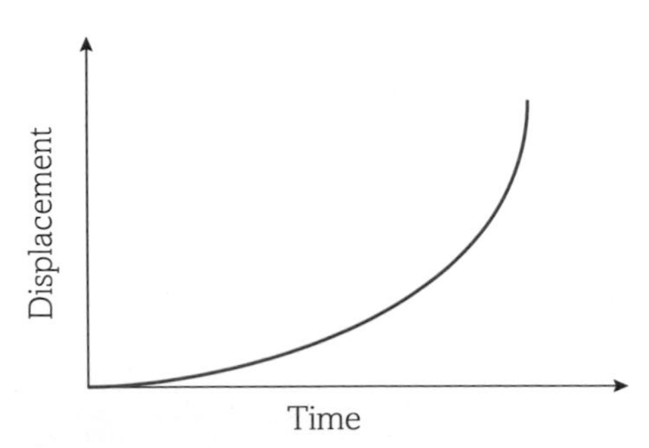

Figure 1 Graph of displacement against time for constant acceleration.

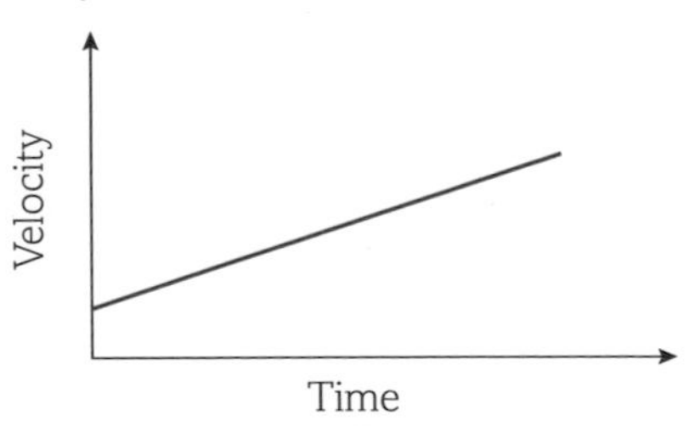

Figure 2 Graph of velocity against time for constant acceleration.

Time/s	Time interval/s	Velocity/m s^{-1}	Change in velocity/m s^{-1}	Distance fallen/m	Distance fallen in each 1 s interval/m
0.00	0.00	0.00	0.00	0.00	0.00
1.00	1.00	9.81	9.81	4.91	4.91
2.00	1.00	19.62	9.81	19.62	14.71
3.00	1.00	29.43	9.81	44.15	24.53
4.00	1.00	39.24	9.81	78.48	34.33
5.00	1.00	49.05	9.81	122.63	44.15

Table 1 Increase in time, velocity and distance for a rock falling under gravity.

Time/s	Time interval/s	Velocity/m s^{-1}	Change in velocity/m s^{-1}	Distance travelled/m	Distance travelled in each 1 s interval/m
0.00	0.00	0.00	0.00	0.00	0.00
1.00	1.00	3.00	3.00	1.50	1.50
2.00	1.00	6.00	3.00	6.00	4.50
3.00	1.00	9.00	3.00	13.50	7.50
4.00	1.00	12.00	3.00	24.00	11.50
5.00	1.00	15.00	3.00	37.50	13.50

Table 2 Velocity, change in velocity and distance travelled by an accelerating car during the first 5.0 seconds of horizontal motion.

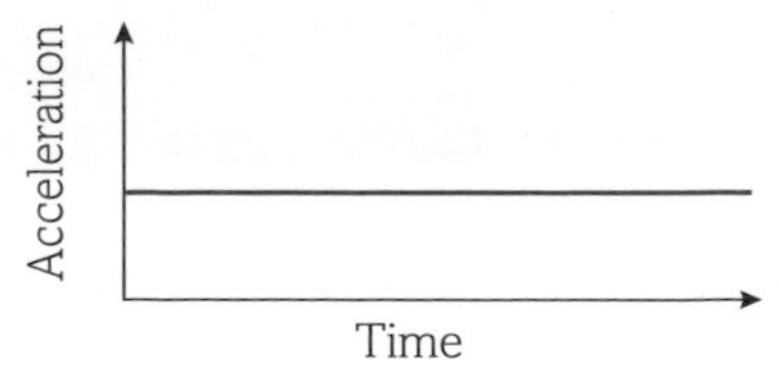

Figure 3 Graph of acceleration against time for constant acceleration.

For a body moving with constant acceleration, the distance travelled from a starting point plotted against time will be a parabola (Figure 1). Its velocity will vary in a linear fashion with time (Figure 2), and its acceleration will be constant (Figure 3). The calculations involved in interpreting such events requires the ability to work from algebraic equations.

Equations of motion

When dealing with constant acceleration over a given time, we always use the standard algebraic symbols shown in Table 3.

Symbol	Quantity	S.I. unit
s	displacement	metre (m)
u	velocity at the start	metres per second ($m\,s^{-1}$)
v	velocity at the end	metres per second ($m\,s^{-1}$)
a	acceleration	metres per second per second ($m\,s^{-2}$)
t	time interval	second (s)

Table 3 Standard algebraic symbols for equations of motion.

The equations of motion that use these symbols are often referred to as the 'suvat' equations. However, their correct name is 'equations of motion' or 'kinematic equations'. You need to be able to apply these four equations to any kinematics problem.

The four equations are:

$$v = u + at$$
$$s = ut + \tfrac{1}{2}at^2$$
$$v^2 = u^2 + 2as$$
$$s = \tfrac{1}{2}(u + v)t$$

LEARNING TIP

- The symbol Δ is known as Delta and signifies the change in a quantity.
- Learn the four suvat equations.
- Write the word 'suvat' on your working page when doing a question and list what information is provided and what value you are being asked to calculate.
- Ignore the terms you do not need to find.
- Check your answer by using a different equation if possible. For example, if you have calculated v by using $v = u + at$, then work it out again using $v^2 = u^2 + 2as$ (if possible) – and don't forget to take the square root of v^2 to get v.
- Don't be caught out by 'hidden' values – for example if a body 'starts from rest' then u is zero.
- For vertical motion, the acceleration due to gravity (g) is used. This means that a is 9.81 $m\,s^{-2}$ if falling and a is −9.81 $m\,s^{-2}$ if it has been thrown up in the air. A positive value of g is used if the object is moving in the same direction as the gravitational acceleration.

WORKED EXAMPLE

A motorcycle is travelling at a constant speed of 5 $m\,s^{-1}$. It passes through a set of traffic lights and accelerates at a rate of 1.8 $m\,s^{-1}$ for the next 15 seconds.

(a) Draw a velocity–time graph of the motion of the motorcycle based on this information.

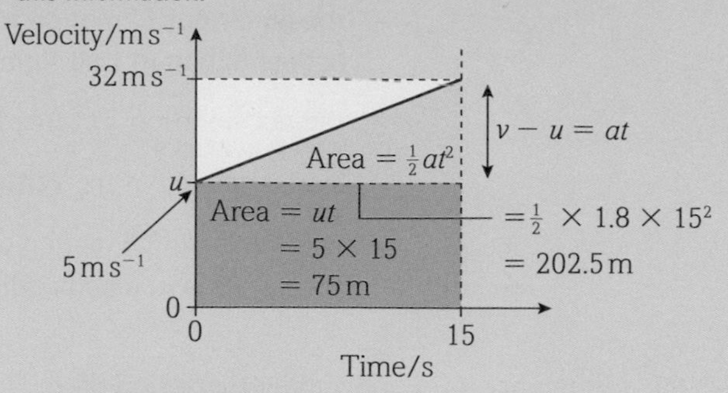

Figure 4 Velocity–time graph for a motorcycle.

(b) What is the final velocity of the motorcycle after the 15 second period?

$v = u + at$
$= 5\,m\,s^{-1} + (1.8\,m\,s^{-2} \times 15\,s) = 32\,m\,s^{-1}$

(c) How far will the car have travelled in this time?

$s = ut + \frac{1}{2}at^2$
$= (5 \times 15) + (\frac{1}{2} \times 1.8) \times 15^2$
$= 75 + (\frac{1}{2} \times 1.8 \times 225) = 277.5\,m$

(d) Use a different equation to check your answer to part (c).

Using $s = \frac{(u + v)}{2} \times t$

$s = \frac{(5 + 32)}{2} \times 15$
$= 18.5 \times 15 = 277.5\,m$

Questions

1 A car, starting from rest, reaches a top speed of 16 $m\,s^{-1}$ in a time of 4 seconds. Calculate:
(a) its acceleration, assuming it to be constant
(b) the total distance the car travels in this time.
(c) Use different equations to check your answers to (a) and (b).

2 A lorry increases its speed at a steady rate from 22 $m\,s^{-1}$ to 29 $m\,s^{-1}$ in 70 seconds.
(a) What is its acceleration?
(b) How far does it travel while accelerating?

3 A ball is thrown vertically upwards with an initial velocity of 14 $m\,s^{-1}$. Calculate:
(a) its maximum height reached
(b) the time taken to reach this height
(c) the total time until the ball is caught again by the thrower.
(d) What assumptions have you made when answering parts (a) to (c)?

4 At the scene of a traffic accident the police notice that a car has taken 28.0 m to stop. The assumed acceleration of the car is −8.0 $m\,s^{-2}$. What value does this assumption give for the speed of the car just before the accident?

5 Explain why you could not use the equations of motion to accurately describe the motion of a rocket that was going up into the night sky.

Free fall and projectile motion

By the end of this topic, you should be able to demonstrate and apply your knowledge and understanding of:

* the equations of motion for constant acceleration in a straight line, including the motion of bodies falling in a uniform gravitational field without air resistance
* the acceleration g of free fall
* independence of the vertical and horizontal motion of a projectile
* two-dimensional motion of a projectile with constant velocity in one direction and constant acceleration in a perpendicular direction

'Free fall' is the acceleration of a body under the action of a gravitational field, with air resistance and buoyancy being ignored. Objects of different masses fall at the same rate under the influence of gravity.

Why do objects of different masses fall at the same rate?

Consider two objects dropped from the same height at the same time. One has a mass of 1000 kg and the other a mass of 1 kg. The gravitational force experienced by each of these masses is calculated using the equation $F = m \times g$, where 'g' is the gravitational field strength (you will study this further in Topic 3.2.2). So the 1000 kg mass will experience a force that is 1000 times larger than the force on the 1 kg mass – hence we may think that the 1000 kg mass will accelerate 1000 times more than the 1 kg mass. However, acceleration is dependent on two things: force and mass. While the force on the 1000 kg mass is 1000 times larger, its inertia (the resistance of an object to changes in its motion) is also 1000 times larger. This is because acceleration of an object is directly proportional to the force acting on it, but inversely proportional to its mass. This is shown below in Figure 1.

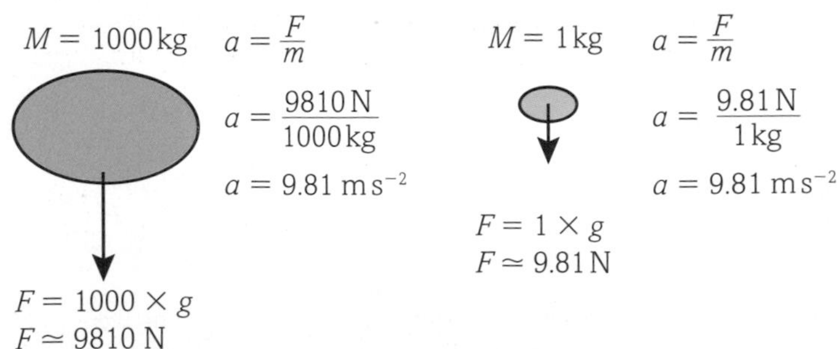

Figure 1 The relationship between acceleration, force and mass.

The value of g changes slightly depending on where it is measured on the Earth's surface. At the North Pole $g = 9.8322\,\text{m s}^{-2}$; in Singapore, near the Equator, it is $9.7803\,\text{m s}^{-2}$. The value of g depends on altitude (distance from the centre of the Earth), local variations in the density of the rocks beneath the Earth's surface and latitude. In this topic we will assume that the value of g is constant, and that air resistance has a negligible effect. We will assume that g has a value of $9.81\,\text{m s}^{-2}$.

Objects thrown upwards

An object thrown vertically *upwards* always has a constant acceleration $-g$, meaning that the acceleration vector is in the opposite direction to the velocity. A graph illustrating the subsequent motion is shown in Figure 2. Notice that the gradient of the graph is negative.

Look at the points P to T on the graph. What is happening at these points?

- P – the object is thrown upwards with an initial upward velocity of $23.0\,\text{m s}^{-1}$.
  The velocity decreases by $9.81\,\text{m s}^{-1}$ each second, as shown by the values on the velocity axis on the graph.
- Q – the object has slowed to $13.2\,\text{m s}^{-1}$ under the influence of gravity.
- R – the object has reached its maximum height and has stopped, momentarily.
  The acceleration it experiences is still equal to g. This is shown by the gradient of the graph, which is constant throughout.
- S – the object now has a velocity of $-16.2\,\text{m s}^{-1}$. The negative value is needed because velocity is a vector and the direction of movement is now downwards, not upwards.
- T – the object returns to the thrower who catches it at $23.0\,\text{m s}^{-1}$. In reality this would not be the case, because some of the kinetic energy of the object is always converted to other forms and the final velocity of the ball would be lower than $23.0\,\text{m s}^{-1}$.

Considering vertical and horizontal motion together

In the Middle Ages, soldiers and archers attempted to address the problem of horizontal and vertical motion, working out theories for the flight of cannonballs and arrows.

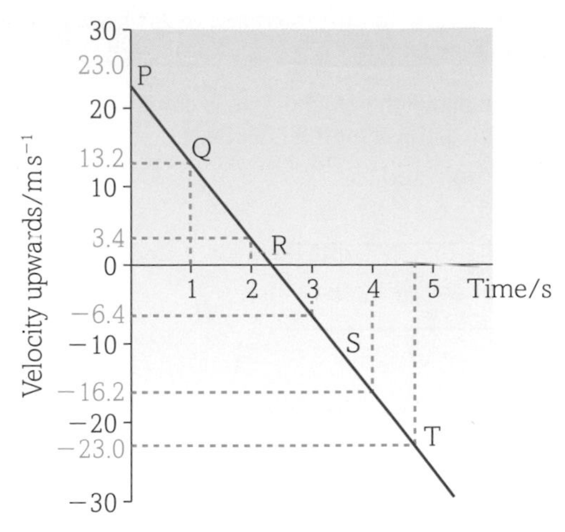

Figure 2 Velocity–time graph for an object thrown vertically upwards.

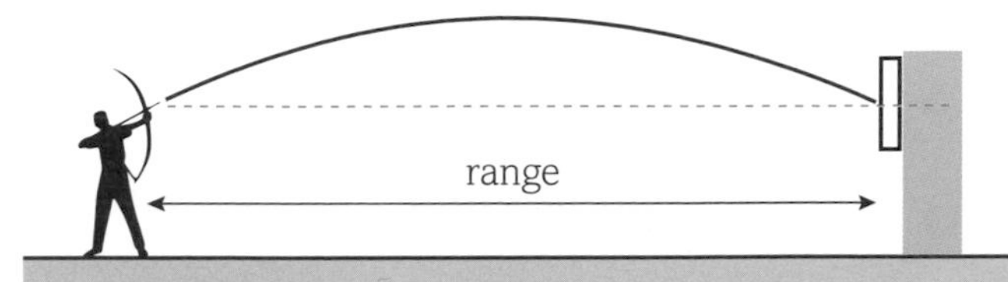

Figure 3 The flight of an arrow – notice how the parabolic shape of the path taken by the arrow is symmetrical, meaning that its horizontal component of velocity remains constant at all times throughout its flight (we assume there are no horizontal forces acting on the arrow and so no horizontal acceleration).

When a batsman hits a cricket ball for six, the ball has both vertical and horizontal components of motion (Figure 4).

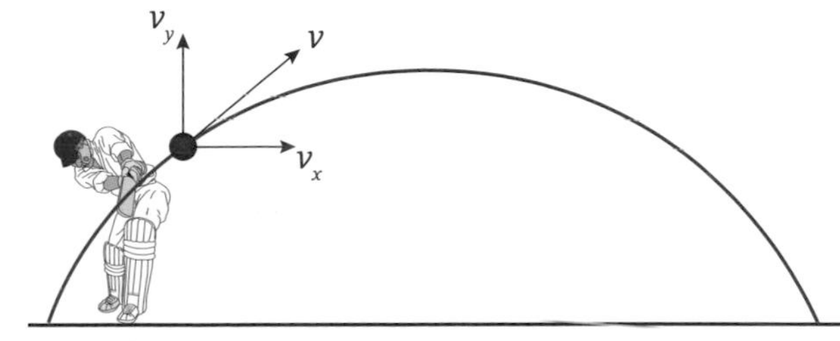

Figure 4 Hitting a cricket ball.

In the absence of air resistance, the horizontal velocity component of any projectile remains constant while it is accelerating downwards.

This effect is illustrated by Figure 5. All the horizontal arrows are the same length, indicating constant horizontal velocity. The vertical arrows show a downward acceleration. When answering questions on free fall you need to deal with horizontal movement and vertical movement entirely separately. You must also be careful with + and − signs. The worked example shows how this information can be used.

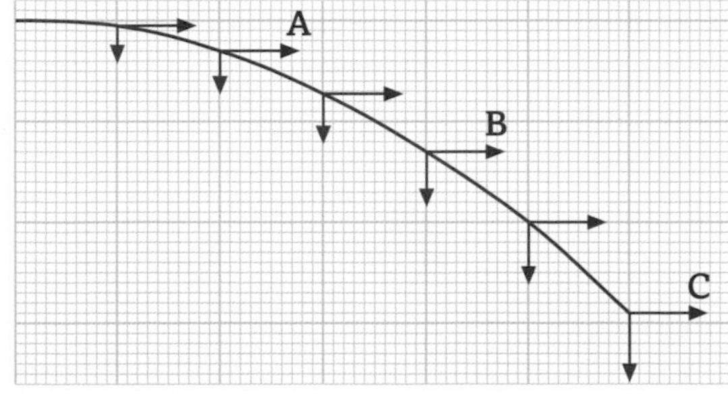

Figure 5 Constant horizontal velocity but vertical acceleration due to the force of gravity acting on the ball.

WORKED EXAMPLE

A cricket ball is thrown with a velocity of 36 m s^{-1} at an angle of 39° to the horizontal.

(a) What height above the throwing point does the ball reach?

(b) How long will it take to fall back to the level at which it was thrown?

(c) What horizontal distance will the ball travel during this time?

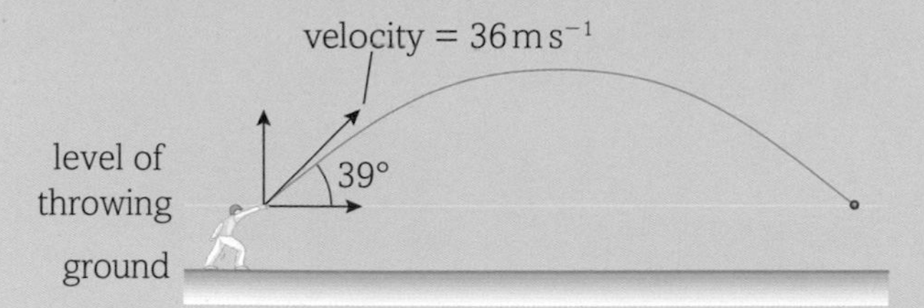

Figure 6 Throwing a cricket ball.

Answers

The first step in solving this is to resolve the starting velocity into horizontal and vertical components (this is explained in Topic 2.3.3).

horizontal component = 36 cos 39° = 28.0 m s^{-1}

vertical component = 36 sin 39° = 22.7 m s^{-1}

(a) Using $v^2 = u^2 + 2as$, where s is the maximum height at which $v = 0$, gives:

$0 = 22.7^2 + (2 \times -9.81) \times s$ and $s = \frac{515.29}{19.62} = 26.3$ m.

(b) Use $s = ut + \frac{1}{2}at^2$ for the entire vertical movement. The vertical displacement is therefore zero because it is back to the level at which it started.

This gives $0 = 22.7t + \frac{1}{2}(-9.81t^2)$. Rearranging gives:

$2 \times 22.7 = 9.8t$ and $t = \frac{45.4}{9.81} = 4.63$ s.

(c) During all this time the horizontal velocity has remained constant, so the horizontal distance travelled = 28.0 m s^{-1} × 4.63 s = 129.6 m. This is 130 m to 2 significant figures.

LEARNING TIP

One final point to mention is that you should not keep rounding numbers when going through a lengthy calculation – your final answer could be some way out. Any rounding of numbers should be done when quoting answers rather than in the course of calculations. A good rule of thumb is to quote all the figures you are certain about and one about which you are uncertain – but no more. In the above example, you were given two significant figures in the question so you can be reasonably sure of two significant figures in your answer. The third figure is doubtful.

When in doubt about the number of significant figures to use, use three. You are most unlikely to be wrong by more than one significant figure.

LEARNING TIP

Remember – the horizontal and vertical components of motion are at right angles to each other, and so are totally independent of each other. Regardless of how quickly or slowly the ball is moving horizontally when thrown, it will hit the ground at the same time as a ball that is dropped vertically at the same time.

Questions

1 Two balls are dropped from the same height of 30 m above the ground. One is dropped vertically from rest; the other is launched sideways at 5 m s^{-1}. Which ball will hit the floor first? Explain your answer.

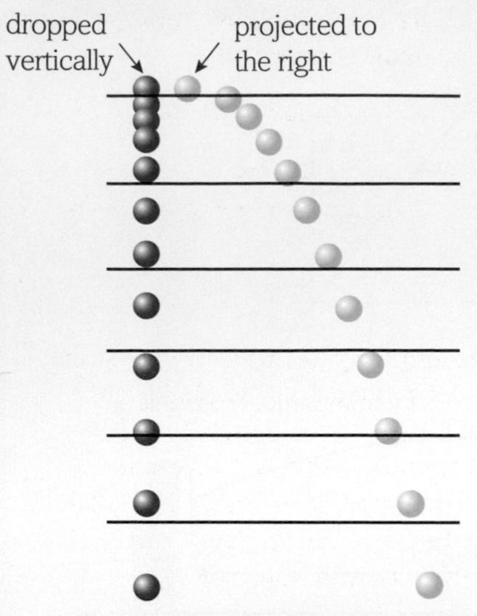

Figure 7

2 Sketch a diagram to show the path of a ball thrown at 48 m s^{-1} at an angle of 72° to the horizontal. Include the velocity vector and its components at the throwing point.

(a) (i) Add to your sketch the velocity vector at the highest point reached by the ball. What is its magnitude?

(ii) Add to your sketch the velocity vector at 3.0 s into the flight of the ball.

(b) (i) Calculate the horizontal and vertical components of the velocity at 3.0 s.

(ii) Calculate the velocity at 3.0 s.

3 A ball is thrown vertically upwards with an initial velocity of 24 m s^{-1}. Calculate:

(a) (i) the maximum height

(ii) the time taken to return to the thrower.

(b) What are the maximum height, the maximum horizontal distance travelled, and the time taken for the ball to hit the ground if the ball was thrown with the same initial velocity and the ball leaves the thrower's hand 1.8 m above the ground?

4 A teacher plans to demonstrate measuring the speed of an air rifle pellet. He intends to fire the pellet horizontally into a block of wood. Both the rifle and the target will be fixed to a board. He will measure the distance from the rifle to the block and the small distance that the pellet will drop under gravity during its flight. Assume that the pellet travels at 150 m s^{-1} and the distance travelled is 12 m to predict his possible results.

(a) Calculate the time of flight.

(b) Taking $g = 9.8$ m s^{-2} calculate the distance fallen when the pellet enters the block.

(c) Discuss whether the experiment is feasible (assume that the experiment is carried out behind a safety screen).

5 A motorcycle stunt rider, moving at constant speed, takes off horizontally from a launch point 2.0 m above the ground, as shown in Figure 8. He lands on the ground 7.7 m away as shown.

(a) By considering his vertical motion only, show that the time taken to reach the ground after he has taken off is about 0.6 s. Neglect the effects of any resistive forces. Take $g = 9.8$ m s^{-2}.

(b) Calculate the horizontal velocity in m s^{-1} at the point at which he leaves the launching point.

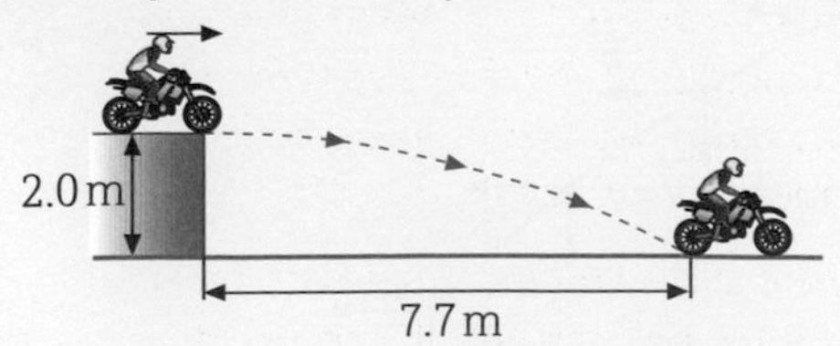

Figure 8

Measurement of *g*

By the end of this topic, you should be able to demonstrate and apply your knowledge and understanding of:

- techniques and procedures used to investigate the motion of objects
- techniques and procedures used to determine the acceleration of free fall using a trapdoor and electromagnet arrangement or light gates and a timer

Investigating motion

Velocity and acceleration in the laboratory can be investigated by using motion sensors or light gates connected to a data logger. A light gate measures the time it takes for a card attached to a model vehicle or falling object to move between two points (Figure 1) while a motion sensor records displacement at regular intervals and can be used to create graphs of displacement against time or velocity against time. Both of these methods allow greater accuracy in measuring time or displacement than when using a stopwatch, as reaction time is eliminated.

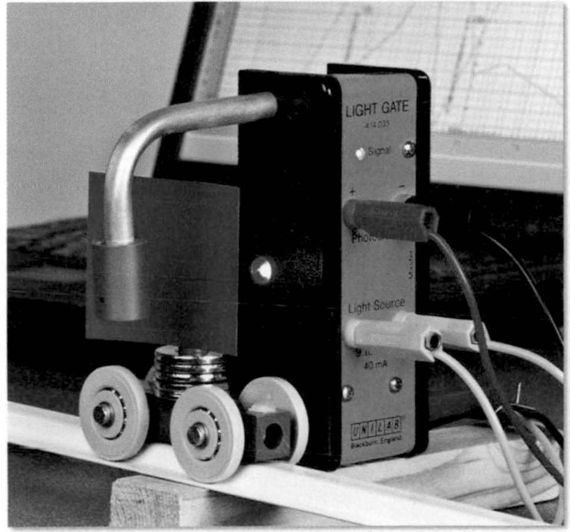

Figure 1 As the trolley passes through the light gate, the card cuts the light beam.

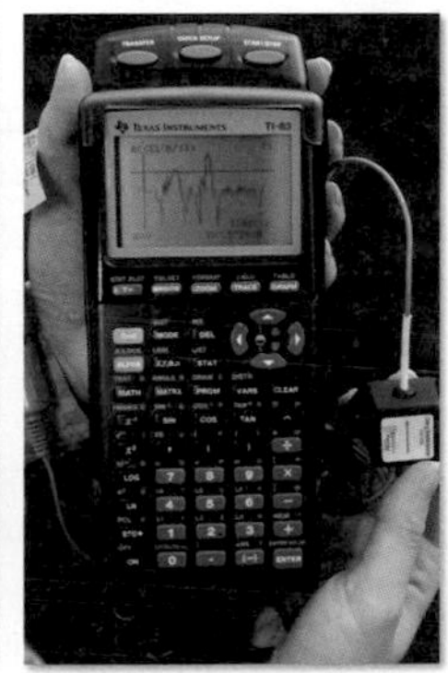

Figure 2 Acceleration can be measured directly using an accelerometer.

Experimental methods used for determining the acceleration of free fall

Provided that an object is relatively close to the Earth's surface and does not reach its terminal velocity, we can calculate the size of the acceleration due to gravity accurately.

Two types of approach can be used to measure the acceleration of free fall:

- direct approaches, such as timing a falling ball and working out the acceleration due to gravity using the equations of motion.
- indirect approaches, such as measuring the time taken for a pendulum to complete a full swing, because the motion of a pendulum is dependent on the value of *g*.

In both cases, the value of *g* is calculated by rearranging an equation that contains a term for the acceleration due to gravity.

INVESTIGATION

The trap door and electromagnet method for determining *g*

The apparatus used in this experiment, shown in Figure 3, consists of an electromagnet that just supports a steel ball. When the current through the electromagnet is switched off, the ball starts to fall and simultaneously an electronic timer is triggered. The ball falls onto a trap door and the timer is stopped. The distance, *s*, that the ball drops is measured with a ruler at this height and the time, *t*, for the fall is taken from the timer. The experiment should be repeated several times to find a mean value for *t*. More readings can then be taken at different heights, *s*.

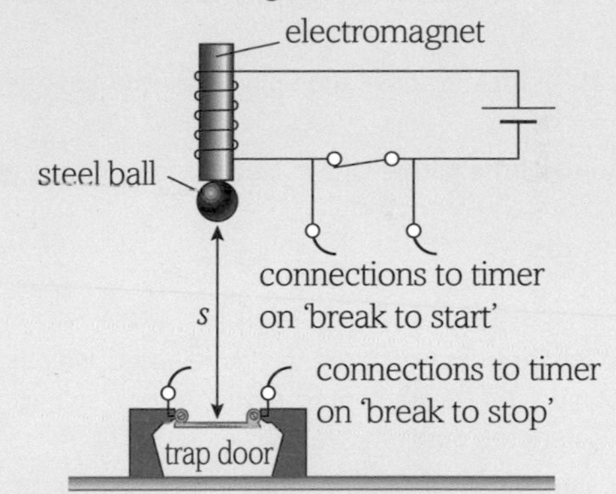

Figure 3 Trap door and electromagnet method.

Using $s = ut + \frac{1}{2}at^2$, *u* is zero because the ball starts from rest and $a = g$. This gives $s = \frac{1}{2}gt^2$ and so $g = \frac{2s}{t^2}$. For any pair of values of *s* and *t* we can substitute the measured values *t* into this equation to obtain a value for the acceleration due to gravity. Alternatively, we can obtain a value by plotting a graph – see Figure 4. A graph of *s* against t^2 will have a gradient of $\frac{g}{2}$.

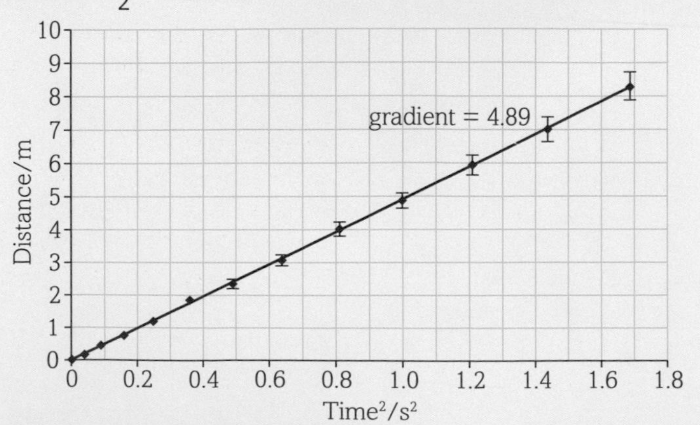

Figure 4 When we plot a straight-line graph, it must have the form $y = mx + c$. Because $s = \frac{1}{2}gt^2$, the *y*-axis must have the values of *s*, the *x*-axis must have the values of t^2, the gradient must be $\frac{1}{2}g$ and the intercept must be zero. So working out the gradient and doubling it gives a value for *g*.

Note the following:

- If the electromagnet current is too high there will be a delay in releasing the ball after the current is switched off and the clock is triggered. You must adjust the current in the electromagnet so that it only just supports the ball.
- If the distance of fall is too large, or the ball is too small, air resistance might have a noticeable effect on the speed.
- The third source of uncertainty concerns measuring the distance of fall. Make sure you measure the distance from the bottom of the ball accurately when held on the electromagnet to the top of the trap door.

INVESTIGATION

The light gates method for determining *g*

We can use a light gate and data logger to measure the time taken by a piece of card to travel through the light gate as it falls. Two small pieces of Blu-tack™ can be added at the lower corners of the card to improve the stability of the card as it falls. The average velocity of the card as it passes through the light gate is found from $v = L/t$, where L is the length of the card and t is the transit time recorded by the digital timer for to the card to travel through the light gate. The velocity may be calculated automatically by the data logging software.

We also use a ruler to measure the vertical height of the card above the light gate, s.

We hold the card vertically above the light gate and then release the card.

The acceleration a of the falling card is calculated using the equation of motion:

$$v^2 = u^2 + 2as$$

Taking $u = 0$, $2g = \frac{v^2}{s}$

By varying the height from which the card is released, we obtain a set of values for v and s. Plot a graph of v^2 (on the y-axis) against s (on the x-axis) and calculate the gradient, which is equal to $2g$.

The method assumes that the card's velocity is constant as it passes through the light gate. The effects of this error can be minimised by measuring the height fallen by the card to a line drawn horizontally across the middle of the card, rather than the lower or upper edge of the card.

Questions

1 A metal ball has been falling towards Earth for 3.5 seconds from rest.

(a) What is its acceleration?

(b) What is its velocity?

(c) How far has it fallen?

2 It takes the squirrel 2.4 s to react when it hears the coconut start to fall. Will it escape being hit by the coconut?

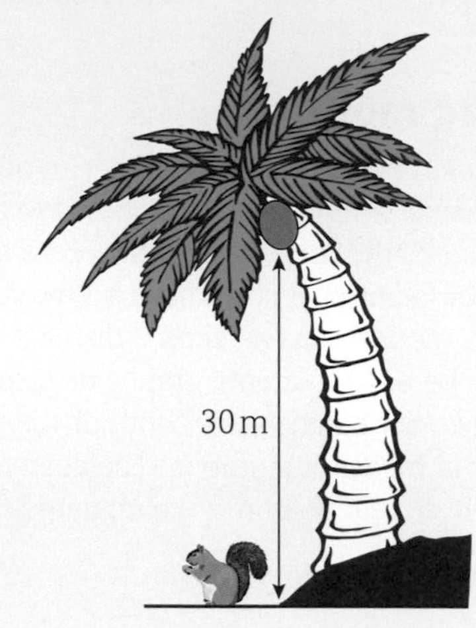

3 Suggest how video techniques could be used to determine g.

4 (a) For the experimental results shown in Figure 4, calculate the value of the acceleration of free fall, g.

(b) Estimate the percentage uncertainty in this value, by finding the gradients of the lines of maximum and minimum gradient.

(c) Suggest reasons why the calculated value is not exactly $9.81\ \text{m s}^{-2}$.

(d) State the likely measurement uncertainty in the values of s and t for this experiment.

(e) From your answer to (d), find the percentage uncertainty in a calculated value of g.

(f) Finally, state the experimental value of g, with an uncertainty value.

5 For the experimental method described on this page to measure g, using light gates, describe the sources of measurement uncertainties. Estimate the sizes of the absolute uncertainties and use these to determine the percentage uncertainty in the calculated value of g.

6 When measuring g using a light gate, errors can occur if the falling weight hits the light gate. Describe how the position of the light gate could be checked to avoid this error.

Car stopping distances

By the end of this topic, you should be able to demonstrate and apply your knowledge and understanding of:

- reaction time and thinking distance
- braking distance and stopping distance for a vehicle

Introduction

Every day, millions of cars use the roads and motorways of the UK. Speed limits range from 30 mph in built-up areas to 70 mph on dual carriageways and motorways, although local councils can impose their own speed restrictions, for example 20 mph near schools. Even slow-moving vehicles can cause injury to both drivers and pedestrians, although the risk will increase as speed, and other variables, become more significant.

Stopping distances

The total **stopping distance** of a car depends on two other distances:

- **thinking distance** – the distance the car travels between the driver seeing the hazard and applying the brakes
- **braking distance** – the distance the car travels between applying the brakes and coming to rest.

Total stopping distance = thinking distance + braking distance.

Thinking distance and braking distance are both affected by the condition of the driver, the car and the roads, as shown in Table 1.

Factors that increase thinking distance	Factors that increase braking distance
higher speed	higher speed
tiredness	poor road conditions (icy or wet)
alcohol and drugs	poor condition of brakes
distractions	poor condition of tyres
age of driver	mass of car (more people or luggage in the car)

Table 1 Factors which affect thinking distance and braking distance.

Thinking distance and braking distance

Both thinking distance and braking distance increase with speed. For a constant reaction time and for a typical car in good condition, the thinking distance increases by 3 m for every 10 mph increase in the car's speed (Figure 1). The thinking distance is calculated using the equation:

$$\text{thinking distance (m)} = \text{reaction time of driver (s)} \times \text{speed of car (m s}^{-1}\text{)}$$

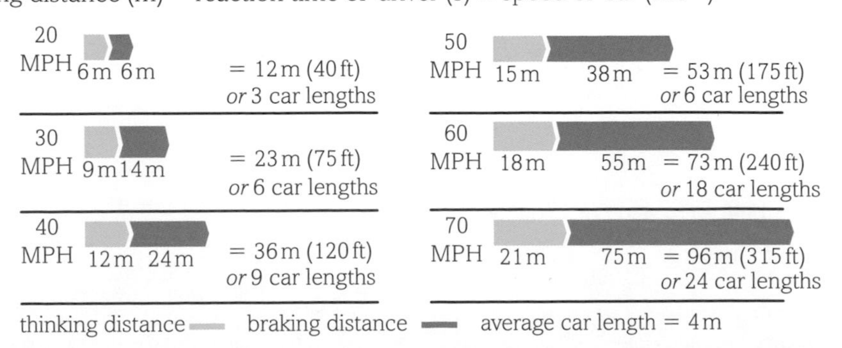

Figure 1 Typical stopping distances (contains public secter information licensed under the Open Government Licence v3.0).

Thinking distance

The thinking distance is calculated using the equation:

thinking distance (m) = reaction time of driver (s) × speed of car ($m\,s^{-1}$)

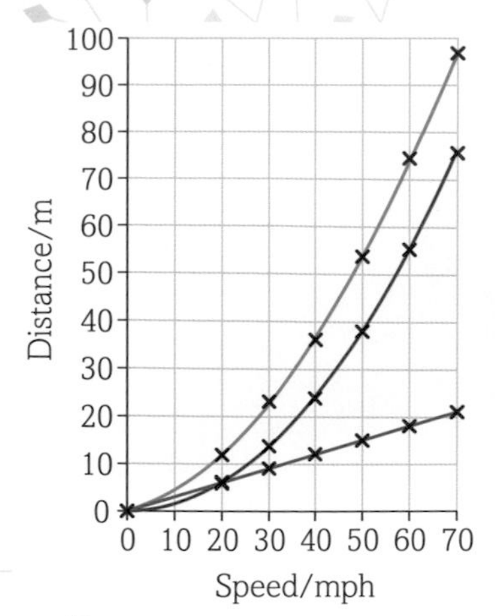

Figure 2 How thinking, braking and stopping distances vary with speed.

WORKED EXAMPLE 1

Using Figures 1 and 2 calculate:

(a) the reaction time of the driver if 1 mph = 0.447 $m\,s^{-1}$

(b) the thinking distance for a driver travelling at 45 mph

(c) the thinking distance for a driver travelling at 80 mph

(d) the speed of a car where the driver's thinking distance is 19.2 m.

Answers

(a) Using any of the values shown in Figure 2, along with the stopping distance equation, we get a value for the driver's reaction time of 0.67 s.
For example, at 50 mph, the reaction time will be $\frac{\text{thinking distance}}{\text{speed}}$, which is $\frac{15\text{ m}}{(50 \times 0.447)\text{ m s}^{-1}} = 0.67$ s. This is true for any other values substituted into the equation.

(b) At 40 mph the thinking distance is 12 m, at 50 mph the thinking distance is 15 m. The relationship is linear so at 45 mph it will be 13.5 m – halfway between the two.
Alternatively, using the equation we obtain
thinking distance = 0.67 × 45 × 0.447 = 13.5 m.

(c) Thinking distance = 0.67 × 80 × 0.447 = 24 m.
Alternatively, extrapolating the graph in Figure 2 will lead to a value of 24 m at a speed of 80 mph.

(d) Rearranging the equation we obtain speed = $\frac{\text{thinking distance}}{\text{reaction time}}$,
leading to speed = $\frac{19.2\text{ m}}{0.67\text{ s}}$, which equals 28.7 $m\,s^{-1}$)

Braking distance

Unlike thinking distance, the relationship between speed and braking distance is not linear. As the speed of the car increases, the braking distance increases by more and more. This may be understood by considering the energy changes that take place during braking.

When a car brakes, the kinetic energy of the car is transferred to thermal energy in the brakes. The work done by the friction force (see Topic 3.3.1) is equal to the transfer of energy:

decrease in kinetic energy = work done = force applied by brakes × distance moved by car

or $\frac{1}{2}mv^2 = Fd$, showing the braking distance is directly proportional to the square of the car's initial speed.

WORKED EXAMPLE 2

A car of mass 1850 kg has good-quality brakes and is travelling on a good road surface. A total braking force of 12 kN is applied to bring the car to rest. (10 mph = 4.47 $m\,s^{-1}$)

Calculate:

(a) (i) the braking distance at a speed of 20 mph
(ii) the braking distance at a speed of 40 mph.

(b) Compare these values.

Answers

(a) (i) $\frac{1}{2}mv^2 = Fd$
$= \frac{1}{2} \times 1850 \times (2 \times 4.47)^2 = 12\,000 \times d$
So $d = 6.2$ m
(ii) $\frac{1}{2} \times 1850 \times (4 \times 4.47)^2 = 12\,000 \times d$
So $d = 24.6$ m

(b) Doubling the speed of the vehicle causes the braking distance to increase by a factor of four. This is because the braking distance is directly proportional to v^2 and not v.

- 1 mile is equal to 1609.3 metres
- 1 hour is equal to 3600 seconds.

So 1 mile per hour = $\frac{1609.3 \text{ metres}}{3600 \text{ seconds}}$ = 0.447 metres per second.

To convert from miles per hour (mph) to metres per second ($m\,s^{-1}$), simply multiply the mph value by 0.447.

For example, 50 mph = 50 × 0.447 = 22.4 $m\,s^{-1}$

Questions

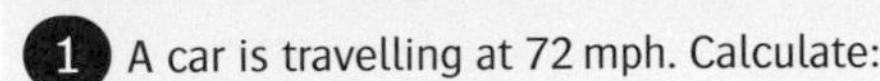

1 A car is travelling at 72 mph. Calculate:

(a) its speed in $m\,s^{-1}$

(b) the distance it will travel, in metres, in 2 minutes

(c) its total stopping distance.

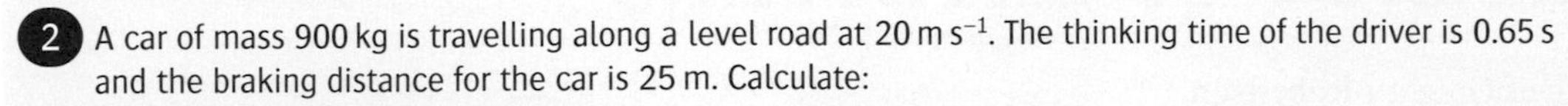

2 A car of mass 900 kg is travelling along a level road at 20 $m\,s^{-1}$. The thinking time of the driver is 0.65 s and the braking distance for the car is 25 m. Calculate:

(a) the overall stopping distance

(b) the initial kinetic energy of the car

(c) the average braking force of the car.

3 The table shows data from the Highway Code. Four spaces in the table are labelled P–S.

Car speed/$m\,s^{-1}$	5	10	15	20	25	30
Thinking distance/m	**P**	7	**Q**	14	**R**	21
Braking distance/m	2	8	**S**	32	50	72

(a) Use the table to calculate the reaction time of the driver. Then work out the values of **P**, **Q** and **R**.

(b) Use the table to:

(i) find a relationship between car speed and braking distance

(ii) show that the deceleration of the car is 6.25 $m\,s^{-2}$

(iii) work out the value of **S**.

(c) What is the total stopping distance at 15 $m\,s^{-1}$?

(d) For a car travelling at 27 $m\,s^{-1}$ (60 mph) what is:

(i) the thinking distance

(ii) the braking distance

(iii) the overall stopping distance?

4 Which of the following statements are true? Correct those that are false by rewriting them making the necessary corrections.

- Doubling the speed of a car doubles the momentum and the kinetic energy of the car.
- If you increase the speed of your car from 10 mph to 30 mph then you are going to cause three times more damage in an accident.
- Thinking distance and speed obey a linear relationship.
- Braking distance and speed are directly proportional.
- Increasing the car speed by a factor of 2 increases the braking distance by a factor of 4.

FALLING FROM THE SKY

Bodies falling through a fluid will eventually reach a terminal velocity. In this activity we will consider the motion of a human body falling in air and analyse the factors that lead to it reaching a terminal velocity.

TERMINAL VELOCITY OF SKYDIVERS

In April 1987, during a jump, the skydiver Gregory Robertson found that his colleague Debbie Williams had been knocked unconscious in a collision with a third skydiver and was unable to open her parachute. Robertson, who was well above Williams at the time and who had not opened his parachute for the 13,500 ft (4100 m) plunge, reoriented his body head-down so as to minimise his effective cross-sectional area, which is perpendicular to the direction of fall. He could thereby increase his velocity to 200 mph so that he could catch up with her. He then went into a horizontal spreadeagle position to reduce his velocity to 135 mph (in the spreadeagle position, a skydiver encounters maximum drag force and thereby falls with minimum terminal velocity of ~ 135 mph) so that he could grab her. He opened her parachute and then, after releasing her, opened his own barely ten seconds before impact. Williams received extensive injuries due to her lack of control on landing, but survived.

[…] When a body falls through the air fast enough so that the air becomes turbulent behind the body, then the magnitude of the drag force, D, is

$$D = \frac{1}{2} C\rho A v^2 \qquad (1)$$

where C is the drag coefficient, ρ the density of air and A the effective cross-sectional area […]

D gradually increases from zero as the speed of the body increases, until

$$\frac{1}{2} C\rho A v^2 = Mg \qquad (2)$$

where M is the mass of the body and g the acceleration due to gravity. Then, the terminal speed, v_t is given by

$$v_t = \left(\frac{2Mg}{C\rho A}\right)^{1/2} \qquad (3)$$

[…] we find the maximum value of A of a skydiver by modelling him as follows […]

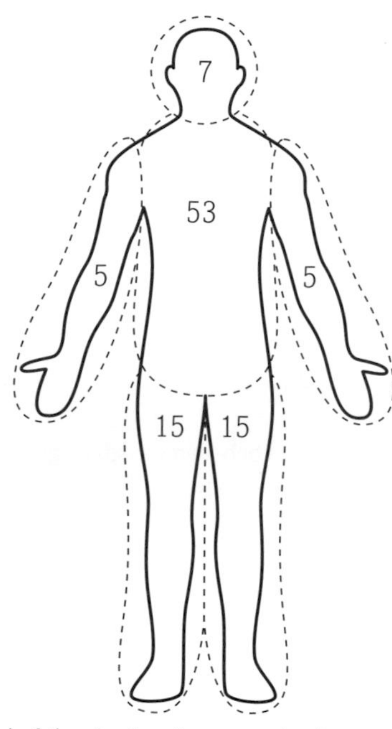

Figure 1 This model of the skydiver is assumed to have a spherical head and cylindrical arms, legs and trunk. Also, if the mass of the skydiver is 100 kg, the percentage masses of the various parts of the body are as indicated.

[…] The total projected area for our model skydiver will be

$$A = 0.0472(MH)^{1/2} + 0.0021\,M^{2/3} \qquad (4)$$

[…] This gives us the desired expression for the terminal velocity as (cf equation (3))

$$v_t = (85g/C\rho)^{1/2}\,M^{1/8}\ \mathrm{m\,s^{-1}} \qquad (5)$$

Sources

Halliday, D., Resnick, R. and Walker, J. (1993) *Fundamentals of Physics*. Wiley.
Agrawal, D.C. (2000) Terminal velocity of skydivers. *Physics Education* vol, 35, no. 4, pp. 281–283.

Where else will I encounter these themes?

1.1 | 2.1 | 2.2 | 3.1 | YOU ARE HERE | 3.2 | 3.3

Let us start by considering the nature of the writing in the article. The article above was taken from the publication *Physics Education*, which is produced by the Institute of Physics (IOP).

1. **Consider the article and comment on the type of writing that is being used, and who it is for. For example, think about whether the writing style is that of a scientist reporting his or her findings to other researchers or is it the style appropriate for a newspaper? Does the report try to explain, persuade or describe? Are the findings open to interpretation by others? How might you change the article to make it more suitable for a younger or less-informed audience?**

We will now look at the physics that is in the article. Do not worry if the physics content or the mathematics is challenging at this stage. You can always return to the article later in your course, once some of the related topics have been studied in more depth. Use the timeline at the bottom of the page to help you put this work in context with what you have already learned and what is ahead in your course.

2. **If you did not reach a terminal velocity, how fast would you hit the ground if jumping from 13 500 feet above the Earth's surface? Assume $g = 9.8\ \mathrm{m\,s^{-2}}$.**
3. **Why would physicists need to model a skydiver – how does this help?**
4. **It can be argued that the mass of the skydiver has little effect on their terminal velocity. How is this evident in the article?**
5. **What are the factors that affect terminal velocity?**

Altitude/km	ρ/kg m^{-3}	v_t /m s^{-1}
0	1.22	45.2
1	1.14	46.8
2	1.06	48.5
3	0.98	50.4
4	0.90	52.6
5	0.82	55.1
6	0.74	58.1
7	0.66	61.5
8	0.58	65.6
9	0.50	70.6
10	0.42	77.1

Table 1 The variation of the density of air and the terminal velocity of a skydiver of mass 80 kg with altitude.

6. **Estimate what Debbie Williams' terminal velocity would have been, based on the model used here or interpolating from the data in Table 1.**
7. **How do the speeds of diving birds compare with the speeds of skydivers? What is responsible for this?**
8. **How far does a person have to fall to reach terminal velocity?**

To what extent is estimation used in the development of the model shown here? Is a model like this a perfect representation of a falling skydiver?

DID YOU KNOW?

On March 24, 1944, 21 year old Flight Sergeant Nicholas Stephen Alkemade was a member of No. 115 Squadron RAF and was flying to the east of Schmallenberg, Germany, when his plane was attacked by enemies, caught fire, and began to spiral out of control. Because his parachute was destroyed by the fire, Alkemade opted to jump from the aircraft without one, preferring his death to be quick, rather than being burnt to death. He fell 18 000 feet (5500 m) to the ground below. His fall was broken by pine trees and a soft snow cover on the ground. He was able to move his arms and legs and suffered only a sprained leg.

Activity

Prepare briefing notes for a TV producer, a non-scientist who is going to commission a sports personality to present a TV programme on extreme sports, and needs to be briefed beforehand on the background and what is achievable.

3.1 Practice questions

1. Which statement is correct when analysing graphs in the area of kinematics? [1]

   A Acceleration is the gradient of a displacement–time graph.

   B Velocity is the area beneath a displacement–time graph.

   C Displacement is the area beneath a velocity–time graph.

   D Acceleration is the area beneath a displacement–time graph.

2. A car accelerates uniformly from rest along a straight road and experiences negligible frictional forces. The car travels 15 m in the second second of its journey. How far does it travel in the fifth second? [1]

   A 15 m

   B 35 m

   C 45 m

   D 75 m

3. A ball is thrown upwards with an initial velocity of $18\,m\,s^{-1}$. Which of the following is a good estimate of its maximum height? [1]

   A 12 m

   B 16 m

   C 18 m

   D 24 m

4. A car passes through traffic lights as they turn from red to green at a speed of $12\,m\,s^{-1}$. A motorcyclist, starting from rest, accelerates from the traffic lights in the same direction as the car at $1.5\,m\,s^{-2}$. After what distance from the lights will they meet? [1]

   A 16 m

   B 120 m

   C 168 m

   D 192 m

[Total: 4]

5. Figure 1 shows a graph of velocity v against time t for a skydiver falling vertically through the air.

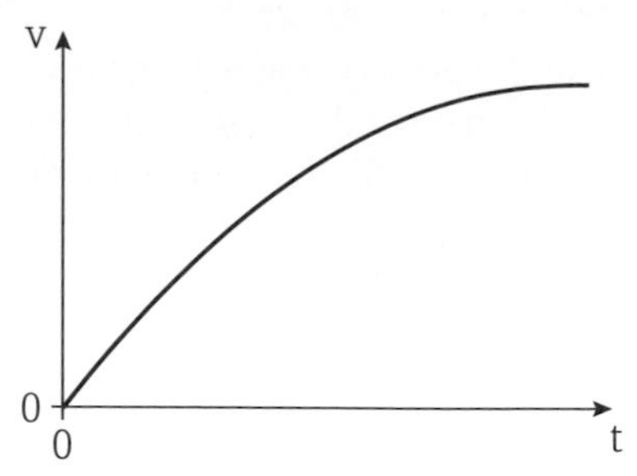

Figure 1 A velocity–time graph.

State how you can use **Figure 1** to determine the acceleration of the skydiver and describe how the acceleration varies with time. [2]

[Total: 2]

6. Calculate the acceleration of a cyclist who travels 256 m from her starting point, from a position of rest, in at time of 8 s. [2]

[Total: 2]

[Q22(a), H156/01 sample paper 2014]

7. A rock is dropped from a cliff of height 200 m. At the same time, an identical rock is launched horizontally from the top of the same cliff. Calculate:

   - the time taken for the rocks to hit the ground
   - the horizontal distance between them when they hit the ground. [4]

[Total: 4]

8. Figure 2 shows a long-jumper at three different stages during her long jump. The horizontal and vertical components of motion are shown at stages A, B and C of the jump.

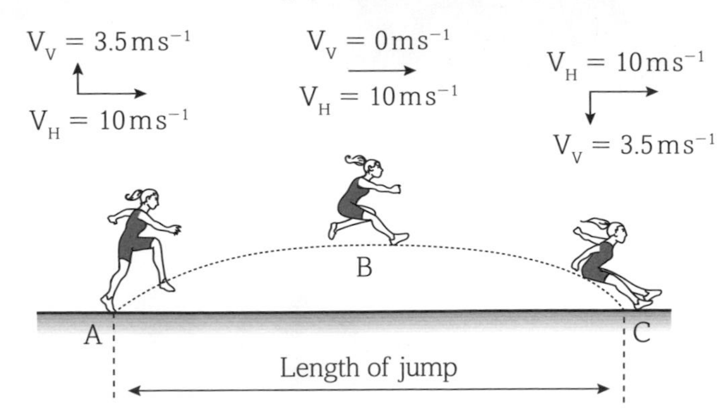

Figure 2

(a) In the model of the long jump shown:

- state any assumptions that have been made, based on the information provided [1]
- account for the change in the vertical component of velocity during the first part of the long jump [2]
- calculate the total time of flight in seconds [3]
- calculate the total length of the long jump. [2]

(b) Explain how increasing the vertical and horizontal components of motion will lead to changes in the length of the long jump. [3]

[Total: 11]

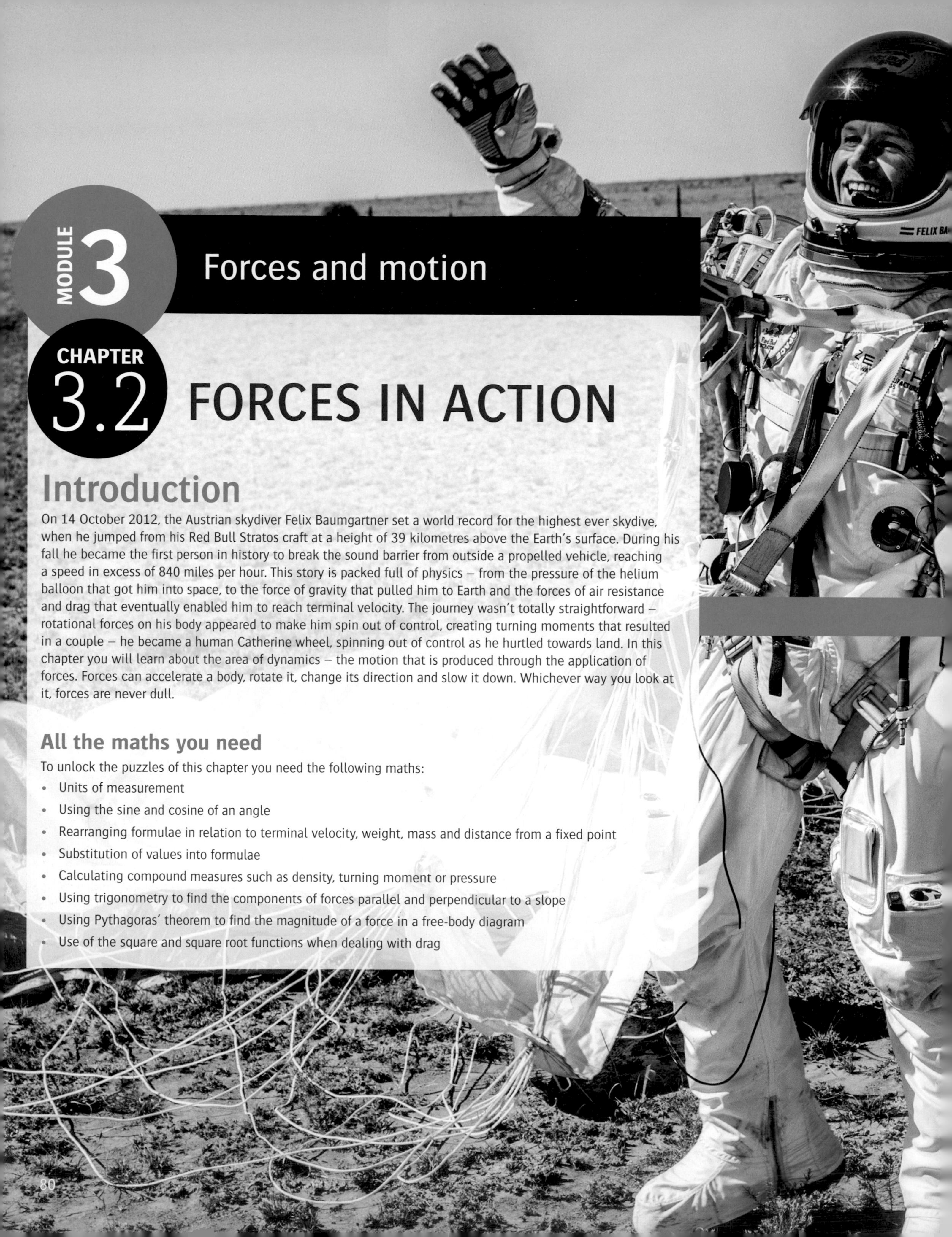

MODULE 3 Forces and motion

CHAPTER 3.2 FORCES IN ACTION

Introduction

On 14 October 2012, the Austrian skydiver Felix Baumgartner set a world record for the highest ever skydive, when he jumped from his Red Bull Stratos craft at a height of 39 kilometres above the Earth's surface. During his fall he became the first person in history to break the sound barrier from outside a propelled vehicle, reaching a speed in excess of 840 miles per hour. This story is packed full of physics – from the pressure of the helium balloon that got him into space, to the force of gravity that pulled him to Earth and the forces of air resistance and drag that eventually enabled him to reach terminal velocity. The journey wasn't totally straightforward – rotational forces on his body appeared to make him spin out of control, creating turning moments that resulted in a couple – he became a human Catherine wheel, spinning out of control as he hurtled towards land. In this chapter you will learn about the area of dynamics – the motion that is produced through the application of forces. Forces can accelerate a body, rotate it, change its direction and slow it down. Whichever way you look at it, forces are never dull.

All the maths you need

To unlock the puzzles of this chapter you need the following maths:

- Units of measurement
- Using the sine and cosine of an angle
- Rearranging formulae in relation to terminal velocity, weight, mass and distance from a fixed point
- Substitution of values into formulae
- Calculating compound measures such as density, turning moment or pressure
- Using trigonometry to find the components of forces parallel and perpendicular to a slope
- Using Pythagoras' theorem to find the magnitude of a force in a free-body diagram
- Use of the square and square root functions when dealing with drag

What have I studied before?

- Forces are measured in newtons
- A force has a size and a direction
- The forces of friction and drag act to slow down the motion of a moving body
- If forces are balanced, a body will remain at rest or move at a constant velocity
- A resultant force will cause a body to accelerate
- For every action, there is an equal and opposite reaction
- A falling body will reach terminal velocity when the weight equals the opposing forces of drag
- The weight of an object is the effect of gravity acting on a mass
- Weight and mass are related by weight = mass × g
- A force acting at a distance from a fixed point will cause a body to rotate

What will I study later?

- How momentum, impulse and force are related (AS)
- The relationship between force, work done, energy transferred and power (AS)
- The nature of forces on materials – Hooke's law, Young's modulus and the subsequent deformation of materials (AS)
- Newton's three laws of motion (AS)
- The nature of the electrostatic force and the force of gravitation (AL)
- Force–distance graphs for a body undergoing motion in a gravitational or electric field (AL)
- The force on a charged particle, and a current-carrying wire, in an electric and magnetic field (AL)
- The nature of centripetal forces on bodies that are involved in circular motion and that are experiencing simple harmonic motion (AL)
- The nature of damping and the forces involved in free and forced oscillations (AL)
- The short range and nature of the strong nuclear force in the nuclei of atoms (AL)

What will I study in this chapter?

- The relationship between weight and mass
- The terms tension, normal contact force, upthrust and friction
- How to draw and interpret free-body diagrams
- The behaviour of a body under the action of a constant force in one and two dimensions
- The factors which affect drag for an object travelling through a fluid and falling in a uniform gravitational field
- Techniques and procedures used to investigate the motion of objects
- Techniques and procedures used to determine the terminal velocity of a body falling in a variety of fluids
- The moment of a force, torque, the torque of a couple and the principle of moments.
- The conditions for a body to be in equilibrium under the action of forces and torques, including the triangle of forces
- The experimental determination of the centre of mass of a body, its significance and its relation to centre of gravity
- Pressure and density
- Use of the equations $p = \frac{F}{A}$ $p = h\rho g$ and $\rho = \frac{m}{V}$
- Archimedes' principle

Force and the newton

By the end of this topic, you should be able to demonstrate and apply your knowledge and understanding of:

- the equation net force = mass × acceleration, $F = ma$
- the newton as the unit of force
- weight of an object, $w = mg$

KEY DEFINITIONS

A **resultant force** is a single force which has the same effect as the sum of all the forces acting on a body.

One **newton** is defined as the force that causes a mass of one kilogram to have an acceleration of one metre per second squared.

Introduction

A force is generally regarded as a push or a pull acting on one or more bodies. We experience forces in our everyday lives, from our weight (the force of gravity that keeps us on the ground) to the forces involved in our transport that allow us to move long distances between our towns and cities. We often use special terms – such as drag, tension, friction, weight and thrust – when using, discussing or calculating forces. Thrust, for example, is the term used for the driving force provided by a jet engine. The thrust will try to move the aircraft forward. However, other forces, such as drag, oppose this forward motion and act in the opposite direction trying to slow the aircraft down. The difference between these two forces is called the **resultant force**. The size of the resultant force will determine the acceleration of the aircraft.

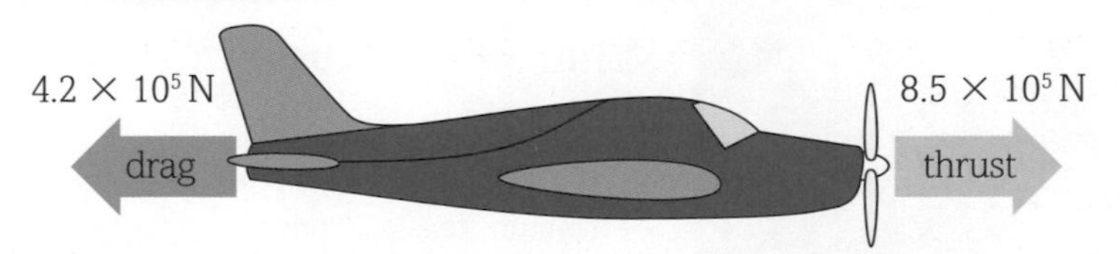

Figure 1 The difference between the thrust and the drag will be the size of the resultant force.

Force, mass and acceleration

The link between these three quantities was first established by Isaac Newton. Previously, people wrongly associated a strong force with a high speed. Newton realised that when an object has no resultant (net) force acting on it, it does not accelerate. So, an object with zero resultant force acting on it will carry on travelling at a constant velocity. An example of this is if you are on a plane travelling at a constant velocity of $280\,\mathrm{m\,s^{-1}}$, the resultant force acting on you is exactly zero. If you are in a lift travelling upwards at a constant velocity, the resultant force acting on you is again zero. In terms of Newton's first law (see Topic 3.5.1), this is true because there is no resultant force, so a body continues to move at a constant speed and will not accelerate. In terms of Newton's second law, using $F = ma$, we can see that the resultant force, F, is zero, so there will be no acceleration.

LEARNING TIP

A resultant force always causes an acceleration. Zero resultant force implies either a constant velocity or the fact that the object is stationary. When using $F = ma$ to work out the acceleration of a moving object, it is always the resultant force that is used for F. For an object with greater mass, the acceleration will decrease if F is constant.

Weight

The force which causes a falling object to accelerate is its weight, due to the pull of gravity. You can calculate the weight of an object using the relationship between force, mass and acceleration.

weight in N = mass in kg × acceleration of free fall in $\mathrm{m\,s^{-2}}$

$$W = mg$$

WORKED EXAMPLE

A go-kart with a rocket attached to it of mass 78 kg (with driver). The go-kart is travelling at a constant speed of 4 m s^{-1}, at which point the rocket fires and the go-kart experiences a driving force of 450 N and drag forces of 280 N. Calculate:

(a) the resultant force acting on the go-kart
(b) the acceleration of the go-kart
(c) the final speed of the go-kart if it continues to accelerate at this rate for 5 seconds.

Answers

(a) resultant force = driving force − drag force = 450 N − 280 N = 170 N
(b) Rearranging Newton's second law:

$$a = \frac{F}{m} = \frac{170}{78} = 2.18\ \text{m s}^{-2}$$

(c) $v = u + at = 4 + (2.18 \times 5) = 14.9\ \text{m s}^{-1}$

The four fundamental forces of nature

There are only four fundamental forces in nature, and they are responsible for all the interactions between bodies and waves in the Universe. Some of these you will be familiar with, whereas others you may not know much about. You do not need to learn this.

- **The gravitational force**

Every particle of matter in the Universe feels the influence of the force of gravitation. This force is directly proportional to the product of the masses of the two bodies and inversely proportional to the square of their separation. In simple terms, as masses get heavier the gravitational force they exert on each other will increase. However, as the distance between their centres of mass increases, the force they exert on each other decreases significantly. Doubling the distance of separation between two masses will cause the force of gravitation to be one-quarter of the original value. The gravitational force holds planets in orbit around the Sun. It is the weakest of the four fundamental forces of nature, is always attractive and has an infinite range.

Figure 2 Hideki Yukawa was responsible for explaining the strong interaction, for which he won the Nobel Prize in 1949.

- **The electromagnetic force**

The electromagnetic force holds atoms and molecules together. The force can be described in terms of the exchange of photons between charged particles or magnetic materials. It can be attractive or repulsive in nature and has an infinite range.

- **The weak interaction**

This is responsible for the process of radioactive decay and is weak in nature. It acts over only a very short range of about 10^{-18} m, which is why we do not experience it for ourselves in our day-to-day lives.

- **The strong interaction**

This is responsible for holding subatomic particles together in the nucleus of atoms and it is by far the strongest of the four forces of nature. This strong force acts over a very tiny range of about 10^{-15} m. Quarks exchange particles called gluons that effectively hold protons and neutrons together.

DID YOU KNOW?

The strong interaction is by far the strongest of the four fundamental forces – its strength is around 100 times that of the electromagnetic force, some 10^6 times bigger than that of the weak force, and about 10^{39} times that of gravitation.

Questions

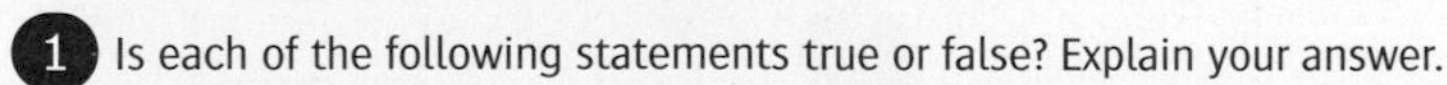

1 Is each of the following statements true or false? Explain your answer.
(a) A body will move at a constant speed if it experiences a resultant force.
(b) The force of gravity is the strongest of the forces of nature.
(c) A parachutist will accelerate downwards after jumping from a plane.
(d) As a car uses up petrol, its acceleration will increase for the same driving force.

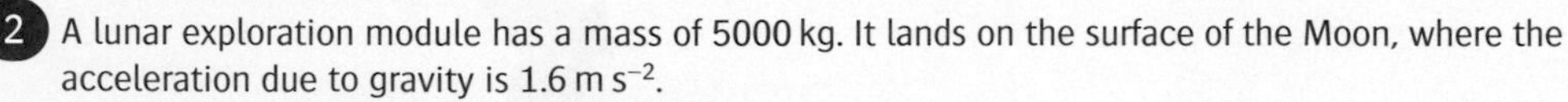

2 A lunar exploration module has a mass of 5000 kg. It lands on the surface of the Moon, where the acceleration due to gravity is 1.6 m s^{-2}.
(a) What is the lunar weight of the module?
(b) What must the thrust of the rocket be for the module to just hover above the surface of the Moon?
(c) The actual thrust of the rocket engine is 14 000 N.
  (i) What is the resultant force on the rocket?
  (ii) What is the initial acceleration of the module?
(d) What will happen to the value of the acceleration as the fuel is burnt up? Explain why.

2 Dynamics

By the end of this topic, you should be able to demonstrate and apply your knowledge and understanding of:

- the terms tension, normal contact force, upthrust and friction
- free-body diagrams

Introduction

The field of **dynamics** is concerned with the motion of bodies *and* the forces that cause that motion to happen. The field of **kinematics**, covered in Chapter 3.1, deals with the motion of bodies *without* any reference to the forces that cause the motion to happen. Together, the fields of dynamics and kinematics make up the study of **mechanics**. Newton's three laws of motion are widely used to describe the behaviour of mechanical systems.

We hear the word 'dynamics' frequently in our everyday lives. For example, there are references to aerodynamics and fluid dynamics – even to people who are dynamic speakers or dynamic sports persons. These words stem from the Greek word *dynamikos* meaning powerful, energetic or forceful. The field of dynamics is also called **kinetics**.

In addition to gravitational, magnetic and electrical forces, there are five main types of force you need to know about: tension, normal contact force, upthrust, friction and drag.

Tension

Tension is experienced by any rope, string, cable or wire that is being pulled, hung, rotated or supported. The force of tension can cause objects to accelerate or to change their shape. Mechanical engineers, architects and scientists need to know about tension – in particular whether or not a material being used to withstand large forces can do so safely before it yields or breaks.

WORKED EXAMPLE

(a) A metal ball with mass m, is stationary and is supported by a metal wire (Figure 1). What is the size of the tension, T, in the wire?

(b) The same metal ball is made to accelerate upwards due to an increase in the tension of the cable. What is the equation that links the tension to the ball's weight and acceleration?

Figure 1 The tension in a stationary, hanging body is equal in magnitude to and opposite in direction to the weight of the body.

Answers

(a) For a stationary ball of mass m, the tension in the wire will be equal in magnitude and opposite in direction to the weight of the ball, hence tension = mass × acceleration g of free fall, or $T = mg$.

Alternatively, $T - mg = 0$. In other words, the difference between the tension in the cable and the weight of the suspended ball is zero, leading to no resultant force on and no acceleration of the ball.

(b) In this example, the *difference* in magnitude between the tension in the cable and the weight of the mass is the resultant force that will be responsible for the upwards acceleration of the ball. In other words:

$$T - mg = ma$$

where a is the acceleration of the ball.

Normal contact force

The **normal contact force**, also known as the **reaction** force or the contact force, is the force that acts perpendicular (at right angles) to the point of contact of a body and the surface with which it is in contact. Some examples are shown in Figures 2–3.

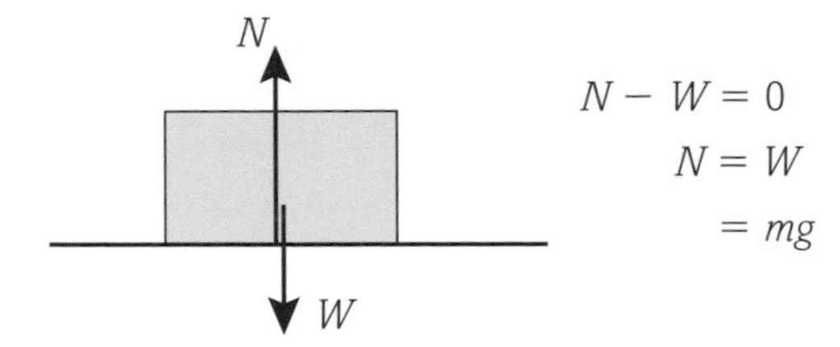

Figure 2 The block, mass m, is in contact with a flat surface, such as a table or floor. The weight of the block acting down through its centre of mass is equal to $m \times g$. The normal contact force acting upwards is equal to this in magnitude but opposite in direction.

Figure 3 The normal contact force acting on the skier is perpendicular to her contact with the slope, but not perpendicular to her weight.

Upthrust

Upthrust is the upwards force that a liquid or gas exerts on a body floating in it due to the water displaced (Figure 4). You will learn more about upthrust in Topics 3.2.7 and 3.2.8.

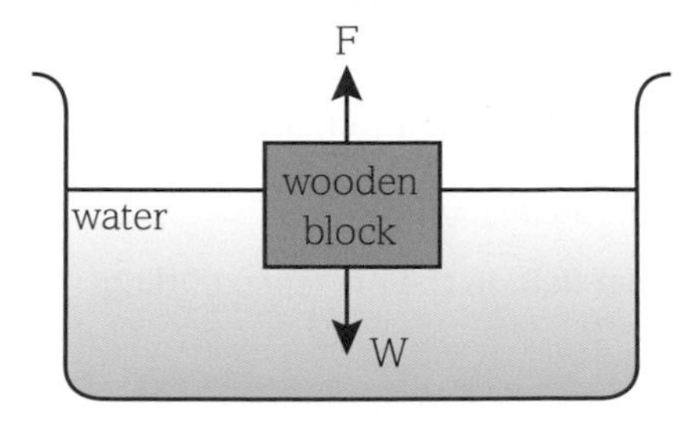

Figure 4 This wooden block floats because the weight of the block is balanced by the upthrust on the object from the water.

A body falling through any fluid will experience both drag and upthrust.

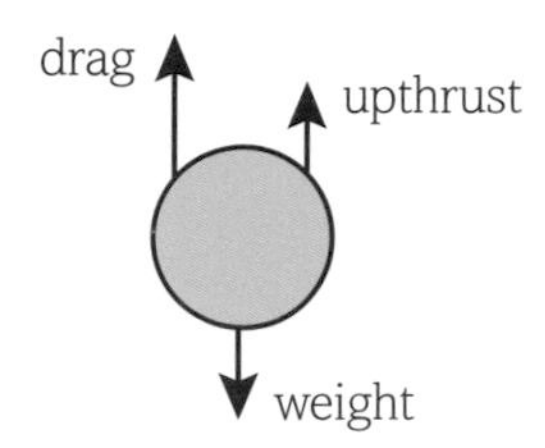

Figure 5 Weight acts downwards on a falling body, whereas the forces of drag and upthrust act upwards opposing its motion. When the magnitude of the body's weight becomes equal to the sum of the drag and the upthrust, the body will move at a constant speed called its terminal velocity.

Friction

The force of friction occurs between two surfaces in contact with one another, and is caused by interatomic and intermolecular forces when surfaces are in contact. Friction opposes motion at the point of contact.

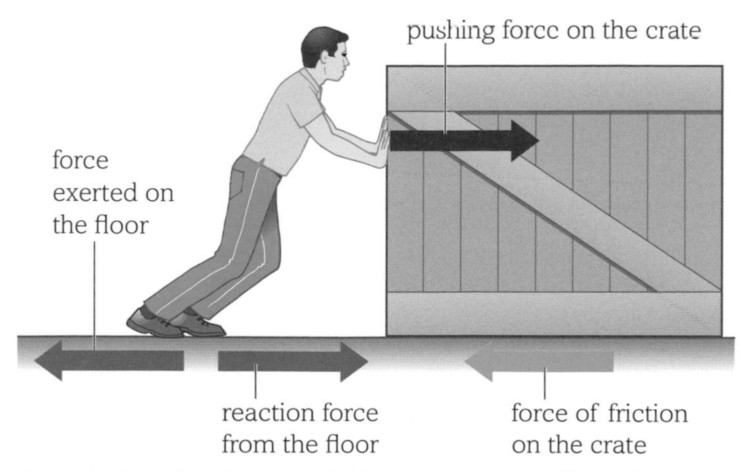

Figure 6 The pushing force provided by the man is opposed by the frictional forces between the floor and the bottom of the crate.

Free-body diagrams

A **free-body diagram** is a labelled drawing (Figure 7) used to show the external forces acting on a body. The diagram should contain:

- the body on which the forces act (often shown as a point)
- the direction of application of each force
- the type of each force
- the size of each force.

If a number of forces act on an object, we can analyse the acceleration produced by resolving all the forces on the free-body diagram along the same two directions. The pair of directions could be horizontal and vertical, or parallel and perpendicular to one of the forces.

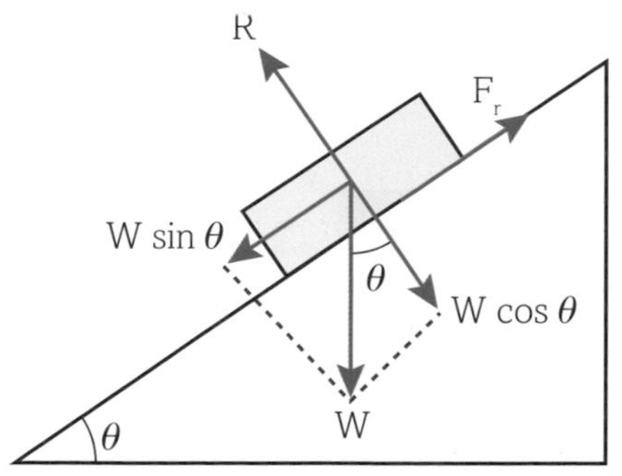

Figure 7 In situations where the normal contact force is acting on a rough inclined plane, the forces will look like this. W is the weight of the body, θ is the angle of inclination of the slope, R is the normal contact force or reaction and F_r is the force of friction.

LEARNING TIP

When drawing a free-body diagram, no scale is needed. Always label the angles to avoid making careless errors in your calculations.

Questions

1. How are the fields of dynamics and kinematics:
   (a) similar to each other
   (b) different from each other?

2. What would be the size of the tension, T, in a cable holding a mass of 3.4 kg if:
   (a) the mass was stationary
   (b) the mass was accelerating upwards at 3.2 m s^{-2}?

3. Draw free-body diagrams showing:
   (a) a ship floating in water and moving at constant speed
   (b) a ship floating in water and accelerating
   (c) a skydiver falling through air
   (d) a skier travelling downhill at an angle of 30° to the horizontal.

4. A man pulls a box of books along the ground by pulling on a strap at 45° to the horizontal.
   (a) Draw a free-body diagram to show the forces acting on the box.
   (b) Resolve the tension in the strap, T, into components along the ground and vertically upwards.
   (c) Write an algebraic equation to represent:
      (i) the resultant force in the horizontal direction
      (ii) the acceleration of the box of books.

5. A concrete block of mass 25 kg is placed on a smooth slope that makes an angle of 30° with the horizontal. Ignoring the effects of friction, calculate:
   (a) the acceleration of the block as it slides down (take $g = 9.81$ m s^{-2})
   (b) the distance the block moves in 3 s.

3.2 (3) Drag and terminal velocity

By the end of this topic, you should be able to demonstrate and apply your knowledge and understanding of:

- drag as the frictional force experienced by an object travelling through a fluid
- factors affecting drag for an object travelling through air
- motion of objects falling in a uniform gravitational field in the presence of drag
- terminal velocity
- techniques and procedures used to determine the terminal velocity of objects in fluids

Figure 1 A car experiences drag as it moves through the air – the higher the velocity the more aerodynamic drag it experiences.

Figure 2 Skydivers experience drag as they fall through air.

Figure 3 At the macroscopic level, this metal looks very smooth.

Figure 4 At the microscopic level, this metal looks rough, which leads to friction when in contact with another surface.

Introduction

A fluid is either a liquid or a gas. When any object such as a car (Figure 1), a ship or a parachutist (Figure 2) moves through a fluid, it experiences the force of drag. The effect of **drag** is always to *oppose* the motion taking place – it will act to slow down the moving object as it is a frictional force. Objects experience aerodynamic drag in air when moving in any direction.

Friction and air resistance

Any objects that are in contact with each other experience a frictional force. Even those that appear to be very smooth (Figure 3) will have many ridges, troughs and indentations when viewed at the microscopic level (Figure 4). Although frictional forces on objects that are in contact act to slow a body down, they do *not* depend on the velocity of the object.

Conversely, the frictional force of drag that an object experiences when moving through a fluid *does* depend on its velocity. It can be shown that:

$F_d = \frac{1}{2}\rho C_d A v^2$, where:

F_d = the force of drag (N)

ρ = the density of the fluid (kg m^{-3})

A = the cross-sectional area of the moving object (m^2)

C_d = the coefficient of drag (no units)

v = the velocity of the moving object (m s^{-1}).

You do not need to remember this equation. The equation can be simplified by writing it as $F_d = kv^2$, which shows that the force of drag is directly proportional to v^2.

Notice that drag is affected by the cross-sectional area of the object and the density of the fluid. When a floating object moves through water, we usually do not notice the aerodynamic drag due to air being much less dense than water.

WORKED EXAMPLE

A car with a cross-sectional area of 4.2 m^2 is moving through air at a speed of 18 m s^{-1}. The drag coefficient is estimated to be 0.28 and the density of air is 1.29 kg m^{-3}. Calculate:

(a) the force of drag acting on the car (give your answer to 1 decimal place)

(b) the acceleration of the car if its mass is 1450 kg and the thrust developed by the engine is 3450 N

(c) the acceleration at a speed of 36 m s^{-1}, assuming no other variables have changed.

Answers

(a) $F_d = \frac{1}{2}\rho C_d A v^2 = \frac{1}{2} \times 1.29 \times 0.28 \times 4.2 \times 18^2 = 245.8$ N

(b) Using Newton's second law:
resultant force = mass × acceleration, or $F_r = ma$
Substituting: (3450 − 245.8) = 1450 × a

So $a = \frac{3204.2}{1450} = 2.2$ m s^{-2}

(c) Because $F_d = \frac{1}{2}\rho C_d A v^2$, we can see that the drag force is directly proportional to the square of the velocity. So, because the velocity has doubled, the drag force increases by a factor of four. This means that the new drag force is 983.2 N. So the new resultant force is 3450 N − 983.2 N = 2466.8 N.

$a = \frac{F_r}{m} = \frac{2466.8}{1450} = 1.7$ m s^{-2}

Terminal velocity and drag

When an object falls from a great height through air, it initially accelerates because the downwards gravitational force is greater than the resistive forces. However, the drag on the object increases as it accelerates. Eventually the drag (upwards) becomes equal to the weight of the falling object (downwards), and so the resultant force on the object is zero – it then continues to fall at a constant velocity. This is called its **terminal velocity**.

Figure 5 shows the variation in velocity for a parachutist. When the parachute is opened there is a sudden increase in the drag force so the resultant force acts *upwards* and the parachutist's speed decreased. As they slow down, the drag force is decreased and eventually a new terminal velocity is reached.

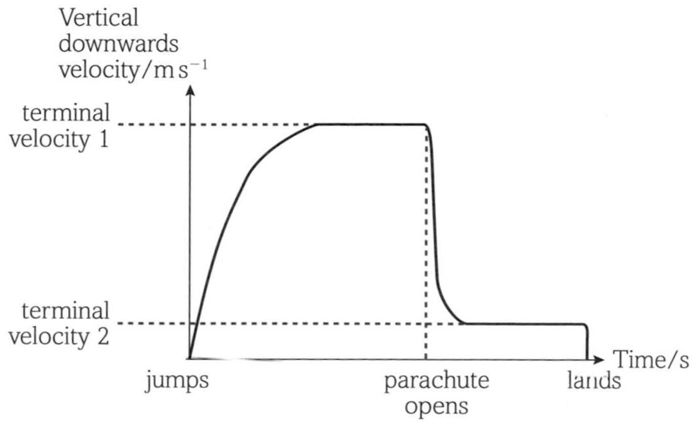

Figure 5 At $t = 0$, the skydiver jumps out of the plane and the only force acting on him is due to his weight. As he accelerates, the frictional force of air resistance increases – hence the reduction in gradient of the graph. At terminal velocity, the forces of weight and drag are balanced and he moves at a speed. Opening the parachute further reduces his velocity and he decelerates rapidly to a second terminal velocity before finally reaching the ground.

INVESTIGATION

Measuring the terminal velocity of a body falling through a fluid

You can do an experiment to determine the terminal velocity of a sphere of a known mass as it falls through a fluid using the apparatus shown in Figure 6. Suitable liquids are wallpaper paste, heavy oil or liquid detergent.

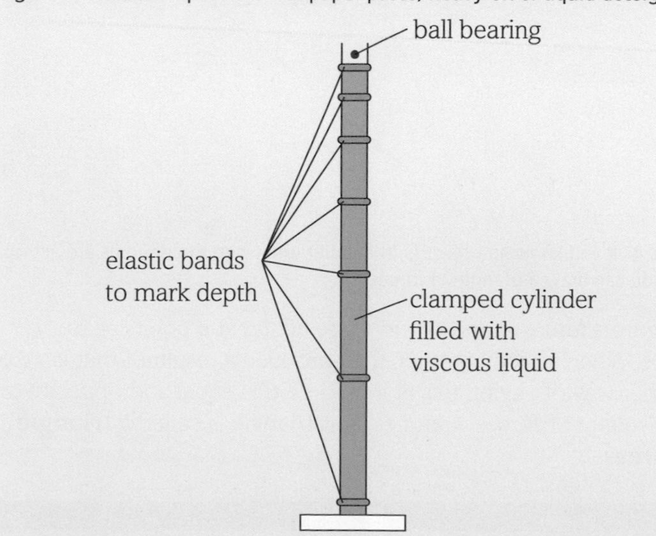

Figure 6 Determining the terminal velocity of a ball bearing falling through a viscous liquid.

Carefully drop a ball bearing into the liquid starting a timer as you do so. At given time periods use the elastic bands to mark the vertical position of the ball.

Once the ball bearing has reached the bottom of the tube, measure the distance between each consecutive pair of elastic bands.

A magnet can be used to take a ball bearing out of the tube to repeat your measurements.

For each time period, measure the distance between consecutive elastic bands, record the time period for each and use this data to calculate the average velocity, v, of the ball.

Once you have several measurements, calculate the mean distance travelled for each time period and use this data to calculate the velocity, v, of the ball for each time period.

Plot a graph of v on the y-axis against cumulative time from the release of the ball, t, on the x-axis. Draw a smooth curve through the points. Identify the time at which the ball reached its terminal velocity and use your graph to determine the 'best' value of terminal velocity.

Questions

1. Explain what is meant by:
   (a) a fluid
   (b) friction
   (c) drag.
2. In the investigation described to measure the terminal velocity of a ball bearing in a viscous liquid, explain how you would use your measurements to show that the ball bearing had reached its terminal velocity.
3. Explain how you would calculate the percentage uncertainty in the measured value of terminal velocity.
4. Explain how, using this apparatus, you could minimise the percentage uncertainty in a single measurement of terminal velocity.
5. Sketch a graph showing how the acceleration of the ball bearing varies with time after its release at the top of the liquid.
6. Discuss what you would expect to happen with an elongated shape of the same mass, giving scientific explanations to support your reasoning.
7. Design an investigation to measure how the terminal velocity of a paper cone varies with its mass. Explain how you will manipulate the independent variable, including how you will keep the other variables constant. Describe what measurements you will take and how the data will be used in order to reach a conclusion.

3.2 ④ Equilibrium

By the end of this topic, you should be able to demonstrate and apply your knowledge and understanding of:

* equilibrium of an object under the action of forces
* condition for equilibrium of three coplanar forces; triangle of forces

Introduction

If the resultant force acting on an object is zero, then the object is said to be in **equilibrium**.

KEY DEFINITION

Objects are in **equilibrium** when all the forces acting on them in the same plane (**coplanar** forces) are balanced – there is zero net or resultant force. In terms of motion, the object is either stationary or is travelling at constant velocity.

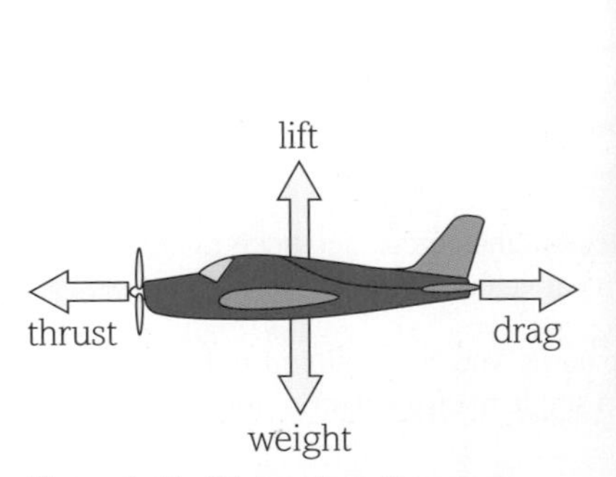

Figure 1 Equilibrium in action.

WORKED EXAMPLE 1

In Figure 2, one of the forces is shown. What are the sizes of the three forces in this example?

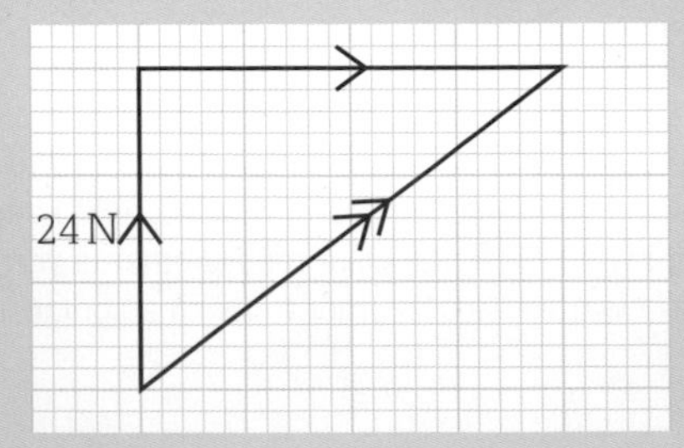

Figure 2

Answer

The force shown by the vector of length 3 cm has a size of 24 N, meaning that 1 cm on the graph is represents a force of 8 N.

If you measure the other two sides of the triangle using a ruler, you will see that they have lengths of 4 cm and 5 cm, so their respective forces are 32 N and 40 N.

You can check this using Pythagoras' theorem using $32^2 + 24^2 = 40^2$.

The triangle of forces

We can work out the resultant of two forces acting on a body in two main ways:

- by scale drawing (Figure 3a, c and d and Worked example 1)
- by calculation from a right-angled **vector triangle** (Figure 3b and Worked example 2).

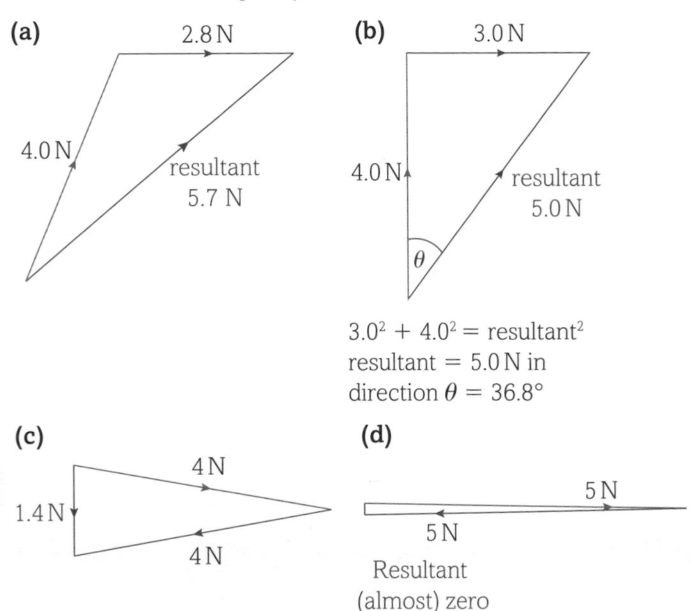

Figure 3 How vector triangles are used to add forces together

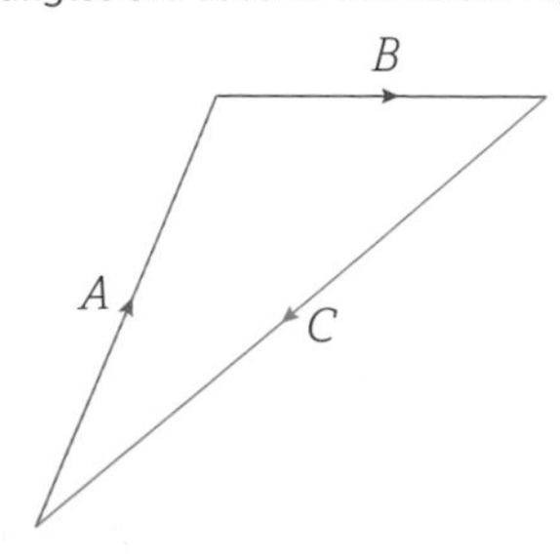

Figure 4 If you know an object is in equilibrium, when three forces are acting on it you can draw a triangle of forces.

In Figure 4 forces A and B and C are acting at a point on an object. When added together, they produce a resultant that is zero. Another way of saying this is that force C is equal and opposite to the resultant of forces A and B. This triangle is called a **triangle of forces**.

KEY DEFINITION

A **triangle of forces** represents the direction and magnitude of three coplanar forces that are acting on an object in equilibrium.

A zero resultant means zero acceleration for an object and such an object is said to be in equilibrium. An object can also be in equilibrium when it is moving with constant velocity. For example, a person at rest in a plane travelling at a constant velocity is in equilibrium.

In the worked example below, you can measure the lengths of each of the lines representing the forces. If you know the length of one of the sides and the scale, then you can work out the sizes of the other forces involved in the system.

WORKED EXAMPLE 2

The climber in Figure 5 has a mass of 80 kg.

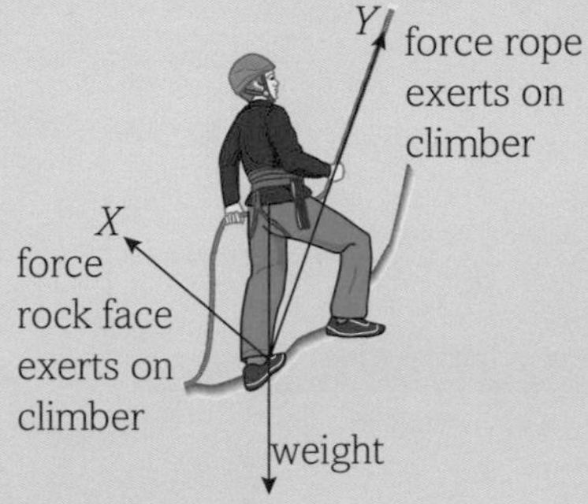

Figure 5 A climber on a rock face.

The force that the rock face exerts on the climber is at an angle of 50° to the vertical. The other force, Y, acting on the climber keeps him in equilibrium and is provided by a rope at an angle of 40° to the vertical. Calculate:

(a) the force, X, the rock face exerts on the climber
(b) the force, Y, provided by the rope.

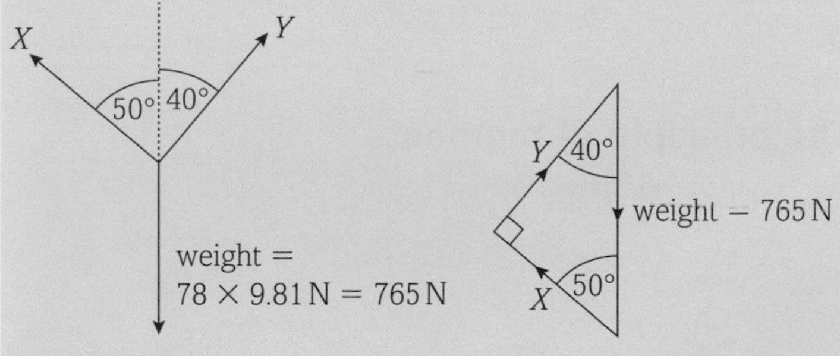

Figure 6 The forces acting on the climber.

Figure 7 Triangle of forces.

Answers

First we work out the weight, W, of the climber:

W = mass × acceleration due to gravity
$= 78\text{ N} \times 9.81\text{ m s}^{-2}$
$= 765$ N (note that we are using 3 significant figures at this stage)

Figure 7 shows the forces acting on the climber in the form of a triangle of forces.

When the sketch diagram has been drawn, add all the given information to it. Note that $50° + 40° = 90°$, so you have a right-angled triangle, which makes it easier to work out the answers.

From the triangle of forces showing the equilibrium arrangement we get:

(a) $X = 765 \times \cos 50° = 492\text{ N} = 490\text{ N}$ to 2 s.f.
(b) $Y = 765 \times \cos 40° = 586\text{ N} = 590\text{ N}$ to 2 s.f.

These answers should be rounded to two significant figures – we are confident in the first two because two significant figures are given in the question. It is only the third that is uncertain.

Questions

1 Use the scale and information provided in the Figure 8 to work out the sizes of forces A–F.

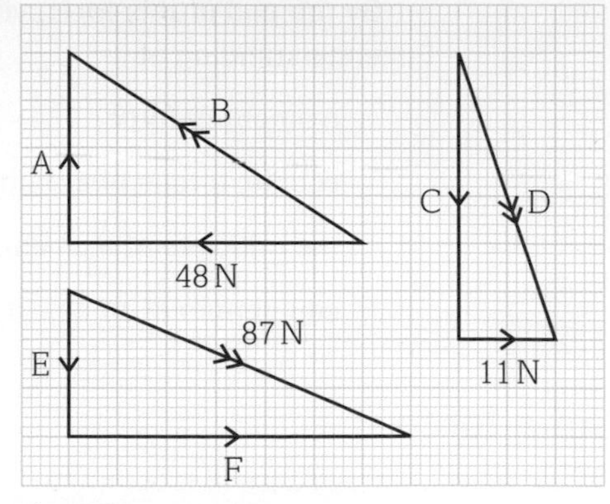

Figure 8

2 A boy and the sledge together (Figure 9) have a mass of 70 kg and are moving at a constant speed down a slope.

(a) Find the value of force R by drawing a closed vector triangle. You know all the angles and one of the lengths.
(b) Use simple trigonometry to find the value of R. How does the value of R calculated here compare with the one you obtained in part (a)?
(c) Calculate the value of the resistive force, S, acting on the boy and the sledge.

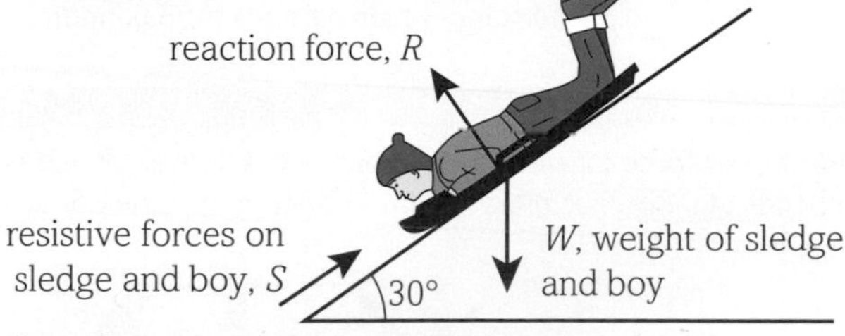

Figure 9

3 A heavy sack of mass 29 kg rests on a surface that slopes at an angle of 30° to the horizontal.

(a) Draw a free-body diagram showing all the forces acting on the sack.
(b) Resolve the forces along the slope and normal to the slope to find the values of the normal contact force and the frictional force.

4 A picture hangs on a wall as shown in Figure 10. It is supported by two strings, each has a tension, T, of 20 N and both are at 30° to the vertical. Use this information to work out the weight of the picture.

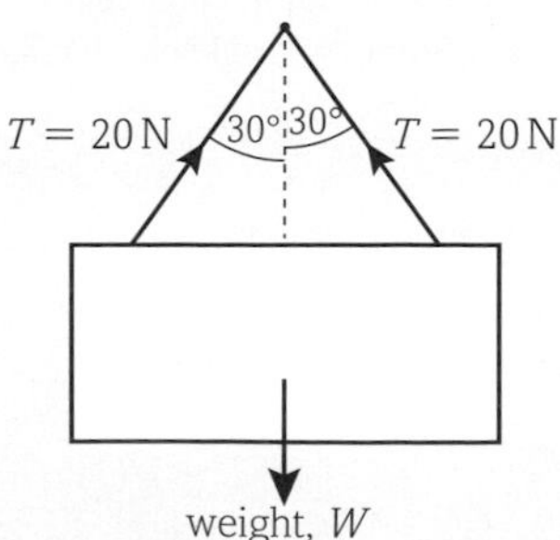

Figure 10

5 Turning forces

By the end of this topic, you should be able to demonstrate and apply your knowledge and understanding of:

* moment of force
* couple; torque of a couple
* the principle of moments
* equilibrium of an object under the action of forces and torques

The moment of a force

When a force acts on an object, it may cause that object to accelerate. A force that acts through the object's centre of mass, parallel to the ground, may cause the object to move in a straight line, increasing its velocity as it does so. We say that work has been done on the object.

Alternatively, a force can be applied to an object causing the object to rotate. The force is often perpendicular to the axis of rotation, leading to a **turning moment** being set up.

The size of the turning moment is calculated using:

turning moment (N m) = magnitude of force (N) × perpendicular distance of force from a fixed point (m)

KEY DEFINITION

The **moment of force** (or **turning moment**) is the product of a force and the perpendicular distance of its line of action from the point or axis.

Figure 1 A force acting on an object may cause it to accelerate in the direction of the force because work is being done on the object – there is no rotation and no turning moment in this case.

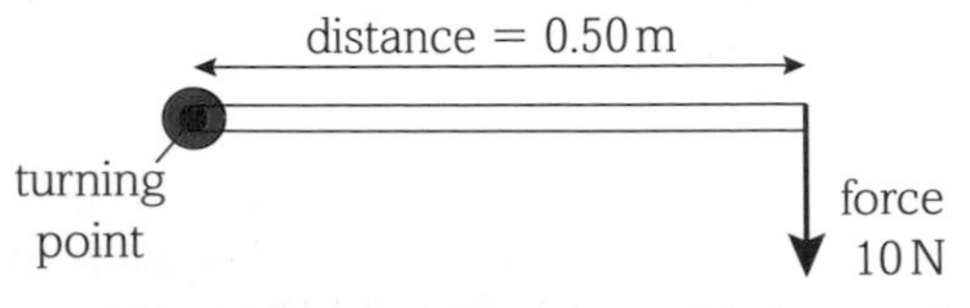

Figure 2 The magnitude of the turning moment will be larger as the size of the force and/or the distance from the fixed point increases.

WORKED EXAMPLE 1

What is the size of the turning moment for:
(a) the spanner (Figure 3)

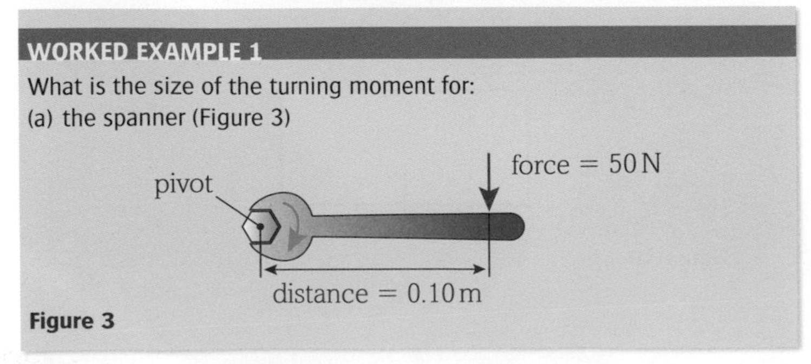

Figure 3

(b) the door (Figure 4)?

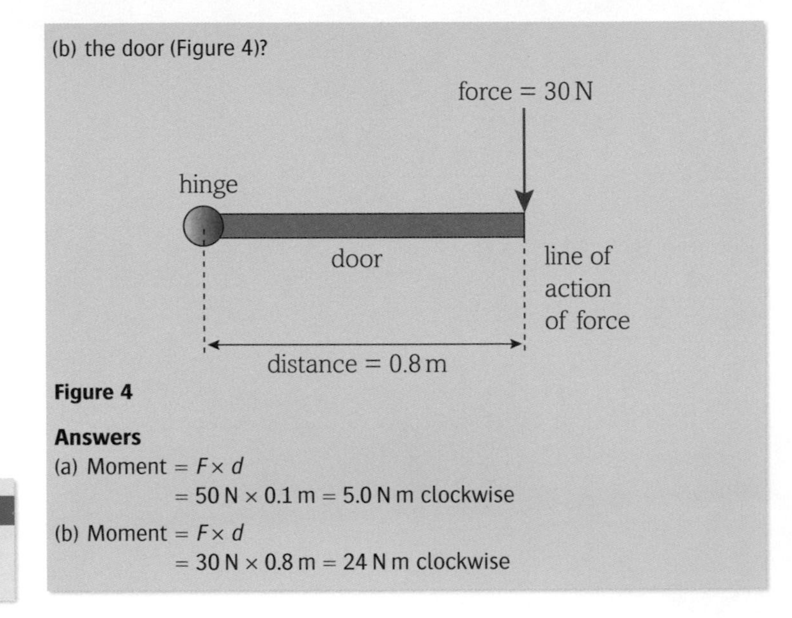

Figure 4

Answers

(a) Moment = $F \times d$
= 50 N × 0.1 m = 5.0 N m clockwise

(b) Moment = $F \times d$
= 30 N × 0.8 m = 24 N m clockwise

The principle of moments

Why does the see-saw shown in Figure 5 balance?

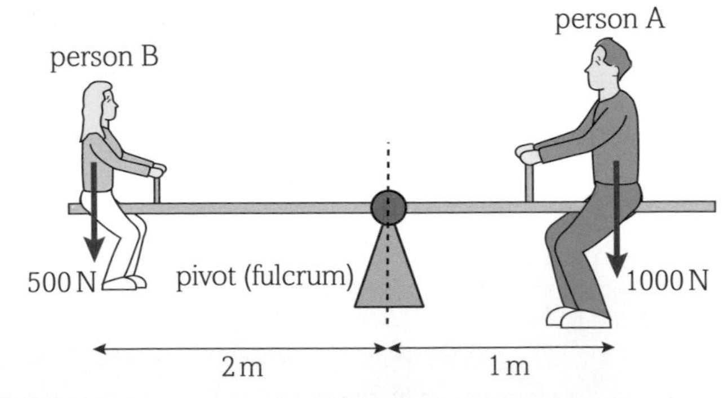

Figure 5

The reason is that the moment acting *clockwise* from the pivot is equal to the moment that is acting *anticlockwise.*

If person A rotates, the see-saw will move downwards in a clockwise direction; if person B rotates, the motion will be anticlockwise.

The moment acting clockwise from person A is:
1000 N × 1 m = 1000 N m

The moment acting anticlockwise from person B is:
500 N × 2 m = 1000 N m

Because the moments acting clockwise and anticlockwise are equal, the see-saw balances. Another term for the see-saw being balanced is that it is in rotational equilibrium. This leads to a key definition in physics called the **principle of moments**.

KEY DEFINITION

The **principle of moments** states that for an object to be in rotational equilibrium, the sum of the clockwise moments must equal the sum of the anticlockwise moments.

WORKED EXAMPLE 2

A see-saw is in rotational equilibrium. Work out the value of the missing force in Figure 6.

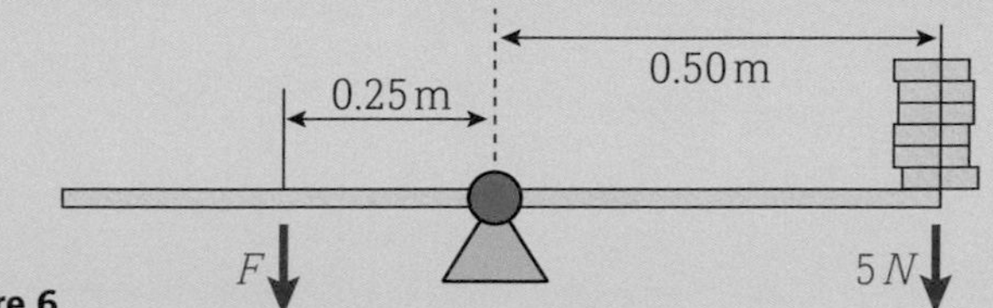

Figure 6

Answer

Clockwise moment = 5 N × 0.5 m = 2.5 N m
Anticlockwise moment = F × 0.25 m

Because the moments must be equal, we get 2.5 N m = F × 0.25 m and the unknown force must be 10 N downwards.

WORKED EXAMPLE 3

Figure 7 shows a human forearm carrying a load of weight 80 N. The weight of the forearm is 20 N.

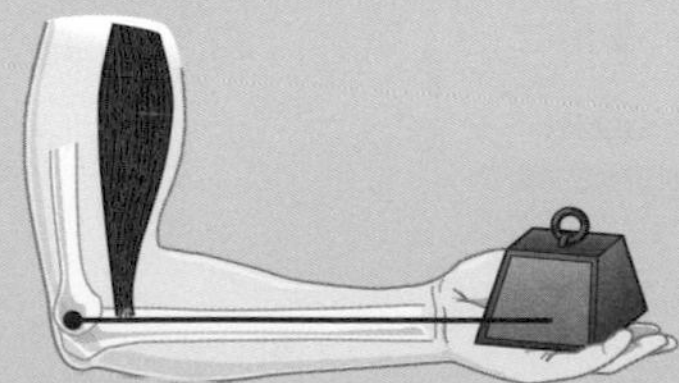

Figure 7

In order to keep the forearm horizontal, the biceps muscle exerts a force, P, on the forearm. An idealised sketch of the forces acting, together with various distances from the elbow, which acts as a pivot, is shown in Figure 8.

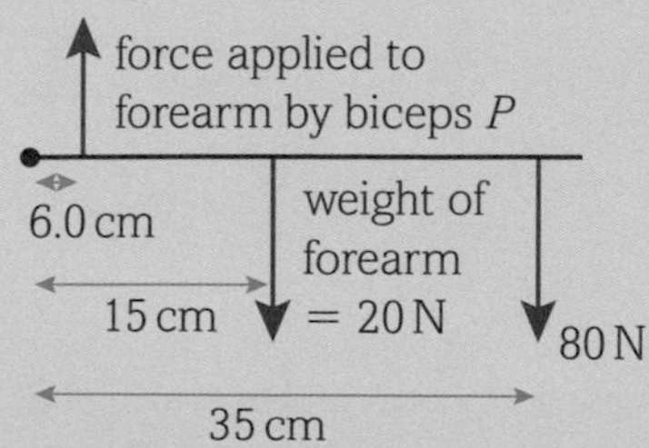

Figure 8

Calculate the force exerted by the biceps.

Answer

Taking moments about the elbow gives:

clockwise moment of load = 80 N × 0.35 m = 28 N m
clockwise moment of forearm = 20 N × 0.15 m = 3 N m
total clockwise moment = 31 N m

For equilibrium, the anticlockwise moment provided by the biceps must be equal and opposite to this, so must also be 31 N m, giving:

$P \times 0.060\text{ m} = 31\text{ N m}$

$P = \dfrac{31\text{ N m}}{0.060\text{ m}}$ = 520 N (2 significant figures) upwards

It is possible to use the principle of moments to work out the weight or mass of a beam by taking moments from fixed points.

WORKED EXAMPLE 4

Look at Figure 9. Calculate the weight of the beam (W) and the reaction force (R_1).

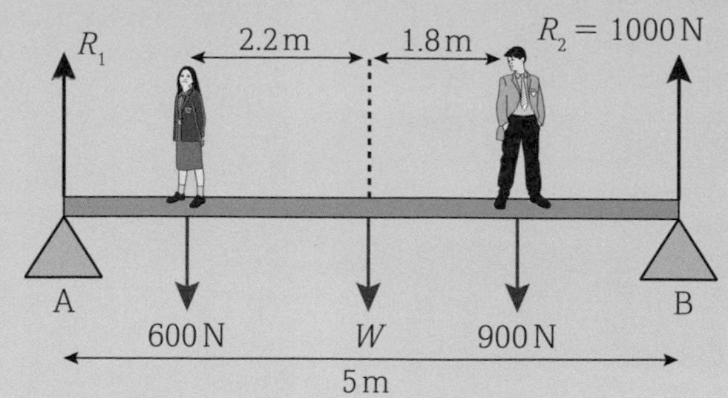

Figure 9

Answer

Taking moments about point A:

(600 N × 0.3 m) + (W × 2.5 m) + (900 N × 4.3 m) = 1000 N × 5 m

Rearranging and simplifying gives W = 380 N

To find R_1 we equate the forces acting vertically upwards and the forces acting vertically downwards. We can do this because if these forces were not balanced then the beam would accelerate in a vertical direction.

$R_1 + R_2$ = 600 N + 380 N + 900 N

We know that R_2 = 1000 N, so substituting this into the equation and rearranging, we get R_1 = 880 N upwards.

LEARNING TIP

There are *two* conditions for equilibrium.

1. The resultant force on the object must be zero.
2. Considering all the forces acting on the object, the sum of the clockwise moments about any point must equal the sum of the anticlockwise moments about that point.

Couples

A **couple** occurs when two equal, antiparallel forces act to produce a rotation – no linear motion occurs. The term 'antiparallel' means that the two forces are parallel to one another, but are acting in opposite directions. The moment or turning effect of a couple is called a **torque**.

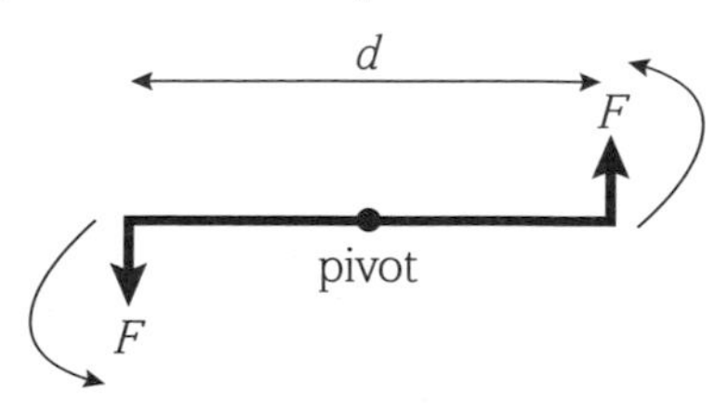

Figure 10 The size of the turning moment produced by the couple is equal to one of the forces multiplied by their separation, $F \times d$.

KEY DEFINITION

The turning moment (**torque**) due to a couple is the product of one of the forces and the perpendicular distance between them. The units are N m.

LEARNING TIP

When calculating the moment of a force, the distance to the pivot must be measured as the perpendicular distance from the line of action of the force to the pivot. In Figure 11, this distance is 1.0 m, so the moment about the pivot is $F \times 1.0$ and not $F \times 1.2$. Alternatively, we can resolve the force into two components along and perpendicular to the handle. Only the perpendicular component produces rotational motion.

WORKED EXAMPLE 5

A man lifts the handles of a loaded wheelbarrow vertically. The axle of the wheel acts as a pivot. The distance d between the man's hands and the axle is 1.2 m as shown in Figure 11.

If the load of 800 N can be considered to act at a horizontal distance of 0.4 m from the axle, find the magnitude of the normal contact force R at the axle when the man holds the wheelbarrow off the ground.

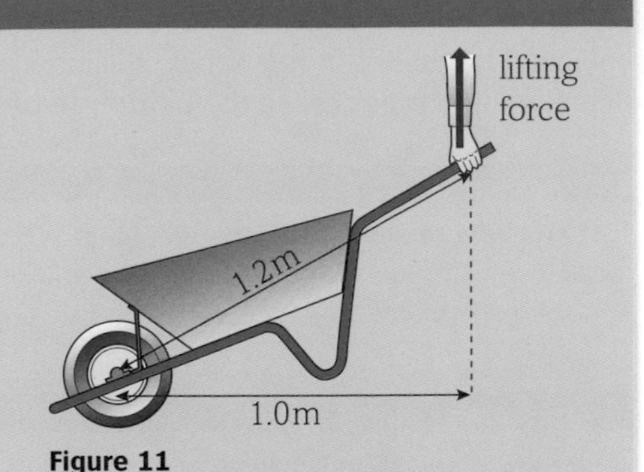

Figure 11

Answer

The wheelbarrow is in equilibrium and not turning about the pivot, so by the principle of moments:

moment of lifting force, F = moment of load, W

$F \times 1.2 = W \times 0.4 = 800 \times 0.4$

$F = 320$ N

Resolving forces vertically,

$F + R = W$

$R = W - F = 800 - 320$

$= 480$ N

Questions

1 State the S.I. units of the moment of a force.

2 Two forces F_1 and F_2 act in the same plane at the centre of mass of an object. The distances of the forces from the centre of mass are d_1 and d_2 respectively. The object is in rotational equilibrium. Copy Table 1 and fill in the missing values.

Force F_1	Distance d_1	Force F_2	Distance d_2
35 N		24 N	8 m
12 kN	35 m		1 m
	3×10^2 m	6550 N	60 m

Table 1 Turning moments, distances and forces.

3 Look at Figure 12 and work out the value of force F.

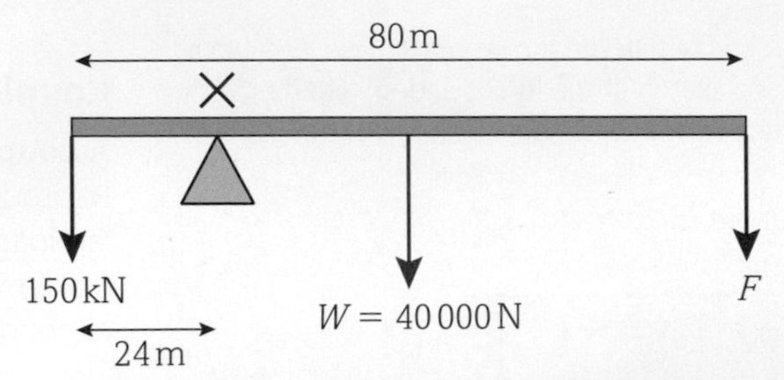

Figure 12

4 By taking moments about point B in the worked example in Figure 9, find the value of R_1 if the weight of the beam was changed to 1450 N.

5 Two equal and antiparallel forces act on a steering wheel. If the forces are both 8 N in size and 25 cm perpendicular from the centre of the steering wheel, what will be the size of the torque due to the couple?

6 Two loads, 10 N and 15 N, are hung from one side of a light beam at distances of 2.5 m and 1.0 m respectively from the pivot. These loads are balanced by two forces on the opposite side of the pivot, a load of 50 N at a distance of 1.4 m from the pivot and an upwards force of 15 N applied at a distance of $1.4 + x$ m from the pivot.

Find the value of x such that the beam is in equilibrium.

3.2 

Centre of mass

By the end of this topic, you should be able to demonstrate and apply your knowledge and understanding of:

- the centre of mass of an object
- the centre of gravity of an object
- simple experiments to determine the centre of gravity of an object

Introduction

The whole mass of an object can be thought of as acting at a single point called its **centre of mass**. If a force acts on the object directly through its centre of mass then the object will move in a straight line with either an acceleration (if there is a net force acting on it) or at a constant speed (if the forces acting on the object are balanced) (see Figure 1).

If the force acting on the object does not act directly through its centre of mass, then the object will move with an added rotational motion. This is due to a turning effect having been introduced by the moment of the external force acting at a distance from the centre of mass (see Figure 2).

A centre of mass can easily be located for shapes that are symmetrical in terms of their shape and the distribution of mass throughout the shape.

For certain shapes, the centre of mass is located at a point where there is no actual matter present – for example, in the case of a hollow circle or a horseshoe. This is because the distribution of mass is completely equal around this point, despite no mass actually being present there.

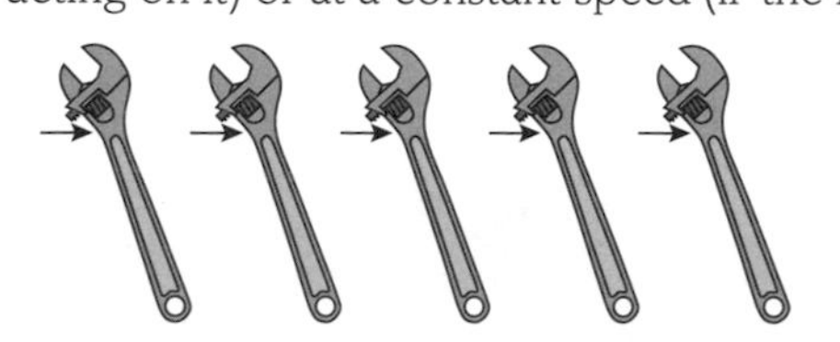

Figure 1 The force acting on this wrench is directly through the centre of mass – if it is free to move it will accelerate in a straight line.

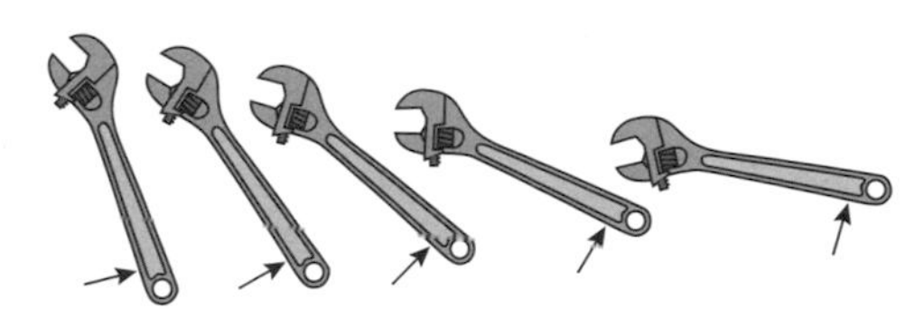

Figure 2 The force on this wrench is not acting directly through its centre of mass, leading to a turning effect that will cause it to rotate.

KEY DEFINITION

The **centre of mass** of an object is the single point at which all of the mass of the object can be assumed to be situated. For a symmetrical body of constant density, this will be at the centre of the object.

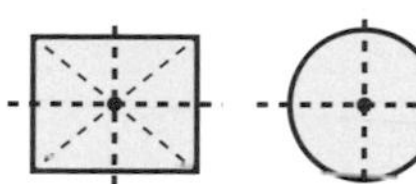
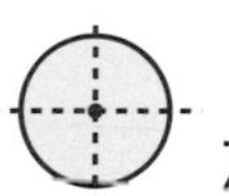

Figure 3 Centres of mass for some symmetrical shapes.

Centre of gravity

The **centre of gravity** of an object is the single point through which the entire weight of the object can be thought to act. The weight of an object can be shown as an arrow pointing vertically downwards through the centre of gravity. The centre of gravity and the centre of mass will correspond for a small body that is placed in a uniform gravitational field.

Figure 5 The centre of mass and the centre of gravity are at the same point for most objects.

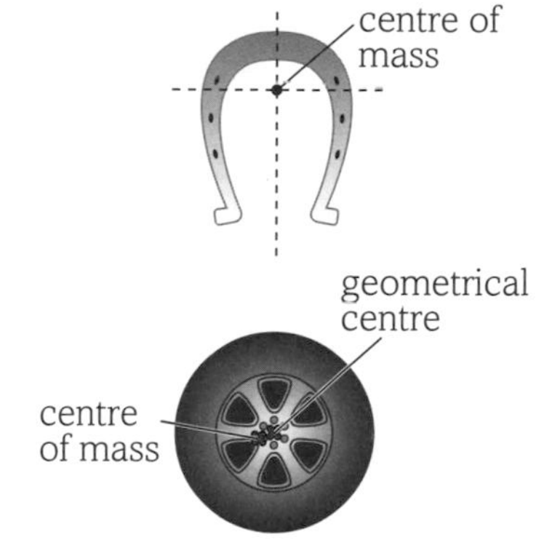

Figure 4 Centres of mass of a car tyre and a horseshoe.

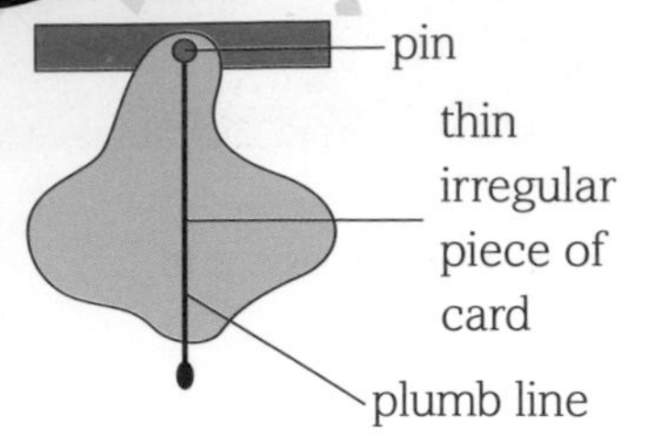

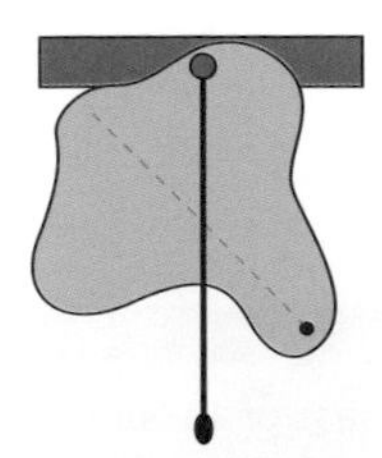

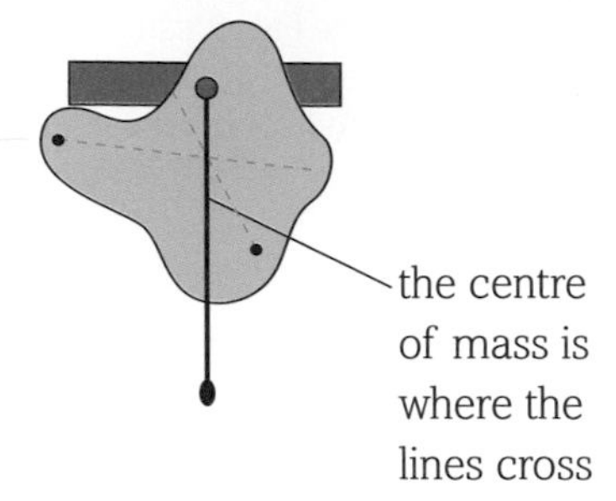

Figure 6 The three lines will lead to a point being formed – or a small triangle.

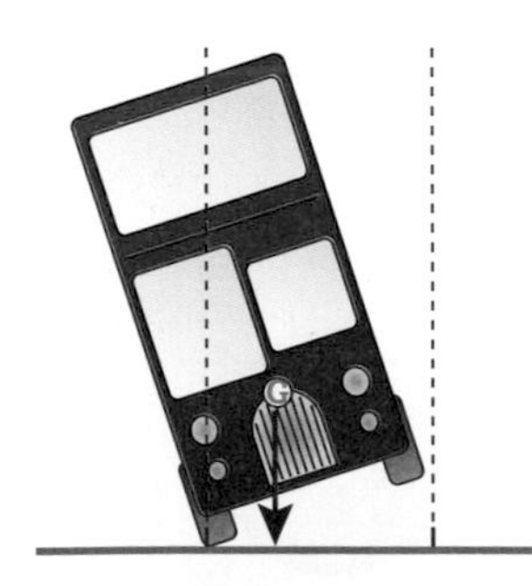

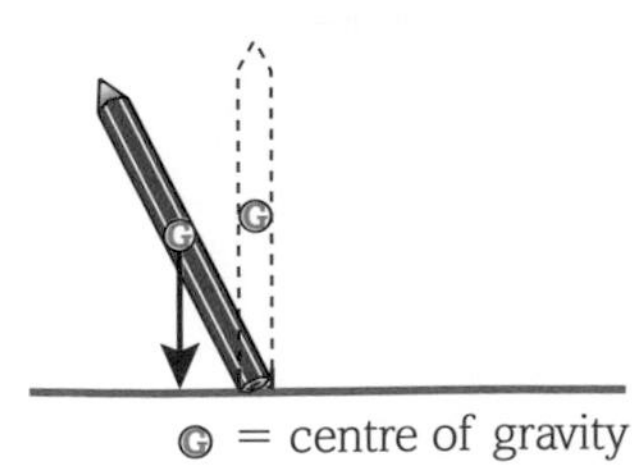

Figure 7 The Formula-1 car is stable; the bus is less stable; and the pencil is unstable.

INVESTIGATION

Finding the centre of mass or centre of gravity of an object

A simple experiment can be done to find the centre of gravity of an irregular shape of uniform thickness, such as a large piece of a jigsaw puzzle. To find the centre of mass or centre of gravity follow these steps.

1. Make three holes in random places on the edge of the shape.
2. Hang the shape from the first hole by placing it through a pin, as shown in Figure 6. Let the shape swing and come to rest. A suspended body will always come to rest with its centre of gravity directly below the point of suspension.
3. Use a plumb line (a piece of string attached to the pin, with a weight tied to the lower end of the string) to draw a vertical line down from the point of suspension to the bottom of the shape.
4. Repeat steps 2 and 3, but from a different hole. There will now be two lines drawn on the shape – where they cross is the centre of mass of the shape.
5. Repeat steps 2 and 3 for a final time from the third hole.

In an ideal experiment, all three lines will cross at a single point – this is the centre of mass. However, in reality the three vertical lines will probably lead to a small triangle and the centre of mass or gravity is assumed to be at the centre of this triangle.

Centre of mass and stability

The stability of cars, lorries and other objects is largely determined by the position of the centre of mass and the base area of the object. Objects that have a low centre of mass and a wide base area are more stable than those with a higher centre of mass and a smaller base area. Objects that are unstable have high positions for their centre of mass and small base areas. An object will become unstable when the object is tilted so that the centre of mass lies outside the width of the base of the object. At this point the object will topple and fall. See Figure 7.

Objects that roll, such as a football or a garden roller, are said to have neutral stability because their centre of mass can never fall outside the base area that is touching the ground.

Questions

1. Some objects have their centre of mass where there is no actual matter present (e.g. a horseshoe). List three objects like this.
2. Sketch a diagram for a metre ruler balanced on the end of your finger in two different positions. In each diagram, mark the centre of gravity and the direction of the weight.
3. Suggest at least two possible sources of error that make it difficult to determine the centre of gravity accurately when using the method shown in the investigation above.
4. Draw the shape shown in Figure 8 on card and use the method shown in Figure 6 to find its centre of gravity.

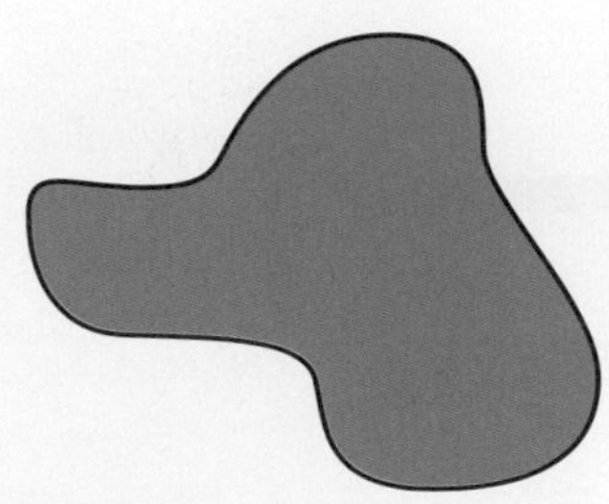

Figure 8

5. Which is more stable – a bicycle or a skateboard? Explain your answer.
6. Successful sportsmen and sportswomen often have to make use of their centre of mass to achieve high levels of performance. Which sports involve the need to understand the importance of centre of mass and what must athletes do to achieve success?

3.2 7 Density

By the end of this topic, you should be able to demonstrate and apply your knowledge and understanding of:

- density; $\rho = \frac{m}{V}$
- upthrust on an object in a fluid, Archimedes principle

Introduction

Figure 1 shows how gold metal can be found in a variety of forms – coins and gold bullion being just two examples. As the volume of gold increases, so will its mass and the amount of money that it will be worth.

Figure 1 Gold comes in a variety of different shapes and sizes, which will have different volumes and masses. However, the density of solid gold remains constant at room temperature.

Regardless of the mass or volume of pure gold that is present, the **density** of gold, in its solid form, will not change. The density of gold is fixed at 19 300 $kg\,m^{-3}$ or 19.3 $g\,cm^{-3}$. This means that every time the volume of some gold increases by 1 cm^3 its mass will increase by 19.3 g.

The density equation is density = mass/volume or $\rho = \frac{m}{V}$, where symbols used are: ρ (the Greek letter 'rho') for density; m for mass; and V for volume.

> **WORKED EXAMPLE 1**
>
> A material is found to have a mass of 350 g and a volume of 180 cm^3. Calculate its density to 2 significant figures.
>
> **Answer**
>
> $\rho = \frac{m}{V}$
>
> $= \frac{350\ g}{180\ cm^3}$
>
> $= 1.9\ g\,cm^{-3}$

> **KEY DEFINITION**
>
> **Density** is defined as mass per unit volume. The S.I. units are $kg\,m^{-3}$.

Calculating the density of regular shapes

The rectangular cuboid in Figure 2 has a mass of 28.4 g and dimensions of 3.2 cm by 2.8 cm by 8.5 cm as shown. We have measured its mass using scales, and the length of each side using a ruler that has a metric scale.

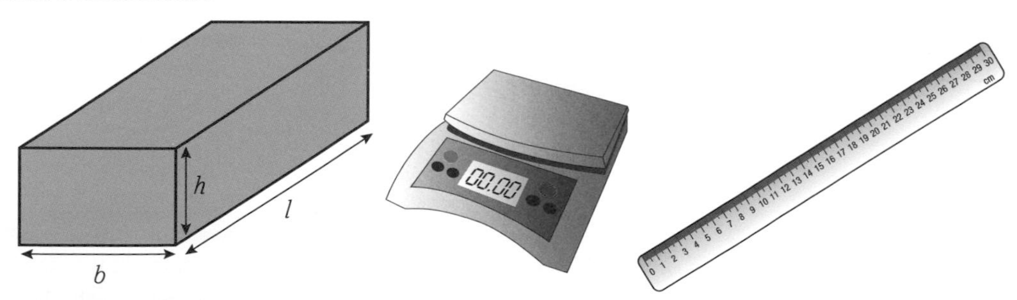

Figure 2 Measuring density.

The volume of the regular cuboid is calculated using volume = length × breadth × height, giving a value of 76.2 cm^3 to 3 significant figures.

So the density of the cuboid is equal to $\frac{28.4\ g}{76.2\ cm^3}$, which is 0.37 $g\,cm^{-3}$.

The volume of a 3D shape is calculated by using the correct mathematical formula for that shape. Examples of the most common regular shapes are given in Table 1.

	Cube	Rectangular prism	Sphere	Cylinder	Cone
Shape	l, l, l	l, h, b	r	r, h	r, h
Volume	$V = l^3$	$V = l \times b \times h$	$V = \frac{4}{3}\pi r^3$	$V = \pi r^2 h$	$V = \frac{1}{3}\pi r^2 h$

Table 1 Shapes and their respective volume formulae.

Calculating the volume of irregular shapes

Unlike the regular shape shown in Figure 2, it is not possible to calculate the volume of an irregular shape using a formula. We need a different way of measuring the volume of the material. The object shown in Figure 3 has an irregular shape, and to determine its volume we have to carefully lower it into a displacement can containing water (so long as the substance is insoluble and does not react with water). The volume of water displaced is equal to its volume. If we know its mass, we can work out the density of the material.

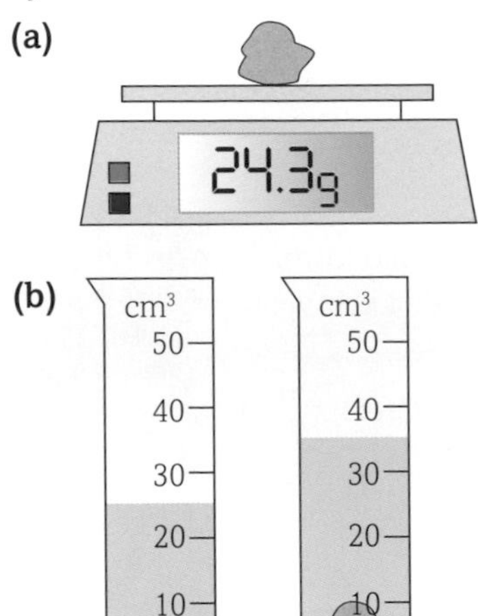

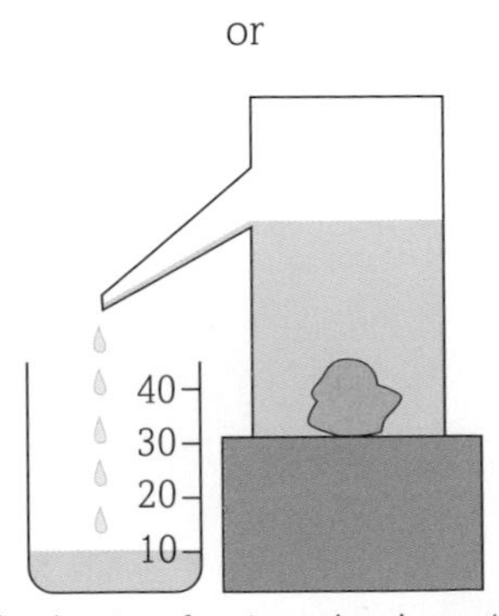

Figure 3 Determining the density of an irregular shape. (a) Weigh the object to find its mass and (b) measure the volume of water displaced to determine the volume of the object.

Densities of commonly found materials

Table 2 lists some common materials along with their densities. The densities of solids and liquids vary very little with temperature or pressure, but the densities of gases can change considerably. The values quoted are at a temperature of 273 K and at 1 atmosphere pressure.

Material	Density/kg m^{-3}
hydrogen	0.0899
helium	0.176
oxygen	1.33
air	1.29
ethanol	789
olive oil	920
water	1000
mercury	13 600
aluminium	2710

Material	Density/kg m^{-3}
silicon	2300
concrete	2400
iron	7870
copper	8930
silver	10 500
gold	19 300
platinum	21 500
osmium	22 500

Table 2 Density of common materials.

LEARNING TIP

You don't have to remember any of these values, but you will find it useful to be familiar with the order of magnitude of the densities of solids, liquids and gases.

DID YOU KNOW?

A ship will float on water because its overall density is less than the density of water. Despite being made of metal which has a density much greater than 1 g cm^{-3}, most of the volume of the ship is air, which has a density much less than that of water.

WORKED EXAMPLE 2

Use the values in Table 2 to calculate:

(a) the mass of 36 m^3 of pure aluminium to 2 significant figures

(b) the volume of 3.6×10^5 kg of pure copper to 1 significant figure.

Answers

(a) mass of aluminium = density of aluminium × volume of aluminium

$= 2710\ \text{kg m}^{-3} \times 36\ \text{m}^3 = 98\,000\ \text{kg}$

(b) volume of copper $= \dfrac{\text{mass of copper}}{\text{density of copper}}$

$= \dfrac{3.6 \times 10^5\ \text{kg}}{8930\ \text{kg m}^{-3}} = 40\ \text{m}^3$

Archimedes' principle

You saw in Topic 3.2.1 that objects floating or submerged in water or any fluid experience an upthrust force due to the displacement of water. The more of the object that is submerged, the greater the upthrust.

A concrete block will sink in water but will float in mercury. This is because the concrete block has a density higher than that of water but lower than that of mercury. Whether an object sinks or floats in a liquid depends on its density relative to that liquid.

KEY DEFINITION

Archimedes' principle states that the upward buoyant force (upthrust) exerted on an object immersed in a fluid, whether fully or partially submerged, is equal to the weight of the fluid that the object displaces.

WORKED EXAMPLE 3

An object floats in water with 20% of its volume above the surface. Find the density of the solid.
(Take the density of water as 1000 kg m^{-3})

Answer

Archimedes' principle states that the weight of water displaced is equal to the weight of the object. As g is constant, the mass of water displaced is equal to the mass of the object.
Therefore, mass of the water displaced = $(0.8 \times V) \times 1000$, where V is the volume of the solid.

Density of the solid $= \dfrac{\text{mass}}{\text{volume}}$

$= \dfrac{0.8 \times V \times 1000}{V} = 800\ \text{kg m}^{-3}$

Questions

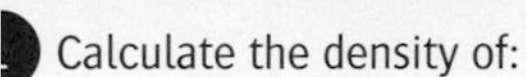

1. Calculate the density of:
   (a) a block of mass 45 g that has a volume of 240 cm^3
   (b) a cube of side 12 cm and mass 580 g
   (c) a rectangular cuboid of dimensions 3 cm, 5 cm and 12 cm and mass 89 g
   (d) a cylinder of height 34 cm, radius 12 cm and mass 45 g.

2. Use the data in Table 2 to work out:
   (a) the mass of 45 m^3 of silver
   (b) the mass of 880 cm^3 of platinum
   (c) the volume of 400 kg of silicon.

3. Use the terms 'mass', 'volume', 'density' and 'temperature' to explain how a lava lamp works.

4. Estimate the mass of air in a typical school laboratory.

5. Explain why an object appears to lose weight when it is submerged in water.

6. A block of wood measures 2.2 m by 2.5 m in plan with a depth of 3.6 m and has a weight of 102 kN. When it is placed in water, how deep will it sink? Take $g = 9.81\ \text{m s}^{-2}$ and the density of water as $1000\ \text{kg m}^{-3}$.

7. An 18 carat gold ring is made from an alloy containing 75% gold with small amounts of other metals including copper and silver. By considering the ring to be a cylinder, find an order of magnitude estimate for the mass of the ring.

3.2 | 8 Pressure

By the end of this topic, you should be able to demonstrate and apply your knowledge and understanding of:

* pressure; $p = \frac{F}{A}$ for solids, liquids and gases
* $p = h\rho g$; upthrust on an object in a fluid; Archimedes' principle

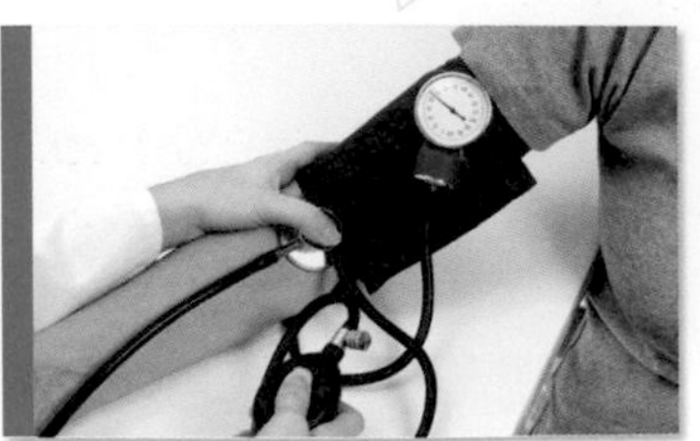

Figure 1 Using a sphygmomanometer to measure blood pressure.

KEY DEFINITION

Pressure is defined as force per unit area at right angles to (normal to) the area.

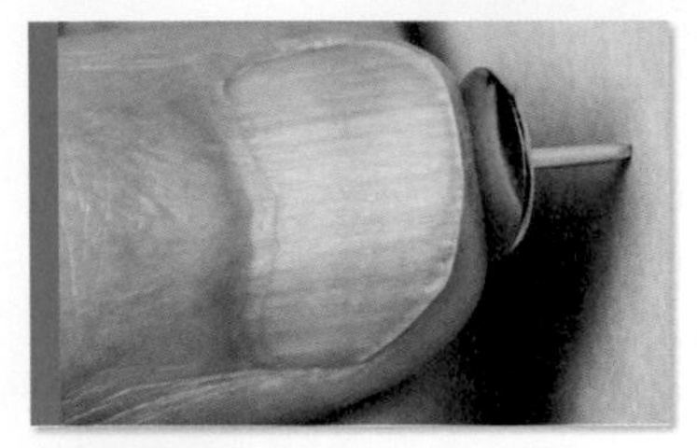

Figure 2 A high pressure is produced by applying a force over a very small area.

Figure 3 The small area of the blades creates a high enough pressure to melt ice and allow the skater to slide.

Introduction

Pressure is defined as force per unit area. It is a vector quantity because it has both a size (magnitude) and a direction. Pressure can be exerted by one object in contact with another, such as when a person stands on the floor or when a drawing pin is pushed into a wall.

Pressure is also exerted by fluids (liquids and gases). We often talk about atmospheric pressure when referring to the weather and we have our blood pressure measured as high blood pressure is a risk for heart disease and strokes. Whenever there is a pressure difference, fluids can flow from an area of high pressure to one at a lower pressure.

The pressure equation is pressure $= \frac{\text{force}}{\text{area}}$ or $p = \frac{F}{A}$

where F is the force normal to an area A.

The S.I. units of pressure are newtons per square metre (N m^{-2}) or pascals (Pa). $1\ \text{N m}^{-2}$ is the same as 1 Pa.

Figures 2–3 show different situations involving pressure. Pressure is highest when:

- the force exerted is large
- the surface area over which the force acts is small.

WORKED EXAMPLE 1

Calculate the pressure exerted by:

(a) a 300 N force acting on a drawing pin with an area of $0.01\ \text{mm}^2$

(b) a man with a mass of 85 kg standing on a plank of wood with an area of $0.45\ \text{m}^2$.

Answers

(a) pressure $= \frac{\text{force}}{\text{area}}$

$= \frac{300\ \text{N}}{0.01\ \text{mm}^2} = 30\,000\ \text{N mm}^{-2}$

Note that $1\ \text{mm}^2$ is $10^{-6}\ \text{m}^2$, so to get a value in Pa (N m^{-2}) we would need to do the calculation:

pressure $= \frac{300\ \text{N}}{(0.01 \times 10^{-6}\ \text{m})^2}$, which is an enormous $3.0 \times 10^{10}\ \text{N m}^{-2}$.

(b) Remember that pressure = force/area, so we need to find the downward force exerted by the man, which is found by using weight = mass × gravitational field strength.

pressure $= \frac{(85\ \text{kg} \times 9.81\ \text{N kg}^{-1})}{0.45\ \text{m}^2} = 1850\ \text{N m}^{-2}$ or 1850 Pa

Pressure at a depth in a fluid

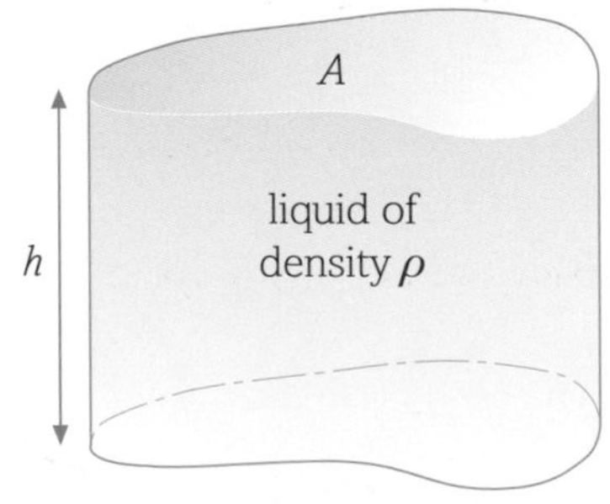

Figure 4 A column of fluid.

Pressure increases with depth in a fluid because of the force exerted by the increased weight of the fluid above.

Imagine a volume, V, of a fluid of constant cross-sectional area, A, and height, h, as shown in Figure 4.

If the density of the fluid is ρ, the mass of the liquid is found by multiplying its density by its volume, giving us a mass of $Ah\rho$.

Because the weight of this mass of fluid is given by $W = mg$, then the formula for the weight of the fluid, F, must be $F = Ah\rho g$.

But pressure $= \frac{\text{force}}{\text{area}}$, so the pressure, p, exerted by a fluid at a depth h is: $p = h\rho g$

WORKED EXAMPLE 2

1. What is the pressure at the bottom of a reservoir containing water to a depth of 30 m? (The density of water is 1 g cm^{-3}.)

Answer

$p = h\rho g$

We know the depth of the water and we can assume that g is 9.81 N kg^{-1}. The density of water is given in g cm^{-3} and this has to be converted to kg m^{-3}. There are 10^6 cm^3 in 1 cubic metre and 10^3 g in a kilogram, so the density value to use will be 1000 kg m^{-3}. This gives:

pressure = 30 m × 1000 kg m^{-3} × 9.81 N kg^{-1} = 294 300 Pa

2. A person suffering from high blood pressure has a systolic pressure of 190 mm Hg. What is this pressure in pascals?

Answer

As before, the pressure is given by $h\rho g$. The values we have here are:

h = 190 mm = 0.190 m; ρ = density of mercury = 13 600 kg m^{-3}; and g = 9.81 m s^{-2}

So p = 0.190 × 13 600 × 9.81 = 25 300 Pa, which is about a one-quarter of normal atmospheric pressure.

DID YOU KNOW?

The reason why the value for the atmospheric pressure is such a large number (about 10^5 Pa) is because above each square metre of land there is a very tall column of air that goes right to the top of the atmosphere. Even though air has a low density, a large volume of it will still have a large mass, and hence a large weight. We do not feel the effect of this in an uncomfortable way because this pressure is balanced by the pressure of the various fluids in our bodies that are acting outwards. Despite being composed of low-density gases, the total mass of the Earth's atmosphere is a staggering 5 × 10^{18} kg.

Floating and Archimedes' principle

Applying what we have learned about pressure, we can now see how equilibrium is established for a floating ship, and what causes upthrust.

A typical loaded oil tanker has a total mass of about 400 000 tonnes. It has a width of 40 m and a length of 500 m. Assuming that the ship has a flat bottom, it must be the pressure of the water on the ship's bottom that is providing the upward force keeping it afloat. For equilibrium:

$$\begin{aligned}\text{force upward (upthrust)} &= \text{weight downward} = mg \\ &= 400\,000\,000\,\text{kg} \times 9.81\,\text{N kg}^{-1} \\ &= 3.92 \times 10^9\,\text{N}\end{aligned}$$

Archimedes' principle states that when an object is floating or submerged in water, the upthrust of the water acting on that object is equal to the weight of the fluid displaced by the object. Therefore:

upthrust = $Ah\rho g$.

This can also be written as:

upward force from displaced water
= pressure × area of bottom of ship:

$3.92 \times 10^9\,\text{N} = h\rho g \times 40 \times 500$

where ρ = density of sea water = 1030 kg m^{-3}

h = distance from the bottom of the ship to the surface:

To find h, rearrange the formula above:

$$h = \frac{\text{upthrust}}{\rho g \times \text{area of bottom of ship}}$$

$$h = \frac{(3.92 \times 10^9)}{(1030 \times 9.81 \times 40 \times 500)} = 19.4\,\text{m}$$

It is clear that this ship requires deep-water harbours for the safe loading and unloading of its cargo.

Figure 5 illustrates the pressure forces acting on a ship. The forces on the sides act horizontally and inwards, tending to squash the vessel. These increase with depth. The forces on the bottom of the ship act upwards and support the vessel.

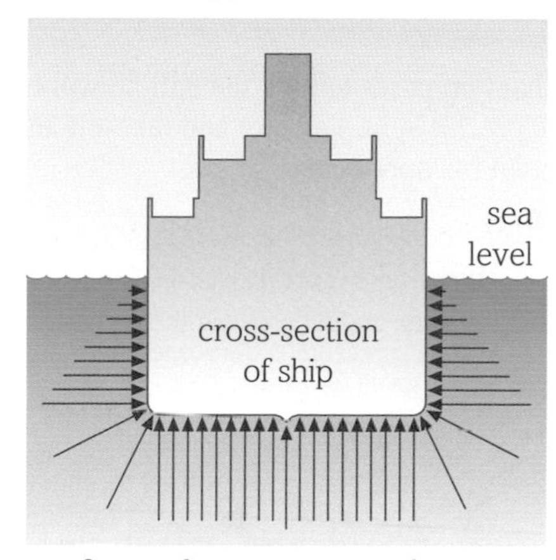

Figure 5 Pressure forces acting on a ship.

Questions

1 Copy and complete Table 1 by filling in the missing values.

Pressure/N m^{-2}	Force /N	Area/m^2
	1200	40
250 000		4.8
1.4 × 10^6	480	
	2.3 × 10^4	1.6 × 10^{-3}
8.6 × 10^9		7.8 × 10^{-5}

Table 1

2 What would be the pressure at the bottom of a 24 m high cylindrical container that has been filled with olive oil (density = 920 kg m^{-3})?

3 In Figure 5, to what depth would the ship sink if it was floating in ethanol (density = 789 kg m^{-3}) instead of water?

4 Estimate the mass of air there is above your head to a height of 1 km in the atmosphere.

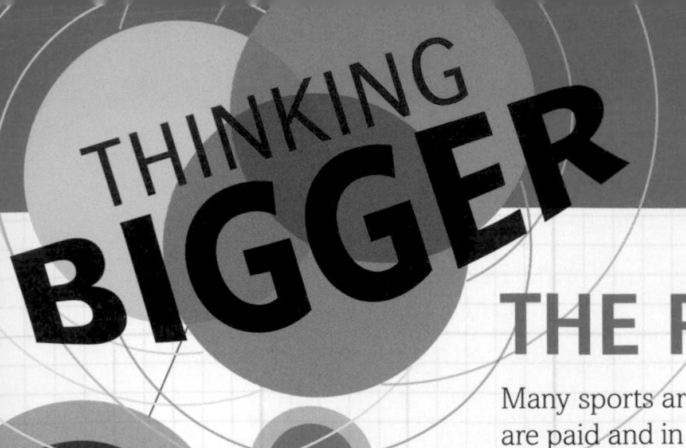

THE PHYSICS OF FOOTBALL

Many sports are becoming more and more professional, both in terms of the money that athletes are paid and in their daily routines, diets, psychological preparation and the use of scientific data and technology to help them improve performance.

In this activity, we will ask how a footballer can bend a free kick and what variables cause this happen.

BECKHAM AS A PHYSICIST?

As David Beckham prepares to kick the ball he works out how to give the ball the precise spin, velocity and take-off angle needed to successfully outwit his opponents.

The Magnus effect

Consider a ball as shown in Figure 1.

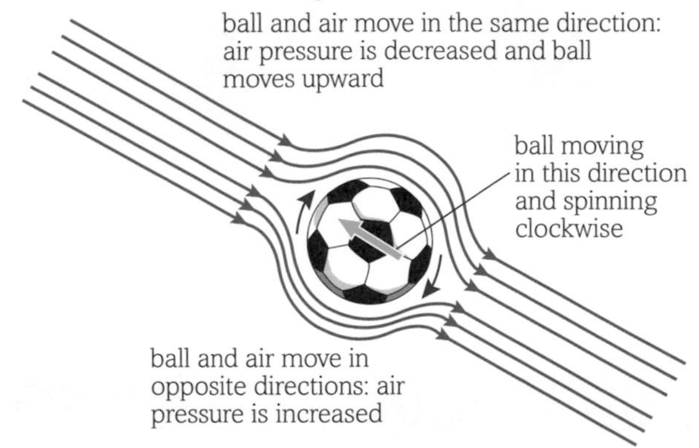

Figure 1 A spinning ball and the Magnus effect.

The air travels faster over the side of the ball where it is moving in the same direction as the flow of the air. The opposite is true of the other side of the ball. According to the Bernoulli principle, faster moving air reduces the pressure. A pressure difference is therefore set up on either side of the ball and this creates a net force, known as the Magnus effect. The force due to the Magnus effect, also referred to as the lift force, is given by

$$F = C_L \rho D^3 f v$$

where C_L is the lift coefficient, ρ is the density of the air ($1.20\,\text{kg m}^{-3}$ at sea level), D is the diameter of the ball (the world governing body, FIFA, state that the circumference of the ball must be between 0.68 and 0.70 m; we can therefore assume a diameter of 0.22 m), f is the spin frequency and v is the velocity of the ball.

The lift coefficient turns out to be a rather complex quantity which needs to be determined experimentally, but a figure of 1.23 meets the requirements of most sporting situations.

How much does, or can, Beckham swing the ball?

First we make the following assumptions (see Figure 2): the free-kick is 25 m away from the goal and is struck with a velocity of $25\,\text{m s}^{-1}$ in such a way as to cause it to swing at $10\,\text{revs s}^{-1}$.

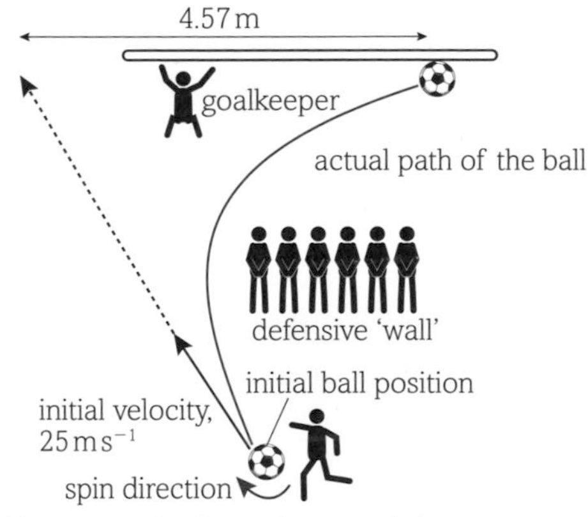

Figure 2 Beckham scores by the application of physics.

[...] Knowing the force on the ball, the acceleration can be calculated from $F = ma$. [...] If we assume the mean value of 0.430 kg then [...] Assuming a flight time of 1 second and applying $s = ut + \frac{1}{2}at^2$, the curve or swing can be calculated.

Source

Ireson, G. (2001) Beckham as physicist? *Physics Education*. vol. 36, no. 1. pp. 10–13.

DID YOU KNOW?

The Brazilian midfielder Didi was the first person to demonstrate, on the international stage, that a football could swerve when kicked. The leather of the 1950s meant that the ball absorbed water and a damp ball would not swerve, but this was not a problem in hot, dry Brazil.

Where else will I encounter these themes?

1.1 | 2.1 | 2.2 | 3.1 | 3.2 | YOU ARE HERE | 3.3

Let us start by considering the nature of the writing in the article. The article above was taken from the publication *Physics Education*, which is produced by the Institute of Physics (IOP).

1. Consider the article and comment on the type of writing that is being used, for example how diagrams and data are used.

We will now look at the physics that is in the article. Do not worry if the physics content or the mathematics is ... course, once some of ... bottom of the page to he... at is ahead in your course...

2. What is t...

3. Estimate ... magnitu...

4. a. Use th... of the ball.

b. Draw ... of the horizont...

5. In this e... does it have on ...

6. The drag ... affect the drag forc...

7. Is it easy ... n your answer.

8. How doe... f a chipped shot, wh... ied?

THANK YOU FOR ANSWERING OUR QUESTIONS

Now we hope we can answer some of yours

opplinsights

Unit B1,
Windsor Place,
Faraday Road,
Crawley,
West Sussex,
RH10 9TF
t: +44 (0) 1293 558955

Your Interview was carried out by: Raheel
Interviewer No: 15026
On Date: 28/02/20

Activity

A football and a... variables that c... to teach an intr... sport as part of ... bout the ... ou could use ... physics of ... ntation.

You could investigate how the design of footballs and golf balls causes a change in the amount that they bend when hit, and the conditions that are needed to get the greatest amount of swing.

3.2 Practice questions

1. A balloon is travelling vertically downwards at a constant acceleration. The upthrust on the balloon is U, its weight is W and it experiences air resistance F.
   Which statement is correct? [1]
   A $F + W > U$
   B $W + U > F$
   C $F > W + U$
   D $W > U + F$

2. The forces acting on a car are shown in Figure 1. Which of the following statements is true if the car is starting from rest? [1]

Figure 1

   (i) The initial acceleration of the car will be $2\ \text{m s}^{-2}$.
   (ii) The size of the drag force will increase as the car accelerates.
   (iii) The car will eventually reach a maximum speed / terminal velocity.

   A (i), (ii) and (iii)
   B only (i) and (ii)
   C only (ii) and (iii)
   D only (i)

3. Look at Figure 2. What mass is needed at point X to make the beam balance? [1]

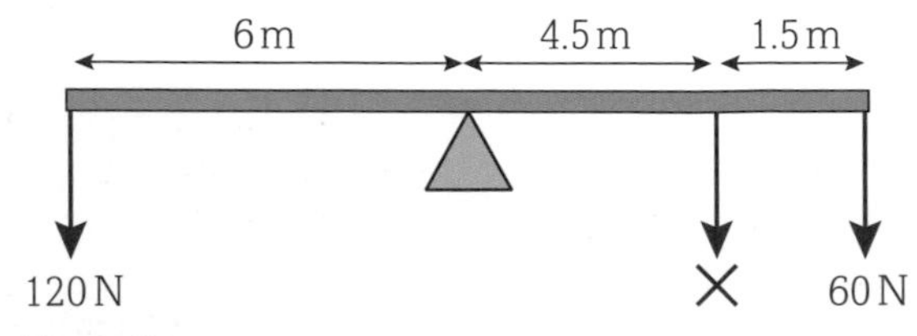

Figure 2

   A 4 kg
   B 8 kg
   C 16 kg
   D 40 kg

4. At what speed will the truck in Figure 3 be travelling down the slope, two seconds after it is released from the position shown?

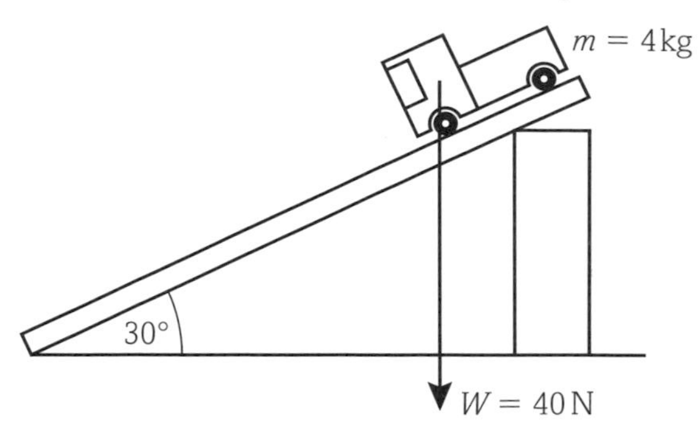

Figure 3

   A $0\ \text{m s}^{-1}$
   B $2\ \text{m s}^{-1}$
   C $6\ \text{m s}^{-1}$
   D. $10\ \text{m s}^{-1}$

[Total: 4]

5. The man in Figure 4 has a mass of 78 kg. The total area of his shoes in contact with the floor is $140\ \text{cm}^2$ and his body has a volume of $0.08\ \text{m}^3$. The value of g on Earth is $9.81\ \text{N kg}^{-1}$.

Figure 4

   Calculate:
   (a) the man's weight [2]
   (b) his density [2]
   (c) the pressure he exerts on the ground. [2]

[Total: 6]

6. Look at Figure 5 below.
   (a) Calculate the size of the torque on the steering wheel. [2]

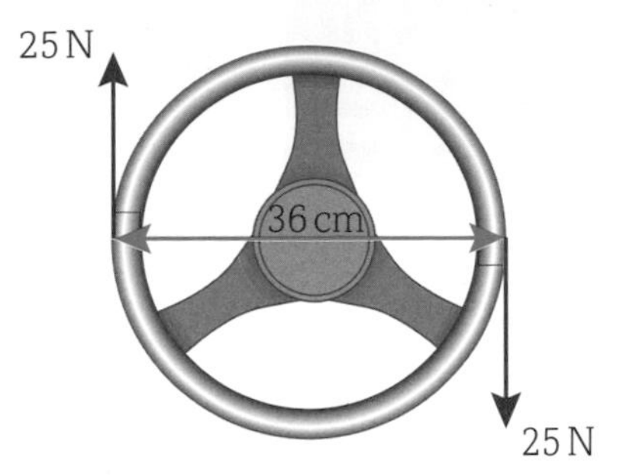

Figure 5

   (b) State two ways in which the torque on the steering wheel could be increased. [2]

[Total: 4]

7. A girl is travelling in a lift. She has a mass of 51 kg.

Figure 6

   (a) Explain why the value of R is 500N when she enters the lift. Take $g = 9.8\text{ m s}^{-2}$ [2]
   (b) What are the magnitudes of W and R, in N, when:
      (i) the lift is rising at a steady speed of 1.2 m s^{-1} [2]
      (ii) the lift is rising at a steady acceleration of 1.2 m s^{-2} [2]
      (iii) the lift is descending at a steady acceleration of 1.2 m s^{-2}? [2]

[Total: 8]

8. A child sits on a swing and is pulled by a horizontal force, P, so that the chains make an angle of 35° with the vertical. This is shown in Figure 7.

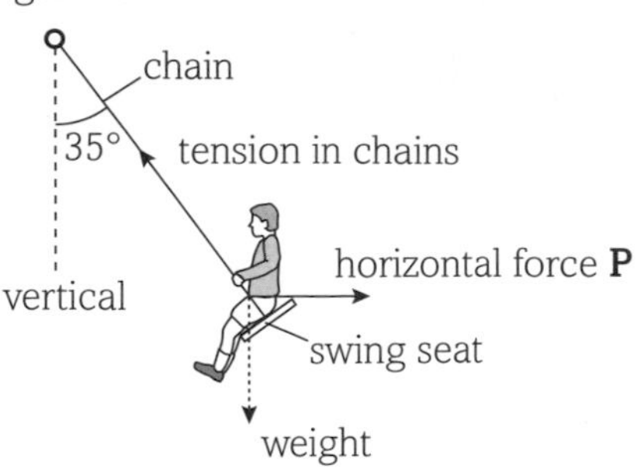

Figure 7

The combined mass of the child and the swing is 26 kg.

   (a) From the information provided, calculate:
      (i) the combined weight of the child and the swing [2]
      (ii) the tension in the chains [2]
      (iii) the size of the horizontal force, P. [2]
   (b) Without calculation, explain what will happen to the following if the size of the angle made by the chains with the vertical increases:
      (i) the combined weight of the child and the seat [2]
      (ii) the tension force in the chains. [2]

[Total: 10]

9. The drag force acting on a body falling through a fluid is related to the velocity of the body by the equation $F = K\rho Av^2$.
   (a) Define the terms ρ and A. [2]
   (b) Show that F has units of kg m s^{-2}. [2]
   (c) Show that K is a dimensionless quantity (it has no units). [2]
   (d) Show that the terminal velocity for a falling body, $v_T = \sqrt{(mg/K\rho A)}$. [2]
   (e) Explain why, when the diameter of a falling spherical drop doubles:
      (i) the cross-sectional area increases by a factor of 4 [1]
      (ii) the weight of the drop increases by a factor of 8 [2]
      (iii) the terminal velocity will increase by a factor of $\sqrt{2}$. [2]

[Total: 13]

MODULE 3 Forces and motion

CHAPTER 3.3

WORK, ENERGY AND POWER

Introduction

In our everyday spoken language we constantly use the words work, energy and power. Adults go to work to earn money, students attend school and college to do school work and homework. We use up our energy when we play sport and we relax and sleep when we have run out of energy. Batteries power our games consoles; the television and newspapers are full of stories of politicians and their quest for power – either gaining it or retaining it. However, despite the common and mundane use of these words, in the realm of physics they have exact meanings which scientists and students need to use carefully and precisely. We have well-defined and precise equations for the calculation of work, energy and power. Work is measured in joules, power is measured in watts. There is equipment that will measure these quantities to extremely high accuracies, and no matter how much we try, we will never be able to create nor destroy these quantities. In addition to this, we always waste energy or power in any conversion from one form to another. In this chapter we will look at the laws that govern the transfer and conservation of energy and power and we will discuss the principles that are fundamental to our understanding of the field of physics.

All the maths you need

To unlock the puzzles of this chapter you need the following maths:

- Units of measurement
- Substitution of values into equations
- Rearranging of formulae
- Use of square root to find velocity from kinetic energy

What have I studied before?

- Work done = energy transferred and work done = force × distance moved
- Work done and energy transferred are measured in joules
- Energy cannot be created nor destroyed, merely converted from one form to another
- Examples of energy forms include kinetic, gravitational potential, chemical potential, elastic potential, light, sound, electrical and nuclear
- When an object falls from a height above the ground, gravitational potential energy is transferred to kinetic energy
- When an object is thrown vertically, kinetic energy is transferred to gravitational potential energy
- In any energy transfer, some energy is always wasted, so no energy transfer is ever 100% efficient
- Power is the energy developed per unit time and is measured in watts
- Energy and power transfers, and their efficiencies, can be shown using Sankey diagrams
- Efficiency = $\left(\frac{\text{useful output}}{\text{input energy}}\right) \times 100\%$

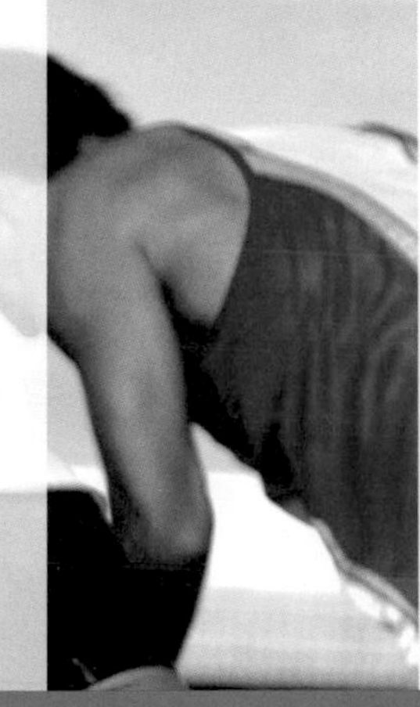

What will I study later?

- The calculation of, and relationship between, electrical power and electrical energy (AS)
- Understanding and calculating the elastic potential energy stored by elastic materials (AS)
- The nature and behaviour of kinetic energy and its conservation during perfectly elastic collisions (AS)
- The nature of the energy transferred or stored in progressive and stationary waves. (AS)
- Photon energy, work function and kinetic energy conservation at quantum level when dealing with the photoelectric effect, excitation and ionisation (AS)
- The energy required to move charged particles and masses in electric and gravitational fields, respectively (AL)
- The notion of internal energy as the sum of random kinetic and potential energies in a system (AL)
- Determination of the specific heat capacity, and specific latent heat, using electrical energy and apparatus (AL)
- The interchange between kinetic and potential energy in oscillatory motion and simple harmonic systems (AL)
- Defining and calculating electrical and gravitational potential and potential energy
- The analysis of the energy stored by a capacitor in equation and graphical form (AL)
- Calculating and evaluating the binding energy and binding energy per nucleon in the processes of nuclear fission and nuclear fusion (AL)

What will I study in this chapter?

- Work done = $Fd \cos \theta$ for a force acting at an angle θ to the direction of motion
- The principle of conservation of energy for a variety of energy transfers
- The different forms of energy, their transfer and conservation
- The transfer of energy is equal to the work done
- Calculation of the kinetic energy of an object using $E_k = \frac{1}{2}mv^2$
- Calculation of the change in gravitational potential energy of a mass, m, in a uniform gravitational field given by $E_p = mg\Delta h$
- The qualitative and quantitative exchange between kinetic and gravitational potential energies
- Power as energy developed per second, measured in watts, W
- Power = force × velocity
- The term efficiency and its calculation for any mechanical system

3.3 

Work and the joule

By the end of this topic, you should be able to demonstrate and apply your knowledge and understanding of:

* the unit joule
* transfer of energy is equal to work done
* $W = F\cos\theta$ for work done by a force

Work

In physics, we define **work** using the equation
work = force applied × distance moved in the direction of the force.

$$W = F \times d$$

Work is a scalar quantity because it has a size (magnitude), but no direction. It can be thought of as the amount of energy transferred from one form to another. For example, work is done when an electric motor transfers electrical energy to kinetic energy, sound, heat energy and gravitational energy.

KEY DEFINITION

The **work** done, or energy transferred is the product of the force and the distance moved by the force in the direction of movement.

WORKED EXAMPLE 1

A man pushes a shopping trolley with a horizontal force of 28 N, moving a distance of 12 m.
(a) How much work did he do on the trolley if there was no friction?
(b) How much work did he do on the trolley if the frictional force was 8 N?

Answers

(a) $W = F \times d = 28 \times 12 = 336$ J
(b) $W = F \times d = 20 \times 12 = 240$ J
So 96 J was done in overcoming frictional forces and this work was not done on the trolley.

Work done and energy transferred

The S.I. unit of work is the **joule** (J), which is equivalent to the newton metre (Nm). One joule of work is transferred when a force of 1 newton causes an object to move a distance of 1 metre in the direction of the force. When we consider horizontal motion, we use the equation $W = F \times d$, with F being the size of the force acting parallel to the direction of movement.

For vertical motion, we also use the equation $W = F \times d$, but we must realise that F is now the weight of the mass in the gravitational field of the Earth, given by $F = m \times g$.

For example, 500 g when a motor lifts a mass, work is done because the electrical energy transferred to the motor is transferred to the kinetic and gravitational potential energy of the mass as it is raised 3.5 m metres off the floor. The total work done in this example is:

$$W = F \times d = mg \times h$$

$$= 0.5\,\text{kg} \times 9.81\,\text{N}\,\text{kg}^{-1} \times 3.5\,\text{m} = 17.2\,\text{N}\,\text{m or } 17.2\,\text{J}$$

WORKED EXAMPLE 2

An object of mass 34 kg is raised from the ground to a shelf 7.5 m above the ground. How much work has been done?

Answer

$W = F \times d$ and $F = mg$, so:
$W = 34\,\text{kg} \times 9.81\,\text{N}\,\text{kg}^{-1} \times 7.5\,\text{m}$
$= 2500$ J or 2500 N m to 2 significant figures

The total work done depends on the size of the applied force and the distance moved in the direction of the force. Typical values are shown in Table 1.

Activity	Typical work value
Lifting an apple to your mouth	$1\,\text{N} \times 0.5\,\text{m} = 0.5\,\text{J}$
Walking upstairs	$750\,\text{N} \times 3\,\text{m} = 2250\,\text{J}$
Pushing a shopping trolley along a supermarket aisle	$150\,\text{N} \times 40\,\text{m} = 6000\,\text{J}$
A crane lifting 40 tonne by 30 m	$1.18 \times 10^7\,\text{J}$

Table 1 Typical work values for different activities.

Remember that forces are vectors and that by resolving a force into its components you can show that there is no component of a force in a direction at 90 degrees to its direction of action (since cos 90° = 0). This means that if a resultant force accelerates a body in a certain direction, the force does no work in a direction *perpendicular* to the direction of motion.

Force at an angle to the direction of motion

Imagine a sailing boat being propelled across the surface of a lake by wind. Due to the relative position of the wind and the sails, the boat does not necessarily always benefit from a maximum force that is parallel to the direction in which it needs to move. When this is the case, we need to consider the **component** of the force that is parallel to the direction in which the boat will move.

In Figure 1, the force of the wind is 400 N, but the force acts at an angle of 36° to the direction of movement of the boat. Using trigonometry, the component of the force of the wind blowing the boat to the right, $F_{||}$, is given by $F_{||} = 400 \times \cos 36°$, or 323.6 N.

If the boat sails for 250 m then the total work done by the wind on the boat is given by:

$$W = F\cos\theta \times d$$

$$= 400\cos 36° \times 250$$

$$= 323.6 \times 250 = 80\,900\,\text{J}$$

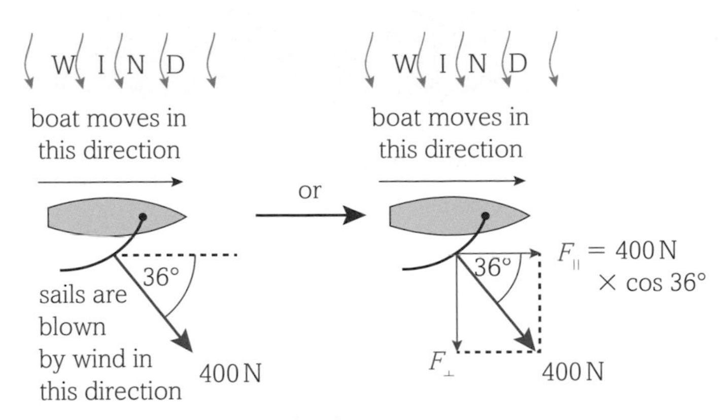

Figure 1 Forces acting on a sailing boat.

WORKED EXAMPLE 3

A man pushes a roller a distance of 50 m with a force of 370 N along the handle of the roller. The handle makes an angle of 38° to the horizontal (Figure 2). How much work has he done?

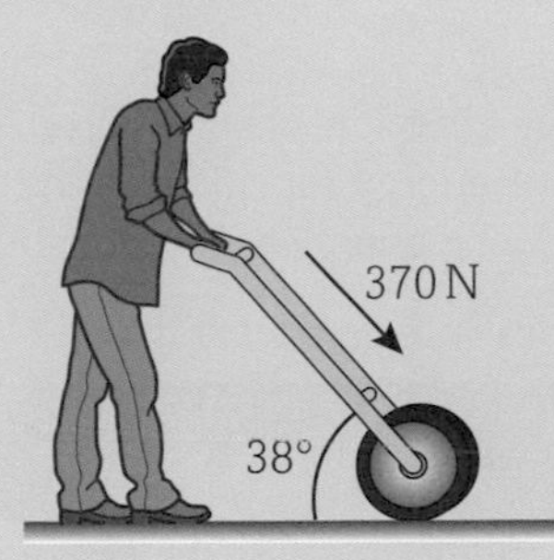

Figure 2 To calculate the work done, you need to know the size of the component of the force that is parallel to the direction of motion.

Answer

$W = F\cos\theta \times d$

$= 370 \times \cos 38° \times 50$

$= 15\,000$ J to 2 significant figures.

WORKED EXAMPLE 4

A barrel of weight 200 N is raised by a vertical distance of 1.8 m by being moved along a ramp at an angle to 25° to the horizontal.

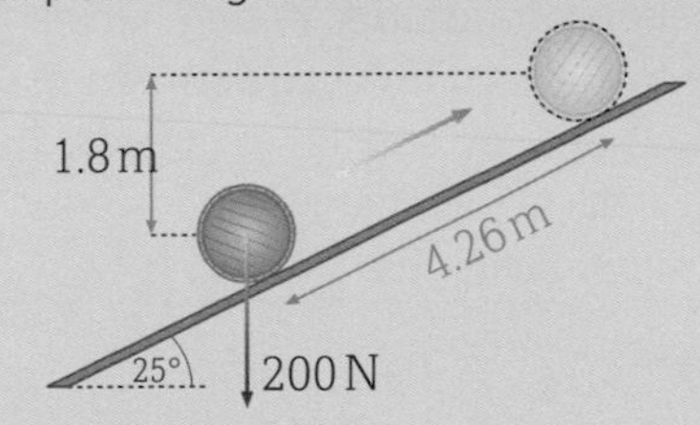

Figure 3

Assuming friction is negligible, calculate:

(a) the work done against gravity

(b) the force needed to move the barrel along the ramp.

Answers

(a) Work done against gravity = 200 N × 1.8 m = 360 J
Notice that we use the distance moved in the direction of the force to calculate the work done –,vertical distance.

(b) The total work done must be the same, but the pushing force acts up the ramp so we use the distance moved along the same direction direction moved parallel to the direction of the slope.
Distance moved along the ramp = 1.8 m /sin 25° = 4.26 m
Force required = 360 J/4.26 m = 84.5 N

Questions

1 A caravan of mass 400 kg is pulled at constant speed along a horizontal road against a drag force of 240 N.

(a) Calculate the work done in pulling the caravan 500 m.

(b) Explain why the weight of the caravan does not enter into the calculation.

2 A barge is pulled by a rope tied to a horse on the towpath of a canal. There is a drag force on the boat in the water. A man on the other bank applies a sideways force using a rope to maintain the course of the barge along the canal. Figure 4 shows the magnitudes and directions of these forces.

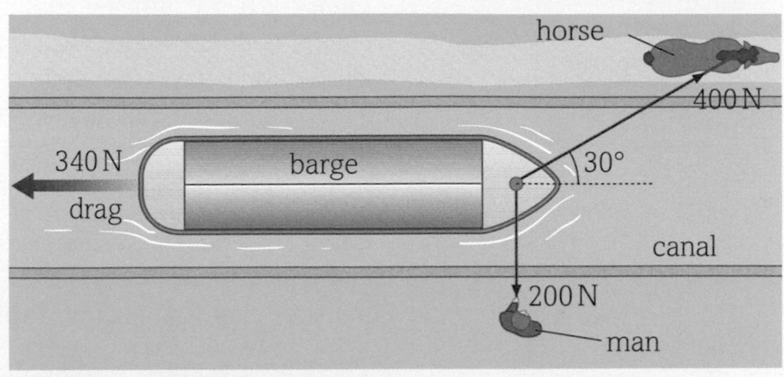

Figure 4

(a) Write down the components of each force along the canal.

(b) How much work is done by each of these forces when the barge moves 10 m along the canal?

(c) How much of the work done on the barge by the horse is not done against the drag force?

(d) Describe how the motion of the barge changes.

3 A cyclist and bike of combined mass 100 kg starts from rest at the top of a slope of length 500 m. The slope is at an angle of 5° to the horizontal and the cyclist rolls down the hill without pedalling. The force of friction acting along the slope is 60 N. Calculate the final kinetic energy of the cyclist at the bottom of the slope.

The conservation of energy

By the end of this topic, you should be able to demonstrate and apply your knowledge and understanding of:

* energy in different forms; transfer and conservation
* the principle of conservation of energy

Energy

Energy is a key principle that underpins physics. Physics is, ultimately, the study of energy – without energy nothing can happen. Energy is the capacity to do work. We come into contact with energy all day every day – the chemical energy stored in food keeps us alive and enables us to do work, the electrical energy that makes our televisions, computers and mobile phones work, the sound energy that allows us to hear, and the nuclear energy stored inside atomic nuclei that we harness to provide power to our homes and businesses.

Energy can be transferred from one form to another. The chemical energy stored in coal is released in the process of combustion – along with other forms of energy, mostly heat and light. A car engine transfers chemical energy stored in diesel or petrol to kinetic energy as a useful form.

There are a number of different forms of energy and these are listed in Table 1.

Form of energy	Equation used to calculate it	Example
Kinetic	$E = \frac{1}{2}mv^2$	A car's engine transfers the chemical energy stored in fuel to kinetic energy.
Gravitational potential	$E = mg\Delta h$	A cyclist in the Tour de France gains gravitational potential energy as he rides up Alpe d'Huez.
Thermal (heat)	$E = mc\Delta T$ (c is the specific heat capacity of the material)	The internal energy of water in a kettle increases as the heating element transfers energy to molecules of water.
Elastic potential	$E = \frac{1}{2} Fe$ (e is the extension of the material)	The work done in stretching an elastic material is stored as elastic potential energy.
Nuclear	$E = mc^2$	The energy transferred during nuclear fission comes from the mass difference between the original nucleus and the smaller fragments. This is called the mass defect. The energy is transferred into thermal and kinetic energies.

Table 1 Energy forms and examples of common energy transfers.

The principle of conservation of energy

In any **closed system**, energy cannot be created or destroyed, it can only be transferred from one form to others. The total energy available before any process is always equal to the energy after the conversion has happened.

KEY DEFINITIONS

A **closed system** is any system in which all the energy transfers are accounted for. Energy or matter cannot enter or leave a closed system.

The **principle of conservation of energy** states that the total energy of a closed system remains constant. Energy can neither be created nor destroyed, it can only be transferred from one form to another.

Figure 1 This car has been filled with fuel, which is a store of chemical energy. Not all of the input energy will be transferred usefully to kinetic energy to turn the wheels – most of it will be transferred to the surroundings by heating the exhaust gases, the surrounding air and the road.

Accounting for all the input energy

However, whenever we transfer energy from one form to another, the process is never perfect – not all input energy is transferred usefully and we say that the process is less than 100% efficient (see Topic 3.3.5). A light bulb is designed to transfer electrical energy to light energy, but the vast majority of the input energy is wasted in heating. The efficiency of bulbs can be as poor as 3%, meaning that 97% of the electrical energy is wasted and only 3% of it is transferred usefully to light energy.

This illustrates the principle of conservation of energy – all of the input energy is accounted for. Similarly, for the Formula 1 car in Figure 1, only some of the energy released by burning fuel is used to increase the kinetic energy of the car. The remaining energy is dissipated to other forms, especially by heating which transfers energy to the car's surroundings, raising the temperature of the surroundings. Note that in this case the closed system refers to the car *and* the surrounding air and road.

DID YOU KNOW?

In the nineteenth century, several eminent scientists could not understand how it was possible not to lose energy. It was the English physicist James Joule who finally proved this principle with a series of meticulous experiments. He measured the rise in water temperature produced when water was stirred by a simple paddle wheel. He designed and made his own thermometers, which were precise to 0.01°C. He repeatedly found that all of the work done by the paddles could be accounted for in the corresponding rise in water temperature.

Figure 2 James Joule.

Questions

1. Give examples of machines that show the following energy transfers:
   (a) chemical to heat
   (b) electrical to sound
   (c) chemical to sound and light
   (d) nuclear to electrical
   (e) gravitational potential to kinetic
   (f) kinetic to electrical
   (g) electrical to kinetic.

2. Use the principle of conservation of energy to explain why, when a catapult is used to launch a paper pellet, the paper does not keep moving for ever.

3. Describe the energy transfers involved when a pole-vaulter does a jump using a flexible pole, from run-up to landing.

4. Explain what is wrong with the statements below.
   (a) '65% of the chemical energy in the fuel was converted to kinetic energy, so 35% of the energy must have been destroyed.'
   (b) 'A pendulum will swing forever in a vacuum, so the conversion of gravitational potential energy to kinetic energy must be 100% efficient.'
   (c) 'The heat energy wasted in power stations can all be reclaimed to heat water so the process will be 100% efficient.'
   (d) 'Any system we consider in energy transfer terms is a closed system.'

3 Potential and kinetic energy

By the end of this topic, you should be able to demonstrate and apply your knowledge and understanding of:

* gravitational potential energy of an object in a uniform gravitational field; $E_P = mgh$
* kinetic energy; $E_K = \frac{1}{2}mv^2$
* the exchange between gravitational potential energy and kinetic energy

Gravitational potential energy

Gravitational potential energy is the energy an object has because of its position. The amount of gravitational potential energy an object has depends on three factors:

- its vertical height above the ground, Δh, measured in metres, m
- its mass, m, measured in kilograms, kg
- the value of the gravitational field strength at that point, g, measured in $N\,kg^{-1}$.

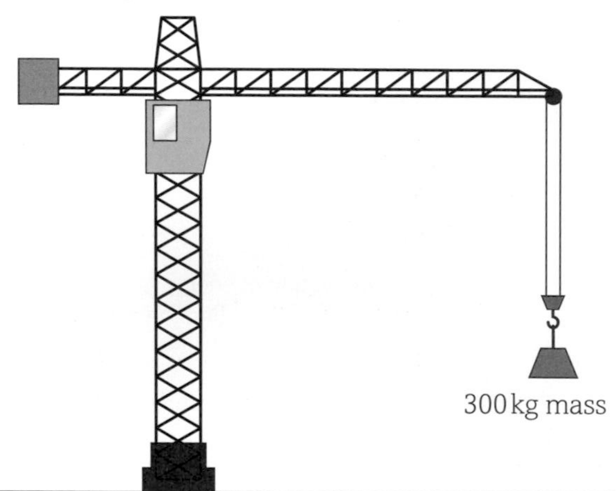

Figure 1 An object's gravitational potential energy depends on its mass, its height and the gravitational field strength.

As each of these quantities increases, so does the object's gravitational potential energy. Doubling any of these factors also doubles the total of amount of gravitational potential energy of the object. Gravitational potential energy is directly proportional to each of the quantities.

As we saw in Topic 3.3.1, the amount of work done is equal to the amount of energy transferred:

work done = energy transferred

and the work done is calculated as force × distance moved in the direction of the force.

When you lift a weight through a change in height Δh, the object gains gravitational potential energy. The amount of gravitational potential energy gained, E_P, equals the work done:

$E_P = \text{work done} = F \times d$

F, the force required to lift a weight, equals mg, so:

$E_P = mg \times \Delta h = mg\Delta h$

WORKED EXAMPLE 1

How much gravitational potential energy E_p is gained by:

(a) a mass of 45 kg that has been carried up stairs of vertical height 20 m on Earth ($g = 9.81\ N\,kg^{-1}$)?
(b) a mass of 45 kg that has been lifted through a vertical height of 20 m on the Moon ($1.6\ N\,kg^{-1}$)?
(c) a girl of weight 540 N who has scaled a climbing wall of height 18 m?

Answers

(a) $E_p = m \times g \times \Delta h$
$= 45\ kg \times 9.81\ N\,kg^{-1} \times 20\ m$
$= 8800\ J$ to 2 significant figures

(b) $E_p = m \times g \times \Delta h$
$= 45\ kg \times 1.6\ N\,kg^{-1} \times 20\ m$
$= 1400\ J$ to 2 significant figures

(c) $E_p = W \times \Delta h$
$= 540\ N \times 18\ m$
$= 9700\ J$ to 2 significant figures

Kinetic energy

The kinetic energy of an object depends on two factors: its mass and velocity.

When an object is accelerated by a force, the object gains kinetic energy. The amount of kinetic potential energy gained, E_K equals the work done:

$E_K = \text{work done} = F \times \Delta s$

where F is the force that produces the acceleration and Δs is the change in displacement caused by the force.

Using $F = ma$: $E_{K.} = ma(\Delta s)$

Using the equation of motion, $v^2 = u^2 + 2as$, we can write

$as = \frac{1}{2}(v^2 - u^2)$

Substituting for as in the expression for $E_{K,}$ we get:

$E_K = \frac{1}{2}m(v^2 - u^2)$

If we consider an acceleration from rest, $u = 0$ so

$E_K = \frac{1}{2}mv^2$

Returning to the pendulum example, we can state:

loss in potential energy = gain in kinetic energy

$mg\Delta h = \frac{1}{2}mv^2$

This leads to an equation that enables us to work out the velocity of an object that has fallen from rest through a height Δh:

$v = \sqrt{2g\Delta h}$

Doubling the mass of a moving object will double its kinetic energy if the speed is kept constant. However, you will notice that the velocity term in the equation is squared. This means that if we keep the mass the same and double the velocity of the object then its kinetic energy will increase by a factor of four. In other words, kinetic energy is directly proportional to mass and directly proportional to the square of the velocity.

Exchange between E_P and E_K

A common example of energy being transferred from one form to another involves the interchange between gravitational potential energy and kinetic energy for a swinging pendulum or a falling object. In accordance with the **principle of conservation of energy**, energy cannot be created nor destroyed, so if drag forces are ignored we can say:

loss in potential energy = gain in kinetic energy.

WORKED EXAMPLE 2

A skydiver with a mass 78 kg jumps out of a plane from a height of 4450 m above sea level. She then accelerates from rest as she falls.
Calculate:
(a) her velocity after falling a vertical distance of 180 m.

Answer

(a) loss of E_p = gain in k.e.

using $mg\Delta h = \frac{1}{2}mv^2$

$v = \sqrt{2g\Delta h} = \sqrt{2 \times 9.81 \times 180}$

$= 59.4\ \mathrm{m\,s^{-1}}$ to 3 significant figures.

In this worked example, we have assumed that all the gravitational potential energy has been transferred to kinetic energy, with none being 'wasted' as heat or sound. In reality, this is not the case and the 'true' value of her speed will be less than the one calculated here.

Figure 2 Skydiving is an example of gravitational potential energy being transferred to kinetic energy.

We can use the equations for the change in potential energy and in kinetic energy to answer questions relating to the transfer of energy between these two forms.

WORKED EXAMPLE 3

A monkey with a mass of 56 kg jumps upwards with a vertical velocity of $12\ \mathrm{m\,s^{-1}}$.

Figure 3

Calculate:
(a) the monkey's initial vertical kinetic energy at take-off
(b) his final vertical kinetic energy once he reaches his maximum height
(c) the vertical height he reaches, assuming that all the kinetic energy is transferred to gravitational potential energy.

Answers

(a) k.e. $= \frac{1}{2}mv^2$

$= \frac{1}{2} \times 56 \times 12^2$

$= 4032\ \mathrm{J}$

(b) His final vertical kinetic energy will be zero because he is no longer moving upwards and has momentarily stopped, at the maximum height, before starting to fall downwards.

(c) $\Delta h = \frac{v^2}{2g}$

$= \frac{12^2}{(2 \times 9.81)}$

$= 7.3\ \mathrm{m}$

Questions

1. Calculate the following gravitational potential energies of the situations described, giving your answers in joules:
   (a) $m = 3.4$ kg, $g = 9.81\ \mathrm{N\,kg^{-1}}$, $\Delta h = 12.6$ m
   (b) $m = 800$ g, $g = 1.6\ \mathrm{N\,kg^{-1}}$, $\Delta h = 0.24$ km
   (c) $m = 2.6 \times 10^4$ kg, $g = 12.8\ \mathrm{N\,kg^{-1}}$, $\Delta h = 235$ km

2. Determine the sizes of the following kinetic energies based on the information provided:
   (a) $m = 1200$ kg, $v = 11.5\ \mathrm{m\,s^{-1}}$
   (b) $m = 9.11 \times 10^{-31}$ kg, $v = 3 \times 10^7\ \mathrm{m\,s^{-1}}$
   (c) $m = 4.2 \times 10^3$ kg, $v = 0.05\ \mathrm{km\,s^{-1}}$

3. A ball of mass 450 g is thrown up in the air with an initial vertical velocity of $7.3\ \mathrm{m\,s^{-1}}$. Assuming there are no energy transfers to other forms, work out:
   (a) its initial kinetic energy
   (b) its final gravitational potential energy
   (c) the maximum height it reaches.

4. What would the maximum height be in question 3(c) if 12% of the initial kinetic energy was transferred to other forms such as heat and sound?

5. Look at Figure 4. Calculate the final speed of the skier, assuming that he starts from rest and that 11% of his initial gravitational potential energy is not transferred to kinetic energy. Take g to be $9.81\ \mathrm{N\,kg^{-1}}$.

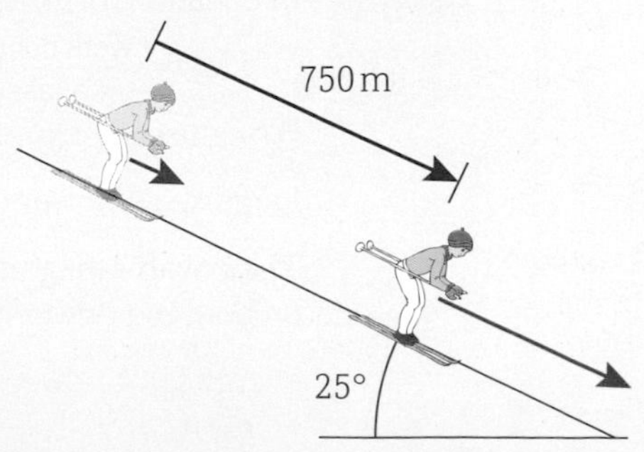

Figure 4

Power and the watt

By the end of this topic, you should be able to demonstrate and apply your knowledge and understanding of:

* power; the unit watt
* $P = Fv$

Power

Power is defined as the rate of doing work and it is measured in watts (W) or joules per second ($J\,s^{-1}$) – it is the rate at which energy is transferred from one form to another. For example, if a light bulb has a rating of 60 W it means that it transfers 60 J of electrical energy to other forms of energy for every second it is turned on.

Consider a sprinter and a long-distance runner (Figure 1). The long-distance runner transfers more energy than the sprinter overall, but the sprinter can transfer energy at a faster rate. This means that the sprinter has a higher power rating than the long-distance runner.

LEARNING TIP

'Rate of' means 'per unit time'. When we state a power value in watts, we are stating how many joules are transferred in each second.

Figure 1 Mo Farah and Usain Bolt.

In equation form, this idea is expressed as:

$$\text{power} = \frac{\text{work done}}{\text{time taken}}$$

Power usually expressed in watts (W) or kilowatts (kW); 1 kW = 1000 W.

Large-scale power generation is expressed in megawatts (MW); 1 MW = 1000 kW = 1 000 000 W.

The power rating of many devices is constant, but the total amount of energy transferred depends on both the power rating and the length of time for which it is used:

energy (J) = power (W) × time (s)

$$E = P \times t$$

If a 100 W light bulb is left switched on for one hour, the electrical energy transferred to the bulb must be $100\,J\,s^{-1} \times (60 \times 60)s = 360\,000\,J$.

WORKED EXAMPLE 1

(a) A 40 W lamp is switched on for 3 hours. How much energy is transferred in this time?

(b) A nuclear power station transfers 3×10^{12} J of energy in 5 minutes. What is its power rating?

Answers

(a) $E = P \times t$
$= 40\text{ W} \times (3 \times 60 \times 60)\text{s}$
$= 4.32 \times 10^5\text{ J}$

(b) $P = \frac{E}{t}$
$= \frac{3 \times 10^{12}\text{J}}{(5 \times 60)\text{s}}$
$= 1 \times 10^{10}$ W or 1×10^4 MW to 1 significant figure

Power can be calculated for any energy transfer, including the human body, as the following example shows.

WORKED EXAMPLE 2

An athlete with a mass of 90 kg performs press-ups at the rate of 50 per minute for a time of 6 minutes. For each press-up he raises his centre of gravity a distance of 24 cm.

(a) Calculate:
- (i) the total work done
- (ii) the power required
- (iii) the energy he transferred, given that the transfer efficiency is 20%.

(b) Use your answers to explain why the athlete gets hot while doing press-ups.

Answers

(a) (i) For one press-up, Δp.e.= mgh
$= 90\text{ kg} \times 9.81\text{ m s}^{-2} \times 0.24\text{ m}$
$= 212\text{ J}$

Number of press-ups in 6 min = 50 × 6 = 300

Total work done = 212 J × 300 = 63 600 J

(ii) Power required $= \frac{63\,600\text{ J}}{(6 \times 60)\text{s}} = 177\text{ W}$

(iii) Only 20% of the energy is transferred to lifting, so the total energy transferred = 63 600 × 5 = 318 000 J

(b) The energy not transferred to lift the athlete's body is transferred as internal energy in his body. So molecules in his body move faster and he feels hotter. Some of the energy he does transfer to raise his body will also be 'wasted' in his muscles when he lowers himself, making him even hotter.
Note that none of these answers will be correct to 3 significant figures. Working is done to this number of figures to ensure that not too much rounding takes place.

The power output of a moving object that is subject to a force can be found using:

power (W) = force (N) × velocity (m s^{-1})

$P = F \times v$

This equation is derived from that for work done:

work done = force × distance (equation 1)

$\text{power} = \frac{\text{work done}}{\text{time taken}}$ (equation 2)

Substituting equation 1 into equation 2 gives:

$\text{power} = \frac{(\text{force} \times \text{distance})}{\text{time taken}}$

But: $\text{velocity} = \frac{\text{distance}}{\text{time}}$, so power = force × velocity.

WORKED EXAMPLE 3

A car engine provides a forward force of 1250 N. If the car is travelling at 24 m s^{-1}, what power is developed?

Answer

power = force × velocity
$= 1250\text{ N} \times 24\text{ m s}^{-1}$
= 30 000 W or 30 kW

Questions

1. A student counts 14 risers on a flight of stairs, each riser being 18 cm high. Her mass is 52 kg. She ran up the stairs three times, taking 3.1 s, 3.0 s and 3.2 s as measured by another student. Calculate her average power while performing this exercise. Do you think that this is a fair way of measuring the mechanical output power of a person?

2. The engine of a car of mass 900 kg produces a driving force of 300 N while travelling at a constant speed of 20 m s^{-1} along a straight level road.
   (a) Suggest a reason why the car is not accelerating.
   (b) Calculate the power required to drive the car forward at 20 m s^{-1}.
   (c) The car comes to a hill with a gradient of 1 in 15, as shown in Figure 2. Calculate the additional power required to maintain the same speed of 20 m s^{-1}.

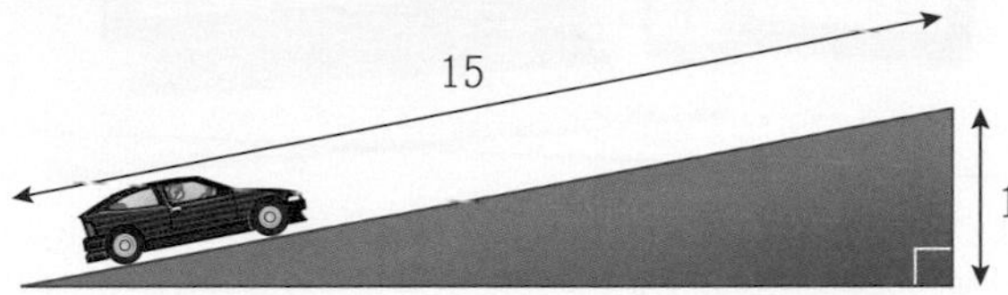

Figure 2

3. To accelerate from rest to 20 m s^{-1} the car in question 2 takes 10 s.
   (a) Calculate the average rate at which kinetic energy is transferred to the car during the acceleration.
   (b) How does your answer to part (a) compare with the power in question 2?
   (c) Explain why the values are different.

4. Calculate the driving force developed by the Lamborghini Aventador LP1600–4 Mansory Carbonado GT in Figure 3 if it accelerates from 0–27 m s^{-1} in 6 seconds. You'll need to do some research to answer this question.

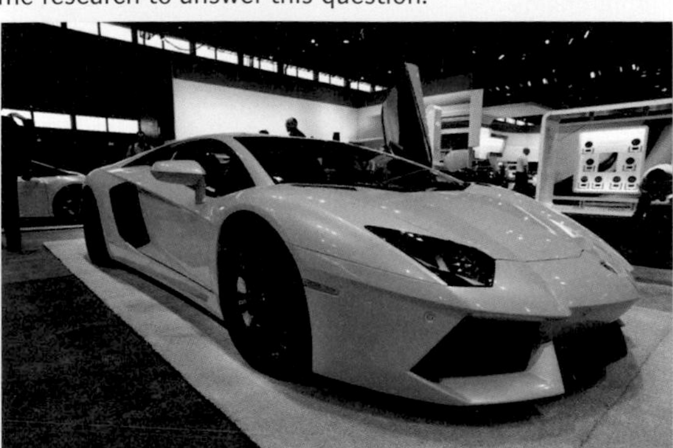

Figure 3

3.3 5 Efficiency

By the end of this topic, you should be able to demonstrate and apply your knowledge and understanding of:

* efficiency of a mechanical system
* efficiency $= \dfrac{\text{useful output energy}}{\text{total input energy}} \times 100\%$

Introduction

Energy of one form or other is often transferred to a different type of energy. For example, the chemical energy stored in cells and batteries is transferred to electrical energy in circuits and eventually, for example, to light and sound from a phone or tablet (Figure 1). Ideally, we want as much of the chemical energy stored in a cell to be transferred to as much of the useful energy as possible.

Figure 1 These devices convert the chemical energy stored in the cells to light and sound as useful forms of energy and 'waste' energy by heating in the electrical circuit.

In accordance with the principle of conservation of energy, that we can never have more output energy than input energy – we cannot simply create energy. More importantly, we can never actually transfer 100% of the input energy to useful output energy – energy is always 'wasted' by transfer to other forms, which are not useful.

In terms of the energy transfers for a device or machine, **efficiency** is expressed as:

$$\frac{\text{useful output energy}}{\text{total input energy}} \times 100\%$$

WORKED EXAMPLE 1

A motor lifts a bucket with a mass of 8 kg a vertical distance of 34 m out of a well. The electrical energy transferred was 4000 J. Calculate the efficiency of the electric motor.

Answer

The input energy = 4000 J

The output energy is the gain in gravitational potential energy:

gain in E_p = mass × gravitational field strength × change in height

$= 8\text{ kg} \times 9.81\text{ N kg}^{-1} \times 34\text{ m} = 2668\text{ J}$

$$\text{Efficiency} = \frac{\text{useful output energy}}{\text{input energy}} \times 100$$

$$= \frac{2668\text{ J}}{4000\text{ J}} \times 100$$

$$= 67\%$$

In terms of the rate of energy conversion, or power of a device or machine, we can also express efficiency as:

$$\text{efficiency} = \frac{\text{useful output power}}{\text{total input power}} \times 100\%$$

WORKED EXAMPLE 2

A light bulb marked 60 W transfers only 12 J of energy per second into light energy. How efficient is the light bulb?

Answer

Input power = 60 W

Output power = 12 W

$$\text{Efficiency} = \frac{\text{useful output power}}{\text{input power}} \times 100$$

$$= \frac{12}{60} \times 100$$

$$= 20\%$$

For every 5 joules of electrical energy that transfer to the bulb per second, 4 joules will be transferred to heat energy!

Device	Energy input	Useful energy output	Typical efficiency
electric motor	electrical	kinetic/potential	85%
solar cell	light	electrical	10%
rechargeable battery	electrical	electrical	30%
electric radiator	electrical	thermal	99%
power station	nuclear	electrical	40%
car (petrol)	chemical	kinetic/potential	55%
car (diesel)	chemical	kinetic/potential	55%
steam engine	chemical	kinetic/potential	8%

Table 1 Examples of common devices for which efficiency is an important consideration.

As can be seen in Table 1, transferring electrical energy to thermal energy is very efficient. Conversely, trying to transfer heat energy or light energy into electrical energy is more difficult, resulting in much lower efficiencies. One of the main challenges for the future is to develop technologies, such as solar cells, in which efficiencies are much higher than they are currently. In truth, 90% of the energy from the Sun is currently being 'wasted' in solar technologies because these devices are only around 10% efficient.

Comparing the efficiencies of power stations

We get electrical energy in our homes from power stations (Table 2). Fuels are used in these power stations to transfer heat energy. The heat energy heats water so that high-pressure steam can be blown onto turbines, which turn generators. The electricity that is produced is then distributed across the country using the National Grid.

Type of power station	Efficiency ranges for generating electrical energy
coal	39–47%
oil	38–44%
natural gas	30–39%
nuclear	33–36%

Table 2 All the methods used for generating electrical energy are less than 50% efficient – they 'waste' more energy than they transfer to useful forms.

WORKED EXAMPLE 3

The electrical energy output of a nuclear power station is 10 900 000 J when operating at its lowest efficiency. Using the data in Table 2 calculate:

(a) the energy input in J

(b) the energy input in MW h.

Answers

(a) $\text{efficiency} = \frac{\text{useful energy output}}{\text{energy input}} \times 100$

Rearranging:

$\text{energy input} = \frac{\text{useful energy output}}{\text{efficiency}} \times 100$

$= \frac{10\,900\,000}{33} \times 100 = 33\,030\,303\text{ J}$

We used the value of 33% (or 0.33) in the calculation here because the question asked for nuclear's lowest efficiency value to be used in the calculation.

(b) $1\text{ MW h} = 1\,000\,000\text{ W} \times 3600\text{ s} = 3.6 \times 10^9\text{ J}$

So, the number of MW h that is equivalent to 33 030 303 J is

$\frac{33\,030\,303}{3.6 \times 10^9} = 9 \times 10^{-3}\text{ MW h}$

Sankey diagrams

Sankey diagrams are usually used to represent power production and energy problems. This is shown in Figure 2 for a typical power station supplying 1000 MW electrical output.

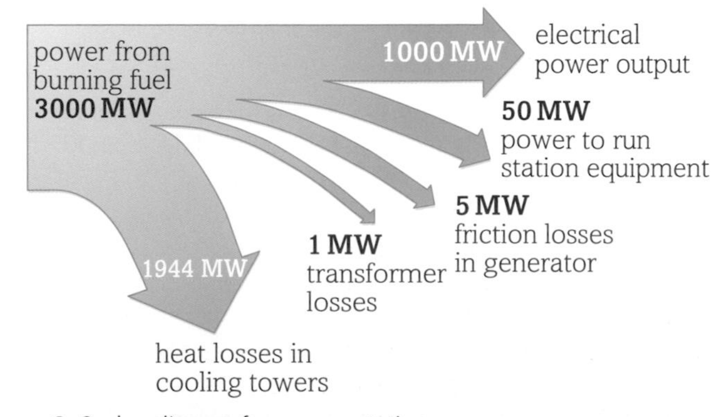

Figure 2 Sankey diagram for a power station.

Before 1000 MW of power can be supplied to customers, further transfer losses occur. For example, the transmission cables used to carry the electricity have resistance and so heat up when the electric current passes through them.

Figure 3 Power transmission lines from a power station.

Questions

1 Copy Table 3 and fill in the gaps.

Device	Efficiency	Input energy/J	Useful output energy/J
refrigerator	48%	4×10^5	
light bulb	7%		43 000
wind turbine	59%	7×10^9	
LED		4×10^2	1.2×10^2
human muscle		34 000	2200
electric motor	73%		2.8 MJ

Table 3

2 Draw Sankey diagrams for:

(a) a washing machine

(b) a television

(c) a lava lamp.

3 A combined-heat-and-power station (CHP) makes use of the hot cooling water to heat the houses and factories near the power station. A CHP station is built on a much smaller scale than a 1000 MW power station, which is designed purely to supply the National Grid with electricity.

For an input of 500 MW of fossil fuel, the CHP station transfers 150 MW of electricity to the Grid and 250 MW to water for local area heating – the rest goes up the chimney or is lost in the generators, etc.

(a) Draw a Sankey diagram for a CHP station.

(b) (i) Calculate the efficiency of the CHP station by assuming that electricity is the only useful output.

(ii) How does this compare with the 1000 MW station considered in Figure 2?

(c) (i) Calculate the efficiency of the CHP station by considering both electrical and local heating as useful output.

(ii) How does this compare with the 1000 MW station?

(d) Explain why it is more efficient to burn a fossil fuel directly at home than to heat the house electrically from the 1000 MW station.

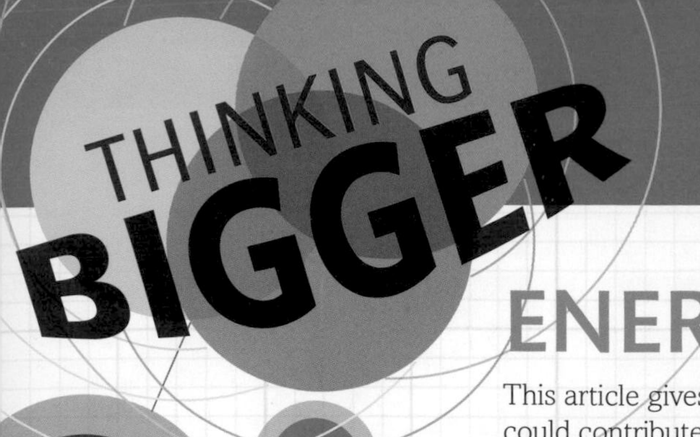

ENERGY SOLUTIONS

This article gives information on some experimental electrical power generation technologies that could contribute to the UK's energy needs in the future.

THIS IS THE NEXT GENERATION OF RENEWABLE ENERGY TECHNOLOGIES

Hi-tech football pitches, wave power and nuclear fusion are helping to move Britain away from 'dirty' fuels towards sustainable energy.

Scientists all over the globe are working to develop sustainable new energy sources to reduce our dependence on dwindling fossil fuel supplies.

In the UK, just 5pc of the nation's energy comes from renewables. The Government has set a target of 15pc by 2020, but progress is slow.

Some sustainable energy sources, such as solar energy, are mature marketplaces, with 60 years of research behind them. Others, such as antimatter, are more experimental.

The science of antimatter is still in its infancy but scientists claim that mixing just half a gram of antimatter with half a gram of matter would create the same energy generated by the Hiroshima bomb.

There are several start-ups developing other ground-breaking technologies for generating electricity, some using methods that seem more Star Trek: The Next Generation than National Grid. We meet three entrepreneurs leading the charge into next-generation renewables.

Turning footsteps into electricity

Youngsters playing on a newly-installed football pitch in one of Rio de Janeiro's most notorious slums are now powering the neighbourhood's street lights with every step.

Their movements across the AstroTurf are converted from kinetic energy into electricity by 200 hidden energy-capturing tiles built by London-based Pavegen.

Founded by Laurence Kemball-Cook in 2009, the company exports its energy-converting tiles to 20 countries across the world. Customers range from infrastructure giants such as Siemens to retail brands Nike and Uniqlo. 'I started this in my bedroom with just a sketch,' Kemball-Cook, 29, tells The Sunday Telegraph. 'Now we employ 30 staff in four offices and we're profitable.'

Pavegen, which converts high footfall areas into pseudo-batteries, and sister company Roadgen, which aims to harvest energy from vehicles on the world's roads, will help to power the cities of the future, says Kemball-Cook. 'We want to take the cost of the technology down to the same price as normal flooring. We're looking to raise investment this year to help us meet that goal.'

Pavegen, which has been shortlisted for this year's UK Private Business Awards, has now passed the £1m turnover mark and is on target to double that this financial year. 'Next year will be pivotal for the company,' says Kemball-Cook, revealing that the company's technology is due to be installed outside the White House next spring in its biggest US project to date.

[...] Wind and solar alternatives are less efficient than Pavegen's technology, and depend heavily on weather and geography, Kemball-Cook claims. 'Human footfall is currently a wasted resource," he adds. "We will become part of the fabric of urban infrastructure.'

The new wave of energy

The World Energy Council has estimated that if the planet's wave power was harnessed, we could generate double the amount of electricity currently produced worldwide. The west coast of Scotland is home to some of the most powerful and consistent waves in the world. Over the past few days, waves have been recorded in the Orkney Islands topping 14 ft.

Sam Etherington, a 24-year-old engineer and founder of Aqua Power Technologies, is testing a new device that captures wave power at a site near the UK archipelago.

[...] Etherington's invention sits on the surface of the water and, unlike other wave power systems, works on a multiaxis basis; it can generate power no matter what direction the waves come from.

Enquiries are flooding in from all over Europe from consortiums keen to install the devices. 'It's very encouraging, given that we're still developing the technology and don't even have a price for them yet,' says Etherington, adding that the devices are likely to cost 'millions'.

[...] Etherington is hoping to have two wave farms deployed in the UK by 2018. [...]

- Source: Burn-Callander, R. (2014) 'This is the next generation of renewable energy technologies', The Sunday Telegraph, 13 December 2014

Where else will I encounter these themes?

1.1 | 2.1 | 2.2 | 3.1 | 3.2 | 3.3 | YOU ARE HERE

Let's start by considering the nature and context of the writing in the article. The text above is an extract from an article published in *The Sunday Telegraph*.

1. a. Discuss who the article is intended for, and comment on the writing style. How does the reporter create impact?

b. What is the source for the facts and data given in the article? How credible are these sources?

Consider the differences in how a newspaper or research report present information and data.

Now let's look at the physics in, or connected to, this article. Don't worry if the physics content is challenging at this stage. You can always return to the article once you have covered other topics later in the book. Use the timeline at the bottom of the page to help you put this work in context with what you have already learned and what is ahead in your course.

2. Which renewable energy resources are most commonly in use currently?
3. Why is only 5% of the UK's energy currently provided by renewable energy resources?
4. How can so much energy be released from only half a gram of matter and antimatter?
5. How do Pavegen tiles work? Draw a diagram to help you explain your answer.
6. Do you think Pavegen tiles are a realistic way of producing electrical power? Explain your reasoning.
7. How does wave power work? How does the wave power technology mentioned here compare with the Pavegen technology?
8. What is the significance of companies such as Siemens, Nike and Uniqlo using Pavegen? What are their respective areas of trade and how would this project appeal to them?
9. What is 'pivotal' about the Pavegen devices being installed outside the White House?
10. Which would you be most likely to want to see being developed – the tiles or the wave power? Explain your answer.
11. Explain why Aqua Power Technologies' prototype wave power generator is more efficient than other devices which orient themselves to waves travelling in one direction.
12. Why are industry consortiums so keen to invest in and install the devices when they have only been tested in large water tanks and no finished device is available?

The equation $E = mc^2$ can be used to calculate how much energy is released when matter and antimatter collide. Find out what the terms and units are that need to be used and then calculate how much energy is released when half a gram of matter comes into contact with half a gram of antimatter.

Activity

As company director, you are making a bid to possible investors for funding for either Pavegen tiles or a wave power farm. A business has agreed to invest £10 million if you can put together a convincing case explaining how the technology you develop will be efficient in terms of energy production and reliability as well as make the company a profit.

Write a 'Dragon's Den' style pitch that includes the underpinning physics, how the investment funds will be used and a compelling, evaluated argument explaining why your pitch and business model is most fit for purpose.

3.3 Practice questions

1. A ball of mass 2.4 kg is thrown upwards with an initial velocity of 16 m s^{-1}. Which of the following statements is true? [1]
   (i) The ball will have an initial vertical kinetic energy of 307.2 J
   (ii) The weight of the ball is given by W = 2.4 × g
   (iii) The kinetic energy of the ball is converted entirely to thermal energy.

   A (i), (ii) and (iii)
   B only (i) and (ii)
   C only (ii) and (iii)
   D only (i)

2. A small electric motor is 20% efficient. Its input power is 9.6 W when it is lifting a mass of 0.50 kg at a steady speed *v*. [1]

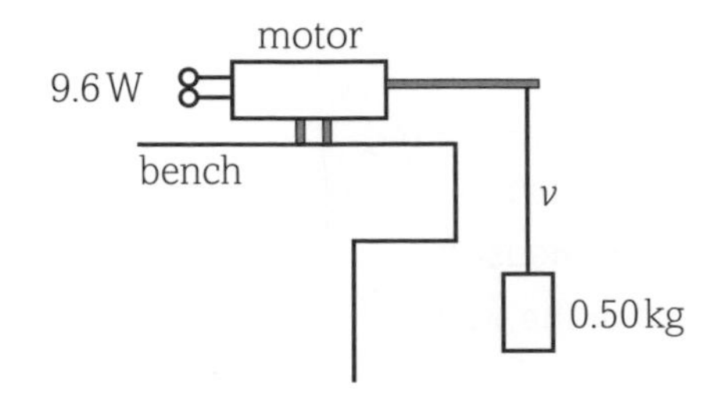

Figure 1

What is the value of *v*?

A 0.39 m s^{-1}
B 2.0 m s^{-1}
C 2.8 m s^{-1}
D 3.8 m s^{-1}

[Q8, H156/01 sample paper 2014]

3. Which of the following is true? [1]
   (i) power = energy × time
   (ii) efficiency = $\frac{\text{energy input}}{\text{useful energy output}}$
   (iii) force × distance = power × time

   A (i), (ii) and (iii)
   B only (i) and (ii)
   C only (i) and (iii)
   D only (iii)

4. A body has a mass of 3.8 kg and a velocity of 6 m s^{-1}. Which of the following is true? [1]
   (i) The body has a kinetic energy of 68.4 J.
   (ii) If the mass was being raised for 8 seconds it would gain 3000 J of gravitational potential energy.
   (iii) If the body was moving horizontally it would have a power of 22.8 W.

   A (i), (ii) and (iii)
   B only (i) and (ii)
   C only (ii) and (iii)
   D only (i)

[Total: 4]

5. Look at the diagram of a pendulum shown in Figure 2.

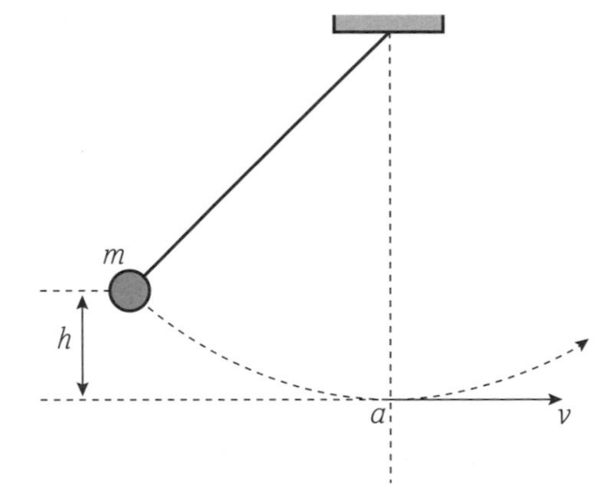

Figure 2

(a) Use Figure 2 to write equations for:
   (i) the potential energy, E_p, when the mass, m, is at the maximum height *h* [1]
   (ii) the kinetic energy, E_k, when the mass passes through its lowest point *a*. [1]

(b) By using the principle of conservation of energy, find an equation for the velocity of the pendulum bob at the lowest point of its swing. [2]

(c) Find values for the maximum potential energy and the maximum kinetic energy if *m* = 500 g and *h* = 80.0 cm. [2]

[Total: 6]

6. A car passes through traffic lights at a constant speed, covering a distance of 240 m in 12 s.

(a) The force acting on the car is 3600 N. Calculate the power generated. [3]

(b) The mass of the car is 1.45 × 10^3 kg. Calculate the kinetic energy of the car during this time. [2]

[Total: 5]

7. Figure 3 shows a conveyor belt for taking young skiers up a slope so that they can ski back down.

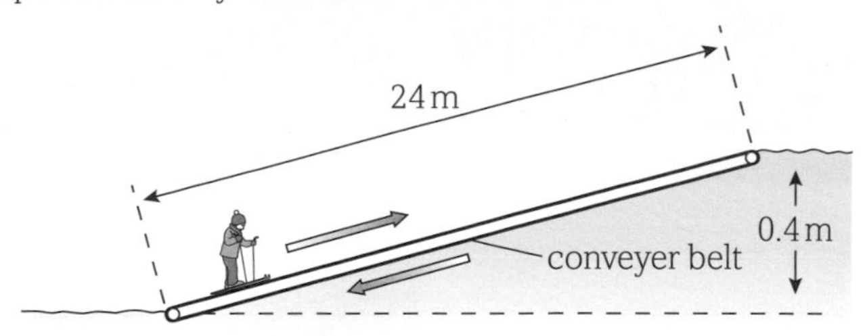

Figure 3

A young skier of mass 28 kg travels up the conveyor belt of length 24 m and vertical height 4.0 m in a time of 52 seconds.

(a) Calculate:

(i) the speed of the young skier on the conveyor belt [2]

(ii) the kinetic energy of the young skier [2]

(iii) the maximum increase in gravitational potential energy of the skier. [2]

(b) The conveyor belt is designed to take a maximum of 12 children at any one time, provided that their mean mass is no greater than 30 kg. Calculate the power needed to lift these children through a height of 4.0 m in 52 seconds. [3]

[Total: 9]

8. An electric motor lifts a 45 kg mass through a height of 37 m in a time of 11.0 seconds.

(a) Assume that the electric motor has an efficiency of 78%. Calculate:

(i) how much electrical energy was required to lift the mass [3]

(ii) the power generated by the electric motor. [2]

(b) Having reached the maximum height, the mass is then dropped onto the ground below.

(i) At what speed will the mass hit the ground? [3]

(ii) What assumptions have you made? [2]

[Total: 10]

MODULE 3

Forces and motion

CHAPTER 3.4

MATERIALS

Introduction

We are surrounded by what would appear to be an almost infinite number and variety of materials. The materials from which our homes are constructed include stone, cement, wood, glass, plastics and ceramics, to name but a few. Each of these materials is used for a specific purpose, e.g. to allow light in to our rooms, to retain heat, to stop flooding, or to conduct electricity. For centuries, people have made use of the natural materials that were present on Earth and we have utilised natural resources and chemicals to make man-made materials for clothing, building, cooking food and in the playing of sports. Materials such as nylon, Teflon and Kevlar are used extensively in many areas of everyday life. Like so many areas of modern day life, the day to day usage of materials-based language has a very specific meaning in the area of physics. The words strong, tough, brittle, hard, elastic, plastic, stress and strain have very specific meanings in the area of materials science and in this chapter you will become familiar with the meanings of these terms and the calculations and equations that allow engineers, architects and physicists to decide why each material is best suited for a particular role.

All the maths you need

To unlock the puzzles of this chapter you need the following maths:

- Units of measurement
- Use of the square function when calculating energy stored by a spring using $E = \frac{1}{2}kx^2$
- Calculating the area of cross-section of a cylindrical wire from πr^2
- Calculating the spring constant from the gradient of a graph
- Calculating the Young modulus from the gradient of a graph
- Calculating or estimating the area under a graph

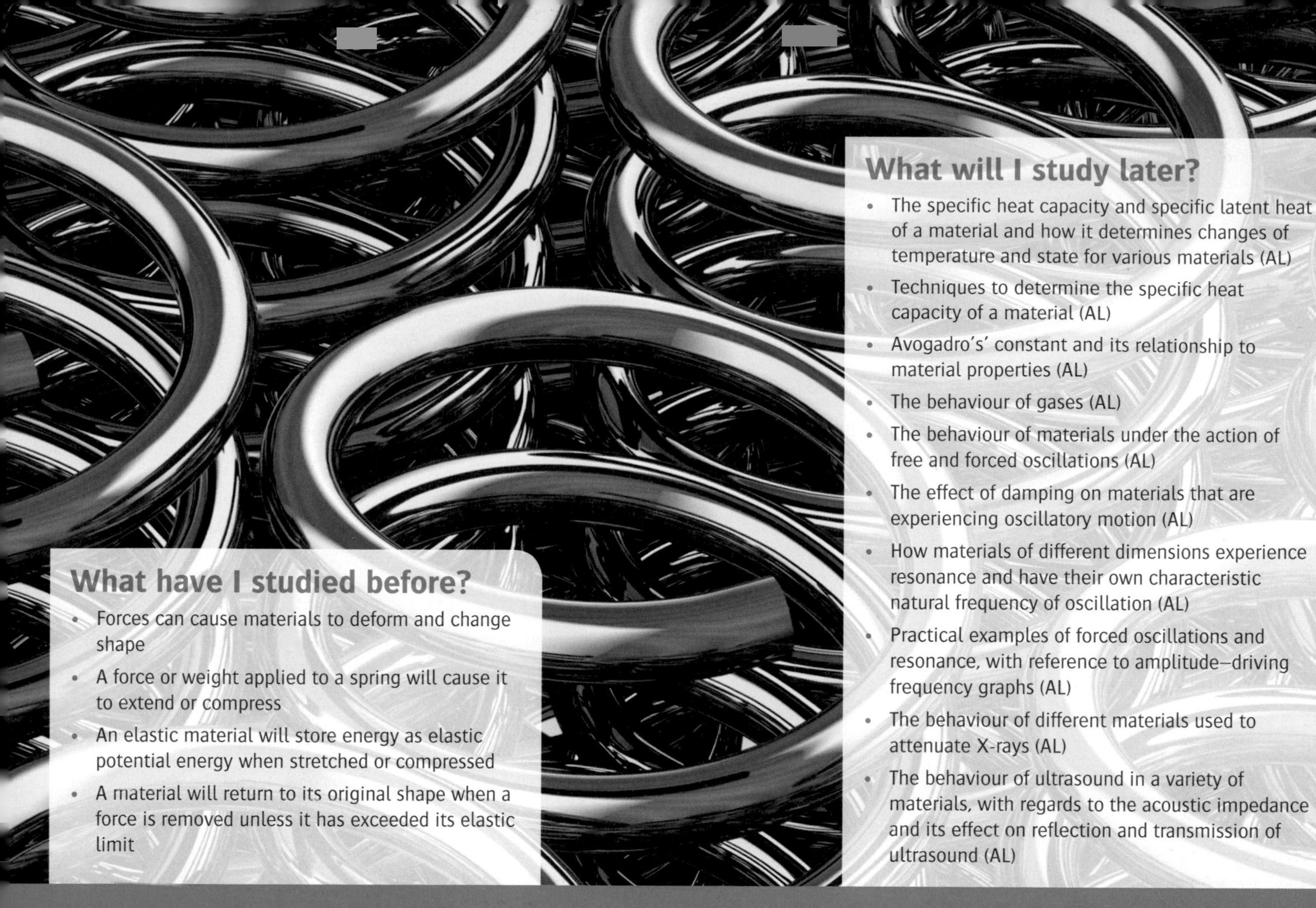

What will I study later?

- The specific heat capacity and specific latent heat of a material and how it determines changes of temperature and state for various materials (AL)
- Techniques to determine the specific heat capacity of a material (AL)
- Avogadro's' constant and its relationship to material properties (AL)
- The behaviour of gases (AL)
- The behaviour of materials under the action of free and forced oscillations (AL)
- The effect of damping on materials that are experiencing oscillatory motion (AL)
- How materials of different dimensions experience resonance and have their own characteristic natural frequency of oscillation (AL)
- Practical examples of forced oscillations and resonance, with reference to amplitude–driving frequency graphs (AL)
- The behaviour of different materials used to attenuate X-rays (AL)
- The behaviour of ultrasound in a variety of materials, with regards to the acoustic impedance and its effect on reflection and transmission of ultrasound (AL)

What have I studied before?

- Forces can cause materials to deform and change shape
- A force or weight applied to a spring will cause it to extend or compress
- An elastic material will store energy as elastic potential energy when stretched or compressed
- A material will return to its original shape when a force is removed unless it has exceeded its elastic limit

What will I study in this chapter?

- The terms extension and compression
- Tensile and compressive deformation of a material, including elastic and plastic deformation
- Hooke's law – that extension is directly proportional to force applied for a spring up until the elastic limit
- An understanding of the force constant, k, and how it determines the stiffness, extension or compression of a spring for a given tensile or compressive force
- The drawing and interpretation of force–extension graphs for a variety of springs, wires and other materials
- That the area beneath a force–extension graph is equal to the work done or elastic potential energy stored in the spring
- The calculation of elastic potential energy using the equations $E = \frac{1}{2}Fx$ or $E = \frac{1}{2}kx^2$
- The terms stress, strain and ultimate tensile strength
- The Young modulus $= \dfrac{\text{tensile stress}}{\text{tensile strain}}$ and the techniques used to determine its value
- Stress–strain graphs for ductile, brittle and polymeric materials

Deformation of materials

By the end of this topic, you should be able to demonstrate and apply your knowledge and understanding of:

* elastic and plastic deformations of materials
* tensile and compressive deformation; extension and compression
* force–extension (or compression) graphs for springs and wires
* techniques and procedures used to investigate force–extension characteristics

Elastic and plastic deformation

Solids are made of huge numbers of tightly packed atoms, so when a solid is compressed (squeezed) or subjected to tension (stretched) the spacing of the atoms in it is altered. The forces between individual atoms are small, but because there are so many of them, a large force is necessary for an appreciable change in the size or shape of the solid. Usually, when the force that is changing the shape of an object is removed, it returns to its original shape.

KEY DEFINITIONS

Elasticity is the property of a body to resume its original shape or size once the deforming force or stress has been removed.
Deformation is the change in shape or size of an object. If the material returns to its original shape then the deformation is elastic; if not, then its deformation is plastic.

It is easy to see that when you sit on a padded chair the cushion changes shape. Less noticeable is the fact that the chair's legs get shorter too – although by a very small amount. But when you stand up, the cushion and the legs return to their original shape. We describe such objects as **elastic**. Other examples of elastic materials are shown in Figure 1.

Figure 1 These materials all change shape when a reasonable force is applied, but will return to their original shape and size when the force is removed.

Plastic behaviour can be thought of as the opposite of elastic behaviour. Instead of returning back to its original shape when the deforming force is removed, the material remains stretched. This happens because the material has exceeded its elastic limit – the force beyond which the atoms can no longer return to their original arrangement. If a spring is exposed to a force that is too large, beyond its elastic limit, it will remain permanently stretched (Figure 2).

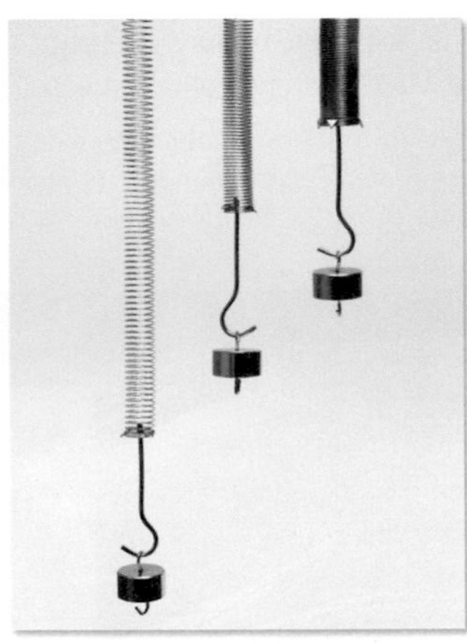

Figure 2 The springs in the middle and on the right returned to their original length when the force was removed – they behaved elastically. The spring on the left shows plastic deformation.

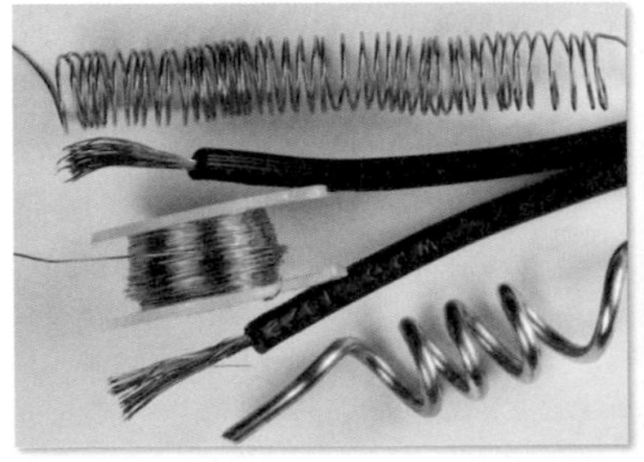

Figure 3 The metals exhibit plastic deformation when stretched beyond their elastic limit. Chewing gum, when wet, exhibits a large range of plasticity extending many times beyond its original length.

Tensile and compressive forces

Tensile forces cause tension in an object. Clearly, for there to be a tension in a fixed, stretched wire it must have equal and opposite forces on it at either end. This will cause the wire to increase in length and have a positive **extension**.

Compressive forces cause **compression** in an object. The forces acting in this situation are also equal and opposite in direction, but they are acting towards each other and will result in a decrease in the length of the spring or a negative extension.

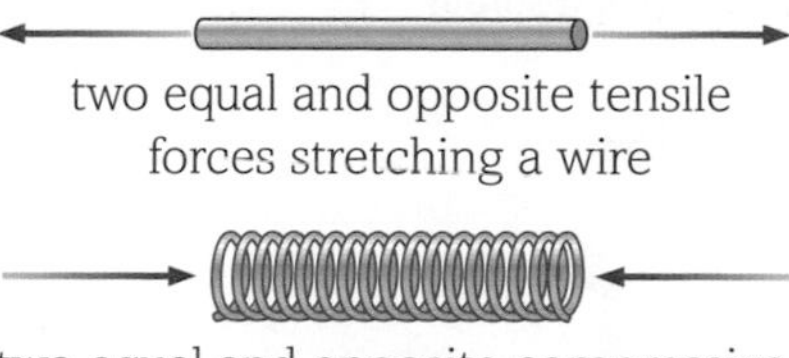

Figure 4 Tensile forces cause a wire to extend; compressive forces cause a spring to compress.

INVESTIGATION

Investigating stretching a wire

Figure 5 shows a long, thin copper wire held firmly in a clamp at one end. The other end supports a weight hanger after passing over a pulley. The hanger must be just heavy enough to keep the wire taut.

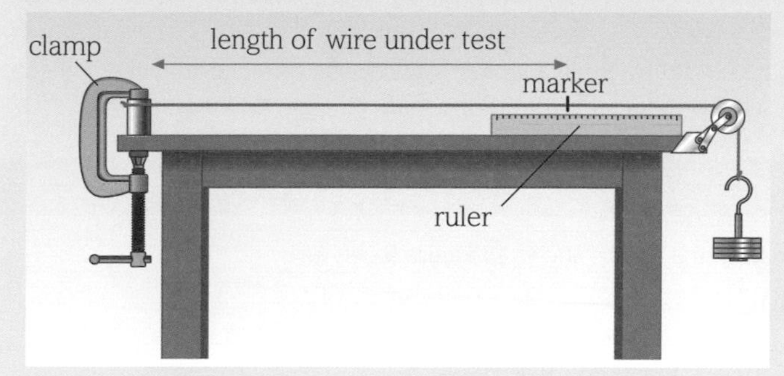

Figure 5 Experiment to measure the stretch of a wire.

A marker is attached to the wire and a ruler is fixed in position below the marker. Table 1 illustrates the position of the marker as weights are added progressively to the hanger. The values of the tensile force are found from the weight, taking $g = 9.81\ \text{m s}^{-2}$

Mass on hanger/kg	Tension in wire/N	Reading on ruler/mm	Extension/mm
0.000	0.00	26.0	0.0
0.050	0.49	29.0	3.0
0.100	0.98	32.0	6.0
0.150	1.47	35.5	9.5
0.200	1.96	38.5	12.5
0.250	2.45	42.0	16.0
0.300	2.94	45.5	19.5
0.350	3.43	48.5	22.5
0.400	3.92	53.0	27.0
0.500	4.90	66.5	40.5
0.600	5.88	87.0	61.0
0.700	6.86	122.0	96.5

Table 1 Results of the stretching experiment.

Note that there is a danger that the wire may snap during this experiment. Eye protection should be worn, and a box should be placed below the hanger – to catch the weight if it falls, and also to ensure that you do not stand directly underneath it.

Describing elastic and plastic behaviour

The graph shown in Figure 6 plots the force of tension against extension for the wire that is being stretched in Figure 5.

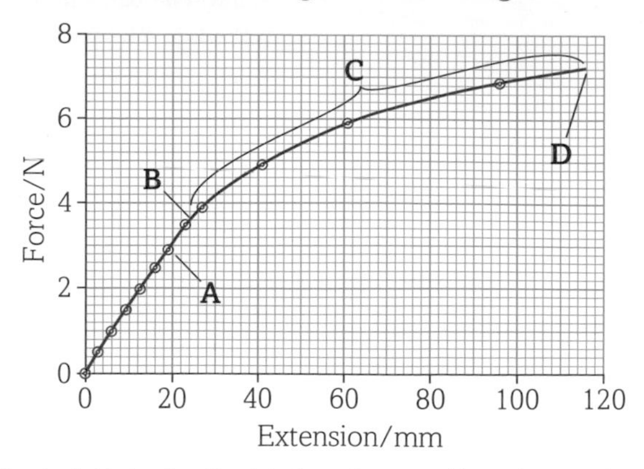

Figure 6 Graph plotted using the data from the stretching wire experiment.

The points labelled A, B, C and D are commonly noted in experiments of this nature and they have names and definitions that you need to be familiar with.

- A – the limit of proportionality. This is the point at which the extension is no longer directly proportional to the force (load). The material will still behave elastically after this point, but not for much more load.
- B – the elastic limit. Beyond this point the material will not show elastic behaviour. Any further load added will lead to plastic deformation and the material will not return to its original shape.
- C – plastic behaviour. This where plastic deformation will be seen when the load is removed and the material will not return to its original shape.
- D – fracture. This is the point at which the material will break. Before this happens, the material undergoes 'creep' where the planes of atoms slide past each other.

Figure 7 shows body panels of cars being pressed from sheet metal. A template of the required shape is attached to the press, which exerts a huge force on a flat piece of steel. The steel must retain its altered shape. The force applied must therefore be sufficiently large to ensure that the steel goes beyond its elastic limit and becomes plastic. Plastic deformation is an integral part of many manufacturing processes.

Figure 7 Presses forming body panels for cars.

INVESTIGATION

Investigating force–extension characteristics for other materials

You will probably have carried out experiments stretching springs, using the simple apparatus shown in Figure 8. A similar method could be used to investigate the *compression* of a spring.

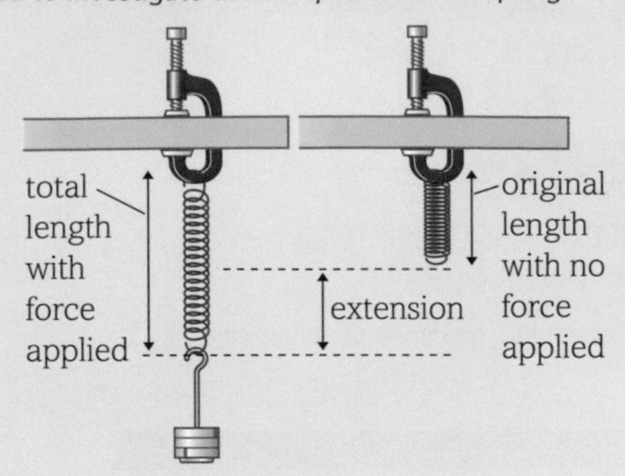

Figure 8

The same safety precautions should be taken as described for the previous investigation. By clamping a ruler vertically alongside the spring, the extension can be measured for different loads. It is a good idea to clamp the ruler so that the 0 cm mark is level with the base of the unstretched spring, allowing extension to be read directly. You can also use a set square to make sure the clamped ruler is vertical in relation to the bench, and to read off the length of the spring against the ruler.

The apparatus can also be used to investigate the stretching of a rubber band, elastic cord or a strip of polythene. You will need to do a trial run to decide a suitable range of values of the load force for each material.

Questions

1. Describe the behaviour of the materials in the graphs shown in Figure 9.

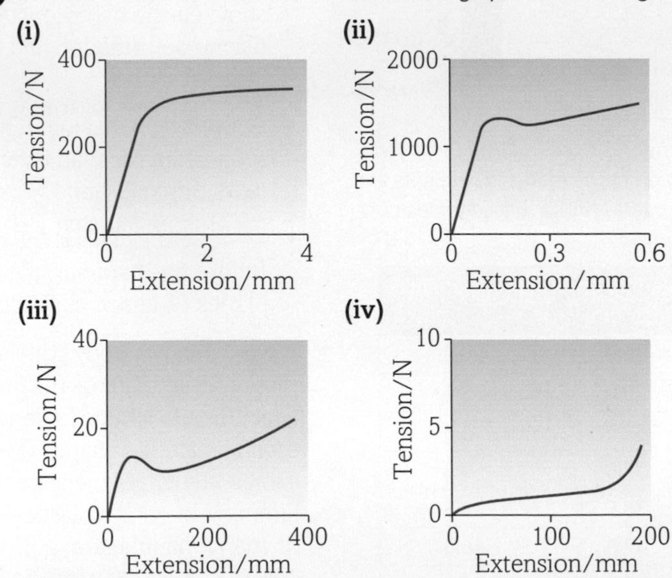

Figure 9

2. Use the terms mentioned in this topic to explain the behaviour of the following materials when undergoing deformation:

   (a) bone (b) chalk
   (c) putty (d) sticky tape
   (e) porcelain (f) lead sheet.

3. In an experiment to investigate the force-extension characteristics of a rubber band, describe and explain the precautions you could take to minimise measurement uncertainties.

4. List one systematic and one random error possible when carrying out an experiment to investigate the extension produced as the load on a spring is increased.

3.4 2 Hooke's law

By the end of this topic, you should be able to demonstrate and apply your knowledge and understanding of:

* Hooke's law and the force constant K of spring or wire; $F = kx$
* force–extension (or compression) graph; work done is area under graph
* elastic potential energy; $E = \frac{1}{2}Fx$; $E = \frac{1}{2}kx^2$

Hooke's law and the force constant

In the previous topic you saw how an elastic deformation usually produces a straight-line graph when the force of tension is plotted against extension. This is summarised by the following statement, known as **Hooke's law**.

KEY DEFINITION

Hooke's law states that the extension of an object is proportional to the force that causes it, provided that the elastic limit is not exceeded.

In equation form Hooke's law becomes:

$F = kx$

where F is the force causing extension x, and k is known as the **force constant** (or spring constant, particularly for springs).

The force constant has the unit newtons per metre – k tells us how much force is required per unit of extension. For example, a k of $6\,\mathrm{N\,mm^{-1}}$ means it takes 6 N to cause an extension of 1 mm. Note that the force constant can only be used when the material is undergoing elastic deformation. When deformation becomes plastic, the force per unit extension is no longer constant.

The work done to stretch a wire

We usually plot the cause of a change (independent variable) on the x-axis of a graph and the effect of that cause (dependent variable) on the y-axis. The graph of tension force against extension plotted in Topic 3.4.1 breaks that convention – but for good reason. If you plot extension on the x-axis, the gradient of the line is equal to the force constant. Figure 1 shows a straight-line graph of tension against extension for the elastic part of a deformation. The gradient $\frac{\Delta F}{\Delta x}$ is the force constant, k.

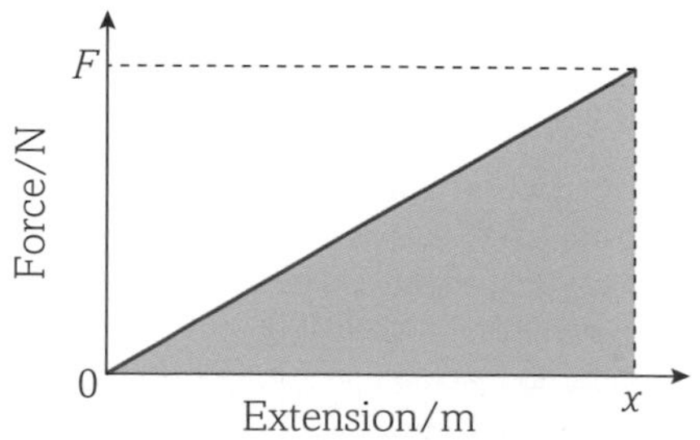

Figure 1 Straight-line graph of tension against extension. Different materials give graphs with different gradients.

The work required to stretch a material depends on the stretching force used and distance moved in the direction of the force, which is the extension produced. The extension produced by tension F is x. The work done to produce this extension is not simply Fx, however, because F is not constant during the extension. We need to consider the average force and not a single maximum value.

The graph in Figure 1 shows Hooke's law – the linear relationship between tension and the extension that occurs during elastic deformation. Provided the limit of proportionality is not exceeded, the tension is proportional to the extension at any point, so the average force needed to produce an extension x is half the final force F. Hence:

work done $= \frac{1}{2}Fx$

Because $F = kx$, the work done can also be expressed as:

work done $= \frac{1}{2}kx^2$

When the force is variable, another way of looking at this is to consider the work done for a small increase in extension, Δx. For a small increase in extension, the tension force is nearly constant. The small amount of work done, $\Delta W = F \times \Delta x$. This is equal to the shaded rectangular area under the force–extension graph in Figure 2.

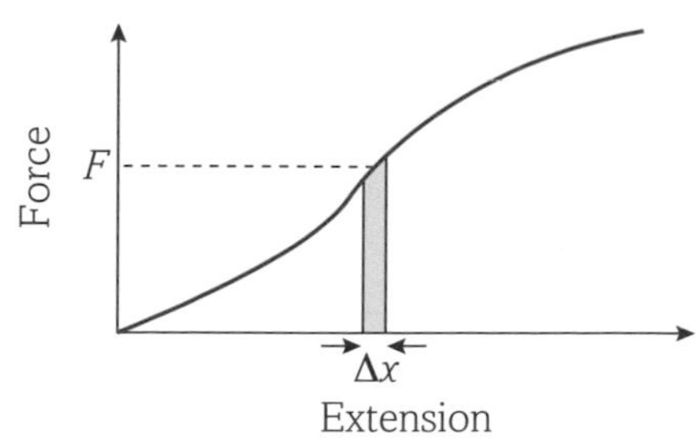

Figure 2 A force–extension graph.

The total work done to reach a final extension can be found by adding up all the areas under the graph.

For an elastic material that obeys Hooke's law this shape is triangular, so the area will be $\frac{1}{2}$ (base × height) or $\frac{1}{2}Fx$ (Figure 1).

Stored elastic potential energy

Work is done on a wire or spring to stretch, or to compress it. The wire itself stores **elastic potential energy**. If the material has been deformed elastically (i.e. the elastic limit has not been reached) this stored energy can be completely released when the load is removed. This applies equally to the extension or compression of the material. Many children's toys incorporate springs that store energy on compression, and that drive the toy forward when released.

In the case of elastic deformations, the elastic potential energy, E, equals the work done, giving:

$$E = \frac{1}{2}Fx = \frac{1}{2}kx^2$$

In general the energy stored in an extension is the area below the line in the force–extension graph.

Energy stored in a plastic deformation

For an elastic material up to its elastic limit, the force–extension graph is the same for loading and unloading. The graph shown in Figure 3 could be produced by stretching a copper wire beyond its elastic limit.

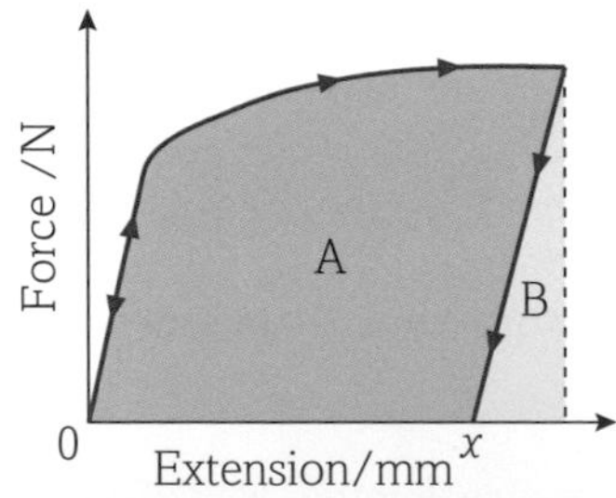

Figure 3 Stretching a wire beyond its elastic limit.

The work done in stretching the wire is given by the area under the graph, area A + area B. If the tension is then reduced to zero, the wire behaves plastically, contracting to a permanent extension x. As the tension is reduced, energy equivalent to area B is released from the wire. The net result of the wire having work A + work B done on it, but only releasing energy B, is that the wire becomes hot to the touch. Hence, if a material deforms plastically the energy recovered on unloading is less than 100% of the work done on the wire.

WORKED EXAMPLE

A car's shock absorbers make a ride more comfortable by using a spring that absorbs energy when the car goes over a bump. One of these springs, fixed next to a wheel, needs to store 250 J of energy when compressed a distance of 10 cm.

(a) What force constant value is required for the spring?

(b) How much energy would be stored if the spring were compressed by 20 cm?

Answers

(a) Because $E = \frac{1}{2}kx^2$

$250 = \frac{1}{2} \times k \times 0.10^2$

$k = \frac{(2 \times 250)}{0.10^2} = 50\,000\ \text{N m}^{-1}$

(b) The extent of compression has been doubled. Provided the spring is still elastic, the energy is proportional to the square of the extension. Twice the extension will therefore cause four times the energy.

Energy stored = 250 J × 4 = 1000 J

Questions

1 A spring that obeys Hooke's law has a force constant of 160 N m^{-1}.

(a) (i) Calculate the extension of the spring, in mm, produced by a force of 4.0 N.

(ii) Calculate the strain energy stored in the spring.

(b) A second identical spring is connected in parallel with the first as shown in Figure 4.

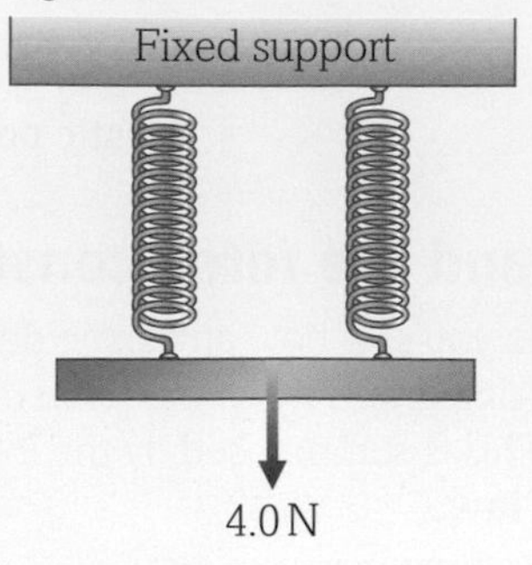

Figure 4

(i) Calculate the extension of each spring, in mm, produced by a force of 4.0 N.

(ii) Calculate the total strain energy stored in the springs.

(c) The second spring is now connected to the end of the first spring to make a single spring of twice the length.

(i) Calculate the extension of each spring, in mm, produced by a force of 4.0 N.

(ii) Calculate the total strain energy stored in the springs.

2 In Figure 5 an open-wound spring is compressed a distance y of 40 mm. When released, the spring rises to a maximum height h of 800 mm. The mass of the spring is 2.0 g.

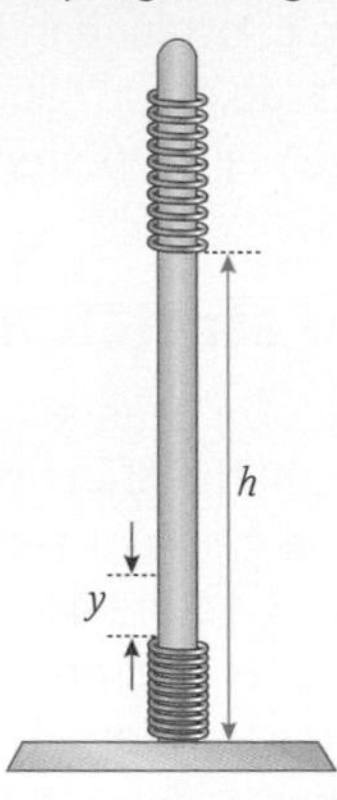

Figure 5

(a) By considering the total change of strain energy to gravitational potential energy, calculate the force constant of the spring.

(b) Plot a graph showing how the maximum height risen by the spring, h, varies with the distance that the spring is compressed, y, up to y = 40 mm.

3 Design an experiment to investigate how the thickness of a rubber band affects the force–extension characteristics. Describe how you will manipulate the variables, what measurements you will take, how you will minimise the effect of errors and how the data will be used in order to reach a conclusion comparing the elasticity of the different band thicknesses.

3.4

3 The Young modulus

By the end of this topic, you should be able to demonstrate and apply your knowledge and understanding of:

* the terms stress σ, strain ε and ultimate tensile strength
* Young modulus = $\frac{\text{tensile stress}}{\text{tensile strain}}$, $E = \frac{\sigma}{\varepsilon}$
* techniques and procedures to determine the Young modulus used for a metal

Introduction

When using Hooke's law, we deal with the force applied to a material and its extension. If the material is elastic, it will return to its original shape when the force is removed. This will happen for any spring, metal wire or rubber material that obeys Hooke's law.

However, the force and extension that we measure for a particular spring are unique for that type of spring. If we change the thickness of the spring or its length then the values for the force needed and the extension observed will also change. This makes it difficult for engineers, metallurgists and other scientists to describe accurately how a particular material will behave when exposed to deformation.

KEY DEFINITIONS

Stress is force per unit cross-sectional area. It has units of N m^{-2} or Pa. It is given the symbol σ.

Strain is extension per unit length. It has no units and is therefore dimensionless. It is given the symbol ε.

LEARNING TIP

Some quantities that we measure in physics will change as the dimensions of the material change. For example, quantities such as the spring constant of a metal wire or the electrical resistance of a metal wire will be different if we change the length or cross-sectional area of the material. We call these quantities 'extrinsic' variables because their value depends on external factors such as length and cross-sectional area.

However, some quantities in physics do not change, even when we change the dimensions of the material. These are called 'intrinsic' variables because they are determined by the material itself and not by its length or cross-sectional area. Examples of these quantities include the Young modulus and resistivity.

Another common example is density. If you increase the volume of a given material, then its mass will increase in direct proportion – double the volume, double the mass. Changing the volume of a material, however, does not change its density, because density is defined as $\frac{\text{mass}}{\text{volume}}$. Density is an intrinsic variable and mass is an extrinsic variable.

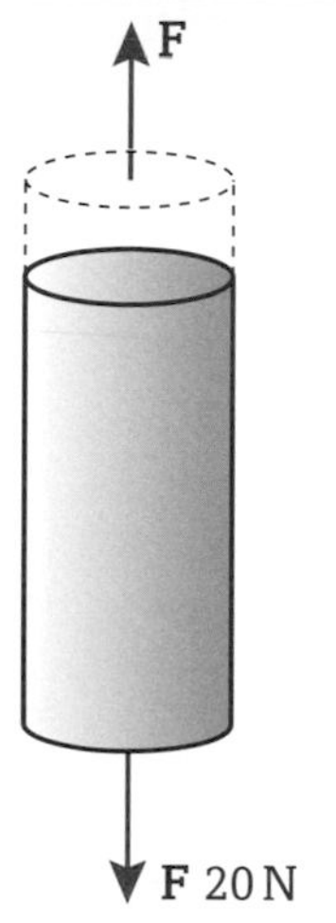

Figure 1 The force applied here is a tension force of 20 N. The cross-sectional area of the wire is calculated using the formula πr^2, because the wire has a circular cross-section.

A more useful way for scientists to compare the behaviour of specific materials is to compare two other quantities called **stress** and **strain**.

Stress and strain

Stress is defined as force per unit cross-sectional area. Stress is therefore expressed in the S.I., unit newtons per square metre (N m^{-2}), also known as the pascal (Pa).

Strain is defined as extension per unit length. This means the change in length of the material caused by the applied force, divided by the original length of the wire. Strain has a value but no units – the units of length cancel when we divide. A strain of 100% (or 1) means that the material has doubled in length.

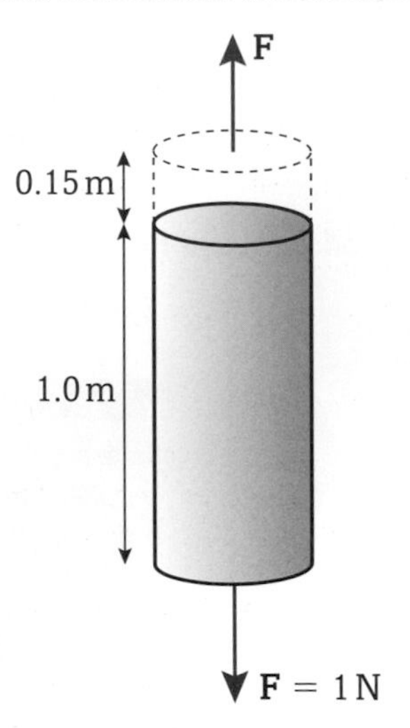

Figure 2 This material shows a strain of 0.15 when a force of 1 N is applied – it has changed from a length of 1.0 m to a length of 1.15 m.

The Young modulus

When a stress is applied to a material it causes strain, and the material will change its length. How much strain is caused depends on the stiffness of the material. A stiff material, such as cast iron, will not alter its shape much when a stress is applied to it, whereas a material such as rubber will undergo a large extension or strain. Materials that are stiff have a large Young modulus.

$$\text{Young modulus} = \frac{\text{tensile stress}}{\text{tensile strain}} = \frac{\sigma}{\varepsilon}$$

Calculating the Young modulus for a material

We can calculate a Young modulus using one of two methods:

- obtain the value from the gradient of a stress–strain graph based on data collected
- calculate the value directly from the equation, using experimental values.

By plotting a stress against strain graph you can obtain a value for the Young modulus of a material, which is a fixed value.

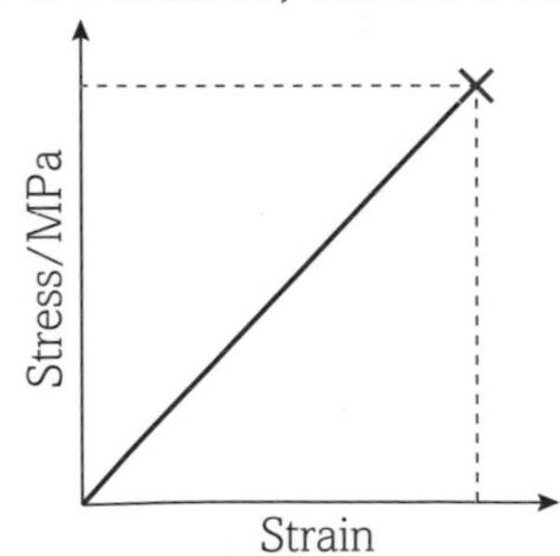

Figure 3 Regardless of the size, length or shape of a spring, the material it is made from will always give a line of constant gradient when behaving elastically. The gradient of this line is the Young modulus for that material.

We know that Young modulus $= \frac{\text{stress}}{\text{strain}}$

We also know that stress $= \frac{\text{force}}{\text{cross-sectional area}}$

and that strain $= \frac{\text{extension}}{\text{original length}}$

This gives us an expression for the Young modulus:

$$\text{Young modulus} = \frac{(\text{force} \times \text{original length})}{(\text{extension} \times \text{cross-sectional area})}$$

So, if the original length of a cylindrical wire is 1.2 m, the total extension is 35 cm, its diameter is 0.4 mm and the total force applied is 25 N:

$$\text{Young modulus} = \frac{(25 \times 1.2)}{(0.35 \times \pi \times 0.0002^2)}$$
$$= 6.82 \times 10^8\ \text{Nm}^{-2}$$

The value of the Young modulus will give you a clue as to what the material is.

Notice how the units of the Young modulus are the same as for stress.

INVESTIGATION

Calculating the Young modulus from a stress–strain graph

The apparatus shown in Figure 4 is known as Searle's apparatus. The diameter of the wire is measured using a micrometer, and the original length of the wire is measured when under tension before any other loads are added.

The force of tension is then increased by adding weights to the right-hand side of the apparatus. Each time a weight is added, the extension is measured. From this data, a graph of stress against strain is plotted, the gradient of which gives the Young modulus of the material of the test wire.

Since the extension produced is very small, a vernier scale is used for its measurement. The reading of the micrometer is noted when the spirit level is levelled and before an extra load is added. After the load is added, the circular scale of the micrometer is rotated until the spirit level is again levelled.

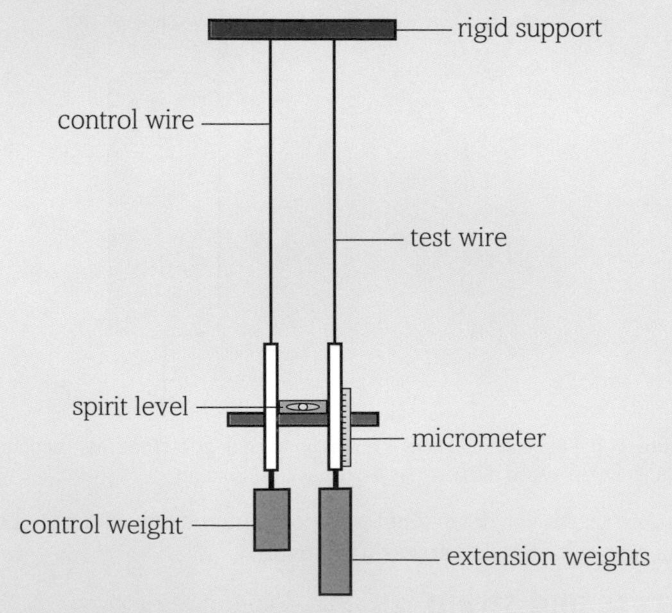

Figure 4 Searle's apparatus can be used to find the Young modulus.

Material	Young modulus/N m^{-2}
diamond	1.2×10^{12}
iron	2.1×10^{11}
copper	1.2×10^{11}
aluminium	7.1×10^{10}
lead	1.8×10^{10}
rubber	2.0×10^{7}

Table 1 The Young modulus for some common materials.

LEARNING TIP

Remember the word 'flea' when you think of the equation Young modulus $= \frac{Fl}{eA}$

Take care when using the Young modulus equation. It is usually safer to calculate stress and strain separately and then calculate $\frac{\text{stress}}{\text{strain}}$

The numerical values are usually large because two large numbers appear on the top of the equation and two small numbers are on the bottom. The very high value for diamond indicates that only a small change in shape occurs for the application of a very high stress. For rubber a large change in shape can be achieved much more easily.

Questions

1. Explain why a Young modulus has the same unit as stress.

2. A wire has a length of 740 mm and a diameter of 12 mm. Calculate:
   (a) the stress when it has a load of 67 N applied
   (b) the strain when it extends by 744 mm
   (c) the Young modulus of the wire, assuming that the stress in (a) caused the strain in (b).

3. Find the Young modulus of the two materials in Figures 5 and 6 and describe their properties.

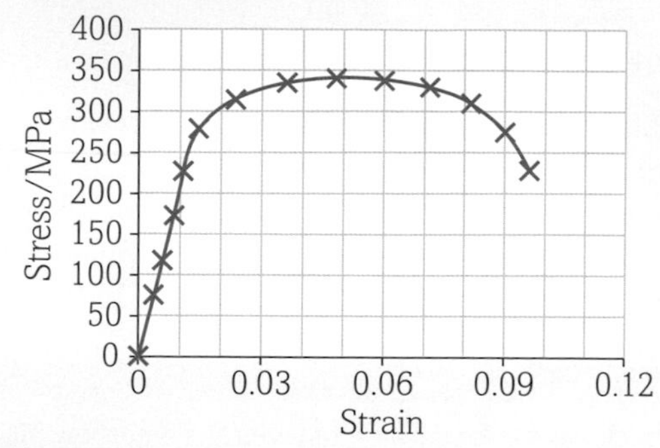

Figure 5

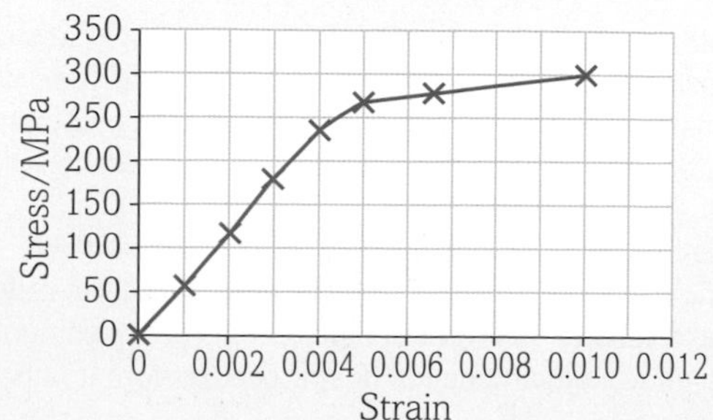

Figure 6

4. A cylindrical wire has an original length of 45 m and a diameter of 80 cm. A force of 1.8×10^5 N is required to increase its length by 1%. What is the Young modulus of the wire?

5. Copy and complete the following sentences, choosing the most suitable words from:

   stress strain tension extension
   Young modulus force constant

   In an experiment to measure the Young modulus of copper, several reels of different gauge (diameter) copper wires are available.
   (a) Whatever the length of wire that is used, a given load attached to the end of the wire will always produce the same .... and ....
   (b) For a given length of wire, the .... will be the same if the strain is the same.
   (c) Whatever the dimensions of the wire, the gradient of the stress–strain graph will always be ....

6. Figure 7 shows the stress–strain graph for a sample of copper wire. The stress in the wire was increased slowly from zero to 2.8×10^8 Pa and then reduced slowly back to zero.
   (a) Use the stress–strain graph to find the Young modulus of copper.
   (b) Why did the strain not return to zero?
   (c) Explain how you could use the graph to estimate the amount of energy that was dissipated in the experiment. Assume that you have data giving the length of the wire and its cross-sectional area.
   (d) Another length of copper wire, which has twice the diameter of the wire used in Figure 7, is stretched until its strain is 2.0×10^{-3}.
      (i) How would the stress–strain graph for this wire compare with Figure 7?
      (ii) How much higher is the tension in the second wire than in the first wire under the same strain?

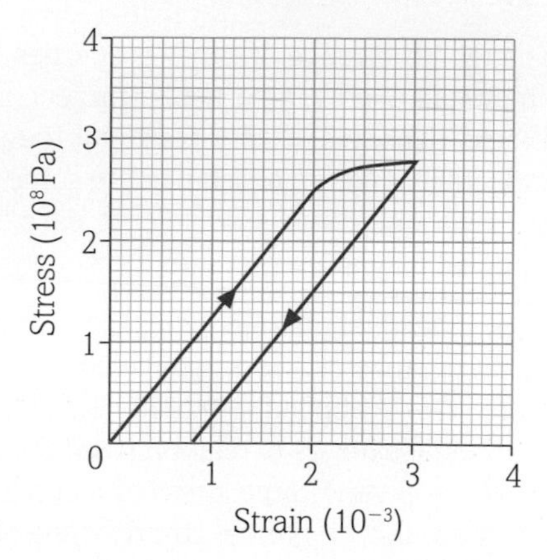

Figure 7

Categorisation of materials

By the end of this topic, you should be able to demonstrate and apply your knowledge and understanding of:

* stress–strain graphs for typical ductile, brittle and polymeric materials

Material variety

We have a vast array of natural and man-made materials at our disposal. Gone are the days when engineers and designers faced a simple choice between wood or metal when planning a construction project. Plastics, first manufactured during the 1930s, are now widely used, while new materials are being designed for highly specific applications.

Alloys of aluminium were traditionally used in the construction of aeroplane wings, for example, but wings are now being manufactured using so-called composite materials that are stronger, lighter and more flexible than any aluminium alloy. In a composite material, several different materials, each with its own advantage, are bonded together.

Figure 1 An aluminium alloy is used to make aircraft wings because it has a low density, is weather resistant, is **malleable**, ductile and relatively soft – it is strong for its density.

Apart from the strength, a range of other properties may affect the choice of materials used in a particular project. Ductility, brittleness, stiffness, density, elasticity, plasticity, toughness, fatigue resistance, conductivity and fire resistance are some of the many properties that may shape choices, while cost, ease of shaping and customer appeal may also enter into the equation.

The properties of individual material types can be illustrated clearly by sketching graphs of stress against strain. Stress is related to the force that is applied to a material, whereas strain is how much the material extends in relation to its original length. For example, if you apply the same stress to a concrete pillar as to an elastic band, you will get a greater strain for the elastic band – that is, the elastic band will extend more for the same force.

Ductile materials

Ductile materials, such as copper, can be drawn out into a wire. Only materials with an extensive plastic region can have their shape altered in this way. Try wrapping the ends of a thin copper wire around two pencils. If you pull steadily on the pencils to stretch the wire, you will feel the plastic flow of the copper. As you pull, the copper wire increases in length, straightens and its cross-sectional area decreases before it eventually breaks (Figure 2).

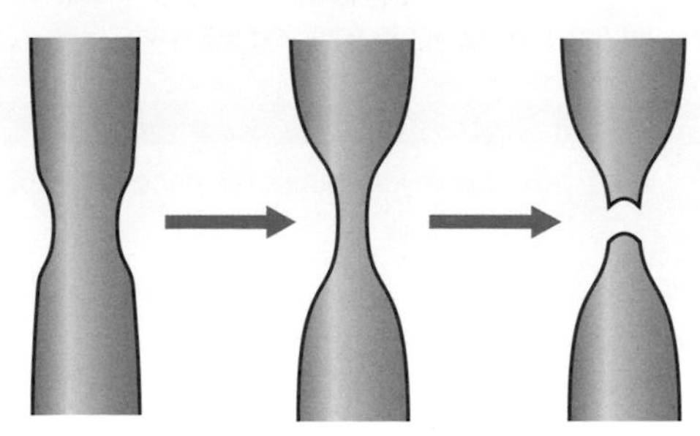

Figure 2 A ductile material such as copper will increase in length and reduce its cross-sectional area – this is known as necking.

The shape of the stress–strain graph for a ductile material is shown in Figure 3. Note that a ductile material has an extensive plastic region within which the material will continue to stretch, even if the stress is reduced. Figure 3 shows the maximum stress that can be applied to the material before it will break. This is known as the ultimate tensile strength of a material.

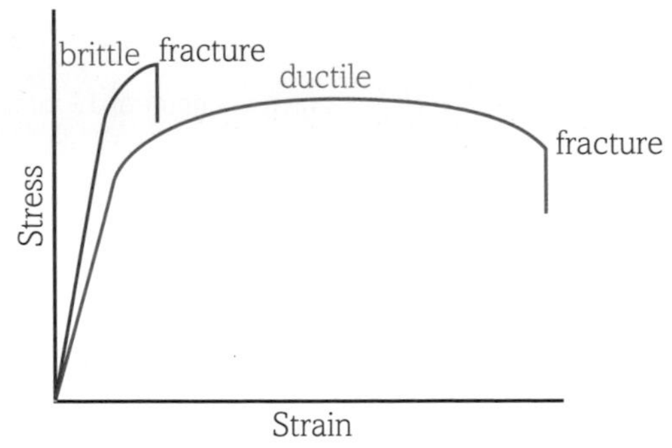

Figure 3 Stress–strain graphs for a ductile material and for a brittle material.

Most metals are ductile, and can be pulled into wires or beaten into sheets. In addition to copper, wires made using steel and aluminium are widely available, while a range of alloys are used in electrical circuit wiring. Silver, gold and platinum are used in both wire and sheet form by manufacturers of jewellery.

KEY DEFINITIONS

A **ductile** material can be drawn into wires and will show plastic deformation under tensile stress before breaking.
A **malleable** material can be hammered or beaten into flat sheets and will show extensive plastic deformation when subjected to compressive forces.
A **brittle** material will break with little or no plastic deformation.
A **hard** material will resist plastic deformation by surface indentation or scratching.
Stiffness is the ability of a material to resist a tensile force.
A **polymeric** material is made of long chains of molecules called polymers.
The **ultimate tensile strength** of a material is the maximum stress it can withstand while being pulled or stretched, before it fails or breaks.

Brittle materials

Figure 3 also shows a stress–strain graph typical of a **brittle** material. Note the contrast between this and the graph for a ductile material. Brittle materials distort very little, but will break or fracture if subjected to a sufficiently large stress. The graph ceases abruptly, and the area under the graph is small, indicating that very little elastic potential energy has been stored by the material. Biscuits and concrete (Figure 4) are both brittle, but illustrate the difference in the amount of stress required to break different materials exhibiting this property. Brittle materials may also be stiff.

Figure 4 Both biscuits and concrete are brittle, but a higher stress is needed to break the concrete pillar.

Polymeric materials

In most stress–strain graphs, the degree of strain is a few percent. For a ductile material it may be as high as 50%, but for certain polymeric materials it may reach 300%. Rubber is a common example – Figure 5 shows a stress–strain graph for a rubber band. A 6 cm rubber band may be stretched to 24 cm quite easily, but is difficult to stretch any further, even if it does not snap. This is due to the nature of rubber molecules, which are arranged in a mass of squashed long chains. When a stress is applied to the rubber, these chains straighten, resulting in a large strain.

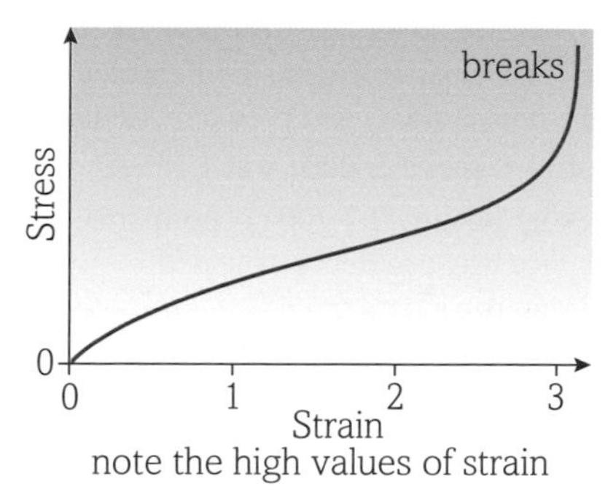

Figure 5 Stress–strain graph for a rubber band.

The flexible properties of rubber have been harnessed in the manufacture of vehicle tyres. Natural rubber becomes weak and sticky when warm, however, so it is subjected to a process called vulcanisation. This involves the addition of impurities, such as sulfur, to bind the chains of rubber molecules together making them harder and stronger – i.e. less easy to stretch.

Questions

1. Here is a list of some mechanical properties of materials – brittle, ductile, elastic, plastic, strong and tough. Choose as many of these properties as are suitable to describe each of the following common materials: bone, cast iron, ceramics, concrete, copper, glass, lead, polyethylene, rubber and wood.

2. Look at the graphs below. Explain how the materials in Figures 6–8 compare based on the terms covered in this spread.

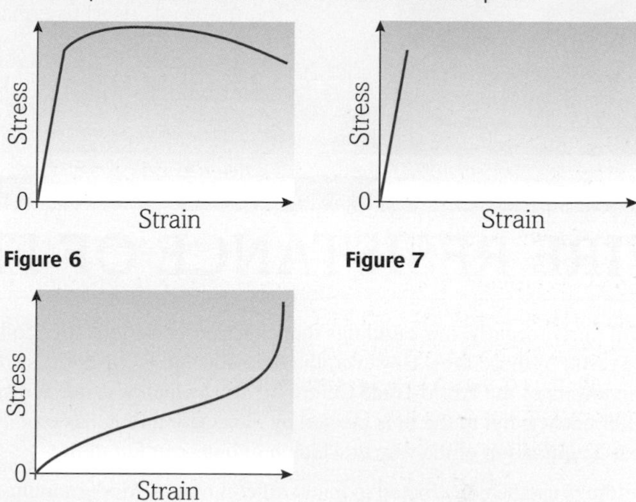

Figure 6

Figure 7

Figure 8

3. Match the statements on the left to those on the right.

This material has a strain of 300%		This material is brittle
After the yield point, the material will show a large strain for a small stress and then break		This material is malleable
This material is stiff and does not extend before it fractures		The extension of the wire is 3 times its original length
I would use this material to make roof tiles		This material is ductile

4. When a rubber band is cooled down it becomes much stiffer but it maintains the same ultimate tensile stress. Imagine that the temperature is such that the strain is half of that shown in Figure 5 for the same stress.
   (a) Sketch the shape of the new graph.
   (b) Suggest an explanation on a molecular scale for such a significant variation of strain with temperature.

5. Describe the properties of the following foods in terms of their material properties:
   (a) chewing gum
   (b) bread dough
   (c) digestive biscuits
   (d) boiled sweets
   (e) fudge.

BUILDING FOR THE FUTURE

Modern buildings are increasingly designed and built with the surrounding environment in mind. Houses and skyscrapers have to be able to resist all sorts of extreme weather and seismic activity, so the materials used to construct them need to be fit for purpose.

In this activity, we address the effect that different materials have on the strength of buildings and their ability to withstand impact and fire, using The Twin Towers disaster of 11 September 2001 as an example.

FIRE RESISTANCE OF FRAMED BUILDINGS

Until fairly recently, few buildings that experienced serious fires collapsed as a result of those fires. However, the dramatic and tragic collapse of the twin towers of the World Trade Centre in 2001, which was not due to the initial impacts but to the fires created by the aviation fuel, has prompted close examination of the way in which buildings can fail in fires.

Buildings can be constructed in many different ways. Modern multi-storey construction tends to use a structural frame, effectively the building's skeleton, to carry the floors. This allows large numbers of storeys in a building and leaves almost complete flexibility to its owners to partition each storey as they wish. Structural frames may be constructed from reinforced concrete or steel. In Britain and North America the great majority of such frames use structural steelwork systems, because these are extremely efficient to build. In composite construction, the most common of the current systems, the concrete floor slabs are continuously connected to the beams of the structural frame, creating a very robust structure in which local loads are shared among a range of members as well as those that directly support them.

All common building materials lose strength when heated to high temperatures. For steel, the change can be seen in the stress–strain curves at temperatures as low as 300 °C. Different types of local failure cause different degrees of risk to a building. Failure of a column by buckling (see Figure 1) is potentially disastrous, because the whole region of the building above it will move downwards with the top of the failing column, and even if the weight of this part of the building could be redistributed to adjacent columns then they may have to be even stronger to kill the huge amount of kinetic energy that they may accumulate.

Figure 1 Consequences of column failure.

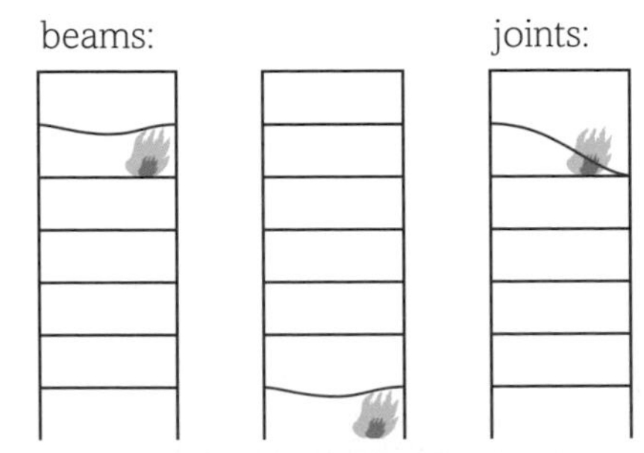

Figure 2 Consequences of failure of beams and connections.

The terrorist attacks on the twin towers of the World Trade centre in New York, followed by their collapse and the deaths of 2800 people, have already had tremendous repercussions in political and international terms, and the aftershocks will undoubtedly continue for some time. On the technical front, one important question has been why the buildings collapsed after they had apparently withstood impacts by Boeing 767 aircraft but after the impact floors had been engulfed in fire.

The main events of 11 September 2001 were:

08.46 WTC1 struck by Boeing 767-200 at approximately 470 mph, centrally on North face between floors 94 and 98.

09.03 WTC2 struck by Boeing 767-200 at approximately 590 mph, off-centre on South face between floors 78 and 84.

09.59 WTC2 collapses. 10.28 WTC1 collapses. 17.20 WTC7 collapses.

Source

Burgess, I. (2002) Fire resistance of framed buildings. *Physics Education*, vol. 37, no. 5. pp. 390–399.

DID YOU KNOW?

Materials that are strong under compression are not necessarily strong under forces of tension. Diamond is incredibly strong under forces of compression whereas carbon nanotubes are incredibly strong under forces of tension. The strongest material in the world is graphene – a one atom thick, ultra-strong material. Pencil lead is made up of many millions of layers of graphene.

Where else will I encounter these themes?

1.1 2.1 2.2 3.1 3.2 3.3

Let us start by considering the nature of the writing in the article. The article above was taken from the publication *Physics Education*, which is produced by the Institute of Physics (IOP).

1. Consider the article and comment on the type of writing that is being used. For example, does the article try to explain, persuade or describe? Is there bias in the article? Are the findings open to interpretation by others?

Cross-reference material with other trustworthy sources to check the validity of what is being stated here.

We will now look at the physics that is in the article. Do not worry if the physics content or the mathematics is challenging at this stage. You can always return to the article later in your course, once some of the related topics have been studied in more depth. Use the timeline at the bottom of the page to help you put this work in context with what you have already learned and what is ahead in your course.

2. What are stress and strain?
3. Figures 3 and 4 show stress–strain graphs for steel and concrete at different temperatures.
4. At what temperature is concrete strongest? Explain your answer.
5. Use Figure 1 to show that the Young modulus of steel falls below one quarter of its maximum value when the temperature exceeds 700 °C.
6. How does the Young modulus vary for:
   a. steel
   b. concrete?
7. Concrete is described as a brittle material.
   a. How is this evident from the graph?
   b. How does this affect the applications for how concrete is used?
8. Which is most problematic – column failure or beam failure? Explain your answer.
9. Why might WTC2 have collapsed first even though it was the second building to receive an impact?
10. Which had most impact on the collapse of WTC1 and WTC2 – fire or impact?

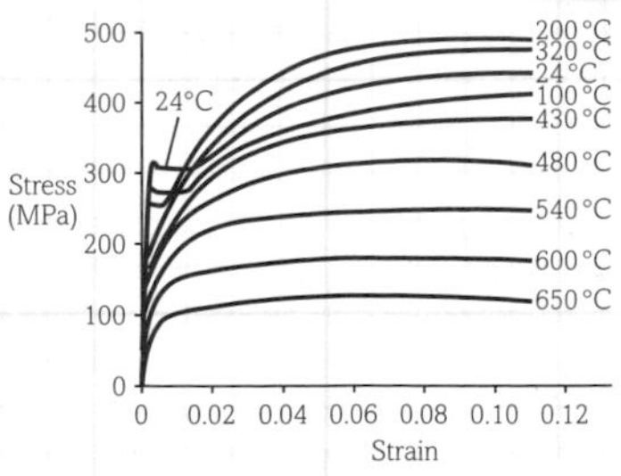

Figure 3 Stress–strain graph for steel.

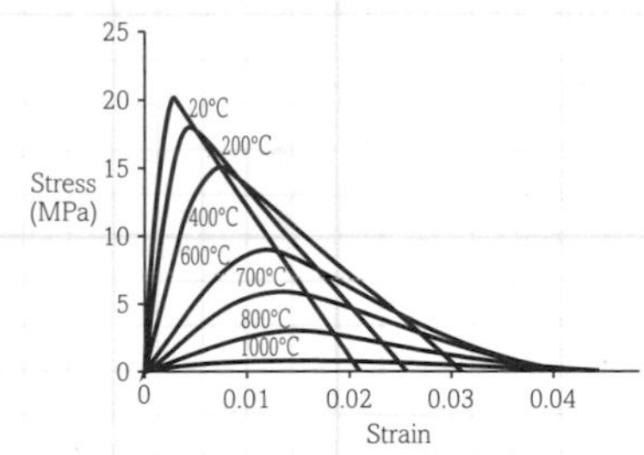

Figure 4 Stress–strain graph for concrete.

Think back to the work you have done relating to Young's modulus and the properties of materials. How would this help to interpret the material provided here.

Activity

Few buildings that experience serious fires collapse as a result of those fires. However, much attention has been centred on the collapse of the twin towers and how the aviation fuel caused them to collapse. Find out:

- how the twin towers were constructed and why they failed;
- why One Meridian Plaza, a building in Philadelphia, did not collapse after a fire in 1991;
- why WTC 7 collapsed even though it had not been hit by a plane.

Choose one of the following activities.

1. Use your findings to write summary notes, for your own reference, of what structural engineers have learned from the collapse of the twin towers and how this may allow engineers to construct safer buildings in the future.
2. Write an inquest report explaining the reasons for the structural failure of the twin towers during the terrorist attacks on 11 September 2001. Your report should refer to the structural designs that could have been implemented so that lives could have been saved and the collapse of the building could have been

Practice questions

1. The graph below (Figure 1) is a force–extension graph for a wire under tension. Which of the following statements is true, based on the graph? [1]

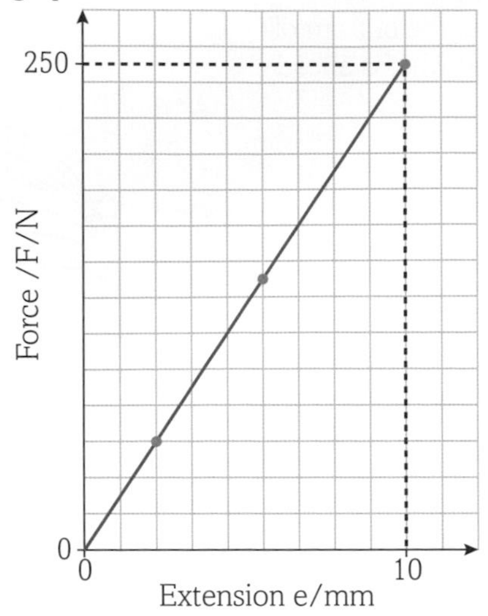

Figure 1

A The gradient of the graph provides us with a value for the elastic energy stored in the wire.

B The wire has passed its elastic limit and is displaying plastic behaviour.

C The spring constant could be expressed as 25 N m^{-3}.

D The spring constant could be expressed as 25 N mm^{-1}.

2. Which of the following statements is/are true? [1]
   (i) The spring constant is constant for a particular metal (e.g. any steel spring).
   (ii) The Young modulus is constant for a particular metal (e.g. any steel wire).
   (iii) The Young modulus is the ratio of stress to strain for any wire.

A (i), (ii) and (iii)

B only (i) and (ii)

C only (ii) and (iii)

D only (iii)

3. A material is hard, has low strain and breaks soon after its elastic limit. What is this material likely to be? [1]

A ductile

B brittle

C plastic

D malleable

4. A spring is stretched so that its extension is 35 cm when a force of 70 N is applied. Which of the following will be true for the spring? [1]

A The spring constant will be 2 N m^{-1}.

B The spring constant will be 20 N m^{-1}.

C The elastic energy stored in the spring will be 12.25 J.

D The elastic energy stored in the spring will be 24.5 J.

[Total: 4]

5. The extension of a metal wire is x when the tension in the wire is F. The table shows the results from an experiment, including the stress and the strain values.

F/N	x/10^{-3} m	stress/10^{7} Pa	strain/10^{-3}
1.9	0.4	1.73	0.20
4.0	0.8	3.50	0.40
5.9	1.2	5.21	0.60
8.0		7.00	0.80
9.0	1.8	7.95	0.90

(a) Complete the table by determining the extension when the force of tension is 8.0 N. [1]

(b) Determine the elastic potential energy stored in the wire when the force of tension is 4.0 N. [2]

(c) Use the table to plot the missing point on the graph of stress against strain as shown in Figure 2. [1]

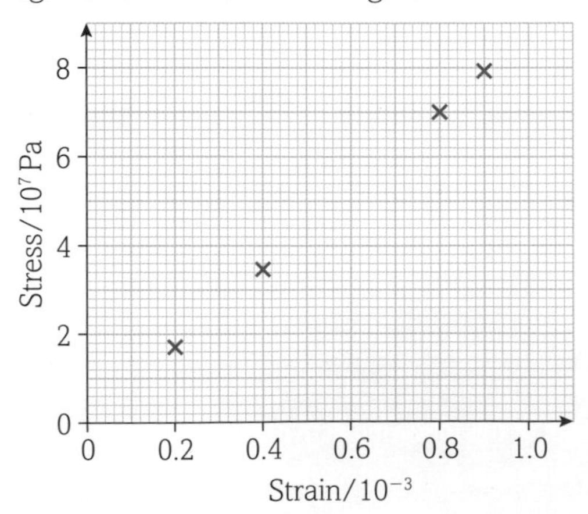

Figure 2

(d) Use the graph to determine a value for the Young modulus of the metal. [3]

[Total: 7]

6. A micrometer screw gauge was used to determine the diameter of a cylindrical wire. The micrometer had a zero error and the value obtained from the reading was greater than the true value for the diameter of the wire. Explain how this will affect the value obtained for the Young modulus compared with its true value. [3]

[Total: 3]

7. Define the following terms:
   (a) ductile [1]
   (b) stress [1]
   (c) strain [1]
   (d) polymeric [1]
   (e) limit of proportionality [1]

[Total: 5]

8. Describe the properties of materials X and Y as shown in Figure 3. [4]

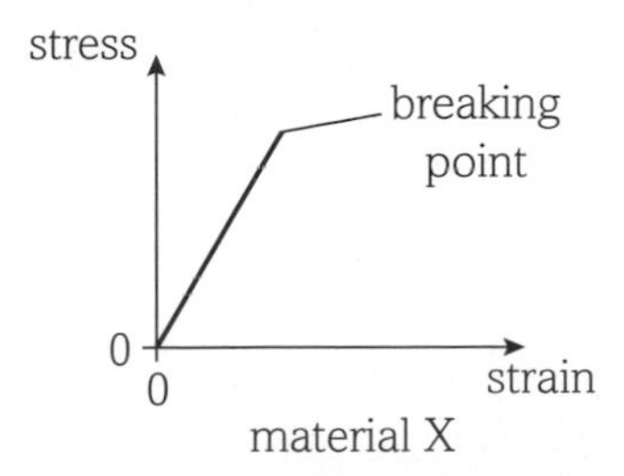

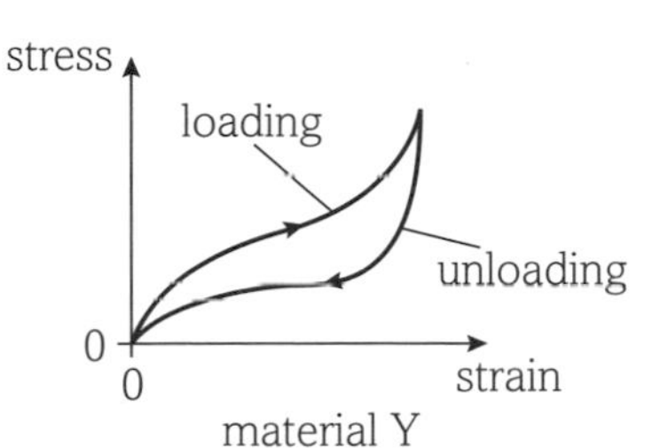

Figure 3

[Total: 4]

9. Explain how you would conduct an investigation to determine the Young modulus of a copper wire of cylindrical cross-section. In your explanation, comment on:
   (a) the apparatus you would need to determine the length and diameter of the wire [3]
   (b) how you would determine tension and extension values [2]
   (c) how you would use an equation or graph to find a value for the Young Modulus [3]
   (d) any systematic errors that may occur and how they could be removed [2]
   (e) a typical value you would expect for the Young modulus of the wire. [1]

[Total: 11]

MODULE 3 Forces and motion

CHAPTER 3.5

NEWTON'S LAWS OF MOTION

Introduction

It is almost impossible to comprehend the enormity of Isaac Newton's contribution to our understanding of physics. Born on Christmas day 1642, Newton contributed more to our understanding of motion, optics, gravitation, thermal physics and mathematics than virtually any other human being, before or since. The NASA moon landings of 1969 were planned with an understanding of motion and gravity based almost entirely on Newton's findings. The three laws of motion that we study today, and that form the bedrock of our understanding of mechanics, still bear his name. He is a true giant in the area of physics, his extraordinary genius equalled only by Albert Einstein. In this chapter we will study the three laws of motion that are named after him. We will, again, visit the vector quantities force, momentum and acceleration and link them to the scalar quantities of mass and kinetic energy. We will discuss the nature of elastic and inelastic collisions and we will discover and examine the term impulse in relation to his laws of motion.

All the maths you need

To unlock the puzzles of this chapter you need the following maths:

- Units of measurement
- Using addition and subtraction to find the magnitude of a resultant force
- Using multiplication to find the resultant force from $F = ma$
- Using multiplication to find the momentum of a moving body from $p = mv$
- Using the square function to calculate and compare kinetic values for elastic and inelastic collisions
- Use of positive and negative values when dealing with vector addition, such as the change in momentum after a collision
- Rearranging formulae to determine momentum, time or impulse
- Finding the area of a graph to determine impulse
- Calculating the gradient of a graph to determine the momentum of a body

What have I studied before?

- The forces of friction and drag act to slow down the motion of a moving body
- If forces are balanced, a body will remain at rest or move at a constant velocity
- A resultant force will cause a body to accelerate
- For every action, there is an equal and opposite reaction
- The greater the momentum of a vehicle, the more damage it can cause
- Momentum is always conserved in any collision or explosion
- Force is the rate of change of momentum

What will I study later?

- The nature of the force of gravitation and the laws of gravitation that explain how every body in the Universe attracts every other (AL)
- The nature of electrostatic forces, including how charged particles can attract and repel each other based on the nature of their charges and the size of their separation (AL)
- Force–distance graphs for a body undergoing motion in a gravitational or electric field (AL)
- The forces experienced by a current-carrying wire in a magnetic field (AL)
- The force on a charged particle in an electric and magnetic field (AL)
- The nature of centripetal forces on bodies that are involved in circular motion (AL)
- Forces involved on bodies that are experiencing simple harmonic motion (AL)
- The nature of damping and the forces involved in free and forced oscillations (AL)
- The short range and nature of the strong nuclear force in the nuclei of atoms (AL)

What will I study in this chapter?

- Techniques and procedures used to investigate the motion and collisions of objects
- Newton's three laws of motion
- Linear momentum, $p = mv$ and the vector nature of momentum
- The relationship between net force and the rate of change of momentum as given by $F = \frac{\Delta p}{\Delta t}$
- The impulse of a force, as calculated by impulse $= F\Delta t$
- The area beneath a force–time graph in connection to the quantity impulse
- The principle of conservation of momentum
- The collisions and interaction of bodies in one-dimensional and two-dimensional motion
- Perfectly elastic collisions in terms of the conservation of kinetic energy
- The nature of inelastic collisions in terms of momentum and kinetic energy

3.5 1

Newton's three laws of motion

By the end of this topic, you should be able to demonstrate and apply your knowledge and understanding of:

- Newton's three laws of motion
- apply the second law using the formula $F = ma$
- how Newton's three laws apply to a range of problems involving motion

Introduction

Newton's three laws of motion can be used to explain how forces cause bodies to remain stationary, move at a constant speed or accelerate. The three laws are:

- Newton's first law – a body will remain at rest or continue to move at constant velocity until an external force acts on it.
- Newton's second law – the resultant force on an object is proportional to the rate of change of momentum of the object, and the momentum change takes place in the direction of the force.
- Newton's third law – if object A exerts a force on object B, then object B will exert an equal and opposite force on object A.

Newton's first law

A body will remain at rest or continue to move in a straight line at a constant velocity unless an external force acts on it.

Consider the car in Figure 1 – it is stationary on a road. The weight of the car acting downwards on the car is balanced by the normal contact force acting upwards. There is no net (or resultant) force acting in any direction on the car. If there was then the car would accelerate in the direction that the net force was acting.

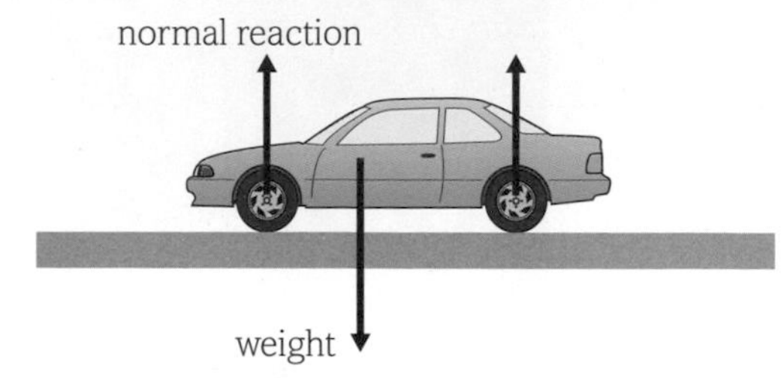

Figure 1 The forces acting vertically on this car are balanced – there are no horizontal forces acting on the car. The car remains at rest.

Contrary to common belief, a moving car does not need to experience a net (resultant) force, provided that it is moving at a constant speed. The drive and resistive forces on the car below are balanced, causing it to move at a constant speed in a straight line.

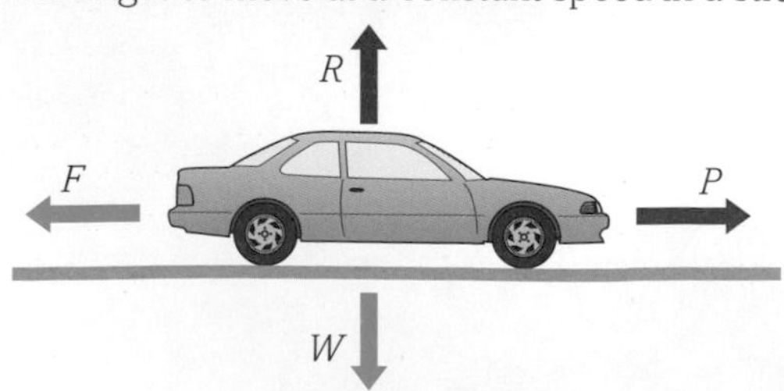

Figure 2 The forces acting on this car are balanced, so the car will move in a straight line at a constant speed – force R = force W; force F = force P.

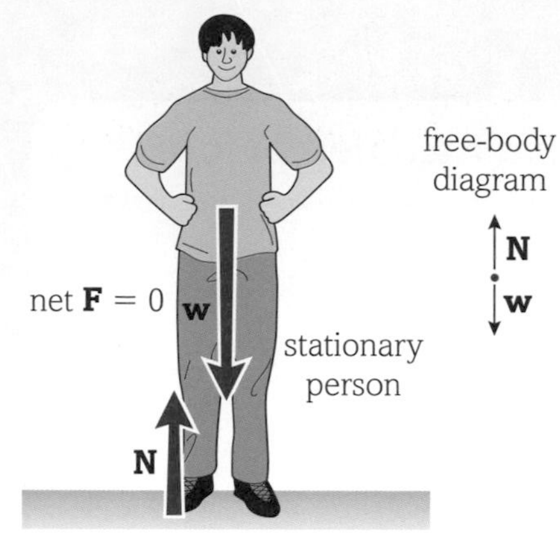

Figure 3 The forces that balance are the weight and normal contact force.

Newton's second law

The resultant force on an object is proportional to the rate of change of momentum of the object, and the momentum change takes place in the direction of the force.

When using Newton's first law, we deal with bodies on which the forces are balanced, resulting in either a stationary body or one that moves at a constant speed. When using the second law, we deal with a *resultant* force that causes an acceleration to occur.

Problems involving the application of Newton's second law usually require the use of the equation $F = ma$, where F is the resultant force, m is the mass of the body and a is the resulting acceleration. The equation $F = ma$ is a special case of the equation defining Newton's second law when mass is constant, as we shall see in Topic 3.5.3.

LEARNING TIP

A force *causes* an acceleration – not the other way round.

WORKED EXAMPLE 1

Consider a ball of mass 4.8 kg moving with an acceleration of $0.8\ \text{m s}^{-2}$. The velocity of the ball will increase by $0.8\ \text{m s}^{-1}$, every second.

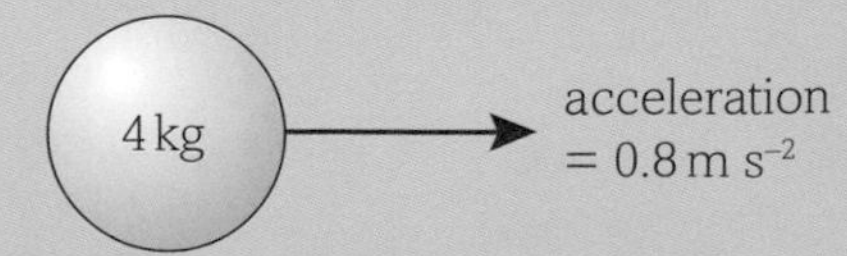

Figure 4 What force is causing this to happen?

Answer

$F = m \times a$

$= 4.8\ \text{kg} \times 0.8\ \text{m s}^{-2}$

$= 3.84\ \text{N}$

More complex second-law problems involve working out the resultant force of a number of forces acting on a body. The *size* and the *direction* of the resultant force have to be known before working out the acceleration of the object using $F = ma$.

WORKED EXAMPLE 2

Determine the size and direction of the acceleration of this vehicle.

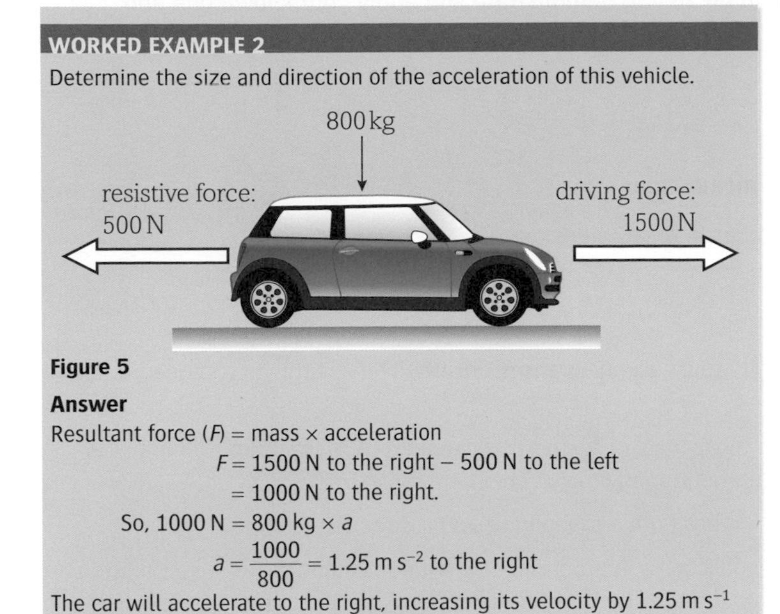

Figure 5

Answer

Resultant force (F) = mass × acceleration

F = 1500 N to the right – 500 N to the left

= 1000 N to the right.

So, 1000 N = 800 kg × a

$$a = \frac{1000}{800} = 1.25\ \text{m s}^{-2} \text{ to the right}$$

The car will accelerate to the right, increasing its velocity by 1.25 m s^{-1} each second until the forces change.

Newton's third law of motion

If object A exerts a force on object B, then object B will exert an equal and opposite force on object A.

Newton's third law often causes confusion. In order not to confuse it with the first law, remember the following requirements for the third law to be used:

- the forces involved must be of the *same type*
- the forces involved must be acting in *opposite directions* on *different objects.*

Consider the example shown in Figure 6. A swimmer is pushing against a wall to propel herself through a pool as she turns to swim in the opposite direction.

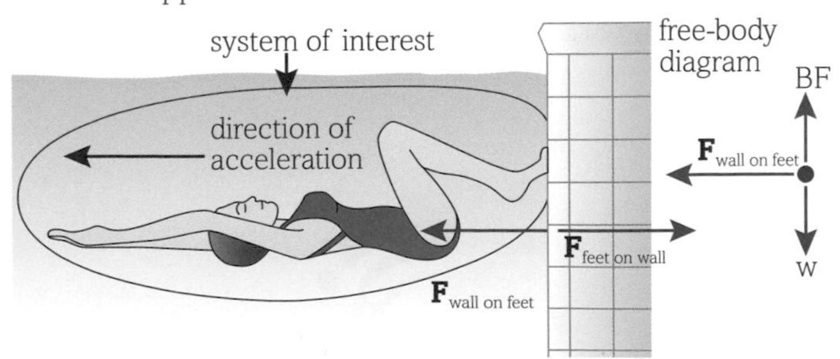

Figure 6 The forces acting on a swimmer.

This is an example of Newton's third law because the forces are of the same type (pushes in a horizontal direction) and they are acting on different bodies. The swimmer applies a force F using her feet on the wall, and the wall applies an equal and opposite force F onto her feet. The free-body diagram in Figure 6 shows that the weight and force of buoyancy are balanced, but a resultant force acts horizontally. Using Newton's second law we can predict that she will accelerate to the left.

How is the example given in Figure 7 *not* an example of Newton's third law?

Although the forces shown are equal and opposite in nature, they do not satisfy the conditions for the problem to be deemed an example of Newton's third law. Firstly, the two forces are both acting on the book – not on different objects. Secondly, the forces are not of the same type – one is gravitational and the other is the reaction force on the book from the desk. This is actually an example of Newton's first law, as opposed to the third law.

The forces on the book are balanced.
the table pushes upward on the book
AS Physics
gravity pulls downward on the book

Figure 7

Questions

1 Complete and fill in the missing values in Table 1.

Mass /kg	Upwards force/N	Downwards force/N	Horizontal force left/N	Horizontal force right/N	Motion
30	250	250	12	48	accelerates to the right at 1.2 m s^{-2}
24	480		200	200	accelerates upwards at 10 m s^{-2}
85	200	200	85	170	
360					remains stationary

Table 1

2 Apply Newton's three laws of motion to the problem depicted in Figure 8. You should:

(a) explain how each of the three laws is involved

(b) calculate any resultant force and acceleration

(c) draw a free-body diagram for system 1 and another for system 2.

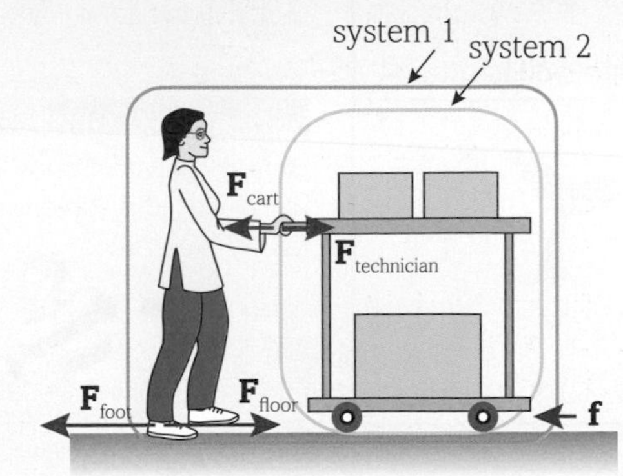

Figure 8 A science technician pushes a cart of demonstration equipment to a school laboratory. Her mass is 65.0 kg, the cart's mass is 12.0 kg, and the equipment mass is 7.0 kg. Calculate the acceleration when the technician exerts a backward force of 150 N on the floor. All the forces that oppose the motion – such as friction on the cart's wheels and air resistance – total 24.0 N.

3.5 (2) Momentum

By the end of this topic, you should be able to demonstrate and apply your knowledge and understanding of:

* linear momentum; $p = mv$
* vector nature of momentum
* the principle of conservation of momentum
* collisions and interactions of bodies in one dimension

electron

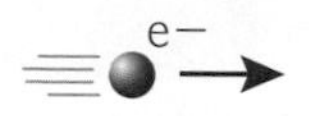

$m = 9.11 \times 10^{-31}$ kg
$v = 10^7\,\text{m s}^{-1}$
momentum $\simeq 10^{-23}\,\text{kg m s}^{-1}$

flea

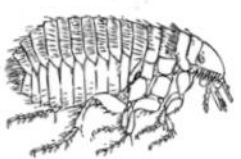

$m = 0.00001$ kg
$v = 10\,\text{m s}^{-1}$
momentum $\simeq 10^{-4}\,\text{kg m s}^{-1}$

rabbit

$m = 5$ kg
$v = 12\,\text{m s}^{-1}$
momentum $= 60\,\text{kg m s}^{-1}$

human

$m = 70$ kg
$v = 8\,\text{m s}^{-1}$
momentum $= 560\,\text{kg m s}^{-1}$

whale

$m = 1.5 \times 10^5$ kg
$v = 6\,\text{m s}^{-1}$
momentum $= 9 \times 10^5\,\text{kg m s}^{-1}$

comet

$m = 5 \times 10^{14}$ kg
$v = 100000\,\text{m s}^{-1}$
momentum $= 5 \times 10^{19}\,\text{kg m s}^{-1}$

Figure 1 Examples of momentum values.

Introduction

Any object that has mass and velocity must also have **momentum**. The key things you need to know about momentum are:

- Momentum = mass × velocity – it has units of kg m s^{-1}. An athlete of mass 86 kg running at a velocity of $11\,\text{m s}^{-1}$ will have a momentum of $86\,\text{kg} \times 11\,\text{m s}^{-1} = 946\,\text{kg m s}^{-1}$.
- Momentum is a vector quantity – it has both a magnitude and a direction.
- Momentum is conserved in any collision or interaction – the total momentum before a collision, interaction or explosion is equal to the total momentum afterwards, provided that no external forces are present.
- Momentum is very closely related to other quantities such as force, impulse and kinetic energy.
- Momentum is directly proportional to both the mass of a moving object and its velocity – doubling either means that the momentum will also double..

KEY DEFINITION

The **principle of conservation of momentum** is the total momentum before a collision is always equal to the total momentum after the collision, provided that no external forces are involved.

In any collision, momentum is always conserved. This makes it easy to calculate the velocities of objects before and after a collision.

LEARNING TIP

Since momentum is a vector , it may be positive or negative, depending upon direction.

WORKED EXAMPLE 1

A rugby player of mass 90 kg is moving at $8\,\text{m s}^{-1}$. He launches a tackle on another rugby player who has a mass of 65 kg and is stationary.

Figure 2 For part (a).

Figure 3 For parts (b) and (c).

Calculate:
(a) the total momentum before the collision
(b) the total momentum after the collision
(c) the velocity of the two players after the collision.

Answers

(a) momentum of rugby player 1 = 90 kg × 8 m s^{-1} = 720 kg m s^{-1}
momentum of rugby player 2 = 65 kg × 0 m s^{-1} = 0 kg m s^{-1}
Total momentum before the collision = 720 kg m s^{-1}

(b) The principle of conservation of momentum states that the total momentum before the collision must be equal to the total momentum after the collision, provided that no external forces are applied. So after this collision the momentum must also be 720 k gm s^{-1}.

(c) momentum before collision = momentum afterwards
720 kg m s^{-1} = combined mass of the rugby players × velocity
720 kg m s^{-1} = 155 kg × v

$$v = \frac{720\ \text{kg m s}^{-1}}{155\ \text{kg}}$$
$$= 4.65\ \text{m s}^{-1}$$

Notice how the velocity of the attacking rugby player has decreased after the tackle due to the increase in mass. This has to happen to conserve momentum.

Momentum as a vector quantity

If a body moving from left to right has a momentum of +250 kg m s^{-1}, then the same body must have a momentum of −250 kg m s^{-1} when it moves in the same way from right to left. Momentum, is a vector quantity. This is important when dealing with momentum calculations that involve explosions or recoil.

WORKED EXAMPLE 2

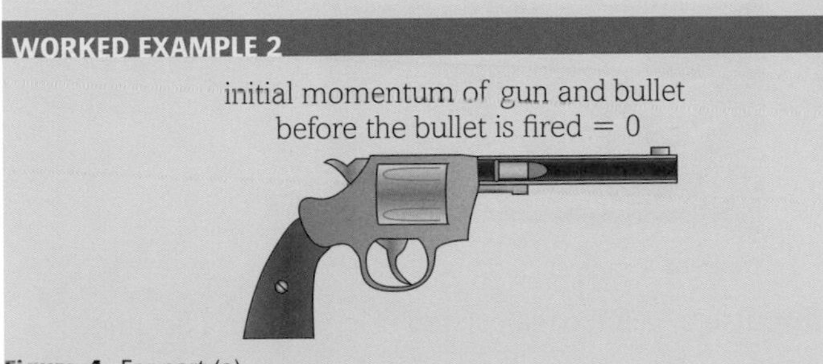

Figure 4 For part (a).

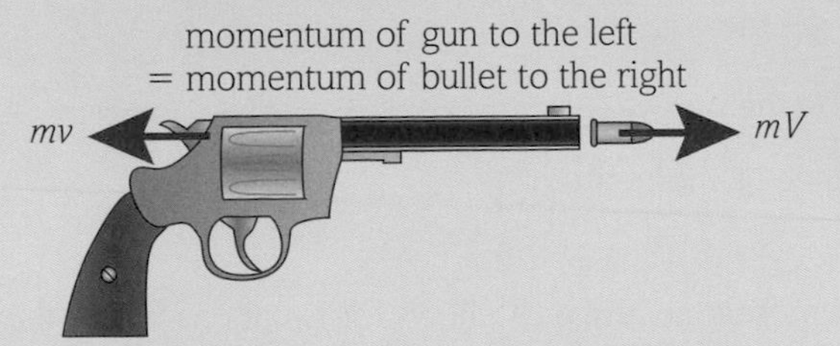

Figure 5 For parts (b) and (c).

A bullet of mass 0.008 kg is shot from a gun with a speed of 240 m s^{-1}. The gun has a mass of 6 kg. Calculate:

(a) the total momentum before the bullet is fired
(b) the total momentum after the bullet is fired
(c) the recoil velocity of the gun.

Answers

(a) momentum = mass × velocity
Nothing is moving before the bullet is fired, so the momentum is zero.

(b) momentum before = momentum afterwards
The momentum after the bullet is fired must also be zero.
... But how is this possible?

(c) momentum before = momentum afterwards
0 kg m s^{-1} = (0.008 kg × 240 m s^{-1}) + (6 kg × v m s^{-1})

$$v = \frac{-1.92}{6}$$
$$= -0.320\ \text{m s}^{-1}$$

The minus sign is the clue – if the bullet moves forward when the gun is fired, the gun must move backwards to conserve momentum. The bullet has a lower mass and higher velocity than the gun, so the gun will move backwards at a lower velocity.

Questions

1. Calculate the momentum in each of the examples below:
   (a) an athlete of mass 75 kg running at 7.5 m s^{-1}
   (b) a car of mass 2.8 × 10^{3} kg moving at 14.5 m s^{-1}
   (c) an electron of mass 9.11 × 10^{-31} kg moving at 2 × 10^{6} m s^{-1}

2. A car of mass 4500 kg is travelling at 12 m s^{-1} when it accidentally hits another car of mass 3400 kg that was at rest with its handbrake off. The first car comes to an immediate stop and its momentum is transferred to the second car. What will be the velocity at which the second car moves off?

3. A train of mass 700 kg travelling at 3 m s^{-1} collides with a stationary train of mass 400 kg. The trains join and move off together. What is their new velocity?

4. Two rugby players, both of mass 80 kg, are run towards one another. The player running from left to right has a speed of 8 m s^{-1}. The player running from right to left has a speed of 6 m s^{-1}. What will be their velocity when they tackle each other head on and move off as one mass?

5. A basic air rifle fires slugs of mass 0.50 g at a velocity of 160 m s^{-1}.
   (a) Suppose the air rifle has a mass of 0.80 kg and is free to move when a slug is fired. Calculate the speed with which it recoils.
   (b) Normally the rifle is held by a person. What mean force does the person experience to prevent the rifle recoiling more than 0.50 mm when it is fired?
   (c) The recoil force of rifles used in the First World War was in excess of 100 N. The military now use recoil-free rifles. Expanding gas in the barrel chamber pushes the bullet forwards, and some of the gas escapes from the chamber. Suggest how this gas can 'absorb' the recoil.

6. An astronaut can propel himself during a space walk by emitting pulsed jets of gas backwards from his backpack.

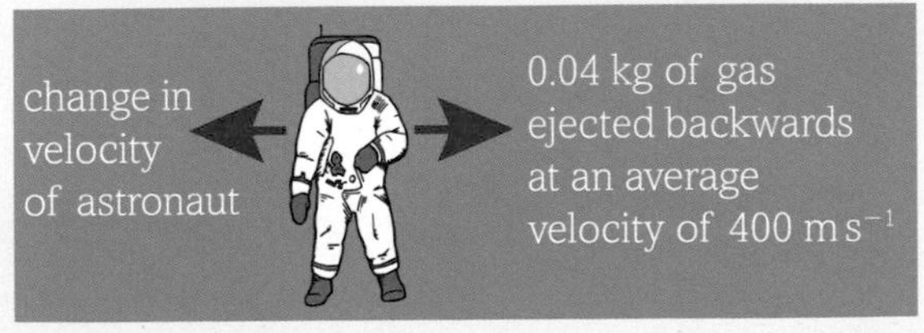

   (a) Explain why the astronaut, initially at rest, moves forward when a jet of gas is emitted from the backpack.
   (b) The backpack expels 0.04 kg of gas in one pulse at an average velocity of 400 m s^{-1}. Calculate the momentum of the pulse of gas.
   (c) The mass of the astronaut and his equipment is 120 kg. Calculate the increase of the velocity of the astronaut.

Momentum, force and impulse

By the end of this topic, you should be able to demonstrate and apply your knowledge and understanding of:

* impulse of a force, impulse = $F\Delta t$
* net force = rate of change of momentum; $F = \frac{\Delta P}{\Delta t}$
* impulse is equal to the area under a force–time graph

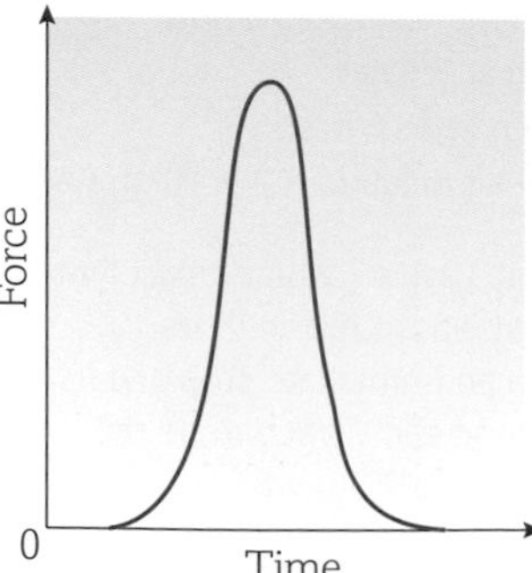

Figure 1 Force–time graph for a smooth impact.

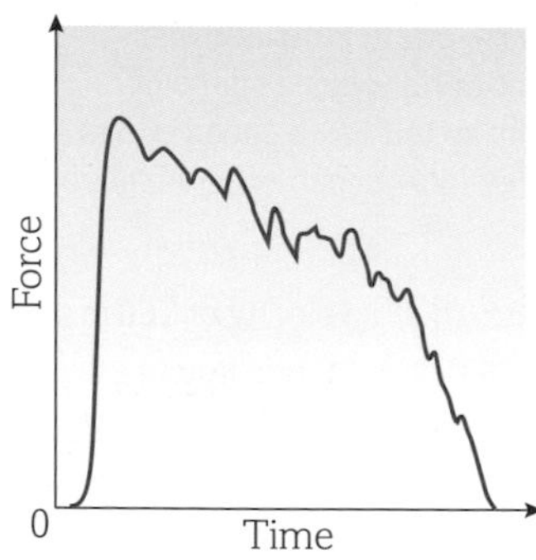

Figure 2 Force–time graph for an uneven impact.

Impulse

Situations arise when the force on an object is, for a short time, large and variable. For example, the sudden force acting on a tennis ball when hit by a racket, a nail hit by a hammer, an object dropped onto the floor or a car colliding with a gatepost.

Typical force–time diagrams of such impacts are shown in Figures 1 and 2. In Figure 1 the impact force increases smoothly to a maximum and then decreases smoothly to zero. This type of impact occurs with a tennis ball and racket, where the force might rise to several hundred newtons during the few milliseconds of contact.

The photograph in Figure 3 shows the distortion that takes place during impact. Immediately after impact, the tennis ball is distorted, and the distortion increases rapidly to a maximum before decreasing back to zero as contact is lost. The overall effect is to change the velocity of the ball considerably.

The photo of the car-test collision in Figure 4 is more complex. Different parts of the car hit the obstacle at slightly different times, so the force–time graph for this will be the one shown in Figure 2.

Figure 3 A tennis racket hitting a ball.

Figure 4 A car test crash.

It is useful to define a quantity called the **impulse** of a force, calculated using $F \times \Delta t$. The units of impulse are N s.

From Newton's second law of motion, we get the equation $F = ma$.

We also know that $a = \frac{(v - u)}{t}$, so combining the two, we get $F = \frac{m(v - u)}{t}$ or $F = \frac{(mv - mu)}{t}$

In other words: net force = the rate of change of momentum

This can be written as $F = \frac{\Delta P}{\Delta t}$

This is a more general statement of Newton's second law, as it allows for situations where the mass is variable, for example, rocket propulsion.

Rearranging this equation we get $Ft = mv - mu$. Hence the units of impulse are also the same as those for momentum, kg m s^{-1}.

KEY DEFINITION

Impulse is the product of a force F and the time Δt for which the force acts.

WORKED EXAMPLE 1

A ball of mass 1.5 kg is travelling towards a wall at 3 m s^{-1}. It hits the wall and bounces back with a speed of 2 m s^{-1}. The ball is in contact with the wall for 0.2 s. Calculate:
(a) the change in momentum (b) the impulse (c) the force exerted on the wall.

Answers

(a) momentum before = $1.5 \times 3 = 4.5$ kg m s^{-1}
momentum afterwards = $1.5 \times (-2) = -3$ kg m s^{-1}
change in momentum = $4.5 - (-3) = 7.5$ kg m s^{-1}

(b) The impulse equals the change in momentum – it is equal to 7.5 kg m s^{-1}.

(c) force = $\frac{\text{change in momentum}}{\text{time the force acts}} = \frac{7.5\text{ kg m s}^{-1}}{0.2\text{ s}} = 37.5\text{ N}$

The area beneath a force–time graph

Consider a theoretical situation in which a steel ball is acted on by a force of 500 N for a time of 4.0 ms. A force–time graph for this is shown in Figure 5. The shaded area beneath the graph has a height of 500 N and a width of 0.0040 s. The area of the rectangle is 500 N × 0.0040 s = 2.0 N s. The answer gives the impulse of the force. The change in the momentum of the ball is therefore 2.0 N s.

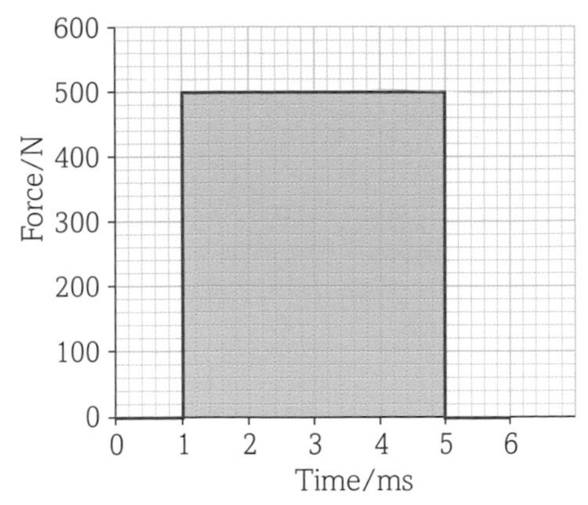

Figure 5 When the force is variable, the impulse is $\Delta F \times \Delta t$. The area beneath the force–time graph is equal to the impulse.

WORKED EXAMPLE 2

A ship of mass 8.0×10^6 kg (8000 tonnes) travelling slowly towards a jetty collides with a wooden protection post and is brought to rest. The braking force provided by the post varies with time in the way shown in Figure 6. How fast was the ship travelling before it hit the post?

Answer

The area beneath the graph is the first thing to determine. Figures 6(a), 6(b) and 6(c) show three ways of doing this.

Method (a) involves counting squares.

The total number of small squares = 25 + 29 + 10 + 63 + 100 = 227

1 small square represents 10 000 N × 0.5 s = 5000 N s

so 227 small squares represent 1.14×10^6 N s.

Method (b) uses a rectangle judged to have the same area as the graph.

This has area = 130 000 N × 9.5 s = 1.23×10^6 N s.

Method (c) uses two triangles and a rectangle judged to have the same area as the graph.

The area here is $(\frac{1}{2} \times 3.5 \times 130\,000) + (5 \times 130\,000) + (\frac{1}{2} \times 4 \times 130\,000) = 1.14 \times 10^6$ N s.

You can see that all the values obtained are of the same order of magnitude. Counting small squares is the most accurate of the methods, so we will proceed using the value it gives.

The area beneath the graph represents the impulse, which is equal to the change in momentum.

Therefore, $1.14 \times 10^6 = mv = 8.0 \times 10^6 \times v$, giving $v = 1.14 \times 10^6 / 8.0 \times 10^6 = 0.12$ m s^{-1}.

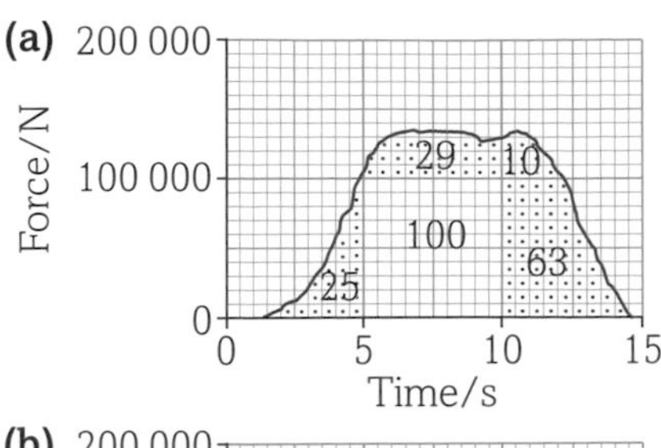

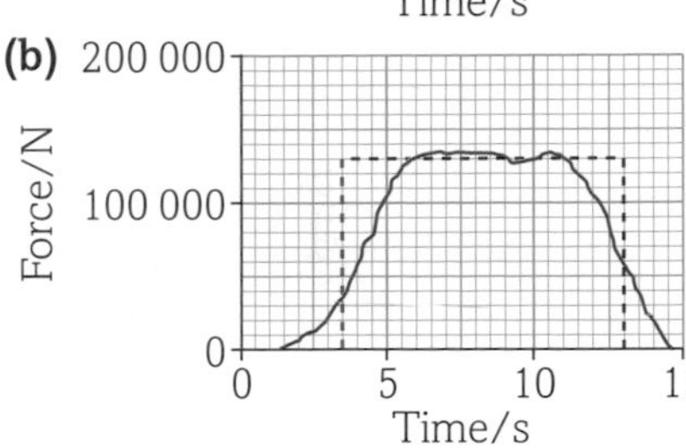

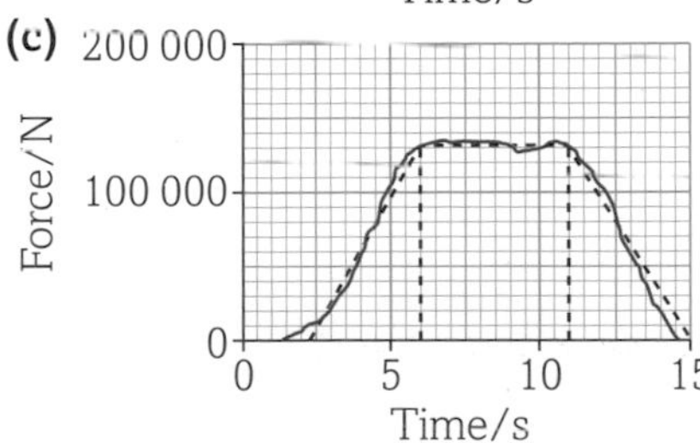

Figure 6 Three ways of measuring the area beneath a graph.

Questions

1. For the example of the steel ball described in worked example 1, use Figure 5 to calculate the ball's final speed assuming the ball has a mass of 0.1 kg and started from rest.

2. A ball of mass 0.8 kg hits a wall at a velocity of 6 m s^{-1}. It rebounds from the wall with a speed of 4.8 m s^{-1}. It is in contact with the wall for 0.08 s. Calculate the impulse and the force exerted on the ball by the wall.

3. Figure 7 shows how the force exerted by a golf club on a ball varies during the time of contact.
   (a) Show that the impulse given to the ball is about 2 N s.
   (b) Calculate the speed of the ball (mass 0.045 kg) initially at rest, at the instant it leaves the club.
   (c) What is the mean accelerating force?
   (d) Imagine a horizontal line drawn on Figure 7 level with the value of the force in part (c) to the maximum time of contact. Is the area within the curve above the line equal to the area within the curve but below the line? Explain your answer.

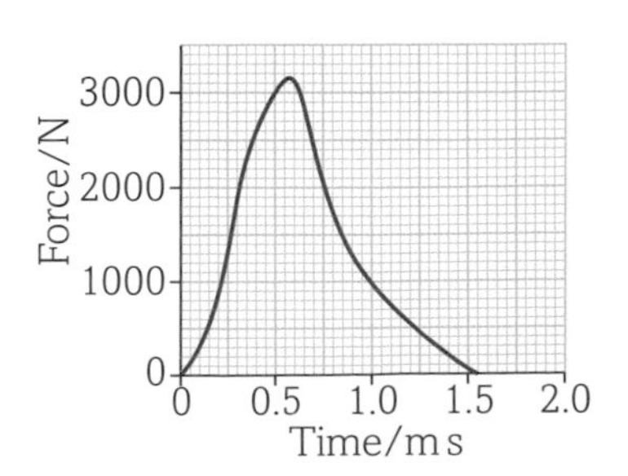

Figure 7

Elastic and inelastic collisions

By the end of this topic, you should be able to demonstrate and apply your knowledge and understanding of:

* collisions and interaction of bodies in one dimension and in two dimensions
* perfectly elastic collision and inelastic collision

Elastic and inelastic collisions

Momentum is always conserved in collisions and explosions.

An **elastic** collision is one in which the momentum and the kinetic energy are conserved – no energy is transferred to other forms such as heat or sound.

If a collision is **inelastic** it means that momentum is conserved but some of the kinetic energy is transferred to other forms in the collision.

DID YOU KNOW?

momentum = mass × velocity
kinetic energy = $\frac{1}{2}$ × mass × velocity^2

WORKED EXAMPLE 1

A mass of 4 kg is moving at 6 m s^{-1} when it collides with a stationary second mass of 4 kg (Figure 1). The first mass stops and the second mass moves off (Figure 2) with a velocity of 6 m s^{-1}. Is this collision elastic or inelastic?

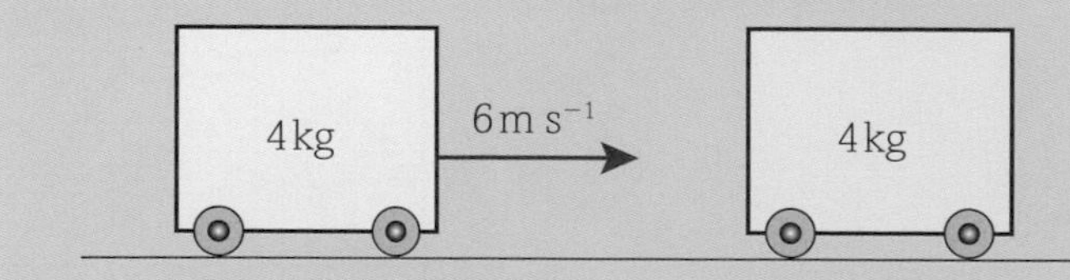

Figure 1 Before the collision.

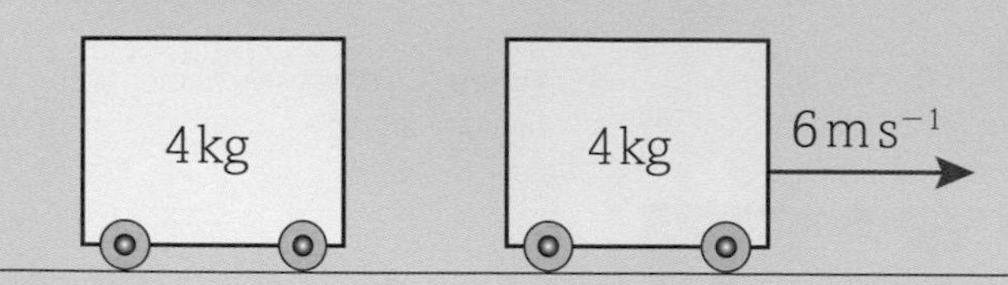

Figure 2 After the collision.

Answer

We have to consider the momentum and the kinetic energy of both bodies, both before and after the collision.

Total momentum before the collision = 4 kg × 6 m s^{-1} = 24 kg m s^{-1}

Total momentum after the collision = 4 kg × 6 m s^{-1} = 24 kg m s^{-1}

As expected, momentum is conserved – it has to be!

Total kinetic energy before collision = $\frac{1}{2} \times 4 \times 6^2 = 72$ J

Total kinetic energy after the collision = $\frac{1}{2} \times 4 \times 6^2 = 72$ J

As both momentum and kinetic energy are conserved, the collision is elastic.

It is important to realise that there are no truly elastic collisions in nature – energy is always transferred to other forms. However, collisions between gas atoms are usually taken to be perfectly elastic.

In all cases, total energy is always conserved, even if kinetic energy is not.

WORKED EXAMPLE 2

A toy bus of mass 1.8 kg travelling at 2.5 m s^{-1} collides with a toy car of mass 800 g travelling in the same direction at a velocity of 2.0 m s^{-1}. The velocity of the bus after the collision is 2.2 m s^{-1} in the same direction.

Calculate:

(a) (i) the new velocity of the car
(ii) the total kinetic energy before the collision
(iii) the total kinetic energy after the collision.

(b) Explain why kinetic energy is not conserved.

Answers

(a) (i) Momentum is always conserved, so:
momentum before collision
$= (1.8 \times 2.5)$ kg m s^{-1} + (0.8×2) kg m s^{-1} = 6.1 kg m s^{-1}
momentum after collision
$= (1.8 \times 2.2)$ kg m s^{-1} + $(0.8 \times v)$ kg m s^{-1}
$0.8v = 6.1 - 3.96$
$v = \frac{2.14}{0.8} = 2.675$ m s^{-1}

(ii) Kinetic energy before = kinetic energy of bus + kinetic energy of car
$= (\frac{1}{2} \times 1.8 \times 2.5^2) + (\frac{1}{2} \times 0.8 \times 2.0^2)$
$= 5.625 + 1.6 = 7.225$ J

(iii) Kinetic energy after = kinetic energy of bus + kinetic energy of car
$= (\frac{1}{2} \times 1.8 \times 2.2^2) + (\frac{1}{2} \times 0.8 \times 2.675^2)$
$= 4.356 + 2.862 = 7.218$ J

(b) The kinetic energy after the collision is less than the kinetic energy before the collision because some has been transferred to other forms of energy, such as heat and sound.

Collisions in two dimensions

When dealing with collisions in a straight line, we consider the total momentum of the objects involved before and after the collision along the *x* axis. From the law of conservation of momentum, the total momentum before the collision must be equal to the total momentum after the collision along this axis.

For collisions which do not take place in a straight line, we have to ensure that the momentum in two perpendicular directions is conserved. Note that two-dimensional problems will only be assessed at A level.

WORKED EXAMPLE 3

A 60 kg mass travels at 12 m s^{-1} to make a head-on elastic collision with a 40 kg mass travelling in the same direction at 7 m s^{-1} (Figure 3).

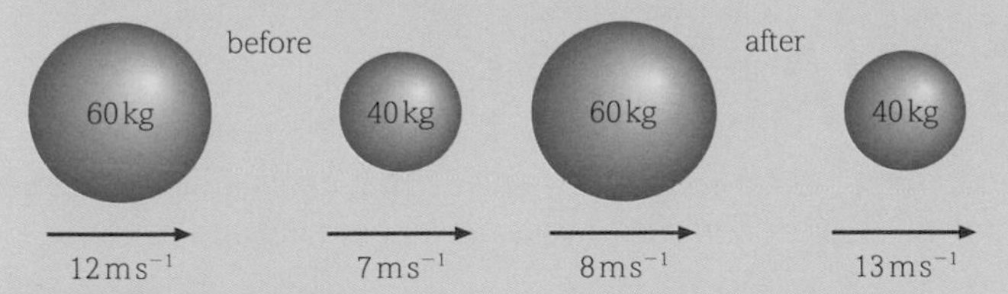

Figure 3

Show that after the collision the 60 kg mass travels forward at 8 m s^{-1} and the 40 kg mass forward at 13 m s^{-1}, and that there is no loss of kinetic energy.

Answer

As with all collision questions, start with conservation of momentum.

Momentum before collision = momentum after collision.

$(60 \times 12) + (40 \times 7) = (60 \times 8) + (40 \times 13)$ (each term is a mass in kg multiplied by a velocity in m s^{-1})

giving $720 + 280 = 480 + 520$, and both sides of the equation come to 1000 N s.

For an elastic collision the total kinetic energy is the same before and after the collision.

Kinetic energy before collision = kinetic energy after the collision

$(\frac{1}{2} \times 60 \times 12^2) + (\frac{1}{2} \times 40 \times 7^2) = (\frac{1}{2} \times 60 \times 8^2) + (\frac{1}{2} \times 40 \times 13^2)$
(each term in joules)

giving $4320 + 980 = 1920 + 3380$, with both sides equal to 5300 J.

Notice what has happened to the velocities. The large mass was catching up on the small mass at a rate of 12 m s^{-1} − 7 m s^{-1} = 5 m s^{-1} before the masses collided. After the collision, the small mass moved away from the large mass at a rate of 13 m s^{-1} − 8 m s^{-1} = 5 m s^{-1}. This is always the case for a perfectly elastic head-on collision: the velocity at which the objects approach one another equals the velocity with which they separate. A simple, familiar illustration is when a ball falls and hits the Earth – it bounces back at the same speed, provided it is a 'perfect' ball.

WORKED EXAMPLE 4

A 3000 kg van, travelling from west to east at 20 m s^{-1} collides with a 1000 kg car travelling due north at 30 m s^{-1}. The vehicles lock together on impact and then move off as one body. Find the velocity of the vehicles after the collision.

Answer

The component of momentum in any given direction must be conserved. We will consider the component in the east direction (the positive x direction) and the component in the north direction (the positive y direction).

Let the final velocity of the vehicles be v, with components v_x and v_y where v makes an angle θ with the positive x direction.

In x direction: 3000 kg × 20 m s^{-1} + 0 = (3000 kg + 1000 kg) × v_x, which simplifies to $v_x = 15$ m s^{-1}

In y direction: 0 + 1000 kg × 30 m s^{-1} = (3000 kg + 1000 kg) × v_y, which simplifies to $v_y = 7.5$ m s^{-1}

Using Pythagoras' theorem, $v^2 = v_x^{2+} v_y^2$, so $v_x = 16.8$ m s^{-1}

We also need to calculate the direction of v, $v_x = v\cos\theta$ and $v_y = v\sin\theta$

so $v_y/v_y = 30\,000/60\,000$ and $v_y/v_y = \tan\theta$

Hence $\tan\theta = 0.5$, and $\theta = 26.6°$

LEARNING TIP

Errors in calculations are usually caused by getting confused about either the positive and negative directions, or the direction in which angles are measured.

Always draw and label a diagram for the situation before and after the collision. This will ensure that you choose the correct angle values to use when working out the components of momentum in the two perpendicular directions. Remember also to indicate directions by using a sign convention.

Questions

1. Explain the difference between an elastic collision and an inelastic collision.

2. Explain how momentum is conserved when a grenade explodes.

3. A ball of mass 3.0 kg, moving at an initial velocity of 5.6 m s^{-1} collides elastically with a stationary metal ball of mass 1.6 kg. After the collision, the 3.0 kg mass moves off at an angle of 34° to its initial direction, with a speed of 1.3 m s^{-1}. The 1.6 kg mass moves off at an angle α to the x-axis. Find the values for the velocity of the 1.6 kg mass and the angle that it makes with the x-axis.

4. Two objects slide over a frictionless horizontal surface. The first object, mass $m_1 = 5$ kg, is propelled with speed $v_1 = 4.5$ m s^{-1} toward the second object, mass $m_2 = 2.5$ kg, which is initially at rest. After the collision, both objects have velocities which are directed at $\theta = 30°$ on either side of the original line of motion of the first object.
   (a) What are the final speeds of the two objects?
   (b) Is the collision elastic or inelastic?

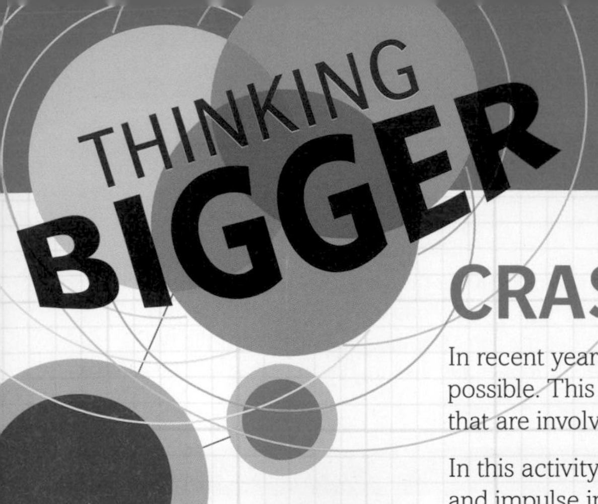

CRASH BANG WALLOP!

In recent years, much research and development has taken place to ensure that cars can be as safe as possible. This has involved a detailed study of crashes and the physics of kinematics and dynamics that are involved.

In this activity we ask ourselves – what are the considerations that need to be made regarding forces and impulse in car safety?

FORCES AND IMPULSE

In past progress toward safer cars, devices like seat belts were aimed at protecting you in a crash. Today, new safety technology is moving toward preventing an accident from happening at all. 'We are seeing a rapid shift from passive safety technology to active safety technology in modern cars,' said senior analyst Karl Brauer of Kelley Blue Book.

A lawsuit over rearview cameras and new ratings for collision warning and automatic braking systems highlight the trend toward accident prevention gear. Public interest groups last week sued the U.S. Department of Transportation for delaying federal requirements that all new cars have rearview cameras, which let a driver see what is behind a vehicle before backing up. The Insurance Institute for Highway Safety also has expanded its testing programs to rate systems that sense an imminent collision and warn the driver and, in some cases, apply the brakes automatically.

Backup cameras

By letting a driver see what is behind his vehicle, this feature can avoid particularly tragic accidents where parents or neighbours back over small children. These 'backover' accidents cause an average of 292 deaths and 18,000 injuries a year, according to the National Highway Traffic Safety Administration. About 44 per cent of those killed are children under five. This is especially an issue for large vehicles like SUVs, pickups and minivans.

Under congressional mandate to improve rear visibility, NHTSA and the Department of Transportation have repeatedly postponed the rule that all new cars must have backup cameras after automakers objected to the potential expense. Now consumer advocacy groups, including Consumers Union and Public Citizen, are suing the department to speed up this requirement.

Crash avoidance systems

First luxury cars and now mid-priced vehicles have added systems that warn a driver when a front-end crash is imminent and, in some cases, apply the brakes automatically. Kelley Blue Book's Brauer believes this is the most essential safety technology since seat belts and rollover-avoiding stability control.

[...] The Insurance Institute tested 74 luxury and moderately priced 2013 and 2014 cars and SUVs that offer a forward collision warning system with or without autonomous braking. These systems often are optional rather than standard equipment. 'Front crash protection systems can add a thousand dollars or more to the cost of a new car,' said IIHS chief research officer David Zuby. 'Our new ratings let consumers know which systems offer the most promise for the extra expense.'

The Institute rated vehicles in three categories – superior, advanced and basic. Each rating depends on the equipment offered and its performance in braking tests at 12 mph and 25 mph to simulate city driving.

[...] Whatever the outcome of the litigation over these technologies, backup cameras – already in 77 percent of 2013 models – will be required on all new vehicles within a few years. And front crash warning and avoidance systems, in the pattern of all past safety features, will eventually spread even to less expensive cars. Some of the radar and video technology involved are even steps along the road to the driverless car.

Source

Jerry Edgerton (2013) 'Car safety features that avoid accidents', http://www.cbsnews.com/news/car-safety-features-that-avoid-accidents/ (last accessed 1 February 2015)

DID YOU KNOW?

Forces on people need to be reduced in car crashes to reduce the likelihood of injury, but sometimes in sport large forces are desirable. It is often not possible to even see the ball when a top tennis player serves. Tennis players can often only tell where the ball is by the sound it makes on the strings as it leaves the racquet.

A boxer's punch was recently measured as having an impact of 3500 kg – or like being hit by an unloaded 17-seater minibus.

Where else will I encounter these themes?

1.1 2.1 2.2 3.1 3.2 3.3

Let us start by considering the nature and context of the writing in the article. The article above was taken from an online article on the *Moneywatch* section of a US TV website.

1. **Consider the article and comment on the type and purpose of writing that is being used. For example, think about whether this is a scientist reporting his or her findings, or a report based on secondary information. Does the report try to explain, persuade, or describe? Is there bias in the article? Are the findings open to interpretation by others? What are the words or terms in the article that would influence you when determining who the article is intended for?**

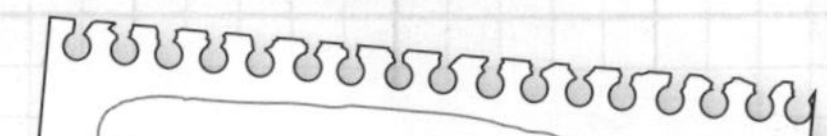

To what extent is an article like this objective and trustworthy? How might the claims made in this document be subject to bias?

We will now look at the physics that is in the article. Don't worry if the physics content or the mathematics is challenging at this stage. You can always return to the article later in your course, once some of the related topics have been studied in more depth. Use the timeline at the bottom of the page to help you put this work in context with what you have already learned and what is ahead in your course.

2. **Define the terms force, momentum and impulse.**
3. **Write a word equation that connects impulse, momentum and force.**
4. **Explain what is meant by the sentence 'We are seeing a rapid shift from passive safety technology to active safety technology in modern cars'.**
5. **A car of mass 1460 kg hits a tree at a speed of 18.4 m s^{-1} and bounces straight back with a speed of 2.8 m s^{-1}. Calculate:**
   a. **the change in momentum of the car**
   b. **the force exerted on the tree if the car was in contact with the tree for 15 m s.**
6. **Explain why momentum must be a vector if velocity is a vector.**
7. **A 2200 kg SUV travelling at 94 km h^{-1} (26 m s^{-1}) can be stopped in 21 s by gently applying the brakes, in 5.5 s in a panic stop, or in 0.22 s if it hits a concrete wall. What is the average force exerted on the SUV in each of these stops?**

Figure 1 How does the car's momentum change?

Activity

As one of the design engineers working on a new high-speed car, your manager has asked you to make a presentation to the Board of Directors. You can assume that they all studied physics at school, but may have left school many years ago. Since the car is designed to travel at high speeds, the car must contain all of the standard safety features, plus a new design of your own that will make the driver and the passengers 'ultra-safe'. Your presentation needs to contain the following elements and must be no longer than 10 slides in length.

- Definitions of force, impulse and momentum and the relationship between them.
- Equations of motion.
- Worked examples of calculations relating to force, impulse and momentum.
- Reference to car safety, including the different types of safety device and how they relate to forces and impulse.
- Your design artwork.
- An explanation of your new innovative design and an explanation of how the physics behind it will make the passengers 'ultra-safe'.

3.5 Practice questions

1. Which of the following is most closely a statement of Newton's second law of motion? [1]
   A For every action there is an equal and opposite reaction.
   B A body will continue to move at a constant speed or velocity until an external force acts upon it.
   C Momentum is conserved in any collision.
   D Force = mass × acceleration.

2. A ball of mass 500 g moves towards a wall at a speed of $4\,m\,s^{-1}$ and rebounds at a speed of $3\,m\,s^{-1}$ after being in contact with the wall for 200 m s. Which of the following statements are true? [1]
   (i) The momentum of the ball before the collision is $2\,kg\,m\,s^{-1}$.
   (ii) The change in velocity is $1\,m\,s^{-1}$.
   (iii) The change in momentum is $3.5\,kg\,m\,s^{-1}$.
   A (i), (ii) and (iii)
   B only (i) and (ii)
   C only (ii) and (iii)
   D (i) only

3. A trolley of mass 3000 kg is moving at $8\,m\,s^{-1}$. It collides with a trolley of mass and they move off together. Which of the following statements are true? [1]
   (i) The collision is perfectly elastic.
   (ii) Momentum is conserved in the collision.
   (iii) Kinetic energy is conserved in the collision.
   A (i), (ii) and (iii)
   B only (i) and (ii)
   C only (ii) and (iii)
   D (i) only

4. Which of the following statements is false? [1]
   A Impulse is equal to the area under a force–time graph.
   B Impulse is the change in momentum.
   C Impulse is the rate of change of momentum.
   D Impulse is a vector quantity.

[Total: 4]

5. The Saturn V rocket used in NASA's space programme had a mass of 3.04×10^6 kg. It took off vertically with a thrust force of 3.40×10^7 N.
   (a) Show that the resultant force on the rocket is about 4×10^6 N. [3]
   (b) Calculate the initial acceleration. [2]
   (c) After 150 s the rocket reached a speed of $2390\,m\,s^{-1}$. Calculate its average acceleration. [2]
   (d) Suggest why the initial acceleration and average acceleration are different. [1]

[Total: 8]

[*Q11, 6PH01/01 Jan 2009 (Edexcel)*]

6. (a) What is meant by Newton's first law of motion? [2]
   (b) Newton's third law identifies pairs of forces.
   (i) State two ways in which the forces in the pair are identical. [2]
   (ii) State two ways in which the forces in the pair differ. [2]

[Total: 6]

7. The forces acting on a car are shown in Figure 1.

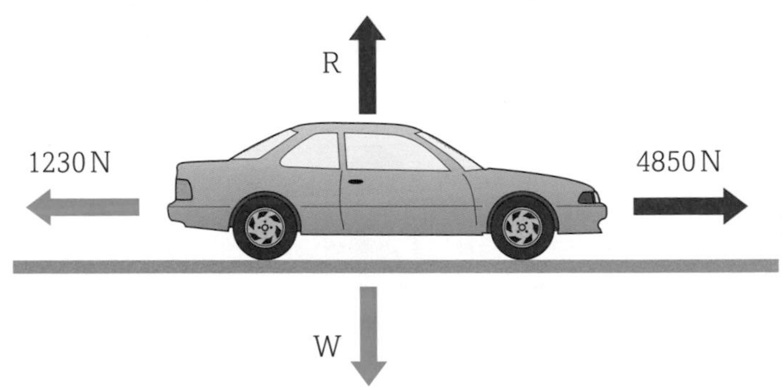

Figure 1

Calculate the size of:
   (a) the car's acceleration [2]
   (b) the change in the car's momentum if it comes to rest in 1.8 seconds. [2]

[Total: 4]

8. Use Newton's first and third laws to explain why the air coming out of the balloon causes the car to move. [3]

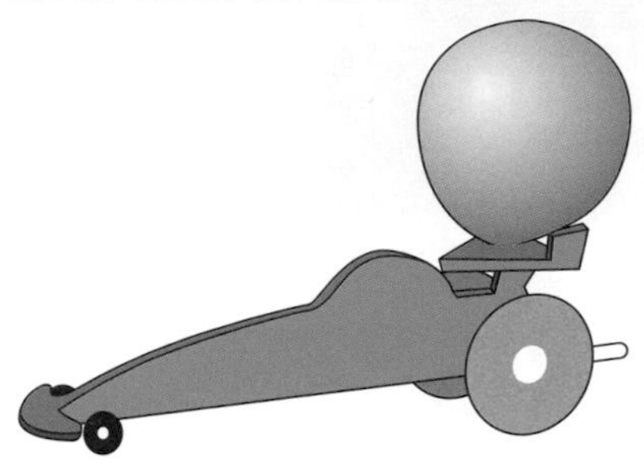

Figure 2

[Total: 3]

9. Figure 3 shows three passengers sitting on a train that is travelling at a high speed in the direction shown. Seat belts are not used on trains. With reference to one of Newton's laws of motion, explain why seat C is the safest seat for a passenger to be sitting on in the event of a rapid deceleration. You may assume that the seats all remain fixed firmly to the floor and do not break. [4]

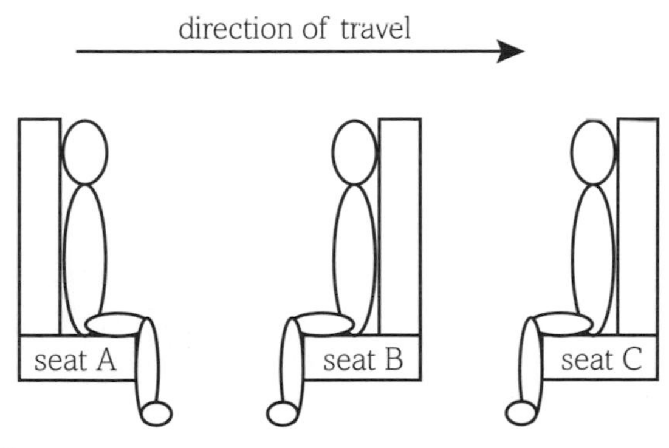

Figure 3

[Total: 4]

[Q14, 6PH01/01 Jan 2012 (Edexcel)]

MODULE 4 Electrons, waves and photons

CHAPTER 4.1

ELECTRICITY: CHARGE AND CURRENT

Introduction

It would be quite difficult to manage in our daily lives without electricity. Our homes, businesses and schools require it for heating and lighting, amongst other things, and our industries rely on it for the manufacture of products such as cars. Communications via phone, computer and television would cease and the food in our fridges and freezers would soon become inedible without electricity to power them. Much of this wonderful energy resource is brought to us from the plethora of power stations that are situated around the UK – powered, mostly, by coal, oil, natural gas or nuclear fuel. The vast network of generators, transformers, pylons and substations ultimately bring us this remarkably clean and reliable form of energy. However, the energy used in our homes is alternating current or a.c., which is produced when a magnetic field rotates inside a coil of wire. In this chapter we are going to look at the essentials of electric current and charge, which are essential when dealing with electrical circuits. We will also consider the mechanisms by which conductors, semiconductors and electrolytes conduct.

All the maths you need

To unlock the puzzles of this chapter you need the following maths:

- Units of measurement
- Rearranging equations to make other quantities the subject of the formula and substituting values
- Use of standard form, e.g. relating to the charge on the electron 1.6×10^{-19} C

What have I studied before?

- Electric charge will flow around a circuit as electrical current if an energy source and electrical conductors are present
- Electrical current is measured with an ammeter, placed in series with the component through which the current is flowing
- The unit of electrical current is the ampere (A)
- The unit of electrical charge is the coulomb (C)
- Electrical circuits may contain components including a cell, ammeter, bulb, voltmeter, switch and conductors
- Conventional current is from positive to negative, electron flow is from negative to positive
- Electrons are the charge carriers in metals, ions are the charge carriers in electrolytes

What will I study later?

- Electrons, and other charged particles, will be accelerated by an electric field (AS)
- Electrons inhabit fixed energy levels in atoms, and excitation can lead to them increasing their energy or leaving the atom via the process of ionisation (AS)
- The behaviour and motion of charged particles in electric and magnetic fields (AL)
- Coulomb's law – the behaviour of charged particles in terms of their respective charges and their separation (AL)
- Electric potential and electric field strength and the scalar and vector properties (AL)
- The force on a current carrying wire in the presence of a magnetic field (AL)
- Fleming's left hand rule and right hand rule (AL)

What will I study in this chapter?

- Electrical circuits are shown via the use of internationally agreed circuit symbols
- Electric current is defined as the rate of flow of charge, where $I = \frac{\Delta Q}{\Delta t}$
- The coulomb, C, is the unit of charge
- The charge on the electron is 1.6×10^{-19} C. All charges are multiples of this
- Current is the flow of electrons in metals and ions in the process of electrolysis
- Conventional current is from positive to negative, whereas electron flow is in the opposite direction
- Charge and energy in circuits behave in conjunction with Kirchhoff's laws
- Electrons have a drift velocity and the current is related to this via the equation $I = nAve$
- Conductors and semiconductors differ in terms of the number of charge carriers they have per unit volume

 Electric circuit components

By the end of this topic, you should be able to demonstrate and apply your knowledge and understanding of:

* circuit symbols
* circuit diagrams using these symbols
* conventional current and electron flow

Introduction

The whole subject of electricity is a relatively large and diverse field. During your course, you will come across terminology and concepts relating to electric charge and current, electric power, energy and resistance – as well as a variety of electrical circuits and the rules that govern their behaviour. In this topic you will cover some of the key ideas that will be expanded further when you meet them later in this book. Some of the material here you will have already met in your GCSE work.

Electric circuits require a number of things if they are to work effectively. These include:

- An **energy source** – a source of **electromotive force (e.m.f.)** so that an electric current can flow in the circuit. As you will learn in Topic 4.2.1, an e.m.f. is measured in volts (joules per coulomb) and is a measure of the amount of energy given to each coulomb of charge when chemical energy in a cell is being transferred to electrical energy in a circuit. Despite its name, e.m.f. is not a force, although it does supply energy to a circuit. The source of the e.m.f. could be a cell or a battery. It could also be a power pack that is plugged into the mains. In circuits containing metal wires, the electric current is carried by negatively charged particles called electrons.
- A **closed path** or **complete circuit** – electrons need to flow in a complete loop from a negative electrode to a positive electrode. A current will not flow if there is an open circuit – if there is a gap where a conducting material is not present.
- **Electrical components** – these can be used to: switch a circuit on and off; act as sensors and respond to changes in the environment; change the amount of current flowing in a circuit; alter the potential difference across components; or transfer electrical energy to other forms of energy.

Electrical circuit components

The most commonly used electrical components, and a description of their respective uses, are shown in Table 1. The table also makes reference to how the electrical resistance of each of these components has an impact on the behaviour of the component in a circuit.

Component	Symbol	Purpose of component	Change in electrical resistance
Junction of conductors		Allows an electric current to split at a junction.	Typically low because it is a conductor. Increases as the temperature increases.
Conductors crossing, no connection		Allows an electric current to flow in a circuit.	Typically low.
Switch		Turns an electric current on and off.	Low when the switch is closed and very high when the switch is open because there is an air gap.
Cell		Provides the circuit with a source of energy or e.m.f.	Will either be described as 'negligible' or will have low internal resistance.
Battery		Provides a circuit with a source of energy or e.m.f. A battery is two or more cells.	As for cells, although the internal resistance will increase as more cells are added in series.
Terminals		Provides a circuit with a source of energy or an e.m.f.	Low.
Lamp		Transfers electrical energy to light energy as a useful form.	Will increase as the current increases.
Fixed resistor		Controls the amount of electric current flowing in a circuit or through a component.	Fixed, and dependent on the material that the resistor is made from. Resistors are normally made from semiconductors such as silicon or germanium.

Component	Symbol	Purpose of component	Change in electrical resistance
Variable resistor		Controls the amount of electric current flowing in a circuit or through a component.	Changes based on the setting of the slider or dial.
Fuse		Safety device – will melt or 'blow' if the current gets too high.	Typically low and dependent on the length and thickness of the fuse wire and the material from which it is made.
Heater		Transfers electrical energy into thermal energy as a useful form.	Typically high.
Ammeter	A	Measures the amount of electrical current flowing in a circuit or through an electrical component.	Very low.
Voltmeter	V	Measures the size of the potential difference (e.m.f.) across an electrical component.	Very high.
Thermistor		Responds to the temperature of the environment and changes its resistance as a consequence.	Changes in response to the temperature of the surroundings.
Diode		Enables a current to flow in only one direction. Often used to protect delicate components from having large currents flow through them.	Low in the forward bias arrangement; very high in the reverse bias arrangement.
Light-emitting diode (LED)		Like a diode, allows electrical current to flow through it in only one direction. When current flows through it, it emits light.	Low in the forward bias arrangement; high in the reverse bias arrangement.
Light-dependent resistor (LDR)		Changes the current flowing in a circuit or though a component when the light intensity levels change.	Decreases as the light intensity increases.

Table 1 Electric circuit components.

Conventional current and electron flow

When electricity was first discovered, it was proposed that the current in a circuit flows from the positive terminal to the negative terminal. This is called **conventional current**. In reality, the charge carriers in metals are electrons that flow from the negative terminal to the positive terminal – this is called **electron flow**. We still use conventional current today, even though the actual flow of electrons is in the opposite direction.

Usually the direction of current flow does not matter. However, in the case of a diode or an LED the arrangement of the circuit is very important, because these components will allow current to flow only when they are arranged in forward bias. This is shown in Figure 1.

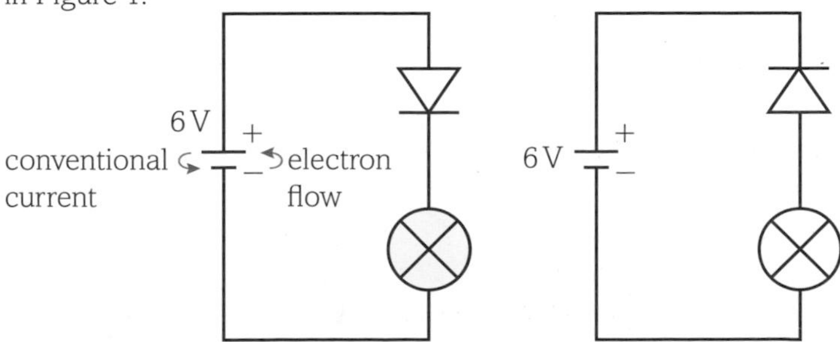

Figure 1 Conventional current and electron flow.

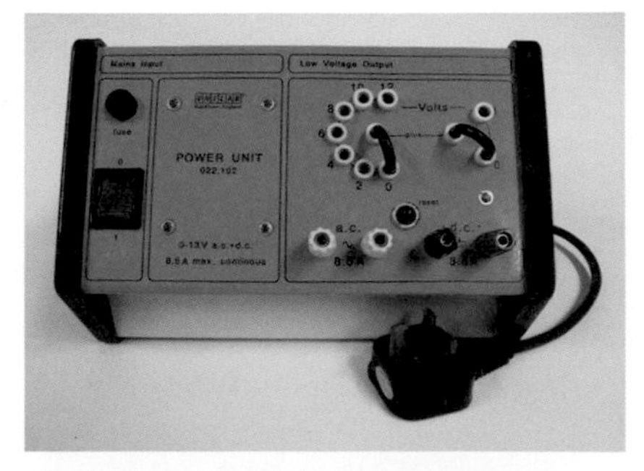

Figure 2 These electrical circuit components are all used to provide an energy source (e.m.f.) to a circuit.

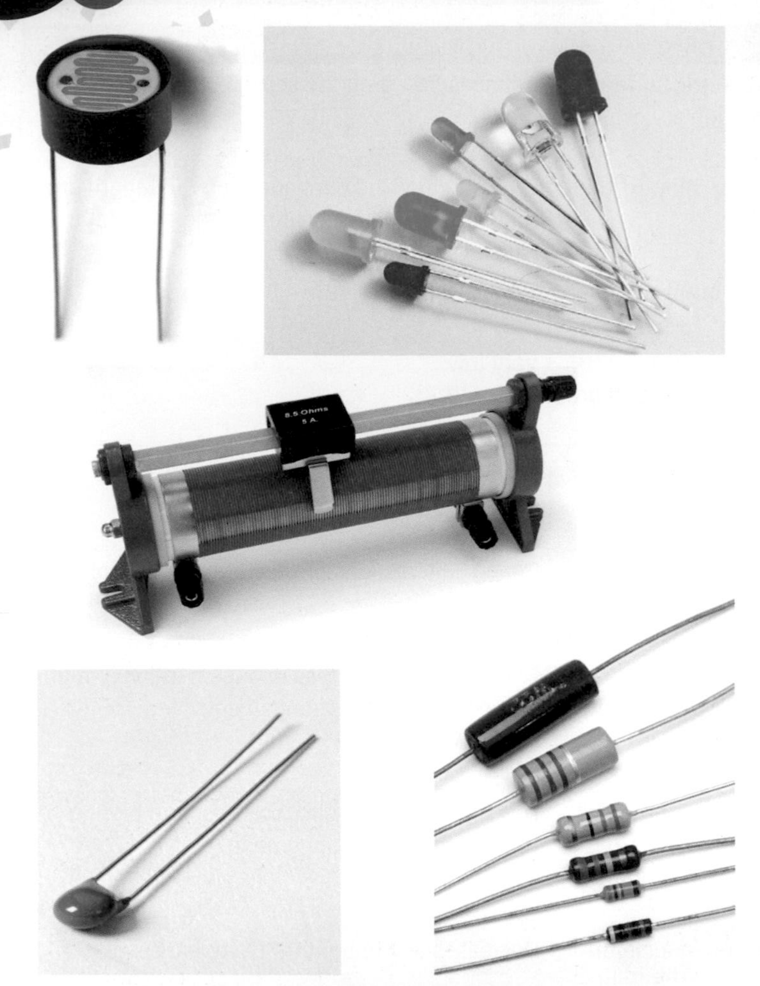

Figure 3 These components all change the current flowing in a circuit and will, therefore, change the potential difference across components in the circuit. Their resistance may be fixed or it may be variable, either by manual control or because of temperature or light intensity changes in the environment.

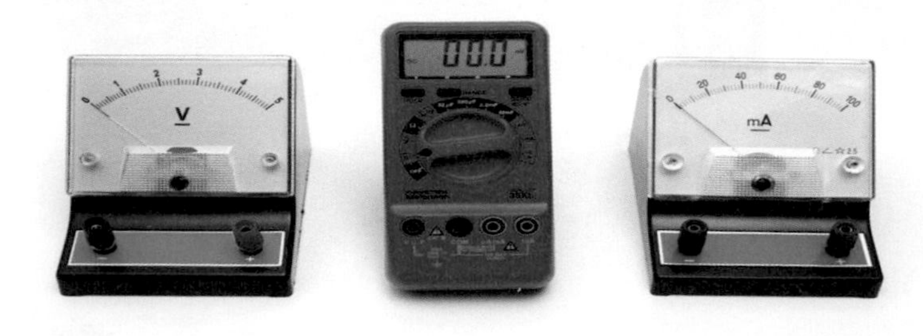

Figure 4 Ammeters and voltmeters measure the sizes of currents and potential differences in circuits.

Questions

1. State the names of the electrical components whose resistance is affected by:
   (a) temperature
   (b) light intensity.

2. Explain why no electric current will flow through the lamps in each of these three circuits.

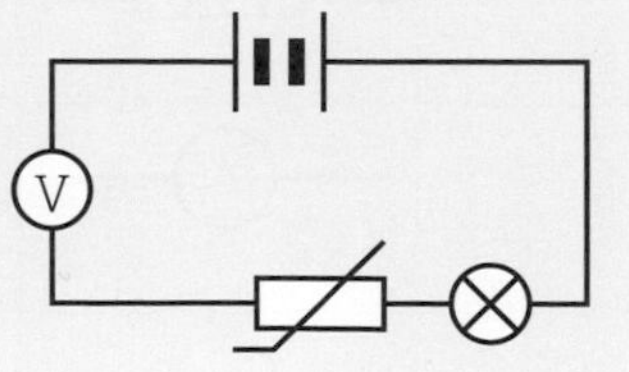

Figure 5

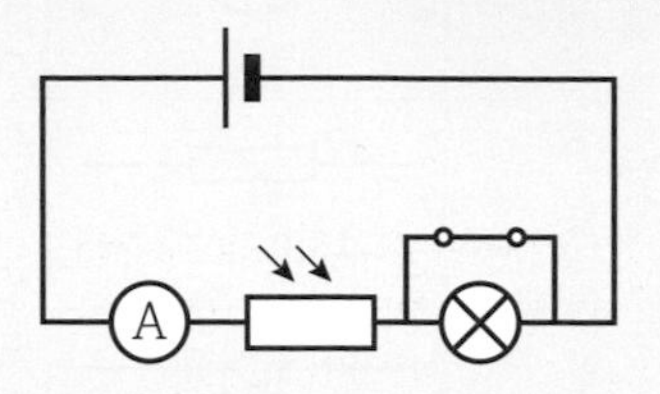

Figure 6

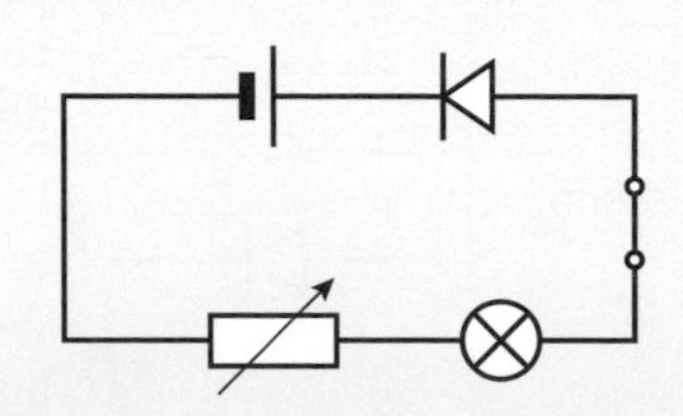

Figure 7

3. In which of the two circuits in Figures 8 and 9 will the buzzer be loudest? Explain your answer.

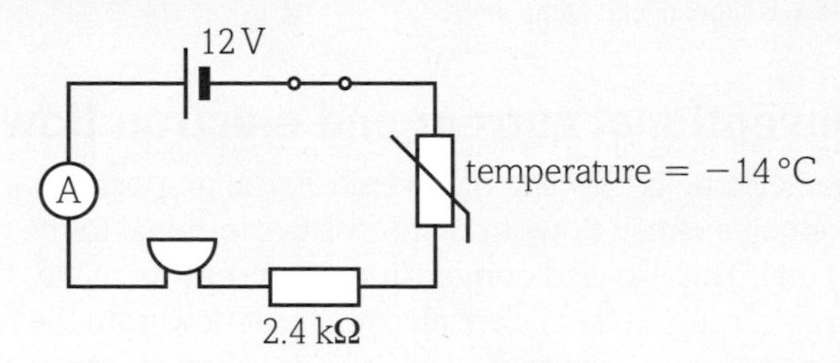

Figure 8

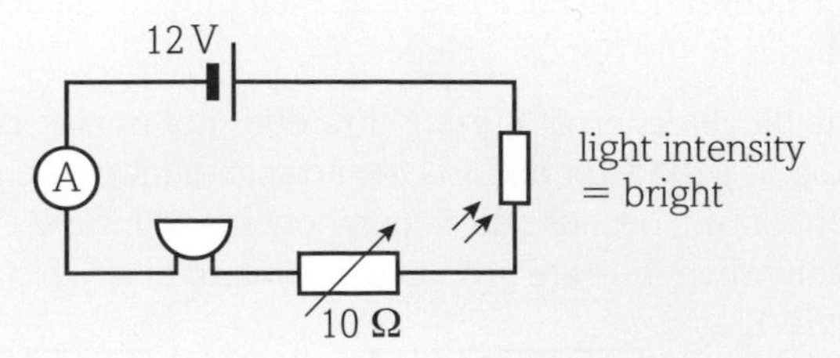

Figure 9

2 Electric current and charge

By the end of this topic, you should be able to demonstrate and apply your knowledge and understanding of:

* electric current as rate of flow of charge; $I = \frac{\Delta Q}{\Delta t}$
* the coulomb as the unit of charge
* the elementary charge e equals 1.6×10^{-19} C
* net charge on a particle or an object is quantised and a multiple of e
* current as the movement of electrons in metals and movement of ions in electrolytes
* Kirchhoff's first law; conservation of charge

Introduction

'Charge' is a property that certain particles may have and it exists in two forms – positive charge and negative charge. From a basic understanding of atomic structure (Figure 1), we know that atoms are neutral overall. They have the same number of positive charges as negative charges. Positively charged particles called protons are located in the nucleus, whereas the negatively charged particles called electrons orbit the nucleus.

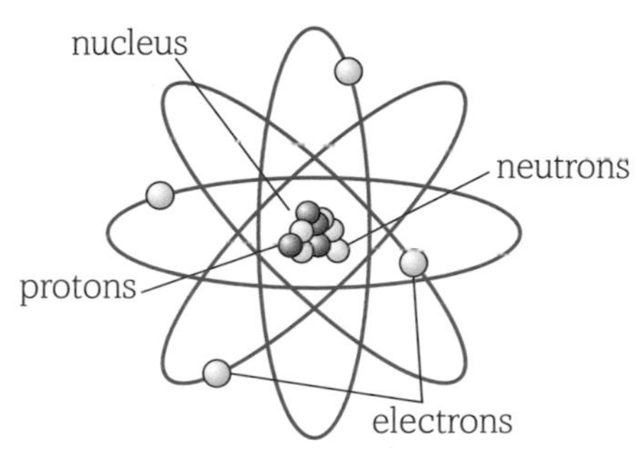

Figure 1 All atoms are neutral overall, meaning they have no net charge – the protons and electrons in any given atom are always equal in number and opposite in charge.

Atoms can become charged by gaining or losing electrons (Figure 2). If an electron is ejected from an atom, due to a collision with another electron or a photon of light, then the atom will become positively charged overall – it is now a positive ion. However, the total charge of any system is always conserved so the net quantity of charge present before or after the movement of any charge is always the same.

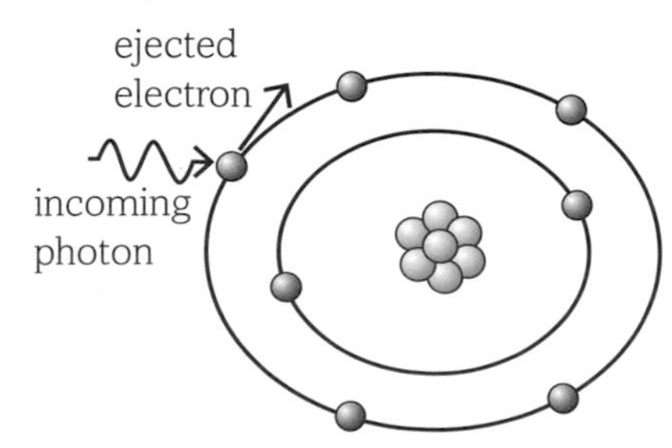

Figure 2 Photons of light can eject electrons from their orbits.

Electric current is the rate of flow of charge, $I = \frac{\Delta Q}{\Delta t}$. An electric current consists of a flow of charged particles.

In any metals – for example copper, gold, lithium, mercury and titanium – the flow of charge is made up of electrons. When copper, for example, is examined in detail, it is found to be crystalline. This means that the metal ions are closely packed and arranged in long, neat rows in what is called a crystal lattice structure. The copper atoms have many delocalised electrons that are free to move randomly. These electrons are also known as free electrons or conduction electrons (see Figure 3). It is these electrons that make metals such good conductors of electricity. When conducting electricity, the conduction electrons drift slowly along the wire.

Some solutions, called **electrolytes**, conduct an electric current. However, the charge carriers in electrolytes are not electrons; they are positive ions and negative ions. Electrolytes are formed when ionic compounds, such as copper sulfate, are dissolved in water. In some conductors the charge is carried by ions.

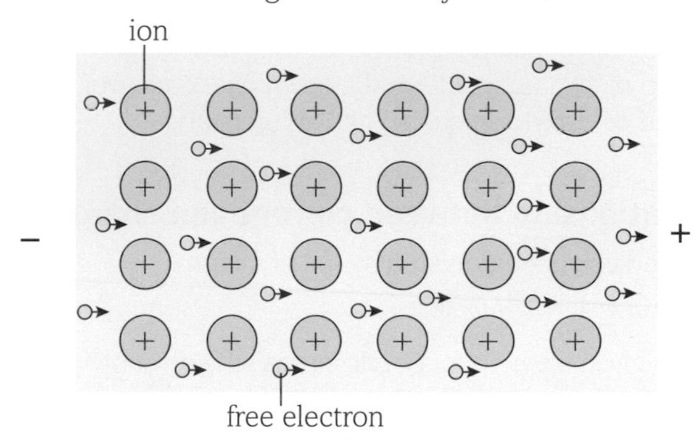

Figure 3 Electrical conduction in a metal.

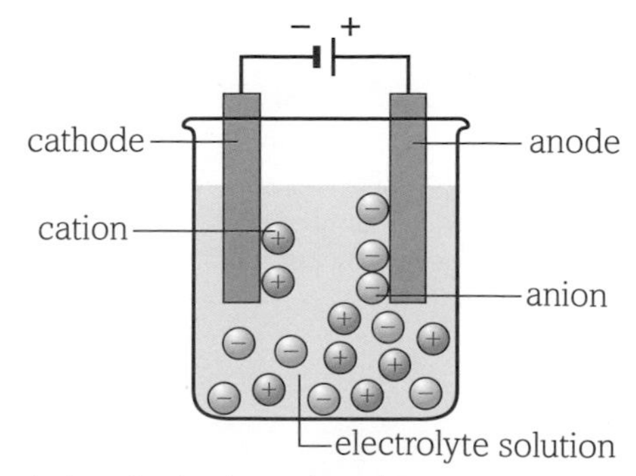

Figure 4 Electrical conduction in an electrolyte.

Kirchhoff's first law

Because charge is always conserved, the sum of the charges flowing into a circuit junction must be equal to the sum of the charges leaving it – charges cannot be created nor destroyed. This law is true for charge, so it must also be true for current – the rate of flow of charge. If, for instance, a current of 8 A flows into a junction, then a total of 8 A must flow out of it. This is shown in Figure 5.

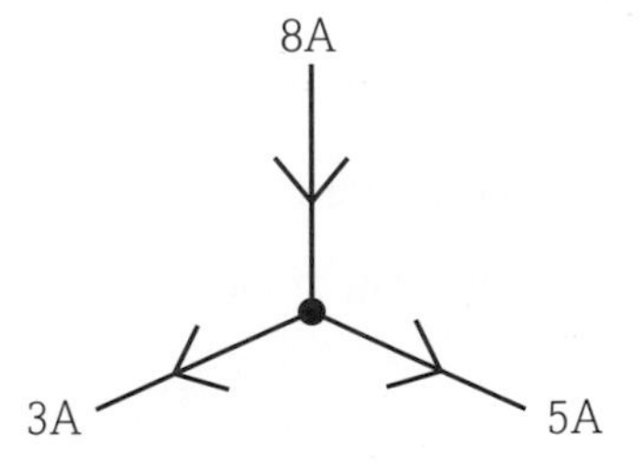

Figure 5

The unit of charge is the coulomb (C), named after the French physicist Charles-Augustin de Coulomb. One coulomb is defined as the quantity of charge that passes a fixed point in one second when a current of one ampere is flowing.

1 C = 1 As

The unit of electric current is the ampere, and it is the S.I. unit on which all other electrical units are based.

The elementary charge *e*

The charge on an electron, e, is -1.6×10^{-19} C. The charge on a proton is $+1.6 \times 10^{-19}$ C. When particles such as oil droplets become charged, they always contain a whole number of electrons, so their net charge will always be a multiple of e. Strangely, particles called quarks that make up the proton and the neutron have fractional charges – an 'up' quark has a charge of $+\frac{2}{3}e$ and a 'down' quark has a charge of $-\frac{1}{3}e$. Protons are composed of two up quarks and a down quark; neutrons are composed of two down quarks and an up quark.

The relationship between current and charge

Charge and current are related by the equation:

charge (C) = current (A) × time (s)

Charge is given the symbol Q, current has the symbol I and time the symbol t, leading to: $Q = I \times t$.

If changes to the current and time are involved, then we use the equation: $\Delta Q = I \times \Delta t$.

WORKED EXAMPLE 1

What charge has passed if:

(a) a current of 3.2 A flows for 25 seconds?
(b) a current of 400 mA flows for 5 minutes?
(c) a current of 850 μA flows for 2 hours?

Answers

(a) $Q = I \times t$
$= 3.2\text{ A} \times 25\text{ s} = 80\text{ C}$
(b) $Q = 0.4\text{ A} \times 300\text{ s} = 120\text{ C}$
(c) $Q = 8.5 \times 10^{-4}\text{ A} \times 7200\text{ s} = 6.12\text{ C}$

WORKED EXAMPLE 2

What time has elapsed if a constant current of 5 A provides a charge of 4000 C to a lamp?

Answers

$\Delta Q = I \times \Delta t$, so $4000 = 5t$, giving a time of 800 s.

If we know the charge on a particle, e, then the number of charged particles flowing past a point each second, n, and the time that has elapsed, t, then we can work out the total charge, Q, that has passed a point using $Q = net$. This means we can also calculate the size of the current using $I = ne$.

WORKED EXAMPLE 3

The charge on one electron is 1.60×10^{-19} C.

(a) What is the current through a cathode-ray tube in which 5.0×10^{14} electrons pass per second?
(b) What time is required for the total charge through the tube to reach 1.0 C?

Answers

(a) Charge on 5.0×10^{14} electrons $= n \times e \times t$
$= (5.0 \times 10^{14}) \times (1.60 \times 10^{-19}\text{ C}) \times 1\text{ s}$
$= 8.0 \times 10^{-5}\text{ C}$

This is the charge in coulombs flowing per second and so the current through the cathode-ray tube is 8.0×10^{-5} A or 80 μA.

(b) $Q = I \times t$. So $t = \frac{Q}{I} = \frac{1.0}{8.0 \times 10^{-5}} = 12\,500\text{ s}$

Questions

1. How much charge has passed through a lamp if a current of 1.6 A flows through it in 10 minutes?
2. Calculate the average current in a wire when a charge of 240 C passes in 60 s.
3. When a small torch is switched on, the current drawn from the cell is 0.25 A.
   (a) Calculate the charge passing a point in the bulb filament when the torch is switched on for 15 minutes.
   (b) How many electrons drift past the point in this time? The charge on one electron is 1.6×10^{-19} C.
   (c) For what length of time would the bulb need to be left on for 5.4×10^{12} C of charge to have passed through it?
4. A rechargeable AA cell, for use in a digital camera, is labelled 2300 mA h. How much charge does it hold when fully charged?
5. A spark travels across a gap between two spherical conductors that are 40 cm apart. The current is 4.0 mA. For electrons travelling at $3.0 \times 10^{7}\text{ m s}^{-1}$ (one-tenth of the speed of light), calculate the number of electrons in the spark at any instant.

3 Electron drift velocity

By the end of this topic, you should be able to demonstrate and apply your knowledge and understanding of:

* mean drift velocity of charge carriers
* $I = Anev$, where n is the number density of charge carriers
* distinction between conductors, semiconductors and insulators in terms of n

The crystalline structure of metals

When copper is examined in atomic detail, it is found to be crystalline (Figure 1). This means that the atoms are packed together closely and are arranged in long, neat rows called a crystal lattice structure. Copper atoms have electrons that are free to move randomly throughout the lattice. These delocalised electrons are also called conduction electrons. It is these electrons that make metals such good conductors of electricity. When conducting electricity, the conduction electrons drift slowly along the wire.

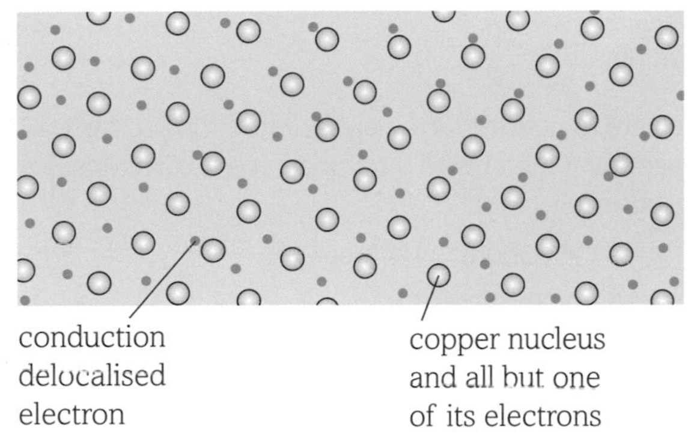

Figure 1 Copper contains many positively charged ions arranged in a regular array, surrounded by a 'sea' of electrons.

The movement of electrons through a metal can be described as random when there is no potential difference across the ends of the wire. This means that there will be no overall net movement of the electrons, as shown in Figure 2.

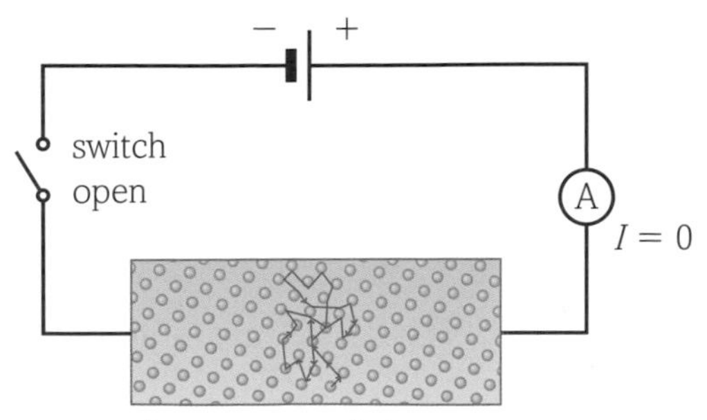

Figure 2 Although the electrons in a metal at room temperature are moving very quickly, their motion is random and there is no overall movement of charge – this means that no current is flowing.

Applying a potential difference across the ends of the copper wire causes a current to flow in the wire (Figure 3). The random motion is still present, but now there is also a net movement of the electrons in one direction, from the negative potential towards the positive potential.

We saw in Topic 4.1.2 that an electric current is a flow of charged particles. Although the presence of many conduction electrons in a metal's lattice allows the flow of an electric current, the electrons drift slowly through the copper lattice. This is because they can move only a small distance before colliding with a copper atom. We can work out the mean drift velocity of these electrons if we know the number density of copper atoms, n, which can be calculated from the density of copper and its atomic mass.

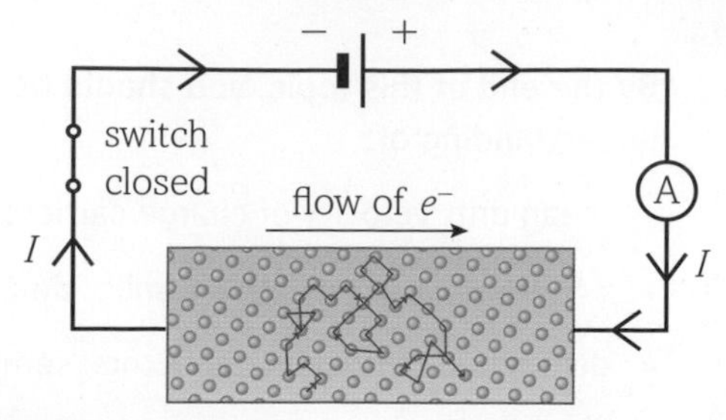

Figure 3 A potential difference across the ends of the metal causes electrons to drift slowly from the negative terminal to the positive terminal, resulting in an electric current.

Calculating electron drift velocity

The size of the electric current flowing in a wire can be calculated using Ohm's law (as we will see in Topic 4.2.3), from which we obtain $I = \frac{V}{R}$.

However, we can also calculate the current, I, by considering the properties of the wire itself. Consider a section of a cylindrical metal wire through which a current is flowing. To derive an equation, we shall use the following:

I = the current flowing through the conductor in amps (A)

L = the length of the conductor in question (m)

A = the cross-sectional area of the wire (m^2)

e = the charge on the electron, which is 1.602×10^{-19} (C)

n = the number of free electrons per unit volume (m^{-3})

v = the mean drift velocity of the electrons ($m\,s^{-1}$).

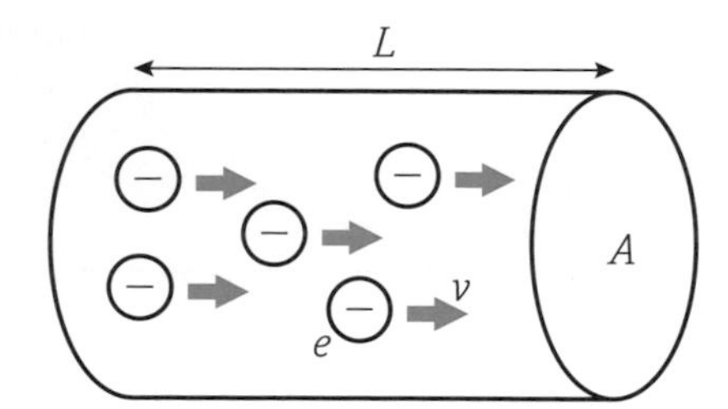

Figure 4

From Figure 4, we can see that:

volume of wire section = length × cross-sectional area = AL

So the number of free electrons available in this volume will be nAL. The total amount of charge that is free to move in this volume is $nALe$.

Because time $= \frac{\text{distance}}{\text{velocity}}$, the time taken for all the charges to travel length L will be $\frac{L}{v}$

Hence the rate of flow of charge will be:

$$\frac{\text{amount of charge in volume of wire}}{\text{time taken for the charge to travel the length L}} = \frac{nALe}{(L/v)},$$ which simplifies to $nAev$.

The rate of flow of charge is the current, so we end up with $I = nAev$.

This is easily rearranged to give an equation in which v, the mean drift velocity, is the subject of the formula:

$$v = \frac{I}{(nAe)}$$

WORKED EXAMPLE

A copper wire has a cross-sectional area of 2.0 mm^2 and carries a current of 6.0 A. The number of free electrons in copper per unit volume is 8.0×10^{28} m^{-3} . Calculate the mean drift velocity of the electrons in this sample of wire.

Answer

We know that $v = \frac{I}{(nAe)}$. The values of *I*, *n* and *e* are all in the correct S.I. units, but the area must be converted from mm^2 to m^2. There are 1000 mm in a length of 1 m, so there will be 1 000 000 mm^2 in an area of 1 m^2.
To convert mm^2 to m^2 we need to divide by 10^6, which give us a value of 2.0×10^{-6} m^2.
Substituting the values into the drift velocity equation gives:

$$v = \frac{6.0}{(8.0 \times 10^{28} \times 2.0 \times 10^{-6} \times 1.602 \times 10^{-19})}$$

$$= \frac{6.0}{25\,632}$$

$= 0.0002$ m s^{-1}, which is a velocity of 0.2 mm s^{-1}.

Comparing conductors, semiconductors and insulators

Metals contain a very large number of free conduction electrons per unit volume and are very good conductors. Conversely, insulators contain very few free conduction electrons and are very poor conductors. In fact, the value for *n* for many insulators is close to zero.

Semiconductors, such as silicon and germanium, have values of *n* between those of conductors and insulators, meaning that they have intermediate conduction properties – neither very good nor very poor in terms of their conducting properties.

Semiconductor materials often undergo doping, where impurity atoms are added to the material. This improves their conducting properties.

Substance	Value of *n*	Type of material
copper	8.0×10^{28}	conductor
aluminium	1.8×10^{29}	conductor
calcium	4.6×10^{28}	conductor
silicon	8.8×10^{9}	semiconductor
germanium	2.0×10^{13}	semiconductor
gallium arsenide	2.0×10^{6}	semiconductor
polythene	close to zero	insulator

Table 1 The values of *n* for a variety of conductors, semiconductors and insulators.

Questions

1. Calculate the current flowing in a copper wire of radius 3.0 mm and length 45.0 cm by using information available to you above.
2. In an experiment using a small wafer of a semiconductor with an area of cross-section 8.2×10^{-6} m^2, the mean drift velocity of the electrons is 80 m s^{-1} when the current in the wafer is 3.6 mA. Show that the number density (*n*) of electrons in the wafer is more than one thousand million times smaller than the value for copper.
3. Use the internet to find out why semiconductors are used in the microprocessors of computers?
4. A current of 0.2 A flows in a gold wire of diameter 1.0 mm. The electron number density for gold is 5.9×10^{29} m^{-3}. What is the mean drift velocity, in m s^{-1}, of the electrons?
5. A current of 0.4 A flows in a copper wire of diameter 0.22 mm. Calculate the mean drift velocity, in m s^{-1}, of the electrons, using the electron number density for copper given in Table 1.
6. The wire in question 4 will melt if the current is too large for too long. For a 5 A fuse wire, calculate the maximum mean drift velocity of electrons in the wire.

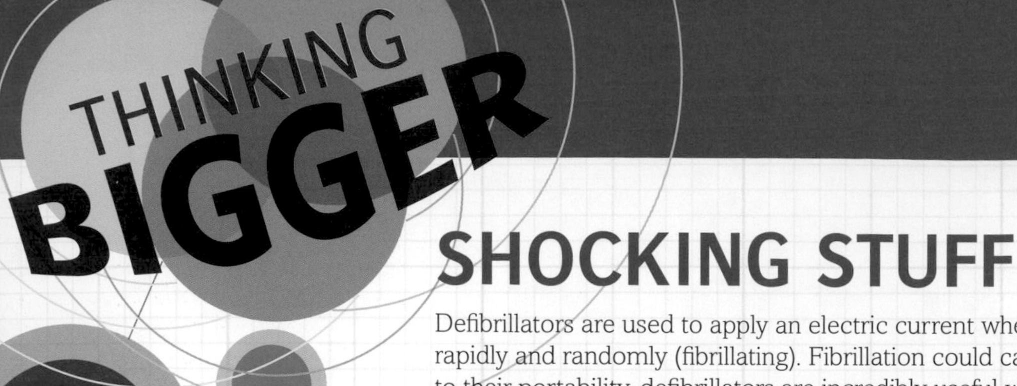

SHOCKING STUFF

Defibrillators are used to apply an electric current when a patient's heart muscle is contracting rapidly and randomly (fibrillating). Fibrillation could cause a sudden stop in blood circulation. Due to their portability, defibrillators are incredibly useful when dealing with serious incidents that need to be dealt with quickly. Service stations and other public places in countries such as France have defibrillators at hand so that a victim can be treated immediately on site.

WHAT ARE DEFIBRILLATORS AND HOW DO THEY WORK?

The most important component of a defibrillator is a capacitor that stores a large amount of energy in the form of electrical charge, then releases it over a short period of time.

[...] Figure 1 shows a defibrillator. When the switch is in position 1, direct current (DC) from the power supply is applied to the capacitor. Electrons flow from the upper plate to the positive terminal of the power supply and from the negative terminal of the power supply to the lower plate. Therefore current flows and a charge begins to build up on each electrode of the capacitor, with the lower plate becoming increasingly negatively charged, and the upper plate increasingly positively charged. As the charge builds up on the plates, it creates a potential difference across the plates (V), which opposes the electromagnetic force of the power supply (E). Initially when there is no charge on the plates, V is zero and it is easy to move electrons onto the plates. As V increases, however, it opposes further movement of electrons, and increasing work must be done to move more electrons onto the plates.

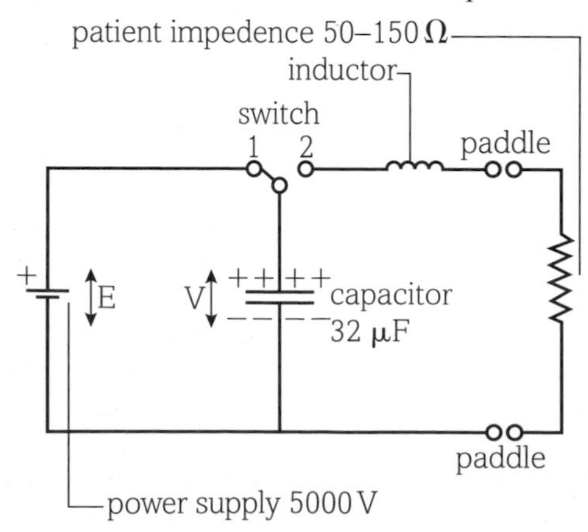

Figure 1

The work done (W) to move charge (Q) through a potential difference V is: $W = VQ$. Charging a capacitor is therefore an exponential process, with a time constant determined by the capacitance and the resistance of the circuit through which the current flows. When V equals E, the current ceases to flow and the capacitor is fully charged. In this example, the amount of charge stored ($Q = CV$) is $32\,\mu F \times 5000\,V = 160\,mC$.

Work must be done against the field to store charge in the capacitor. The charged capacitor is therefore a store of potential energy, which may be released on discharge.

When the paddles are applied to the patient's chest and the switch is moved to position 2, a circuit is completed. Electrons stored on the lower (negative) plate of the capacitor are able to pass through the patient and back to the upper plate. Thus, current flows, stored electrical energy is released, and the potential difference across the plates (V) falls to zero (i.e. the capacitor is discharged). The rate of discharge declines as the potential difference across the plates falls; it is an exponential process with a time constant determined by the capacitance and the resistance of the circuit through which the current flows.

For successful defibrillation, the current delivered must be maintained for several milliseconds. However, the current and charge delivered by a discharging capacitor decay rapidly and exponentially. Inductors are therefore used to prolong the duration of the electric current flow. Inductors are coils of wire that that produce a magnetic field when current flows through them. When current passes through an inductor, it generates a flow of electricity in the opposite direction which opposes current flow as predicted by Faraday's law. This opposition to current flow is called inductance.

Source

http://www.frca.co.uk/documents/4_1_29.pdf (last accessed 1 February 2015)

Where else will I encounter these themes?

1.1 2.1 2.2 3.1 3.2 3.3

Let us start by considering the nature of the writing in the article. The text above was taken from the website Anaesthesia UK – physical principles of defibrillators.

1. **Consider the article and comment on the type of writing that is being used. Who is the audience? Is it for the typical citizen or is it for members of the medical community? Does the text try to explain, persuade or describe at any point? What are the words or terms in the text that would influence you when determining who the article is intended for? How might you change the article to make it more suitable for a less well-informed audience?**

Remember to check the value of prefixes and the correct equation before performing a calculation.

We will now look at the physics that is in the article. Do not worry if the physics content or the mathematics is challenging at this stage. You can always return to the article later in your course, once some of the related topics have been studied in more depth. Use the timeline at the bottom of the page to help you put this work in context with what you have already learned and what is ahead in your course.

2. **What is defibrillation?**

3. **A defibrillator is charged before applying it to a patient. Use your knowledge of charge, current and resistance to explain why different currents might flow in different people's bodies.**

4. **Why is it so difficult to store electrical energy in large quantities?**

5. **Capacitors can store charge better if they contain a dielectric. What is a dielectric and how does it work?**

6. **How much charge could be stored by a capacitor of capacitance 470 μF if it had 5000 V across it?**

7. **What are inductors and why are they used in defibrillator circuits?**

8. **What are the potential hazards associated with using a defibrillator incorrectly?**

9. **On 17 March 2012, the footballer Fabrice Muamba suffered a cardiac arrest and collapsed during a football game. He was given multiple defibrillator shocks and, in total, his heart stopped for 78 minutes. How is it possible for somebody to survive for 78 minutes without a heartbeat?**

10. **When a capacitor is discharged, the rate at which the charge leaves the plates is not linear. Why is this?**

Activity

You have been asked to write a leaflet which is going to be left in the waiting room at hospitals to inform people about defibrillators and how they can save lives. Your leaflet needs to explain:

- how a healthy heart works, using the correct medical terms;
- what happens during a cardiac arrest;
- how a defibrillator works (including ideas about charge and energy);
- how a normal life can be achieved after a heart attack, through an implanted defibrillator.

DID YOU KNOW?

In Australia up until the 1990s it was quite rare for ambulances to carry defibrillators. This changed in 1990 after Australian media mogul Kerry Packer had a heart attack and, purely by chance, the ambulance that responded to the call carried a defibrillator. After recovering, Kerry Packer donated a large sum to the Ambulance Service of New South Wales in order that all ambulances in New South Wales should be fitted with a personal defibrillator, which is why defibrillators in Australia are sometimes colloquially called 'Packer Whackers'.

4.1 Practice questions

1. Which of the following statements is true? [1]

   A Charge is the rate of flow of current.

   B A particle can have a charge of 1.4 e.

   C 20 C of charge pass a point when 10 A flows for 2 hours.

   D 20 C of charge contains 1.25×10^{20} electrons.

2. 4.5×10^{24} electrons flow through a device in 2 hours. What is the current flowing through the device? [1]

   A 1 A

   B 10 A

   C 100 A

   D 1000 A

3. Which of the following is/are true? [1]

   (i) Kirchhoff's first law concerns the conservation of charge.

   (ii) Ions are the charge carriers in electrolytes.

   (iii) A charged particle must have a charge that is a multiple of e.

   A (i), (ii) and (iii)

   B only (i) and (ii)

   C only (ii) and (iii)

   D only (i)

4. Which of the following statements are true? [1]

   (i) Conductors have a greater number density of charge carriers than semiconductors.

   (ii) The drift velocity of electrons is inversely proportional to the cross-sectional area of the wire.

   (iii) The rate of flow of charge is directly proportional to the diameter of the wire.

   A (i), (ii) and (iii)

   B only (i) and (ii)

   C only (ii) and (iii)

   D only (i)

[Total: 4]

5. 7.6×10^{23} electrons flow through a resistor in 12 minutes. Calculate the size of the average current flowing through the resistor in this time. [2]

[Total: 2]

6. A current of 6.4 A flows in a copper wire of length, l, and cross-sectional area πr^2. What will the size of the current become if:

   (a) the length of the wire is doubled [1]

   (b) the cross-sectional area of the wire is doubled [2]

   (c) the radius of the wire is halved [1]

   (d) the material is changed to a material of greater resistivity? [1]

[Total: 5]

7. Copper sulfate solution is an example of an electrolyte.

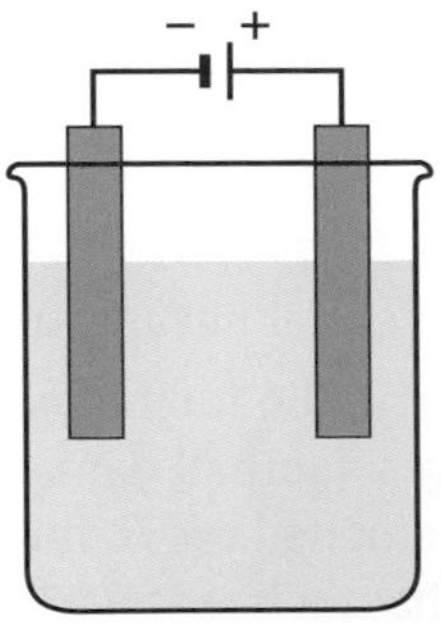

– +

   (a) State:

   (i) the charge carriers present in the electrolyte [2]

   (ii) the charge carriers present in the copper wires. [1]

   (b) Explain how the charges present in the electrolyte get to each of the terminals shown in the diagram and how electric current will therefore be registered on an ammeter. [5]

[Total: 8]

8. (a) Name the charge carriers responsible for electric current:
   (i) in a metal [1]
   (ii) and in an electrolyte. [1]

(b) A copper rod of cross-sectional area $3.0 \times 10^{-4}\,m^2$ is used to transmit large currents. A charge of 650 C passes along the rod every 5.0 s. Calculate:
   (i) the current I in the rod [1]
   (ii) the total number of electrons passing any point in the rod per second [1]
   (iii) the mean drift velocity of the electrons in the rod given that the number density of free electrons is $1.0 \times 10^{29}\,m^{-3}$. [2]

(c) The copper rod in (b) labelled **X** in **Figure 2**) is connected to a longer thinner copper rod **Y**.

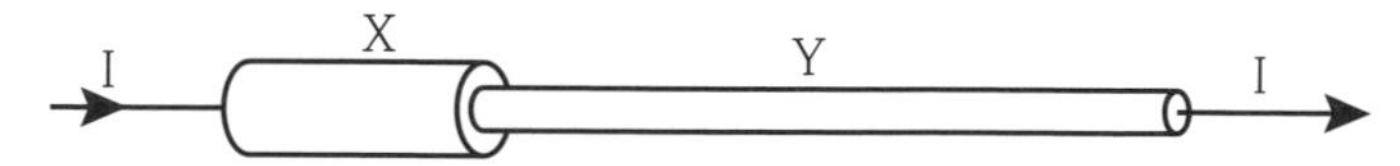

Figure 2

   (i) State why the current in **Y** must also be I. [1]
   (ii) Rod **Y** has half the cross-sectional area of rod **X**. Calculate the mean drift velocity of electrons in **Y**. [1]

[Total: 8]

[Q4, H156/02 sample paper 2014]

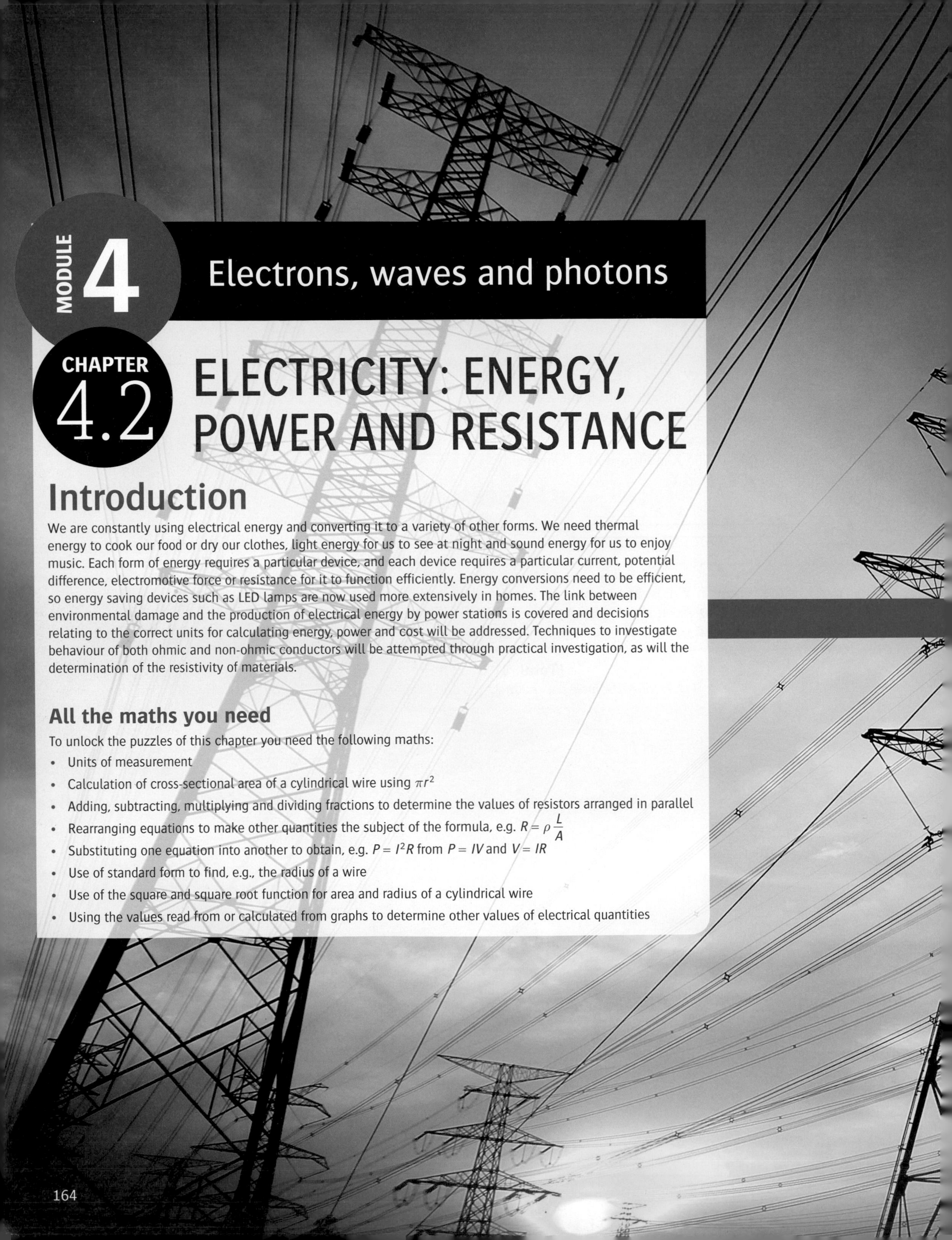

MODULE 4

Electrons, waves and photons

CHAPTER 4.2

ELECTRICITY: ENERGY, POWER AND RESISTANCE

Introduction

We are constantly using electrical energy and converting it to a variety of other forms. We need thermal energy to cook our food or dry our clothes, light energy for us to see at night and sound energy for us to enjoy music. Each form of energy requires a particular device, and each device requires a particular current, potential difference, electromotive force or resistance for it to function efficiently. Energy conversions need to be efficient, so energy saving devices such as LED lamps are now used more extensively in homes. The link between environmental damage and the production of electrical energy by power stations is covered and decisions relating to the correct units for calculating energy, power and cost will be addressed. Techniques to investigate behaviour of both ohmic and non-ohmic conductors will be attempted through practical investigation, as will the determination of the resistivity of materials.

All the maths you need

To unlock the puzzles of this chapter you need the following maths:

- Units of measurement
- Calculation of cross-sectional area of a cylindrical wire using πr^2
- Adding, subtracting, multiplying and dividing fractions to determine the values of resistors arranged in parallel
- Rearranging equations to make other quantities the subject of the formula, e.g. $R = \rho \frac{L}{A}$
- Substituting one equation into another to obtain, e.g. $P = I^2R$ from $P = IV$ and $V = IR$
- Use of standard form to find, e.g., the radius of a wire
- Use of the square and square root function for area and radius of a cylindrical wire
- Using the values read from or calculated from graphs to determine other values of electrical quantities

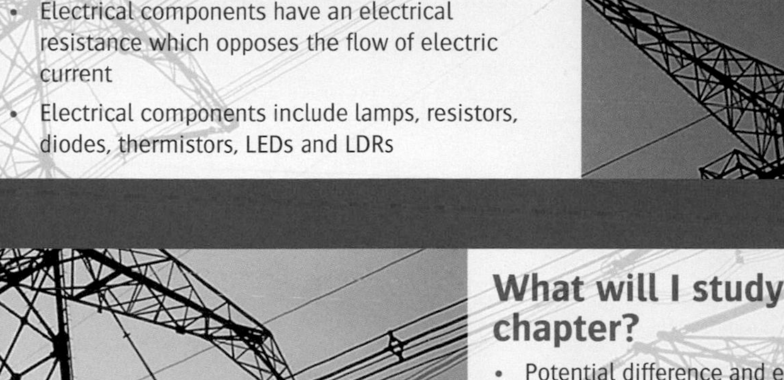

What have I studied before?

- The mains electrical supply in UK homes is 230V a.c.
- The unit of electrical energy that is used by energy companies is the kilowatt-hour
- The cost of electrical energy is calculated based on the kilowatt-hour or kW h
- Electrical energy used in the home is a.c. (alternating current) whereas that used in circuits with cells or batteries is DC.
- Potential difference is measured by connecting a voltmeter across the component
- Electrical energy is measured in joules, electrical power is measured in watts
- Electrical circuits may contain components including a cell, ammeter, bulb, voltmeter, switch and conductors
- Electrical components have an electrical resistance which opposes the flow of electric current
- Electrical components include lamps, resistors, diodes, thermistors, LEDs and LDRs

What will I study later?

- For components arranged in series, the total resistance is the sum of the individual resistances, i.e. $R = R_1 + R_2 + R_3$ (AS)
- For components arranged in parallel, the total effective resistance of the network is given by $\frac{1}{R_T} = \frac{1}{R_1} + \frac{1}{R_2} + \frac{1}{R_3}$ (AS)
- E.m.f., internal resistance and lost volts (AS)
- Kirchhoff's first and second laws (AS)
- The arrangement of resistors in series and parallel to form potential divider circuits (AS)
- Capacitors are used to store electrical energy, which can be calculated using $E = \frac{1}{2}QV$, $E = \frac{1}{2}CV^2$ or $E = \frac{\frac{1}{2}Q^2}{C}$ (AL)
- Capacitors and resistors can be added in parallel or series so that the capacitors can be charged and discharged (AL)
- The exponential laws that govern the charging and discharging of capacitors (AL)
- The force on a current carrying wire in the presence of a magnetic field (AL)
- Fleming's left hand rule and right hand rule (AL)

What will I study in this chapter?

- Potential difference and electromotive force (e.m.f.) are both measured in volts
- Potential difference and e.m.f. are different in terms of the energy transfers involved in a circuit
- Electrical energy, or work done, can be expressed as $W = VQ$, $W = EQ$ or $W = VIt$
- The kinetic energy gained by an electron that has been accelerated through an electric field is given by $eV = \frac{1}{2}mv^2$
- I–V characteristics are different for ohmic and non-ohmic conductors, which can be investigated using practical procedures. The resistance of certain components is temperature-dependent and others vary with light intensity
- Resistivity is a property of a material that can be calculated using $\rho = \frac{RA}{L}$ at constant temperature, but is also temperature dependent.
- Power is the rate of transfer of electrical energy per second. It is measured using the watt and calculated using $P = IV$, $P = I^2R$ and $P = \frac{V^2}{R}$
- The kilowatt-hour is the unit of electrical energy used in the home to calculate the cost of using electrical energy

Potential difference and e.m.f.

By the end of this topic, you should be able to demonstrate and apply your knowledge and understanding of:

* potential difference (p.d.); the unit *volt*
* electromotive force (e.m.f.) of a source such as a cell or power supply
* distinction between e.m.f. and p.d. in terms of energy transfer
* energy transfer; $W = VQ$; $W = EQ$
* energy transfer $eV = \frac{1}{2}mv^2$ for electrons and other charged particles

Introduction

In order for an electric current to flow in a circuit, there must be an energy source – in school laboratories this is often a chemical cell or battery. When no current is flowing through the supply, we have a situation in which there is the maximum possible potential difference across the supply. This is called the **electromotive force** or **e.m.f.** of the supply and is normally associated with energy sources in which chemical energy is transferred to electrical energy.

Components in an electrical circuit transfer the energy carried by electrical charges to other forms of energy. The **potential difference** or **p.d.** across each component describes how much energy per unit charge is being transferred at each component.

Both e.m.f. and p.d. can be measured in either joules per coulomb ($J\,C^{-1}$) or volts (V). There are subtle differences between e.m.f. and p.d.:

- e.m.f. is the energy transferred to each unit charge, so $\text{e.m.f.} = \frac{\text{energy transferred}}{\text{charge}}$
- p.d is the work done by each unit of charge, so $\text{p.d.} = \frac{\text{work done}}{\text{charge}}$

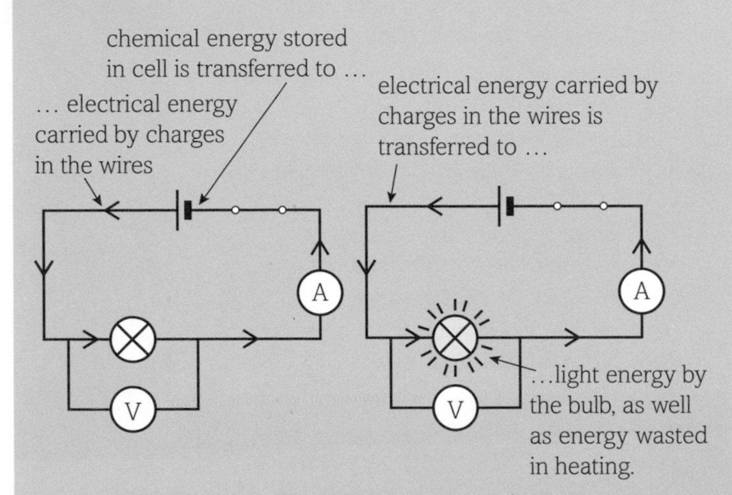

Figure 1 Both e.m.f. and p.d. can be measured in $J\,C^{-1}$ or V, but the natures of the energy transfers are different.

KEY DEFINITIONS

The **e.m.f. (electromotive force)** of a supply is the energy gained per unit charge by charges passing through the supply, when a form of energy is transferred to electrical energy carried by the charges. It is measured in volts (V) or joules per coulomb ($J\,C^{-1}$).

The **p.d. (potential difference)** measured across a component is the energy transferred per unit charge by the charges passing through the component. It is measured in volts (V) or joules per coulomb ($J\,C^{-1}$).

WORKED EXAMPLE 1

1.25×10^{20} electrons are provided with 200 J of energy by a cell. What is the e.m.f. of the cell?

Answer

Remember – one coulomb of charge is equal to 6.25×10^{18} electrons.

$$1.25 \times 10^{20} \text{ electrons} = \frac{1.25 \times 10^{20}}{6.25 \times 10^{18}} = 20 \text{ coulombs of charge}$$

$$\text{e.m.f.} = \frac{200\text{ J}}{20\text{ C}}$$

$$= 10\text{ V}$$

Measurement of potential difference

We use a voltmeter to measure the p.d. across a component. A voltmeter is always connected in parallel across the electrical component.

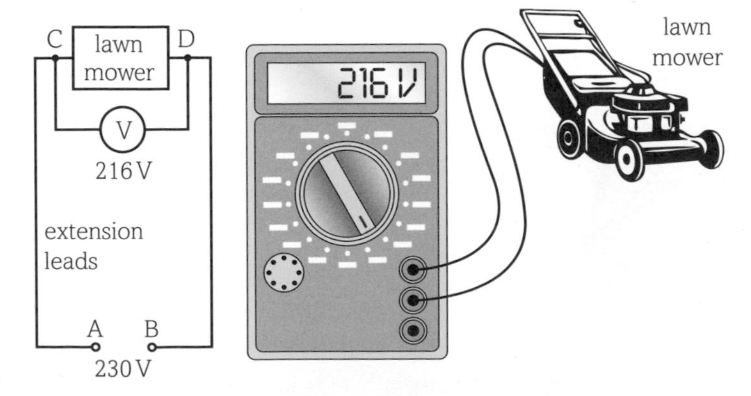

Figure 2 Measuring a p.d.

A voltmeter connected across the mains supply will read 230 V. However, when the voltmeter is connected across points C and D, the p.d. measured is 216 V. Of the 230 joules per coulomb available from the mains supply, 14 joules per coulomb have been transferred to heating the extension lead.

In other words, the circuit starts with 230 J C^{-1}; 7 J C^{-1} is used heating the extension lead on the way to the mower, leaving 223 J C^{-1}; 216 J C^{-1} is used by the mower leaving 7 J C^{-1} to heat the extension lead on the way back.

Energy in circuits

The e.m.f., ε, of a supply such as a solar cell is the energy transferred (W) by the supply when charge Q passes through the cell.

$$\varepsilon = \frac{\text{energy transferred to charge}}{\text{charge}}$$

$$\varepsilon = W/Q \text{ so } W = \varepsilon Q$$

Since the unit of energy is joule and the unit of charge is coulomb, the S.I. unit of e.m.f. is joules per coulomb, also known as volts (V).

$$1\ \text{J C}^{-1} = 1\ \text{V}$$

The p.d. (V) across a component is the energy transferred *to* the component (the work done) when charge Q passes through it.

$$\text{p.d.} = \frac{\text{energy transferred to charge}}{\text{charge}}$$

$$V = W/Q \text{ so } W = VQ$$

The unit of p.d. is also J C^{-1} or V.

Energy transferred to each charged particle

In metals, the charge carriers are electrons. A dynamo, battery or solar cell transfers energy to each electron that passes through. Each electron has a charge of e coulombs, so an energy change of eV takes place when an electron passes through a p.d. V. As we shall see in Topic 4.5.1, the energy transferred when one electron travels through a potential difference of one volt is a derived unit called one electronvolt.

When a potential difference is applied across a conductor, the electrons are accelerated and gain kinetic energy. When an electron is accelerated through a p.d. V, the kinetic energy gained is:

$$\frac{1}{2} m_e v^2 = eV$$

where m_e is the mass of the electron.

In metals, the conduction electrons quickly lose some of this kinetic energy by collisions with the metal ions, as we shall see in Topic 4.3.2.

WORKED EXAMPLE 2

An electric kettle requires 300 000 J to be supplied in 120 s to boil a cup of water. The e.m.f. of the mains supply is 230 V. Calculate:

(a) the charge supplied

(b) the current to the kettle

Answer

(a) The e.m.f. of 230 V means that 230 joules per coulomb is supplied to the charge carriers passing through the supply. 300 000 J are required in total, so the charge supplied is $\frac{300\,000}{230}$ = 1304 coulombs (do not round up figures at this point).

(b) 1304 coulombs in 120 s means a current of 1304 C/120 s = 10.9 C s^{-1} = 10.9 A

Questions

1 Look at the statements in Table 1. Do they apply to e.m.f. or p.d. – or both?

Statement	Applies to e.m.f.	Applies to p.d.
'Measured in volts.'		
'Maximum value across the cell when no current is flowing in the circuit.'		
'Used when describing energy being transferred to electrical energy from another form.'		
'Used when describing the transfer of electrical energy to another form.'		
'Can have units JC^{-1}'		

Table 1

2 What values are recorded by each of the meters in Figure 3?

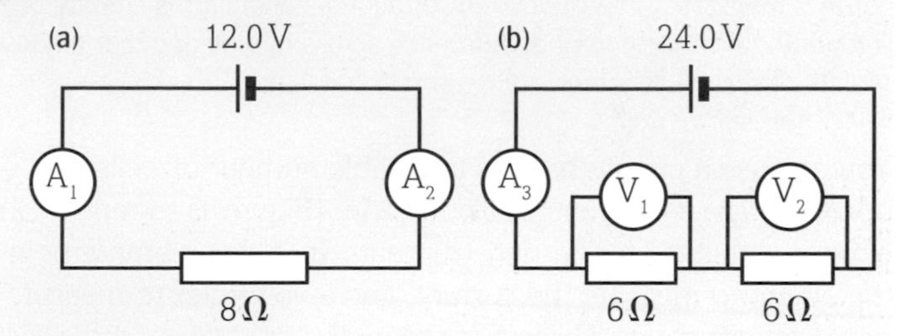

Figure 3

3 The electric circuit in a small torch consists of a battery of e.m.f. 3.0 V and a bulb connected in series. The potential difference across the bulb is 2.7 V when the current in the bulb is 0.3 A.

(a) Calculate how much energy is transferred in one minute:

(i) from chemical energy to electrical energy in the battery

(ii) from electrical energy to thermal energy and light energy in the bulb.

(b) Suggest why the values in part (a) are different.

4 Inside a cell, the energy required to move an electron from the positive to the negative terminal is 2.4×10^{-19} J. Show that the e.m.f. of the cell is 1.5 V.

5 (a) How much energy is transferred when 1.0 C of charge passes through the cell in question 4 above?

(b) How much charge passes through the cell in 1.0 minute?

6 When there is a potential difference of 1.0 V between two points in an electrical circuit, what is the mean amount of energy lost by an electron when moving between these points?

7 When an electron is accelerated across a potential difference of about 15 V it has enough energy to ionise a nitrogen atom, that is, to remove an electron from the atom by collision with it. How many attojoules, symbol aJ, is this? 1 aJ = 1.0×10^{-18} J.

Resistance and Ohm's law

By the end of this topic, you should be able to demonstrate and apply your knowledge and understanding of:

* resistance; $R = \frac{V}{I}$; the unit Ω
* Ohm's law

Ohm's law

In the 1820s, the physicist Georg Ohm discovered that increasing the number of cells in a battery that was connected to a length of wire produced an increase in the current – and that the current was directly proportional to the potential difference across its ends, provided that the temperature of the wire remained constant. This led him to publish the statement we now know as **Ohm's law**.

KEY DEFINITION

Ohm's law states that the current through a conductor is directly proportional to the potential difference across it, provided that physical conditions, such as temperature, remain constant.

You can use a circuit that has a variable number of cells connected to a long length of thin wire (Figure 1) to repeat Ohm's experiment using modern instruments. Include an ammeter in the circuit to measure the current, and a voltmeter to measure potential difference. Using a long length of thin wire and a switch keeps the current at a low value so the heating effect is negligible. Table 1 shows the readings obtained from one such experiment.

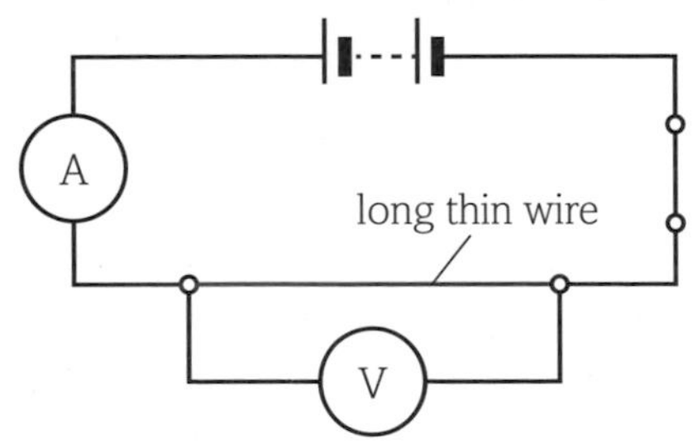

Figure 1 Circuit for Ohm's experiment.

Number of cells in battery	p.d./V	current/mA
0	0	0
1	1.56	3.2
2	3.09	6.5
3	4.57	9.7
4	6.14	12.9
5	7.62	15.9
6	9.12	19.1
7	10.59	22.3
8	12.07	25.5

Table 1 The results show how the voltage and the current change when different numbers of cells are included in a circuit.

A graph plotting these results is shown in Figure 2. It illustrates clearly that the current is proportional to the potential difference. This graph is called the *I–V* characteristic of the wire used in the experiment.

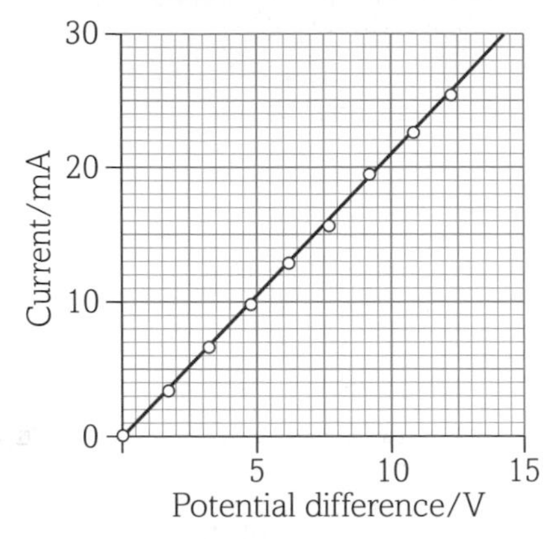

Figure 2 An *I–V* characteristic graph.

Resistance and the ohm

Resistance is defined by the equation:

$$\text{resistance} = \frac{\text{potential difference}}{\text{current}}$$

$$R = \frac{V}{I}$$

The unit of resistance must therefore be 'volts per ampere'. This S.I. unit is known as the *ohm* and has the Greek symbol Ω.

WORKED EXAMPLE

An ammeter, connected in series with a resistor, registers a current of 280 mA when a potential difference of 18 V is applied across the ends of the resistor. What is its resistance? Give your answer to 2 significant figures.

Answer

$$R = \frac{V}{I} = \frac{18\text{ V}}{0.280\text{ A}}$$

$= 64.3\ \Omega$, which is 64 Ω to 2 significant figures.

The factors that affect electrical resistance

From Ohm's law, we know the relationship between resistance, potential difference and current – increasing the potential difference or decreasing the resistance will lead to a greater flow of electric current in a circuit.

However, there are other physical factors present that will affect the resistance of a circuit.

The material the wire is made from

Metals contain some electrons that are free to move throughout the structure under the influence of a potential difference. This means they are good conductors and so the resistance of metals is low. A copper wire of length 1 m and diameter 1 mm has a resistance of $2 \times 10^{-2}\,\Omega$. Plastics and other insulators, which do not have free electrons, have high resistance values in excess of $10^{6}\,\Omega$. See Topic 4.2.4.

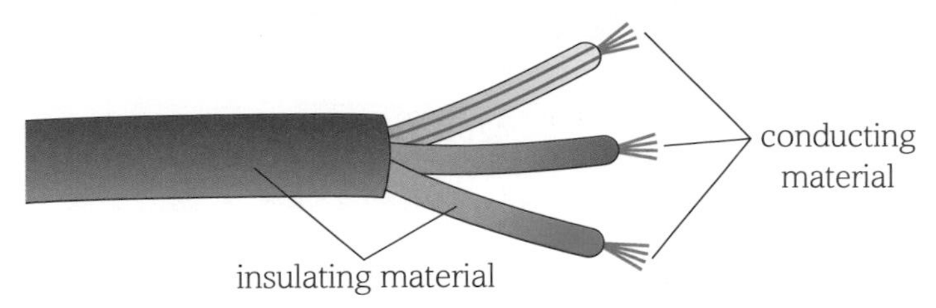

Figure 3 Different materials have different electrical properties – metals are conductors and plastics are insulators.

The length of the wire

As the length of a conductor increases, so does its electrical resistance. This is because the electrons, which are the charge carriers in metals, experience a smaller potential gradient as the wire gets longer. The increased length of the wire means that the size of the 'voltage per metre' value is lower, and this causes a lower drift velocity and hence a smaller current. This means that the resistance of the wire increases – resistance is directly proportional to the length of a wire.

Another view of this is that electrical resistance can be thought of in terms of the likelihood of collisions occurring between the conduction electrons and the fixed vibrating ions in the metal lattice. Increasing the number of collisions between the conduction electrons and the fixed ions causes the electrical resistance to become higher and the current to become lower.

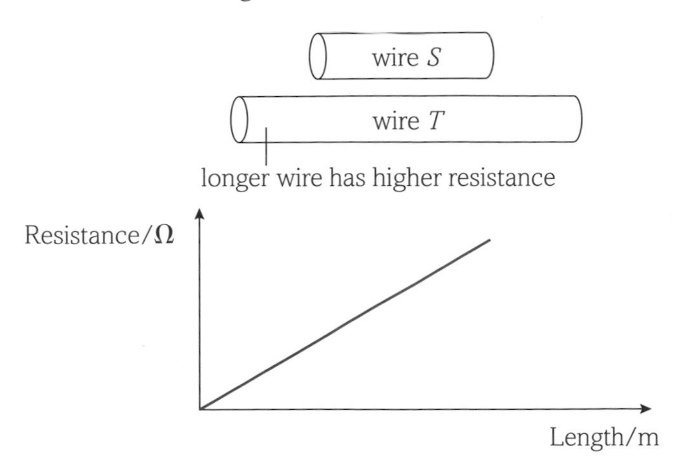

Figure 4 The longer the wire, the higher its resistance.

The cross-sectional area of the wire

Increasing the cross-sectional area of the wire is much like having a number of wires bundled together. There is no increase in the potential gradient, but there is a bigger volume of electrons that are available to flow at the same drift velocity. Resistance is inversely proportional to the cross-sectional area of a wire.

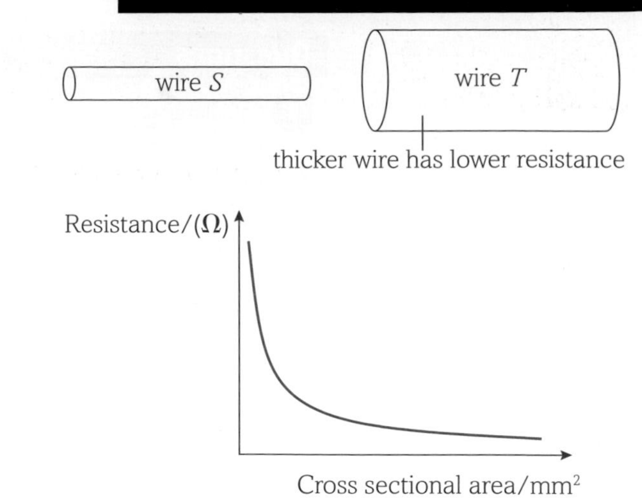

Figure 5 The thicker the wire, the lower the resistance.

The temperature of the wire

When the temperature of a metal wire increases, the average vibrational kinetic energy of the fixed metal ions also increases (Figure 6).

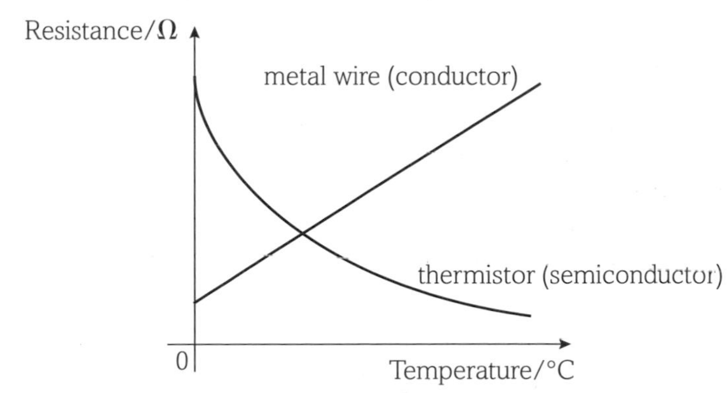

Figure 6 The resistance of a metal wire increases as the temperature increases. The resistance of a thermistor decreases as the temperature increases.

This means that the effective cross-sectional area of the ions increases and it is more difficult for the conduction electrons to pass them. Ultimately this will lead to more collisions between the conduction electrons and the vibrating metal ions. In turn this will result in an increase in the electrical resistance of a metal. However, some materials (made from semiconductors) have structures that lead to a decrease in electrical resistance when their temperature increases. Thermistors (Figure 7) are examples of such devices, and this property makes them very useful in potential divider circuits, which will be covered in Topic 4.3.4.

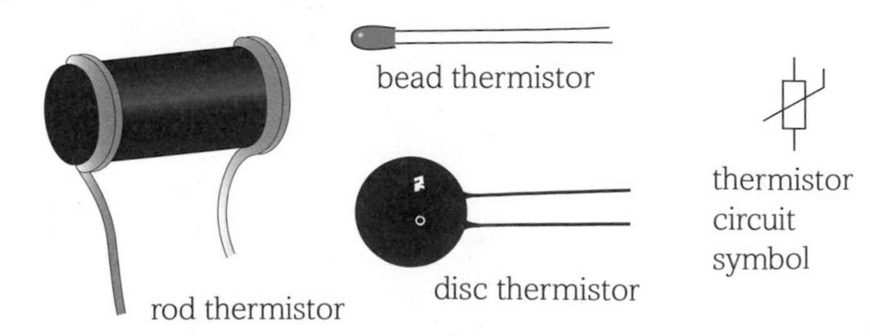

Figure 7 Thermistors come in different forms.

Questions

1 Calculate:

(a) the resistance of a component that requires 120 V across it for a current of 4.8 A to flow through it

(b) the potential difference needed to drive a current of 0.45 A through a resistor marked 47 Ω

(c) the current that flows through a 470 Ω resistor when a p.d. of 12 V is applied across it.

2 Copy and complete Table 2 by filling in the missing values using Ohm's law.

Potential difference/V	Current/A	Resistance/Ω
24.0	0.2	
	0.04	4700
10 000		12×10^6
1.2×10^4	6×10^{-5}	
6.8×10^2		3.6×10^6

Table 2

3 Which of the graphs in Figure 8:

(a) obey Ohm's law?

(b) has the highest resistance when $V = 2.5$ V?

(c) has a resistance of 4.4 Ω when $I = 0.5$ A?

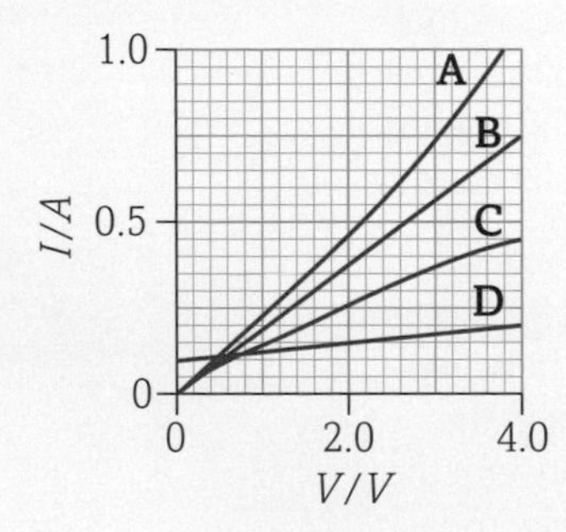

Figure 8

4 Find the values of A, B, C, D and E in the circuits shown in Figure 9:

(a)

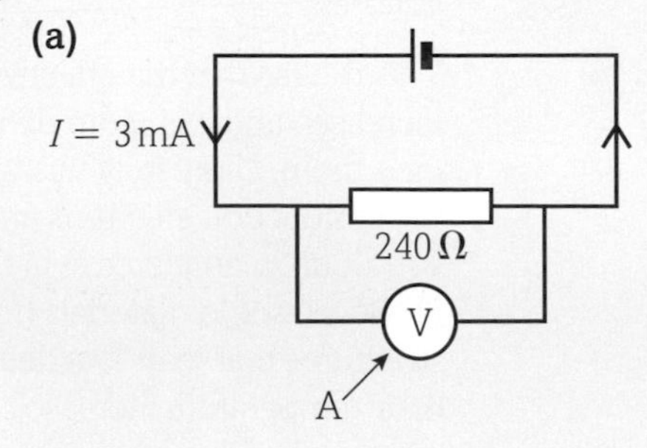

(b)

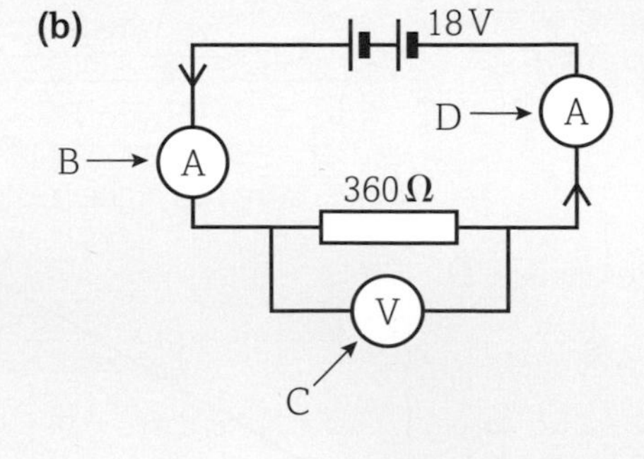

(c)

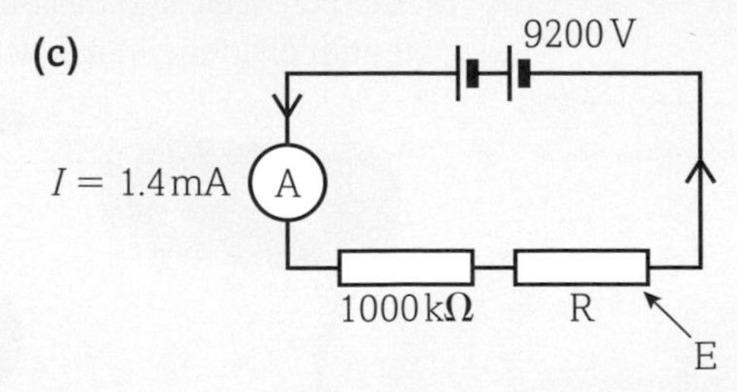

Figure 9

LEARNING TIP

You will need to know the rules for combining resistance in series. This is covered in Topic 4.3.2.

Resistance of circuit components

By the end of this topic, you should be able to demonstrate and apply your knowledge and understanding of:

* *I–V* characteristics of a resistor, filament lamp, diode and light-emitting diode (LED)
* light-dependent resistor (LDR); variation of resistance with light intensity
* techniques and procedures used to investigate the electrical characteristics for a range of ohmic and non-ohmic components

Introduction

The resistance of a metal, measured in Ω, is a measure of its opposition to the flow of current through it.

Electrical resistance is the opposition to the flow of electric current – the higher the resistance of a circuit, the lower the current.

A variety of electrical components can be used in circuits. Their resistance depends on a number of variables, including their internal structure, and this will determine their behaviour in circuits.

I–V characteristic of a resistor

A resistor at constant temperature behaves in accordance with Ohm's law. This means that the current through the resistor is directly proportional to the potential difference across its ends. The resistor behaves in the same way regardless of the direction of the electric current. Conductors that show a direct proportionality between current and potential difference (Figure 2) are called ohmic conductors.

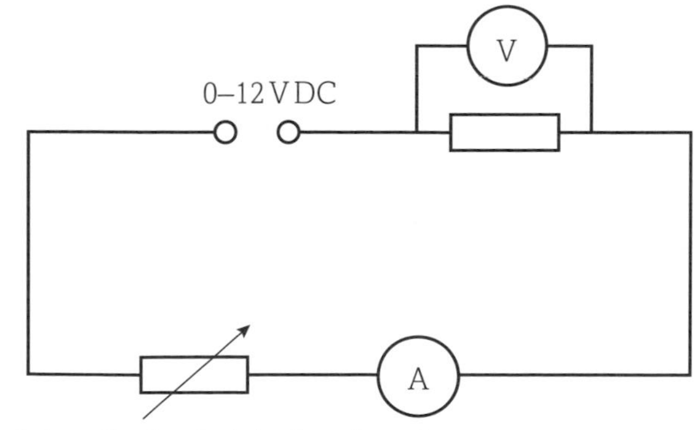

Figure 1 Investigating the behaviour of a resistor.

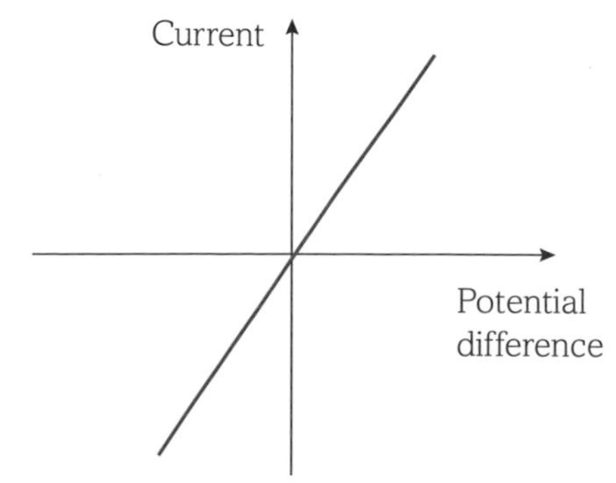

Figure 2 Current against potential difference for a ohmic conductor.

I–V characteristic of a filament lamp

At low temperatures, a filament lamp (Figure 3) behaves in accordance with Ohm's law and the relationship between current and resistance is fairly linear. However, as the current increases the free electrons drift through the metal lattice more quickly. This leads to more collisions between the electrons and the fixed ions in the structure. The collisions cause the metal ions to have a higher vibrational kinetic energy, which reduces the number of electrons that can get past. In simple terms, the resistance of the filament lamp increases as its temperature increases. Like a simple resistor, changing the polarity of the cell has no effect on the behaviour of the filament lamp.

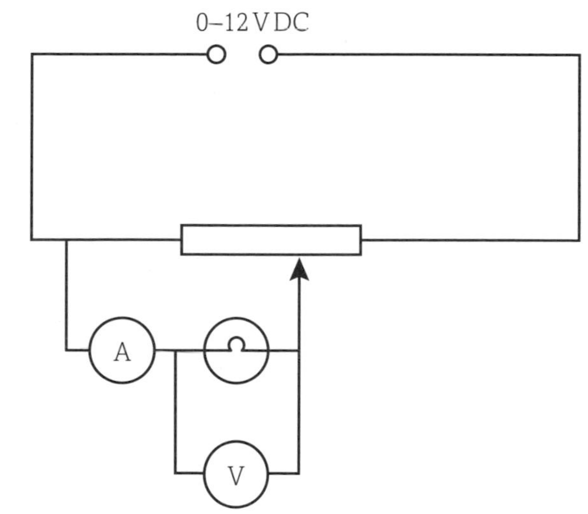

Figure 3 Investigating the behaviour of a filament lamp.

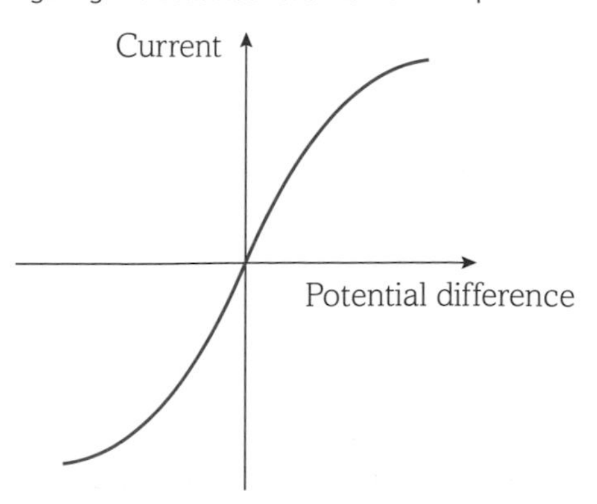

Figure 4 Initially, as the potential difference across the filament lamp increases, the current increases in an almost linear way. At higher potential differences, the gradient decreases as the current increases. A filament lamp is a non-ohmic conductor.

I–V characteristics of a diode or LED

Reversing the direction of the e.m.f. in a simple circuit causes the current to reverse through a resistor or filament lamp. With a diode this is not the case: the diode will allow current to flow through it in only one direction, from positive to negative. This is called the *forward bias* direction. In the *reverse bias* direction, effectively no current can pass and the diode is said to have almost infinite resistance. This can be seen in Figure 6 – reversing the cell polarity leads to no current, regardless of how high the e.m.f.

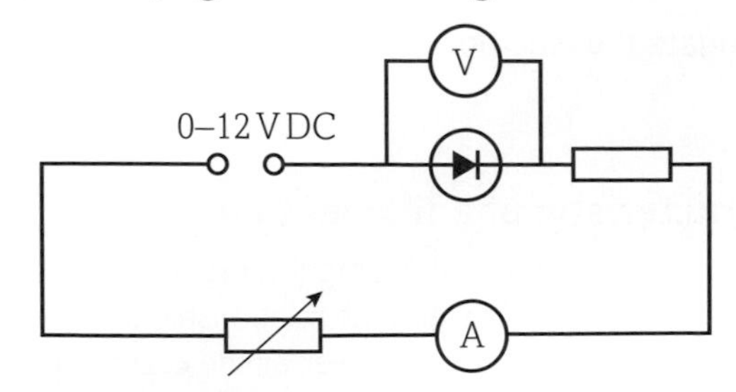

Figure 5 Investigating the behaviour of a diode.

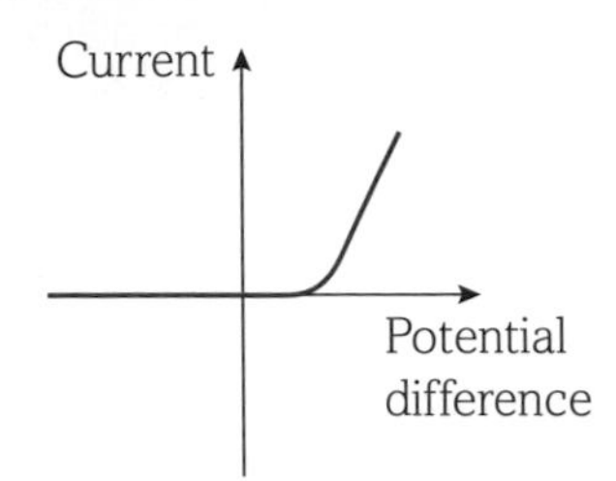

Figure 6 The *I–V* behaviour of a diode.

Diodes are simple devices that allow an electric current to pass in only one direction. Through developments in technology, their usefulness has increased. The current passing through them can now be much larger and some types, called light-emitting diodes (LEDs), emit visible light when current flows through them. You can buy LEDs that act as strong sources of white light or any other colour. They:

- switch on instantly
- are very robust
- are very versatile
- operate with low potentials
- have a long working life.

They are increasingly being used as light sources, and high-powered LEDs are expected to transform electric lighting. 10 000 LEDs can be used in a set of traffic lights compared with a single LED in the on/off indicator of a TV set.

The resistance of thermistors and light-dependent resistors

Some components can be used in circuits designed to respond to external conditions. A thermistor changes its resistance based on the surrounding temperature (Figure 7a), whereas a light-dependent resistor (LDR) has a high resistance when it is dark and a low resistance when light is shining on it (Figure 7b). These components are very useful for switching on/off other components under certain conditions of temperature and brightness.

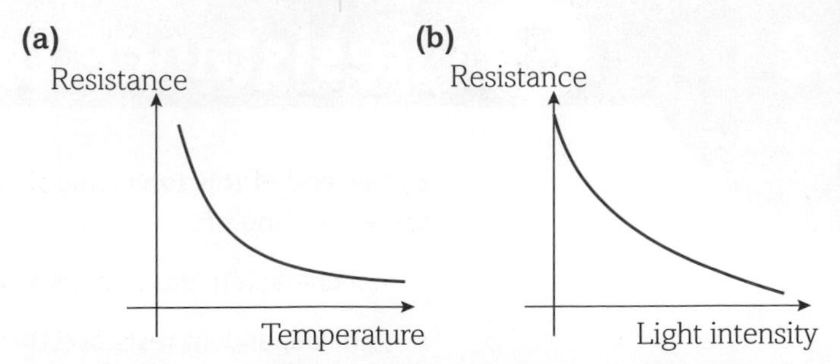

Figure 7 (a) The resistance of a negative coefficient thermistor decreases as its temperature increases, whereas (b) the resistance of an LDR decreases as the intensity of light falling on it increases.

Questions

1 Copy and complete Table 1 by deciding which component or components relate to each of the statements.

Statement	LED	Diode	Filament lamp	Resistor	LDR	Thermistor
ohmic conductor						
emits light						
resistance is temperature dependent						
infinite resistance in reverse bias						
resistance is light dependent						
detects changes in the surroundings						
semiconductor						

Table 1

2 Which of the graphs in Figure 8 go with the components described in this topic?

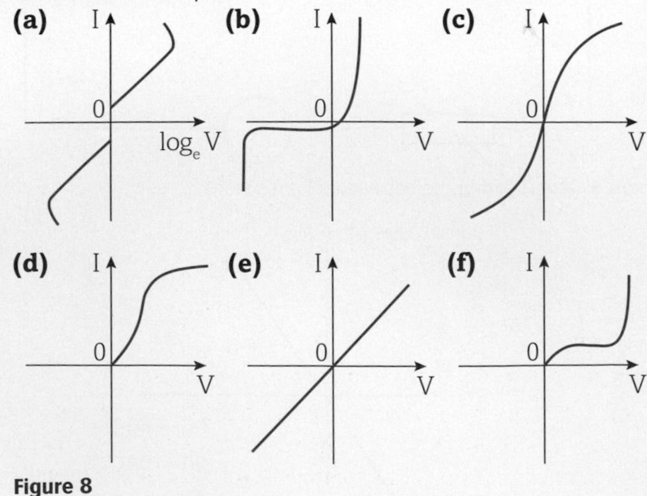

Figure 8

4 Resistivity

By the end of this topic, you should be able to demonstrate and apply your knowledge and understanding of:

* resistivity of a material; the equation $R = \frac{\rho l}{A}$
* techniques and procedures used to determine the resistivity of a metal

Introduction

We know that the resistance of a given material depends on a number of factors – in particular, if we change the length, cross-sectional area or temperature of a material then its resistance will change. For example, increasing the length or the temperature of a copper wire will lead to an increase in its resistance. Conversely, increasing its cross-sectional area lowers its resistance and allows a higher current to flow through it.

Resistivity, on the other hand, is not affected by changing the physical dimensions of a material – it is a fixed value for that material and is called an intrinsic property (see Topic 4.1.3). Even if we change the length or the cross-sectional area of a material, we can never change its resistivity.

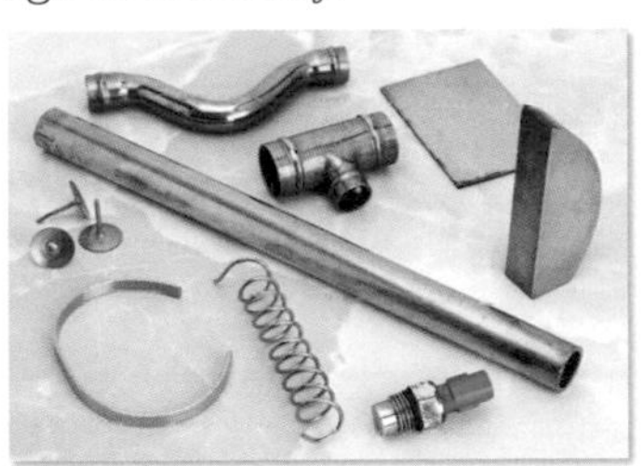

Figure 1 These specimens of copper all have different dimensions. Their respective resistance values are different but they will all have the same resistivity of $1.7 \times 10^{-8}\,\Omega\,m$ at 20° C.

Material	Resistivity/Ω m
silicon	1.0×10^{-2} to 2.3×10^{3}
germanium	1.0×10^{-4} to 10^{1}
nichrome	1.5×10^{-6}
constantan	4.9×10^{-7}
iron	9.7×10^{-1}
tungsten	5.6×10^{-8}
aluminium	2.8×10^{-8}
copper	1.7×10^{-8}
silver	1.6×10^{-8}

Table 1 Resistivities of different materials at 20 °C.

When you know the resistivity of a material at a particular temperature, you can calculate the resistance of a wire form of that material at that temperature. This is because:

- the resistance of a wire is proportional to its length, other factors being constant – the increase is because of the higher number of collisions the conduction electrons encounter as they travel along the wire
- the resistance is inversely proportional to the wire's cross-sectional area.

In algebraic form this gives:

$$R \propto \frac{l}{A}$$

The constant of proportionality that turns this expression into an equation is the resistivity, ρ, leading to the equation for calculating resistivity:

$$R = \frac{\rho l}{A} \text{ or } \rho = \frac{RA}{l}$$

WORKED EXAMPLE

A cylindrical wire of length 135 cm has an average diameter of 0.8 mm. If a current of 2.4 A flows through this wire when there is a potential difference of 16 V across its ends, calculate:

(a) the resistance of the wire in Ω to 2 significant figures

(b) the resistivity of the wire to 2 significant figures.

Answers

(a) $R = \frac{V}{I}$

$= \frac{16\,V}{2.4\,A}$

$= 6.7\,\Omega$

(b) The cross-sectional area of the wire is:

$A = \pi r^2$

$= \pi \times (0.4\,mm)^2 = \pi \times (0.4 \times 10^{-3}\,m)^2$

$= 5.0 \times 10^{-7}\,m^2$

$\rho = \frac{RA}{l}$

$\rho = \frac{(6.7 \times 5.0 \times 10^{-7})}{1.35}$

$= 2.5 \times 10^{-6}\,\Omega\,m$

Using resistivity in the real world

Geological surveys using electrical resistivity meters were carried out in Jacksonville Ohio, USA in 1998. The purpose of these surveys was to investigate if holes (or voids) had been left underground in areas where coal had been mined. Where this is the case there is a risk of roads collapsing due to the weight of traffic above the voids.

DID YOU KNOW?

Resistivity meters are used by archaeologists to scan sites. The data can be used to determine what structures are present underground as well as the nature of any erosion that has occurred in underground rocks. Recent studies near Stonehenge have shown that in the area with a radius of 330 m from Stonehenge there are a further 17 similar henge structures, 50 massive stones, 60 huge pillars and a wooden burial house that is 3000 years older than Stonehenge. Until recently, archaeologists and geologists believed that Stonehenge existed in isolation, but now it is known that it was part of a bigger structure.

INVESTIGATION

Determining the resistivity of a wire

Since resistivity is defined as:

$$\rho = \frac{RA}{l}$$

we can find ρ by making measurements of current and potential difference for different lengths of wire. To minimise errors from the heating effect it is best to adjust the variable resistor to keep a low fixed current flowing through the wire.

If we rearrange the equation we get $R = \frac{\rho l}{A}$. This equation can be compared to $y = mx + c$ (the equation of a straight line).

If we plot a graph of resistance, R, against length, l, of the sample wire then the gradient of the straight-line graph is equal to $\frac{\Delta R}{\Delta l}$. If we multiply the value of this gradient by the cross-sectional area of the wire, A, we obtain a value for the resistivity of the material under test.

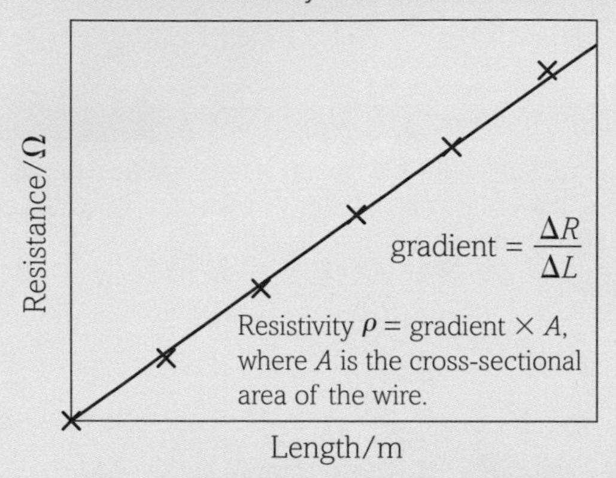

Figure 2

When measuring the diameter of the wire using a micrometer it is important to measure it in a number of different places along the whole length of the wire. This is because the wire is not necessarily perfectly cylindrical in all places and so one value could lead to erroneous results. Also, applying too much pressure when using the micrometer could lead to an 'egg-shaped' wire, from which the value of the cross-sectional area would be incorrect.

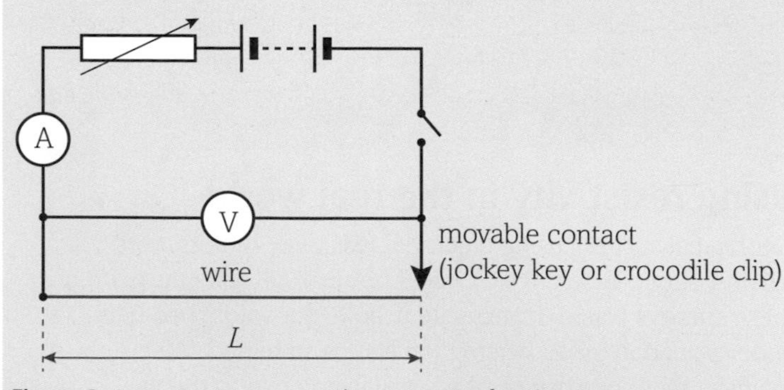

Figure 3 Apparatus to measure the resistivity of a wire.

Questions

1. A wire has a diameter of 2.4 mm and a length of 82 cm. When a potential difference of 12 V is applied across its ends, a current of 1.2 A flows through it. Calculate the resistivity of the wire.

2. For the wire in question 1, explain what would happen to the following quantities if the length of the wire was doubled and its resistance was halved:
   (a) the resistance of the wire
   (b) the current flowing through the wire
   (c) the resistivity of the wire.

3. Match the following resistivity values with the material they describe:

metallic element	$\rho = 3.8 \times 10^{6}\,\Omega\,\text{m}$
an alloy	$\rho = 4.7 \times 10^{-3}\,\Omega\,\text{m}$
a material found in semiconductor circuits	$\rho = 2.2 \times 10^{-8}\,\Omega\,\text{m}$
an insulator	$\rho = 1.6 \times 10^{-6}\,\Omega\,\text{m}$

4. Calculate the resistivity of the material shown in Figure 4 using the information provided.

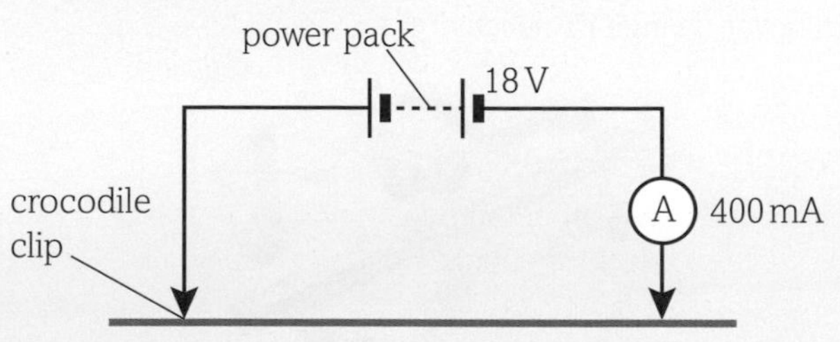

Figure 4

5. The resistivities of silicon and germanium are about 100 Ω m and 10 Ω m respectively at room temperature. The resistivity of carbon in the form of graphite is about 10^{-5} Ω m. Polyethylene has a resistivity higher than 10^{14} Ω m. Compare these values with those for the materials in Table 1. Suggest why they differ so widely.

6. Read these comparisons:
   - Material X is a better electrical conductor than material Y.
   - Material P is a better insulator than material Y.
   - Material S is a better conductor than P, but not as good as Y.

   Arrange materials P, S, X and Y in order of their resistivities, from highest to lowest.

The effect of temperature on resistivity

By the end of this topic, you should be able to demonstrate and apply your knowledge and understanding of:

* variation of the resistivity of metals and semiconductors with temperature
* negative temperature coefficient (NTC) thermistor; variation of resistance with temperature

Introduction

Changing the dimensions of the material will change its resistance. For example, making a copper wire longer and thinner will increase its resistance; making it shorter and thicker will cause its resistance to decrease.

Heating a copper wire will cause its resistance to increase – the extra thermal energy manifests itself as more vibrational kinetic energy in the metal ions, making it more difficult for the free electrons to pass.

Resistivity, on the other hand, is not affected by the length or the cross-sectional area of the material in question. Generally speaking, the resistivity of copper should remain the same, regardless of the dimensions of the copper wire being tested.

Resistivity and temperature

However, resistivity *is* affected by changes in temperature. Table 1 shows the temperature coefficients for a number of different metals and semiconductors.

Material	Resistivity at 20 °C/Ω m	Temperature coefficient
silver	1.59×10^{-8}	0.0038
copper	1.68×10^{-8}	0.0039
gold	2.44×10^{-8}	0.0034
aluminium	2.82×10^{-8}	0.0039
carbon (graphene)	1.00×10^{-8}	−0.0002
nichrome	1.10×10^{-6}	0.0004
silicon	6.4×10^{2}	−0.075
germanium	4.6×10^{-1}	−0.048

Table 1 The resistivity values of materials vary considerably. They are also temperature dependent.

The relationship between the resistivity of a material and its temperature is shown by:

$$\rho_T = \rho_0[1 + \alpha(T - T_0)]$$

where: ρ_T is the resistivity of the material at a temperature, T;

ρ_0 is the resistivity value that is quoted at T_0 (usually room temperature);

α is the temperature coefficient;

T is the temperature of the material; (given in °C or K)

T_0 is the reference temperature at which the resistivity of the material is quoted – we shall take this to be 20 °C or 293 K.

WORKED EXAMPLE 1

The resistivity of silver at room temperature is 1.59×10^{-8} Ω m. What will its resistivity be at a temperature of 80 °C?

Answer

$\rho_T = \rho_0[1 + \alpha(T - T_0)]$

We know that $\alpha = 0.0038$ °C^{-1}, $T = 80$ °C and $T_0 = 20$ °C.

Substituting these values:

$$\rho(\text{at } 80\,°\text{C}) = 1.59 \times 10^{-8} \times [1 + 0.0038 \times (80 - 20)]$$
$$= 1.59 \times 10^{-8} \times 1.228$$
$$= 1.95 \times 10^{-8}\ \Omega\,\text{m}$$

This shows that the resistivity has increased as the temperature increased, which is typical for metals.

WORKED EXAMPLE 2

The resistivity of germanium at room temperature is 0.46 Ωm. What will its resistivity be at a temperature of 40 °C?

Answer

$\rho_T = \rho_0[1 + \alpha(T - T_0)]$

We know that ρ_0 is 0.46 Ω m, $\alpha = -0.480$ °C^{-1}, $T = 60$ °C and $I_0 = 20$ °C.

Substituting these values:

$$\rho(\text{at } 40\,°\text{C}) = 0.46 \times [1 - 0.048 \times (40 - 20)]$$
$$= 0.46 \times 0.04$$
$$= 1.84 \times 10^{-2}\ \Omega\,\text{m}$$

In this example using germanium, the resistivity has *decreased* as the temperature has increased. This is because in some semiconductor materials more electrons are made available for electrical conduction when the material is warmer.

The negative temperature coefficient thermistor

The fact that the resistivity of some materials decreases as they get warmer means that they can be useful in some electric circuits. One example is the *negative temperature coefficient thermistor* (*NTC*). As its temperature increases, its resistance decreases, so the size of the current passing through it increases, allowing other parts of the circuit to operate. This is described further in the topic on potential divider circuits (Topic 4.3.4).

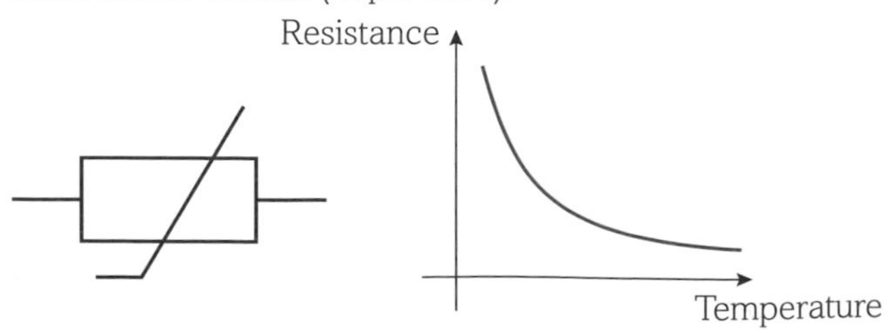

Figure 1 The resistivity and resistance of an NTC thermistor decreases as it gets warmer.

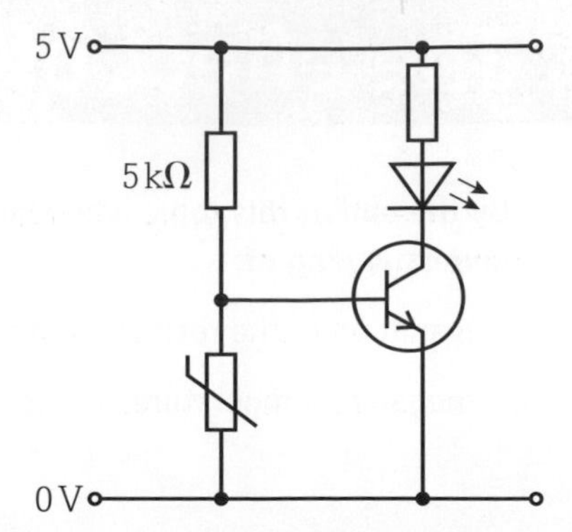

Figure 2 As the thermistor gets warmer, the current through it increases – if this current is more than 5 mA, the transistor will allow a current to flow through an LED and it will light up.

Questions

1. Calculate the resistivities of:
   (a) silver at 30 °C
   (b) silicon at 38 °C.

2. Calculate the difference in the resistivities of copper between 30 °C and 40 °C.

3. Explain why the resistivity of a metal increases as its temperature rises, but the resistance of a semiconductor falls as the temperature rises.

4. Explain the possible functions of the following components.
   (a) a thermistor
   (b) a transistor
   (c) an LED
   (d) a variable resistor
   (e) a 1 kΩ resistor
   (f) a resistor placed in series with the LED.

5. Certain materials, called superconductors, can have zero electrical resistance. Find out about the conditions that are needed for this to happen.

4.2 ⑥ Electrical power

By the end of this topic, you should be able to demonstrate and apply your knowledge and understanding of:

* the equations $P = VI$, $P = I^2R$ and $P = \frac{V^2}{R}$
* energy transfer; $W = VIt$

Introduction

Power is the rate at which energy is transferred from one form to another. The more powerful a device or machine is, the more energy it will transfer per unit time.

The unit of power is the watt, W. If a device has a power of 1 W it converts 1 J of energy from one form to another in one second.

Example	Nuclear power station	Jet aircraft	Formula 1 car	Usain Bolt	Hairdryer	Ant

Power	10^9 W	1.4×10^8 W	1.2×10^6 W	2700 W	1800 W	<1 W

Figure 1 A range of powers.

We calculate the amount of energy transferred, or the amount of work done, using the relationships:

- energy (J) = power (W) × time (s)
- work done (J) = power (W) × time (s)

WORKED EXAMPLE 1

A light bulb transfers 144 000 J of electrical energy to other forms in 1 hour. What is the power of the light bulb?

Answer

$E = P \times t$

so $P = \frac{E}{t}$

$= \frac{144\,000\text{ J}}{(60 \times 60)\text{s}}$

$= 40\text{ J s}^{-1}$ or 40 W

Electrical power

As well as calculating power from the generic equations above, which apply to all types of energy, we can also express electrical power in terms of specific values of current, voltage and resistance in a circuit at any given time. We know that the relationship between work done, charge and voltage is given by:

$W = Q \times V$

We also know that the relationship between charge, current and time is given by $Q = I \times t$.

Substituting for Q in equation 1 gives:

$W = V \times I \times t$

We know that the power, P, is work done, or energy converted, per unit time, so we can show that:

$$P = I \times V$$

We also know, from Ohm's law, that $V = I \times R$, so the three electrical equations for power that we most commonly use are:

- $P = I \times V$
- $P = I^2R$
- $P = \frac{V^2}{R}$

WORKED EXAMPLE 2

An electric toaster is switched on for 3 minutes. It is marked 230 V: 1200 W. Calculate:

(a) the current, I, flowing in the toaster
(b) the resistance, R, of the toaster
(c) the total charge, Q, that flows through the toaster in the 3 minute period
(d) the energy, W, transferred to the toaster.

Answers

(a) $P = VI$: $1200\text{ W} = 230\text{ V} \times I$, so $I = \frac{1200}{230} = 5.2\text{ A}$

(b) $V = IR$: $230\text{ V} = 5.2\text{ A} \times R$, so $R = \frac{230}{5.2} = 44\ \Omega$

(c) $Q = It$: $Q = 5.2\text{ A} \times 180\text{ s}$, so $Q = 940\text{ C}$

(d) $W = Pt$: $W = 1200\text{ W} \times 180\text{ s}$, so $W = 216\,000\text{ J}$

You can do a quick check by calculating the number of joules per coulomb. Here 216 000 J are supplied by 940 C, giving $216\,000\text{ J}/940\text{ C} = 229.8\text{ J C}^{-1} = 230\text{ V}$ (to 2 significant figures), and this is the potential difference of the supply voltage.

Current and voltage in transmitting electrical power across the UK

Electrical power is transferred across the UK using the National Grid (Figure 2).

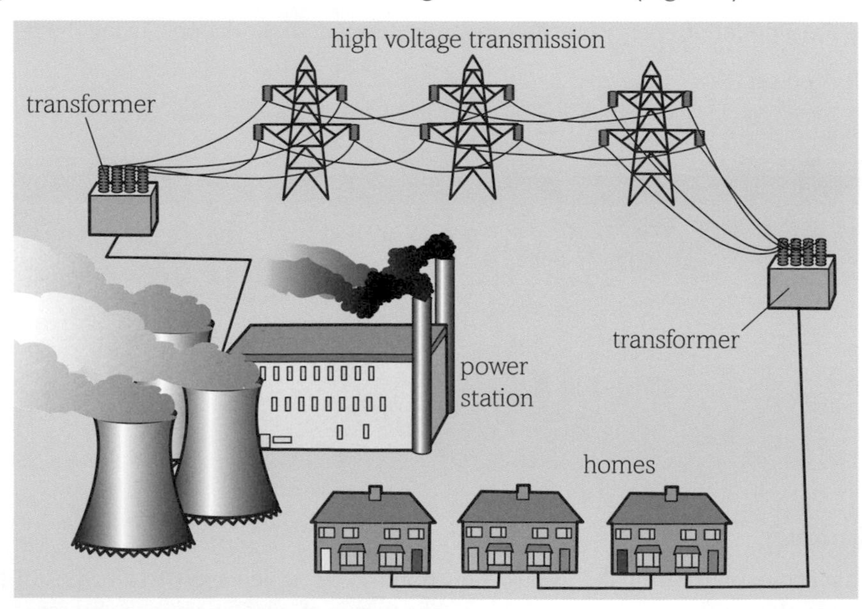

Figure 2 The National Grid.

From the equation $P = IV$, power can be transferred at high voltage/low current, or at low voltage/high current. The choice is to use a low current because the higher the current is, the more energy is 'wasted' to heat, which is calculated using $P = I^2R$. Because the value of current is squared in the equation, doubling the current will increase the amount of energy transferred to heat per second by a factor of four.

WORKED EXAMPLE 3

A nuclear power station transfers 5.2×10^{12} J of energy from nuclear energy to electrical energy in a one hour period. Calculate:
(a) the power of the nuclear power station in watts
(b) the current in the wires when the transmission voltage is 200 kV
(c) the current in the wires when the transmission voltage is 2 kV
(d) the ratio power 'wasted' as heat at 2 kV : power 'wasted' as heat at 200 kV.

Answers

(a) $P = \frac{E}{t}$

$= \frac{5.2 \times 10^{12}\ \text{J}}{3600\ \text{s}}$

$= 1.4 \times 10^{9}\ \text{W}$

(b) $I = \frac{P}{V}$

$= \frac{1.4 \times 10^{9}\ \text{W}}{200 \times 10^{3}\ \text{V}}$

$= 7000\ \text{A}$

(c) $I = \frac{P}{V}$

$= \frac{1.4 \times 10^{9}\ \text{W}}{2000} = 700\,000\ \text{A}$

(d) $P = I^2R$

$= (700\,000)^2 : (7000)^2$

$= 10\,000 : 1$

This means that 10 000 times more heat will be 'wasted' when the current is increased by a factor of 100.

Questions

1 Calculate:
(a) the power of a light bulb that is labelled 240 V/0.25 A
(b) the power of a clock radio that uses a 3 V cell and takes a current of 1.2 mA
(c) the power of a heater labelled 20 V/0.2 Ω.

2 A kettle has a power rating of 2.3 kW and is connected to a 240 V mains supply. Calculate:
(a) the current it takes
(b) the current it would take if its power rating was only 2 kW
(c) how much less energy the 2 kW kettle would transfer compared to the 2.3 kW kettle in a two-minute period.

3 An electrical device is labelled 240 V/3.5 Ω. Calculate:
(a) the current it takes
(b) its power rating
(c) the energy it transfers as heat each second.

4 On a building site, a small crane lifts a load of 800 N through a height of 7.0 m.
(a) Calculate the increase in gravitational potential energy of the load.
(b) The 110 V electric motor on the crane draws 2.0 A from the supply for 60 s.
  (i) Calculate the power rating of the motor.
  (ii) How much electrical energy is transferred?
(c) Calculate the efficiency of the crane motor.

5 A typical power station providing electricity for the National Grid generates electricity at 25 kV. The typical demand on an average winter's day in the north-west of England is 6000 MW.
(a) Calculate the current output from one of the generators.
(b) Transformers raise the transmission voltage to 400 kV. Assuming 100% efficiency in the transformers, show that the current in the transmission line is 15 kA.

Cost of electrical energy

By the end of this topic, you should be able to demonstrate and apply your knowledge and understanding of:

* the kilowatt-hour (kW h) as a unit of energy
* calculating the cost of energy

Introduction

The typical house is filled with a variety of electrical devices, all of which transfer electrical energy to other types of energy. In an ideal world, all this electrical energy would be transferred to other useful forms, but no device is 100% efficient, so some energy is always 'wasted'.

Table 1 shows the power ratings of typical devices used in a kitchen, along with the typical amount of energy that would be used by that appliance during just one day. These devices are used for differing lengths of time. For example, a kettle or toaster is likely to be used for only short periods of time, whereas a fridge would be left on all the time.

Appliance	Power rating/W	Time used/h	Time used/s	Total energy used by appliance/J day^{-1}
kettle	2200	0.5	1800	3 960 000
toaster	1100	0.5	1800	1 980 000
electric oven	3000	2	7200	21 600 000
dishwasher	1250	1	3600	4 500 000
fridge	400	24	86 400	34 560 000
washing machine	2800	2	7200	20 160 000
microwave oven	850	0.5	1800	1 530 000
television	250	4	14 400	3 600 000
radio	10	2	7200	72 000
lights	600	4	14 400	8 640 000

Table 1 Daily domestic electrical energy use.

LEARNING TIP

To calculate energy transferred, use: Energy (J) = power (W) × time (s)

The kilowatt-hour

As you can see in Table 1, if we record energy use in joules, enormous numbers of joules are going to be used each day. This is because the joule is a very small unit of energy, so we will transfer many millions each day. Electricity bills are commonly recorded over a three-month period, so the number of joules recorded on a bill are gigantic – but the cost of one joule of energy is incredibly small.

To make all this easier for the mind to take in, we use another unit called the **kilowatt-hour** (kW h). One kilowatt-hour is the energy used by a 1 kilowatt device in 1 hour when it is on.

We know that energy (J) = power (W) × time (s), so 1 kilowatt-hour is equal to 1000 W being transferred in 3600 s – this means that 1 kW h is equal to 3.6×10^6 J. This is the *unit* used on electricity bills.

KEY DEFINITION

A **kilowatt-hour** (kW h) is 1000 watts for 3600 seconds. It is therefore 3 600 000 J.

Appliance	Power rating/W	Power rating/kW	Time used for/h	Total energy used/kW h
kettle	2200	2.2	0.5	1.1
toaster	1100	1.1	0.5	0.55
electric oven	3000	3.0	2.0	6.0
dishwasher	1250	1.25	1.0	1.25
fridge	400	0.4	24	9.6
washing machine	2800	2.8	2.0	5.6
microwave oven	850	0.85	0.5	0.425
television	250	0.25	4.0	1.0
radio	10	0.01	2.0	0.02
lights	600	0.6	4.0	2.4

Table 2 It is far more sensible, straightforward and manageable to charge consumers using the kW h instead of joules.

WORKED EXAMPLE 1

Calculate how much energy is transferred by an 850 W microwave used for 240 minutes:
(a) in joules
(b) in kW h.

Answers

(a) $E = P \times t$
$= 850\text{ W} \times (240 \times 60)\text{s}$
$= 12\,240\,000\text{ J}$

(b) $E = P \times t$
$= 0.85\text{ kW} \times 4\text{ h}$
$= 3.4\text{ kW h}$

Calculating the cost of electrical energy

When charging for electricity, we use the kW h as the unit. We read the domestic electricity meter to work out how many units have been used by the consumer since the last bill, and then multiply this by the cost of each kW h (Figure 1).

DID YOU KNOW?

To convert from J to kW h, divide by 3 600 000. To convert from kW h to J, multiply by 3 600 000.

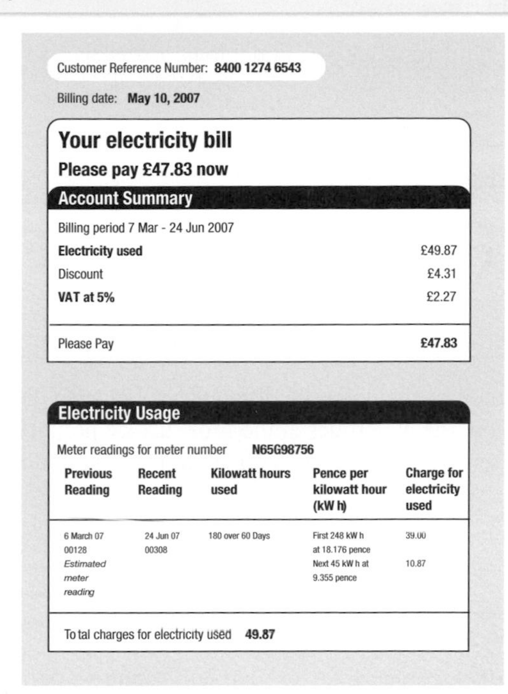

Customer Reference Number: **8400 1274 6543**

Billing date: **May 10, 2007**

Your electricity bill

Please pay £47.83 now

Account Summary

Billing period 7 Mar - 24 Jun 2007

Electricity used	£49.87
Discount	£4.31
VAT at 5%	£2.27
Please Pay	**£47.83**

Electricity Usage

Meter readings for meter number **N65G98756**

Previous Reading	Recent Reading	Kilowatt hours used	Pence per kilowatt hour (kW h)	Charge for electricity used
6 March 07 00128 *Estimated meter reading*	24 Jun 07 00308	180 over 60 Days	First 248 kW h at 18.176 pence Next 45 kW h at 9.355 pence	39.00 10.87

To tal charges for electricity used **49.87**

Figure 1 Part of a typical household electricity bill. The cost to the customer will be the number of kilowatt-hours used by the cost of 1 kW h (unit).

WORKED EXAMPLE 2

Calculate the cost of using:

(a) A 2.4 kW shower for three hours if the cost of 1 kW h is 12p.

(b) An 800 W microwave, a 2200 W kettle and a 3000 W oven for 90 minutes if 1 unit costs 14.7p

(c) An oven transfers 1.2×10^9 J of electrical energy to heat. How much will this cost if 1 kW h costs 14.2p?

Answers

(a) cost = number of kW h × cost per unit
= (2.4 kW × 3 h) × 12p
= 7.2 kW h × 12p
= 86.4p

(b) Total power used = 800 W + 2200 W + 3000 W = 6000 W = 6 kW
cost = number of kW h × cost per unit
= (6 kW × 1.5 h) × 14.7p
= 132.3p = £1.32

(c) To convert from joules to kW h we divide the value in J by 3 600 000:

$$\frac{1.2 \times 10^9 \text{ J}}{3\,600\,000} = 333.3 \text{ kW h}$$

Cost = 333.3 kW h × 14.2p
= 4733p
= £47.33

Note that in this question that we were given a total amount of energy, so a power rating or a time is not needed.

LEARNING TIP

Remember, 'kW h' is a unit of energy, not a unit of power.

Questions

1. Copy Table 3 and then fill in the missing values relating to time, energy used and cost at a price of 12.8p per kW h.

Appliance	Power rating	Time used for	Energy/J	Energy/ kW h	Cost
kettle	2350 W	20 minutes			
shower	10.4 kW			2	
television	0.15 kW		4×10^8		
radio	20 W	2 days			
vacuum cleaner	450 W	25 minutes			

Table 3

2. An electricity meter shows a reading of 60 169.1 kW h. After a further 3 months, the reading increased to 60 586.8 kW h. Calculate:

(a) how many joules have been transferred in this 3 month period

(b) the cost of this if each unit (kW h) costs 11.8p.

ELECTRICAL RESISTIVITY

Resistivity is a property of material that has a fixed value, regardless of the dimensions of the material in question. It is low for metals and high, or infinite, for insulators. This means that it can be useful when used as part of a calibrated scientific instrument.

In this activity, we will consider how the property of resistivity can enable geophysicists to model the substructure of an area of land.

ELECTRICAL RESISTIVITY IN THE CONTEXT OF GEOPHYSICS

Electrical resistivity studies in geophysics may be understood in the context of current flow through a subsurface medium consisting of layers of materials with different individual resistivities. For simplicity, all layers are assumed to be horizontal. The resistivity ρ of a material is a measure of how well the material retards the flow of electrical current. Resistivities vary tremendously from one material to another. For example, the resistivity of a good conductor such as copper is of the order of $10^{-8}\,\Omega$ m, the resistivity of an intermediate conductor such as wet topsoil is $\sim 10\,\Omega$ m, and the resistivity of poor conductors such as sandstone is $\sim 10^{8}\,\Omega$ m. Due to this great variation, measuring the resistivity of an unknown material has the potential for being very useful in identifying that material, given little further information. In field studies, the resistivity of a material may be combined with reasoning along geologic lines to identify the materials that constitute the various underground layers. [...]

Figure 1 shows a general linear electrode configuration for a typical resistivity survey.

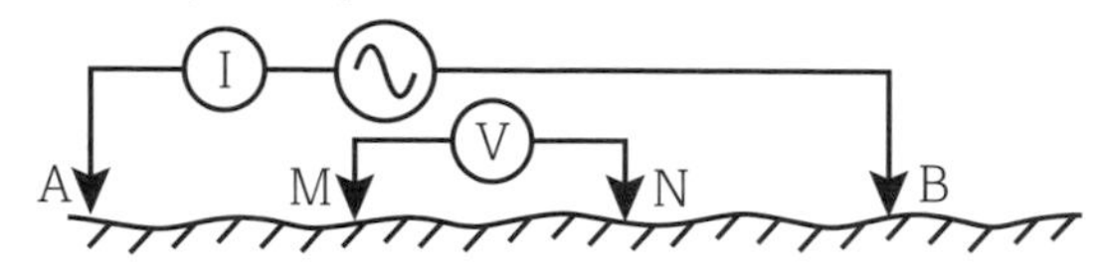

Figure 1 Electrode configuration for a typical resistivity survey.

All four electrodes are chosen to be in a straight line in the present work for simplicity. In general, the electrodes are not restricted to being collinear, although solving the electromagnetic field equations that accompany such arrays becomes more difficult. The ac current source is in series with an ammeter, which measures the total current I going into the ground through the electrodes at points A and B. A voltmeter attached to the two electrodes at points M and N measures the potential difference V between these points. By convention, the electrodes at the four surface points (A, M, N, B) are also named (A, M, N, B). The ratio (V/I) obtained is the apparent resistance for the entire subsurface. [...]

The total potential difference between the electrodes M and N [...] may be rearranged to yield:

$$\rho = \frac{V_{MN}}{I} \times K \tag{1}$$

where K is the 'geometric factor' that will acquire a particular value for a given electrode spacing. For the Wenner array, all of the separations are equal to a constant value a and the Wenner geometric factor assumes the simple form $K = 2\pi a$.

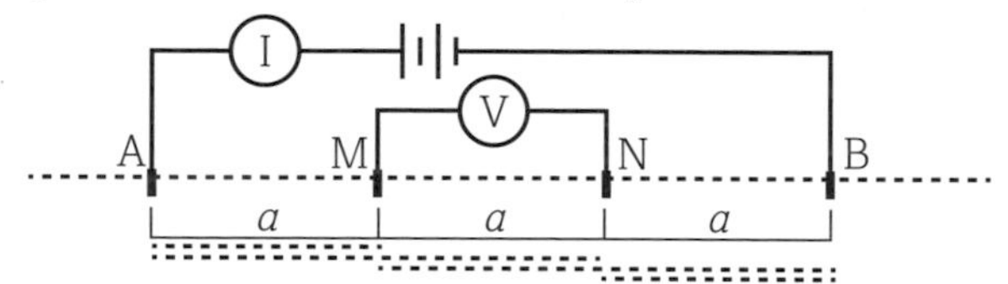

$r_{11} = a \qquad r_{12} = 2a$

$r_{21} = 2a \qquad r_{22} = a$

$$K = 2\pi\left(\frac{1}{r_{11}} - \frac{1}{r_{21}} - \frac{1}{r_{12}} + \frac{1}{r_{22}}\right)^{-1}$$

$$K = 2\pi\left(\frac{1}{a} - \frac{1}{2a} - \frac{1}{2a} + \frac{1}{a}\right)^{-1}$$

Figure 2 Wenner array with geometric factor calculation shown.

Source

Herman, R. (2001) An introduction to electrical resistivity in geophysics. *American Journal of Physics*, 69, 943–952.

Where else will I encounter these themes?

1.1 2.1 2.2 3.1 3.2 3.3

Let us start by considering the nature of the writing in the article. The extract above is from a journal published by the American Association of Physics Teachers (AAPT).

1. Consider the extract and comment on the type of writing that is being used. Who is the audience? What are the words or terms in the article that would influence you when determining who the article is intended for? Does the article try to explain, persuade or describe? Are the findings open to interpretation by others? How might you change the article to make it more suitable for a younger or less-informed audience?

How are the findings in articles like this checked to see if they are correct?

We will now look at the physics that is in the article. Do not worry if the physics content or the mathematics is challenging at this stage. You can always return to the article later in your course, once some of the related topics have been studied in more depth. Use the timeline at the bottom of the page to help you put this work in context with what you have already learned and what is ahead in your course.

2. How is electrical resistivity similar to and different from electrical resistance?

3. What is the relationship between resistivity and conductivity?

4. Why is the resistivity in these studies referred to as 'apparent resistivity'?

5. Show that the quantity K for the geometric factor has units of metres, by considering the homogeneity of units in the equation $\rho = K \times V/I$.

6. If the spacing between electrodes A and M is 30 m, when $V = 8$ kV and $I = 3$ mA then what will the apparent resistivity be of the sample of ground being analysed?

7. Apart from temperature, which factors will influence the magnitude of the apparent resistivity of the soil or rock sample being analysed?

8. What are the advantages and limitations of the Wenner method when trying to determine the substructure of a piece of land using a resistivity model as outlined here?

Activity

You are going to write a short report of no longer than 500 words to present the case for how a local water authority could investigate the potential size of a subsurface aquifer (an underground layer of permeable rock in which water can rest or be transported). Find out more about resistivity methods can be used in groundwater surveys. Your report should describe what resistivity is and how it can be used to determine a model of the structure under the ground. Include limitations of the method and how the predictions from the model could be verified.

DID YOU KNOW?

If you cool some metals to near absolute zero, they have virtually zero resistance. In 1986 it was discovered that the resistance of a brittle ceramic compound – normally an insulator – drops abruptly to zero when the material is cooled below a relatively high temperature of 30 K. An electric current flowing through a loop of wire of this material below its critical temperature can persist indefinitely with no power source.

4.2 Practice questions

1. A potential difference of 1.2 kV is applied across a resistor of resistance 450 Ω. What current flows through the resistor? [1]
   A 375 A
   B 0.375 A
   C 27 A
   D 2.7 A

2. If the potential difference across a resistor doubles, which of the following statements is true? [1]
   A The current will stay the same if the resistance remains constant.
   B The cell will take half as long to lose its energy.
   C The charge passing will be four times as large.
   D The power dissipated as heat will be four times as large.

3. A filament lamp is described as being 240 V, 120 W. The lamp is connected to a supply so that it lights normally.
   Which statement is correct? [1]
   A The charge passing through the filament in one second is 2.0 coulomb.
   B The lamp transfers 120 J for each coulomb of charge passing through the lamp.
   C The lamp has a resistance of 480 Ω.
   D Doubling the potential difference will halve the lamp's power.

   [Total: 3]

4. The wires connecting a 24 W 8 V battery to a series circuit are made from copper. They have a cross-sectional area of $1.2 \times 10^{-7}\,m^2$. The number n of free electrons per m^3 for copper is $8.0 \times 10^{28}\,m^{-3}$.
   (a) Calculate the current in the wire. [2]
   (b) Describe what is meant by the term mean drift velocity of the electrons in the wire. [2]
   (c) Calculate the mean drift velocity, v, of the electrons in the copper wire. [3]

   [Total: 7]

5. Look at the graph shown in Figure 1.
   (a) Explain why this graph shows that the component does not obey Ohm's law. [2]
   (b) Identify the component. [1]

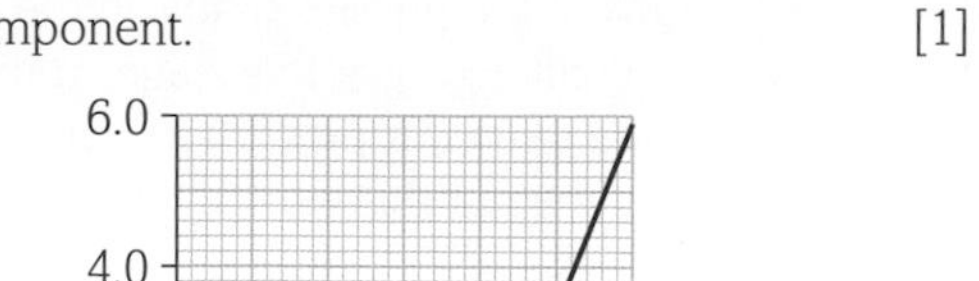

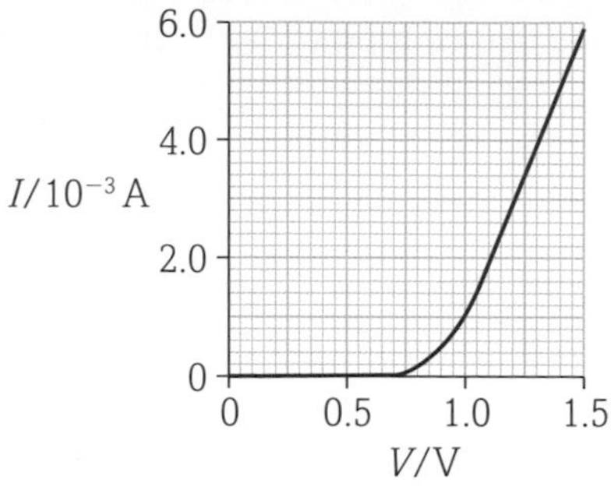

Figure 1

[Total: 3]

6. A current of 2 A is flowing in a filament lamp of resistance 20 Ω. If the current through the lamp is then doubled, explain what will happen to the:
   (a) kinetic energy of the atoms in the wire of the filament lamp [2]
   (b) the resistance of the filament lamp [2]
   (c) the energy dissipated as heat per second. [3]

   [Total: 7]

7. An electric kettle has a power rating of 2.4 kW and is connected to a mains voltage of 230 V a.c. Calculate:
   (a) the current, in A, flowing through the kettle's element when operating under normal conditions [2]
   (b) the resistance of the kettle's element in Ω [2]
   (c) how much energy has been supplied to the kettle if it takes 2 minutes to boil the water inside it [2]
   (d) state an assumption you have made when calculating the answer above. [1]

   [Total: 7]

8. How are the terms electromotive force and potential difference:
   (a) similar [2]
   (b) different. [2]
   [Total: 4]

9. A 350 W heater is left running for 8 hours. Calculate:
   (a) the energy converted by the heater in kW h [2]
   (b) the energy converted by the heater in J [2]
   (c) the cost of using the heater for this time if 1 unit of electrical energy costs 12.8p. [2]
   [Total: 6]

10. The electrical energy supplied to an electric motor can be calculated using the formula $E = VIt$.
    The gravitational potential energy gained when a mass is raised through a height h is given by the formula $E_p = mg\Delta h$.
    (a) State what the quantities are in both of the two equations provided. [2]
    (b) Describe an experiment you would conduct to determine the efficiency of an electric motor. In your description, refer to:
       (i) any equipment you would need with an explanation of why it is needed [4]
       (ii) any laws, principles and equations that you would need to use [3]
       (iii) the nature of any systematic errors that might be present and how they would be removed. [2]
    [Total: 11]

11. An investigation is carried out to determine the resistivity of a piece of nichrome wire.
    (a) State the apparatus that would be needed in order to obtain suitable values for the resistance of the wire. [2]
    (b) Which other readings would need to be obtained so that a value for the resistivity of the wire could be obtained? [2]
    (c) What apparatus would be needed in order for accurate and precise values to be obtained? [2]
    (d) Having obtained a range of values, explain how the resistivity of the wire could be obtained:
       (i) by calculation from an equation [2]
       (ii) by the plotting of a suitable graph. [2]
    (e) The investigation is repeated, but this time the wire is placed under conditions that leads to a significant change in its temperature. Explain what effect, if any, this have on the value obtained for the resistivity of the wire. [2]
    [Total: 12]

MODULE 4 Electrons, waves and photons

CHAPTER 4.3

ELECTRICITY: ELECTRICAL CIRCUITS

Introduction

Essentially, physics is a practical and investigative discipline. The equations relating to charge, current, resistance, energy and potential difference need to be applied to actual circuits. Over recent decades and centuries, electrical circuits have been central to the discovery of some of the most important laws and theories in physics – from the discovery of the Higgs particle to the determination of the Planck constant.

In this chapter we are going to look at the key components of DC electricity and circuits – developing further our understanding of the physical quantities and units in electrical circuits that have been covered earlier in your studies. The areas of electromotive force, internal resistance and Kirchhoff's laws will be covered to enable a detailed understanding of the key principles behind the working of circuits. Opportunities will arise for the planning and implementing of electrical experiments, with a focus on the behaviour of light and temperature-dependent components used in potential divider circuits.

All the maths you need

To unlock the puzzles of this chapter you need the following maths:

- Units of measurement
- Adding, subtracting, multiplying and dividing fractions to determine the values of resistors arranged in parallel
- Rearranging equations to make other quantities the subject of the formula, e.g. $V = E - Ir$
- Use of standard form, e.g. when dealing with small current and large resistance values
- Use of ratio for potential divider calculations
- Use of simultaneous equations to determine e.m.f. and internal resistance
- Finding values from the gradient and y-intercept of a graph, including resistance, resistivity, e.m.f. and internal resistance
- Using the values read from or calculated from graphs to determine other values of electrical quantities
- Applying positive and negative values to the e.m.f. and currents in circuit analysis using Kirchhoff's laws

What have I studied before?

- Electric charge will flow around a circuit as electrical current if an energy source and conductors are present
- Electrical current is measured with an ammeter, placed in series with the component through which the current is flowing
- Potential difference is measured by connecting a voltmeter across the component
- Electrical energy is measured in joules, electrical power is measured in watts
- Electrical circuits are used to convert energy from electrical energy to other forms.
- Electrical circuits may contain components including a cell, ammeter, bulb, voltmeter, switch and conductors
- Electrical components have an electrical resistance which opposes the flow of electric current
- Electrical components include a lamp, a resistor, a diode, thermistor, LED and LDR

What will I study later?

- The force on a current carrying wire in the presence of a magnetic field (AL)
- Fleming's left hand rule and right hand rule (AL)
- Faraday's law and Lenz's law (AL)
- Magnetic flux, magnetic flux linkage and the laws of electromagnetism (AL)
- The generation of an e.m.f. from the rate of change of magnetic flux linkage (AL)
- The generation of a.c. current and a.c. voltage (AL)
- The structure and function of the a.c. transformer (AL)

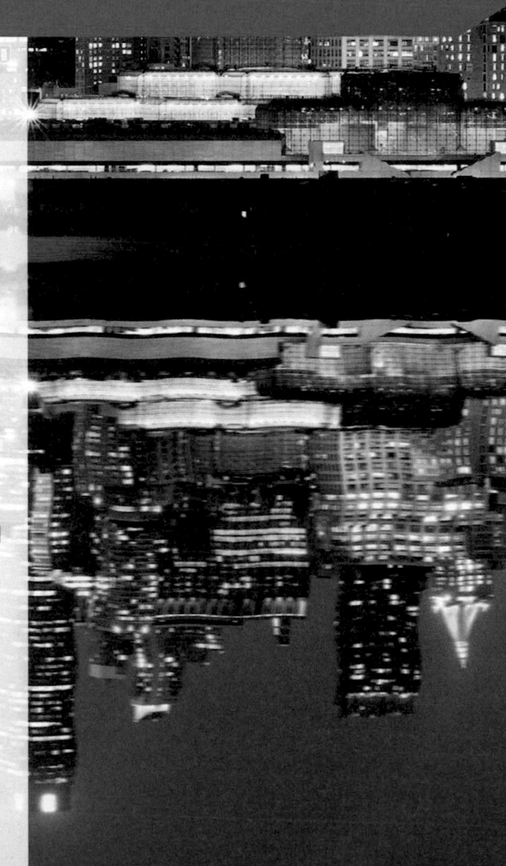

What will I study in this chapter?

- For components arranged in series, the total resistance is the sum of the individual resistances, i.e. $R = R_1 + R_2 + R_3$
- For components arranged in parallel, the total effective resistance of the network is given by $\frac{1}{R_T} = \frac{1}{R_1} + \frac{1}{R_2} + \frac{1}{R_3}$
- Power is the rate of transfer of electrical energy per second. It is measured using the watt and calculated using $P = IV$, $P = I^2R$ and $P = \frac{V^2}{R}$
- Electrical energy transferred, or work done, is calculated using the equation $W = VIt$
- The kilowatt-hour is the unit of electrical energy used in the home to calculate the cost of using electrical energy
- Circuits involving series and parallel arrangements of components and two or more sources of e.m.f. can be solved using Kirchhoff's laws and the resistance equations
- Cells, batteries and other electrical energy supplies experience an internal resistance that is related to the e.m.f. through the equation $E = I(R + r)$
- Techniques can be used to determine the e.m.f. and internal resistance of a cell via calculation or graphically
- A potential divider, or potentiometer, arrangement can be used to share the potential difference across components in a circuit. The potential divider equations are $V_{out} = \frac{R_2}{(R_1 + R_2)} \times V_{in}$ or $\frac{V_1}{V_2} = \frac{R_1}{R_2}$
- Techniques and procedures may be used to investigate, and design, potential divider circuits which may include a thermistor or LDR

Kirchhoff's first and second laws

By the end of this topic, you should be able to demonstrate and apply your knowledge and understanding of:

* Kirchhoff's first law; conservation of charge applied to electrical circuits
* Kirchhoff's second law; the conservation of energy applied to electrical circuits

Kirchhoff's first law

The current at any point in a series circuit has the same value as at any other point. Current (flow of charge) does not get 'used up' by the components in the circuit. The charge carries energy to the components, which transfer it into other forms of energy.

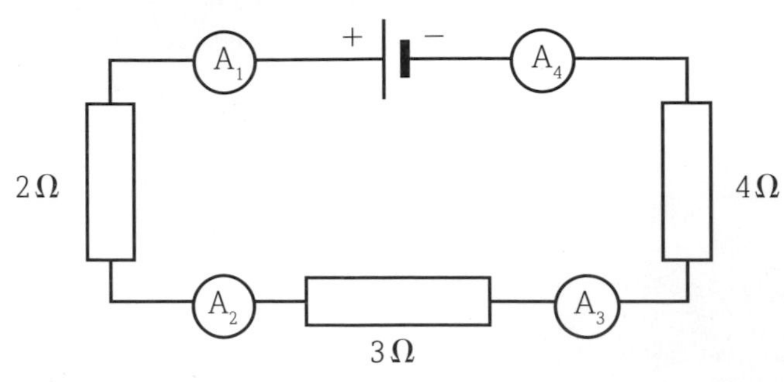

Figure 1 The ammeter readings in a series circuit will be identical because the current is the same at all points.

The sum of the currents entering a junction (or node) in a circuit is equal to the sum of the currents leaving the junction (Figure 2). This is called **Kirchhoff's first law** and is a consequence of the conservation of charge (see Topic 4.1.2).

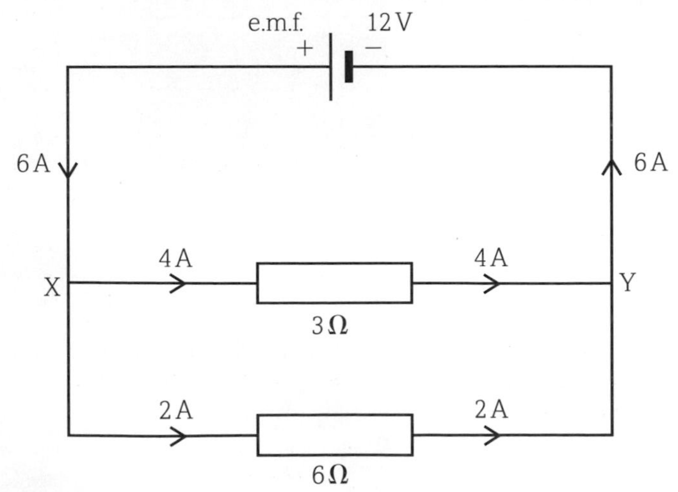

Figure 2 In this parallel circuit a current of 6 A leaves the cell of e.m.f. 12 V. At X it splits, with 4 A going through the 3 Ω resistor and 2 A going through the 6 Ω resistor. The currents recombine at Y, with 6 A flowing back to the cell.

Kirchhoff's first law states that the algebraic sum of the currents entering a junction must be zero. In Figure 3, if we take the currents entering the junction as being positive and the currents leaving the junction as being negative, then we get:

$I_1 + I_2 - I_3 - I_4 - I_5 = 0$

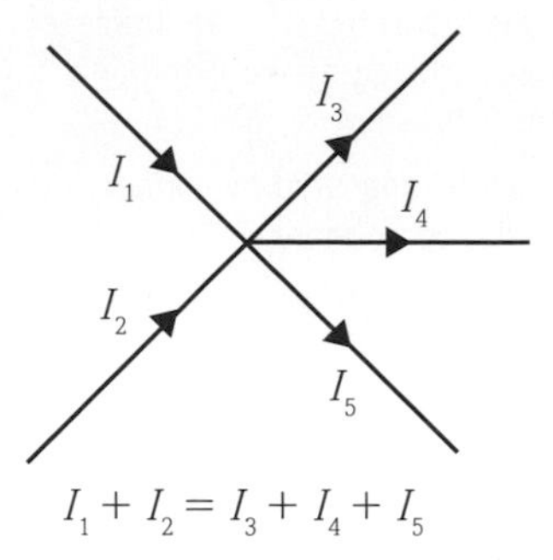

$I_1 + I_2 = I_3 + I_4 + I_5$

Figure 3 Regardless of how many currents enter a junction, Kirchhoff's first law will always apply. Here the sum of the currents flowing into the junction ($I_1 + I_2$) must be equal to the sum of the currents leaving it ($I_3 + I_4 + I_5$).

Kirchhoff's second law

Kirchhoff's second law states that in any closed loop, the sum of the e.m.f. is equal to the sum of the products of the current and the resistance. In equation form, this is:

$\Sigma\mathcal{E} = \Sigma IR$

Kirchhoff's second law is a consequence of the conservation of energy.

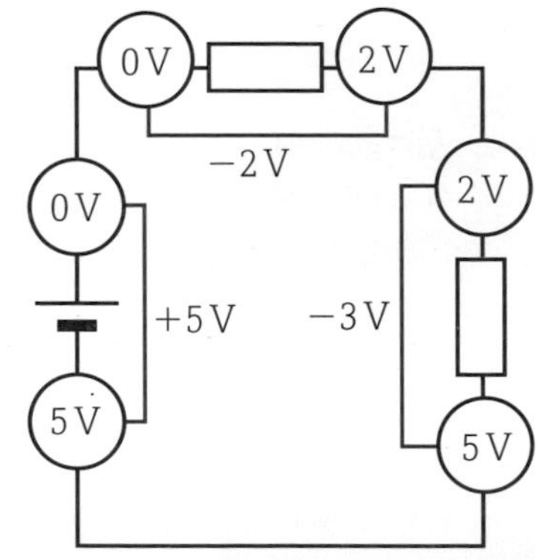

Figure 4 The e.m.f. provided by the cell must be the same as the total drop in potential differences across all of the resistors in the loop.

Kirchhoff's first and second laws involving circuits with two sources of e.m.f.

When using Kirchhoff's first and second laws, you must be careful about the directions that the currents are travelling in. For more complex problems involving two sources of e.m.f. you need to make sure that the current flowing in one direction is taken as positive and the current flowing in the opposite direction is taken as negative. For example, you may decide that the current flowing clockwise is positive, and so any current flowing anticlockwise must be taken as negative.

WORKED EXAMPLE

Look at the circuit shown in Figure 5.

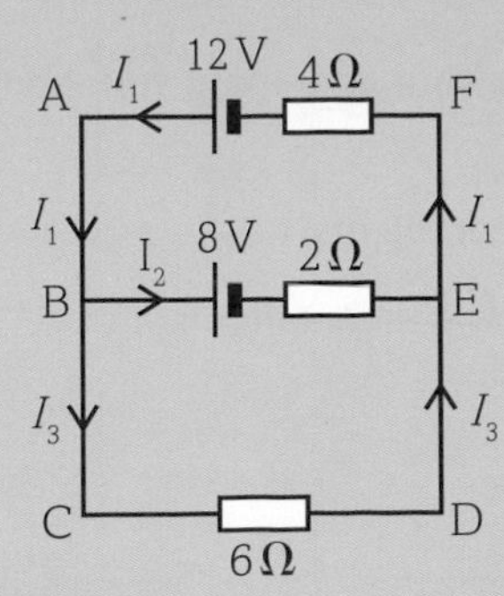

Figure 5

Write equations to describe:

(a) Kirchhoff's first law for the circuit.

(b) Kirchhoff's second law for the big loop ACDFA.

(c) Kirchhoff's second law for the small loop BCDEB.

Answers

(a) At point B, we can see that I_1 enters the node at junction B and that I_2 and I_3 leave the junction.
So, $I_1 = I_2 + I_3$

(b) Going anticlockwise from point A, and applying $\Sigma E = \Sigma IR$ for the big loop ACDFA we get:
$12\text{ V} = (6 \times I_3) + (4 \times I_1)$
This is because the total e.m.f. in the loop is 12 V and the current I_3 travels anticlockwise through the 6 Ω resistor and the current I_1 travels anticlockwise through the 4 Ω resistor.

(c) $8\text{ V} = (6 \times I_3) - (2 \times I_2)$
This is because in the smaller loop BCDEB, the total e.m.f. is 8 V and, having taken the current to be positive in the clockwise direction, we get a positive value through the 6 Ω resistor and a negative direction current (clockwise) through the 2 Ω resistor.

How can we use Kirchhoff's laws to determine the sizes of different currents?

From (b) and (c) in the worked example, we have two equations that show the currents, resistances and e.m.f. in two of the loops in the circuit shown in Figure 5. However, we have three unknowns in the two equations – currents I_1, I_2 and I_3. We can solve these using simultaneous equations if we can reduce this to two unknowns.

The two equations we are dealing with are:

$$12\text{ V} = (6 \times I_3) + (4 \times I_1)$$
$$8\text{ V} = (6 \times I_3) - (2 \times I_2)$$

Instead of having I_2 in the second equation, we can replace it with $(I_1 - I_3)$, the value from rearranging the answer to (a) in the worked example, and we now have:

$$12\text{ V} = (6 \times I_3) + (4 \times I_1)$$
$$8\text{ V} = (6 \times I_3) - 2 \times (I_1 - I_3)$$

The second equation now simplifies to $8\text{ V} = (8 \times I_3) - (2 \times I_1)$

We now have two simultaneous equations, with unknowns I_1 and I_3:

$$12\text{ V} = (6 \times I_3) + (4 \times I_1) \quad \text{equation 1}$$
$$8\text{ V} = (8 \times I_3) - (2 \times I_1) \quad \text{equation 2}$$

Multiplying both sides of equation 2 by 2 we get:

$$12\text{ V} = (6 \times I_3) + (4 \times I_1) \quad \text{equation 1}$$
$$16\text{ V} = (16 \times I_3) - (4 \times I_1) \quad \text{equation 3}$$

Adding equation 1 to equation 3 cancels out the I_1 values, leaving:

$$28\text{ V} = 22 \times I_3$$

This gives us a value for I_3 of 1.27 A to 2 decimal places.

Substituting the values into the other equations allows us to obtain values of I_1 and I_2, which are 1.1 A and −0.17 A respectively. The negative sign for I_2 means that it flows in the opposite direction to the arrow.

Questions

1. Use Kirchhoff's first law to find the missing values in Figure 6.

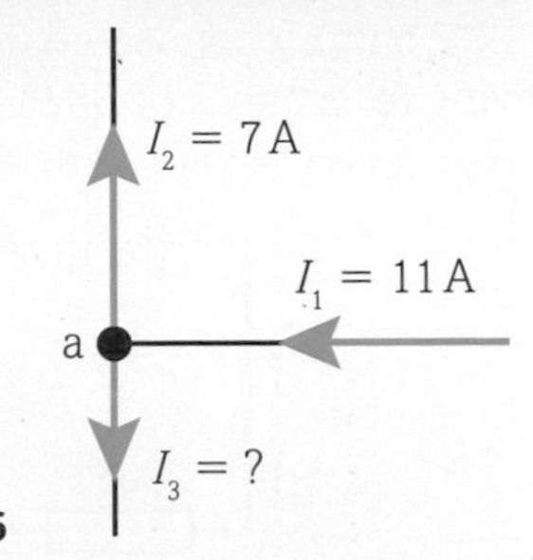

Figure 6

2. Apply Kirchhoff's first law to find the missing value in the circuit shown in Figure 7.

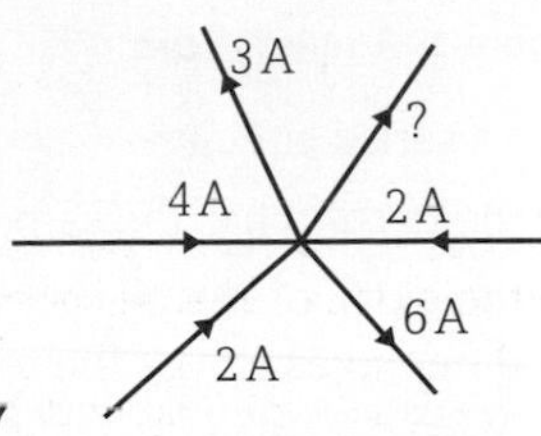

Figure 7

3. By referring to Figure 4, draw the circuit shown in Figure 8 and show that the e.m.f. provided by the cell must be the same as the drop in potential differences across each of the resistors (or lamps as shown here) in the loop.

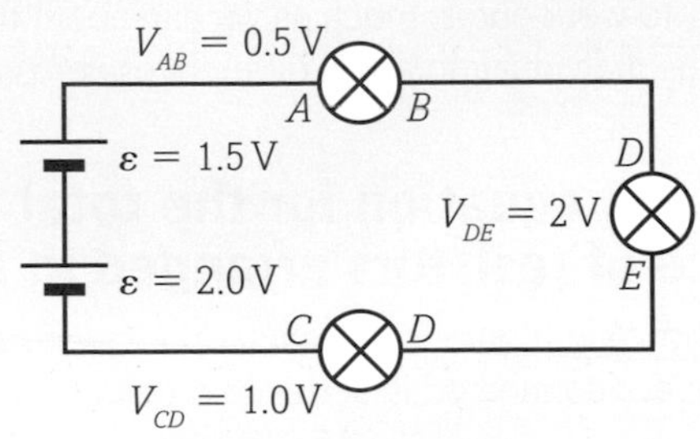

Figure 8

4. Use Kirchhoff's second law to calculate the size of each individual current flowing through resistors R_1, R_2 and R_3 for the circuit shown in Figure 9.

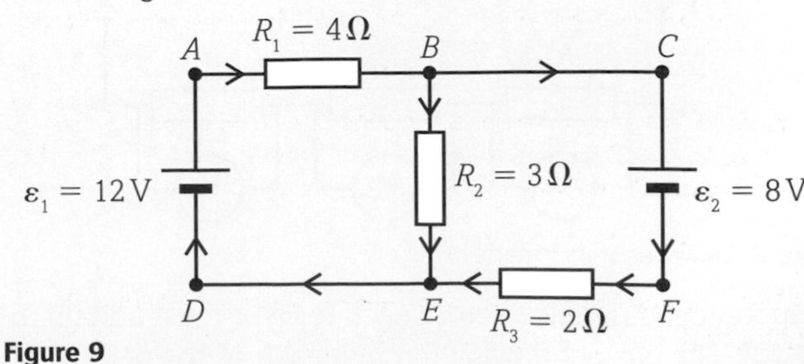

Figure 9

2 Series circuits

By the end of this topic, you should be able to demonstrate and apply your knowledge and understanding of:

* applying Ohm's law to solve a range of series circuit problems
* total resistance of two or more resistors in series; $R = R_1 + R_2 + R_3$
* analysis of circuits with series components

Introduction

A diagram of a **series circuit** is shown in Figure 1. It consists of a 12 V battery connected to three resistors, one after the other, having resistances of 30 Ω, 10 Ω and 60 Ω respectively.

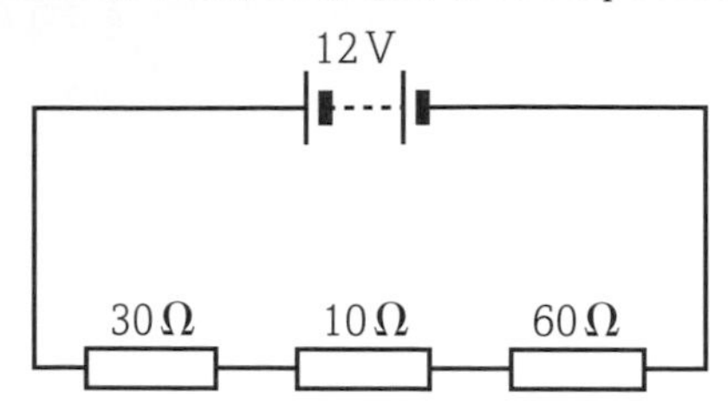

Figure 1 A series circuit.

For a series circuit:

- the current has the same value at any point in the circuit
- the e.m.f. of the cell is equal to the sum of the potential differences across the resistors
- the total resistance of the circuit is equal to the sum of the individual resistors.

When dealing with simple series circuits, we often use the Ohm's law equation $V = I \times R$. We also tend to 'simplify' the circuit to make it easier to work out as much as we can about the current flowing and the individual potential differences across each resistor.

Deriving the equation for the total effective resistance of resistors arranged in series

The circuit shown in Figure 2 contains three resistors, R_1, R_2 and R_3, which are connected in series to a cell of e.m.f. V.

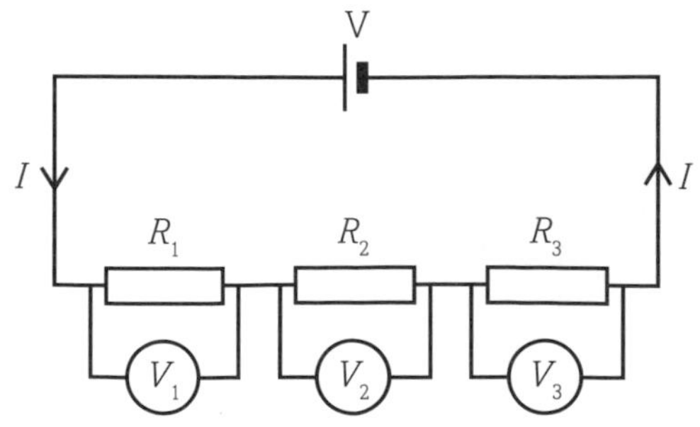

Figure 2 Three resistors in series.

The potential difference across each resistor must add up to the e.m.f., V. The current flowing through each of the resistors must be the same, I.

WORKED EXAMPLE

For the series circuit shown in Figure 1:

(a) Draw a 'simplified' circuit

(b) Calculate the current flowing in the circuit

(c) Calculate the potential difference across each of the resistors.

Answers

(a)

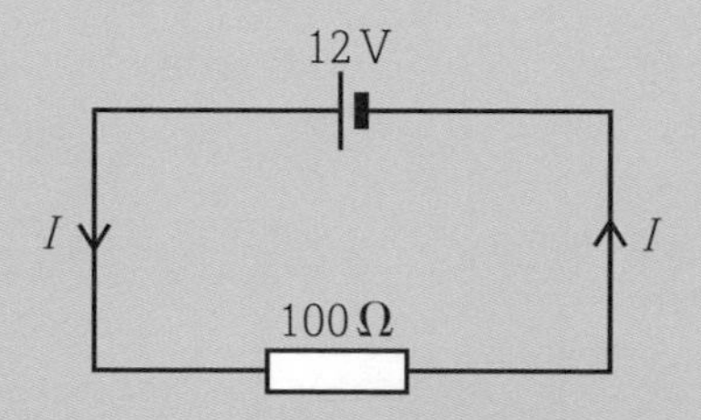

Figure 3

(b) Rearranging the Ohm's law equation gives $I = \frac{V}{R}$, where I is the current flowing, V is the e.m.f. of the cell and R is the total resistance in the circuit.

$$I = \frac{12\,\text{V}}{100\,\Omega}$$
$$= 0.12\,\text{A}$$

This is the value of the current at any point in the circuit. It does not increase or decrease regardless of whether it is travelling through the wires, the resistors or the cell.

(c)

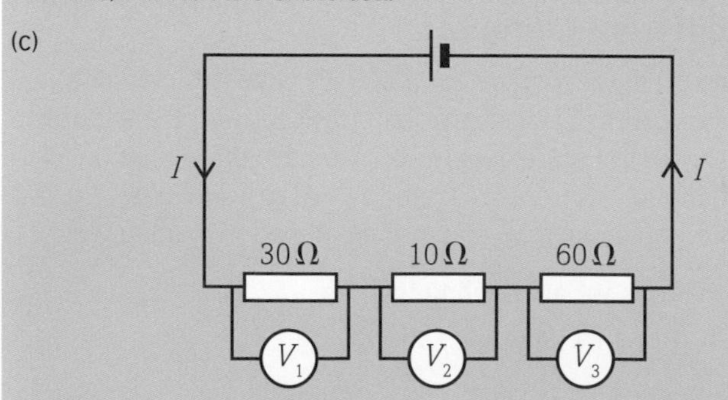

Figure 4

Using Ohm's law $V = I \times R$ on each of the resistors:

$V_1 = I \times R_1$
$= 0.12\,\text{A} \times 30\,\Omega = 3.6\,\text{V}$

$V_2 = I \times R_2$
$= 0.12\,\text{A} \times 10\,\Omega = 1.2\,\text{V}$

$V_3 = I \times R_3$
$= 0.12\,\text{A} \times 60\,\Omega = 7.2\,\text{V}$

Note that the three individual values of 3.6 V, 1.2 V and 7.2 V add to give the cell e.m.f., 12.0 V

From Ohm's law, $V = IR$, we can state that:

$V = V_1 + V_2 + V_3$

So $IR_T = IR_1 + IR_2 + IR_3$, where R_T is the total effective resistance of the series arrangement.

Factorising gives $I(R_T) = I(R_1 + R_2 + R_3)$

So: $R_T = R_1 + R_2 + R_3$

This means that the total effective resistance of a series network of resistors is equal to the sum of the individual resistances.

Ammeters and voltmeters in circuits

Ammeters are always connected in *series* with components to measure the size of the current flowing through them. Ammeters are designed to have close-to-zero resistance.

Voltmeters are always arranged in *parallel* with components in a circuit. They are designed to have a close-to-infinite resistance so that none of the current flows through them.

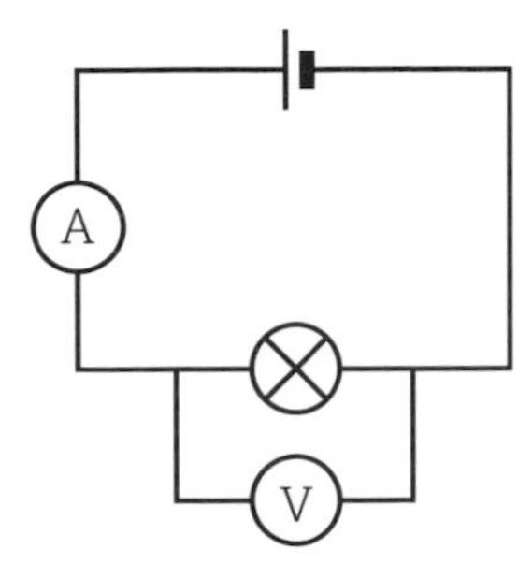

Figure 5 Connecting ammeters and voltmeters in circuits.

Questions

1 Three resistors of resistance 100 Ω, 120 Ω and 80 Ω are connected in series to a 30 V cell. Calculate:

(a) the total resistance in the circuit

(b) the current in the circuit

(c) the potential difference across each of the resistors.

(d) Show that $V = V_1 + V_2 + V_3$.

2 Two resistors of value 50 Ω and 150 Ω are connected in series to a cell of e.m.f. 24 V. Calculate:

(a) the current in the circuit

(b) the potential difference across each of the two resistors.

3 Three resistors of values $0.5R$, R and $1.5R$ are connected in series to a 6.0 V battery. The current in the circuit is 20 mA. Calculate the values of the resistors.

4 The reading on the ammeter in Figure 6 is 1.2 A. How much would the current increase by if the switch was closed?

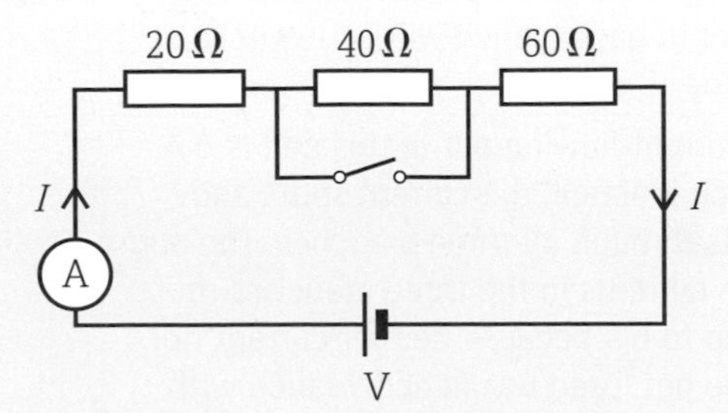

Figure 6

4.3 (3) Parallel circuits

By the end of this topic, you should be able to demonstrate and apply your knowledge and understanding of:

- **total resistance of two or more resistors in parallel;** $\frac{1}{R} = \frac{1}{R_1} + \frac{1}{R_2} + ...$
- **analysis of circuits with components, including both series and parallel**

e.m.f. = 36 V

$I_1 = \frac{36}{12} = 3A$, $R_1 = 12\,\Omega$

$I_2 = \frac{36}{18} = 2A$, $R_2 = 18\,\Omega$

$I_3 = \frac{36}{6} = 6A$, $R_3 = 6\,\Omega$

$I = I_1 + I_2 + I_3$
$= 6A + 3A + 2A$
$I = 11A$

Figure 1 A parallel circuit.

Figure 2 Two resistors connected in parallel.

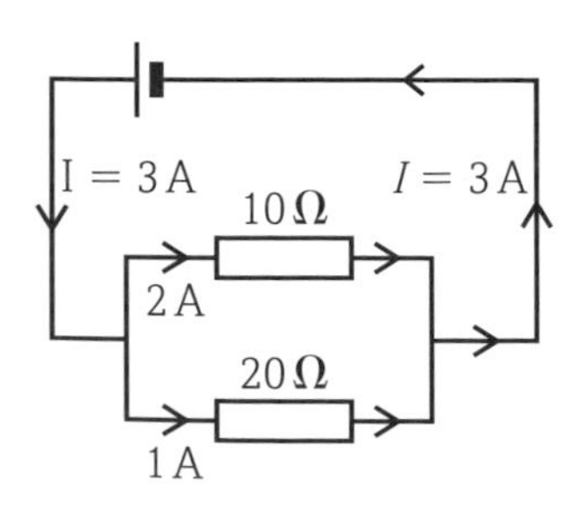

Figure 3 The current flowing through a resistor depends on its resistance.

Introduction

When resistors are connected one after another, as shown in Figure 1 of Topic 4.3.2, we say that they are arranged in series. Resistors that have been connected like this have the same size current passing through them and the potential differences across them add up to the cell e.m.f.. We work out the size of the current flowing through each resistor using Ohm's law.

In a parallel circuit, the resistors are arranged as shown in Figure 1. Instead of being connected one after the other, they are connected in parallel branches.

In a **parallel circuit**, the currents flowing through the resistors are not necessarily the same – the current splits up at the junctions and different amounts of current will flow through a resistor depending on its resistance. The higher the resistance, the smaller the current flowing through it.

In Figure 3, the current splits so that twice as much goes through the 10 Ω resistor. This is because it has half the resistance of the 20 Ω resistor and so will allow twice as much current to flow through it. If we multiply the current passing through each resistor by its resistance, we obtain the size of the voltage across that resistor. So, for both resistors in Figure 3, the voltage across them must be 20 V.

This brings us to an important rule for parallel circuits – the potential difference across the resistors in each branch of a parallel circuit must be the same.

The rules for parallel circuits are given in Table 1.

Rule	Example	Diagram
The components in each branch of a parallel circuit share the same voltage as the cell: $V_{total} = V_1 = V_2 = ...V_n$	Six identical bulbs are connected so that they are in three parallel branches. They are connected to a cell that has an e.m.f. of 12 V. There is a potential difference of 12 V across each branch. The bulb in the first branch receives all 12 V. The two bulbs in the second branch share the 12 V and have a potential difference of 6 V across each. The three bulbs in the third branch each have 4 V across them. Although a current I leaves the cell, the current in each branch will be lower than this.	12 V; 12 V; 6 V, 6 V; 4 V, 4 V, 4 V; I
The total current in a parallel circuit is the sum of the individual branch currents: $I_{total} = I_1 + I_2 + ... I_n$	The current flowing out of this cell is 6 A. At each junction, the current splits and travels through all three branches. The sum of the currents in the three branches must add up to 6 A because neither current nor charge get 'used up', in accordance with Kirchhoff's first law. The size of the current in a branch is inversely proportional to the resistance of that branch. The higher the resistance of the branch, the lower the current flowing through it.	6 A; 2 A; 3 A; 1 A; 6 A

Table 1 How the current and voltage rules differ for series and parallel circuits.

Deriving the equation to calculate resistance when resistors are arranged in parallel

Imagine that we have three resistors arranged in parallel, as shown in Figure 4.

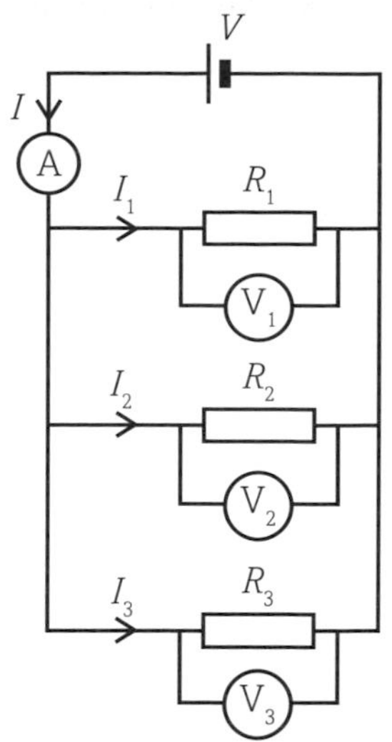

Figure 4

For resistors connected in parallel, we know that the current divides at the junctions, according to the resistance of each branch. The voltage across each branch is the same as the cell voltage.

The two useful equations are:

$I = I_1 + I_2 + I_3 ...$ equation 1

$V = V_1 = V_2 = V_3 ...$ equation 2

Applying Ohm's law, $V = IR$, and rearranging equation 1:

$$\frac{V}{R} = \frac{V_1}{R_1} + \frac{V_2}{R_2} + \frac{V_3}{R_3} ... \quad \text{equation 3}$$

Applying equation 2 to equation 3, we get:

$$\frac{V}{R} = \frac{V}{R_1} + \frac{V}{R_2} + \frac{V}{R_3} ...$$

Factorising:

$$V\left(\frac{1}{R}\right) = V\left(\frac{1}{R_1} + \frac{1}{R_2} + \frac{1}{R_3}\right) ...$$

Cancelling the V terms on both sides:

$$\frac{1}{R} = \frac{1}{R_1} + \frac{1}{R_2} + \frac{1}{R_3} ...$$

WORKED EXAMPLE

For the circuit in Figure 5:

(a) Calculate the effective resistance of the circuit to 2 decimal places.

(b) Calculate the current through the 5 Ω resistor to 1 decimal place.

(c) Work out the current, I, leaving the answer to 1 decimal place.

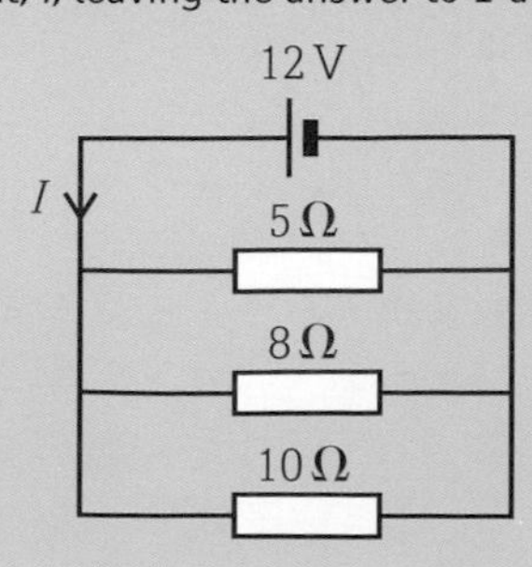

Figure 5

Answers

(a) $\frac{1}{R} = \frac{1}{R_1} + \frac{1}{R_2} + \frac{1}{R_3} ...$

Substituting the values given:

$$\frac{1}{R} = \frac{1}{5} + \frac{1}{8} + \frac{1}{10} = \frac{8}{40} + \frac{5}{40} + \frac{4}{40} = \frac{17}{40}$$

So, $R = \frac{40}{17}$, which is 2.35 Ω to 2 decimal places.

(b) The current through the 5 Ω resistor is found by applying Ohm's Law:

$I = \frac{V}{R} = \frac{12\text{ V}}{5\,\Omega}$, giving a value of 2.4 A

(c) Using Ohm's law:

$$I = \frac{V}{R} = \frac{12\text{ V}}{2.35\,\Omega} = 5.1\text{ A}$$

Questions

1 Find the effective resistance of:

(a) two 12 Ω resistors connected in parallel

(b) three 12 Ω resistors connected in parallel.

2 What do you notice about the size of the effective resistance when you connect resistors in parallel?

3 Find the reading on the ammeters and voltmeters in Figure 6.

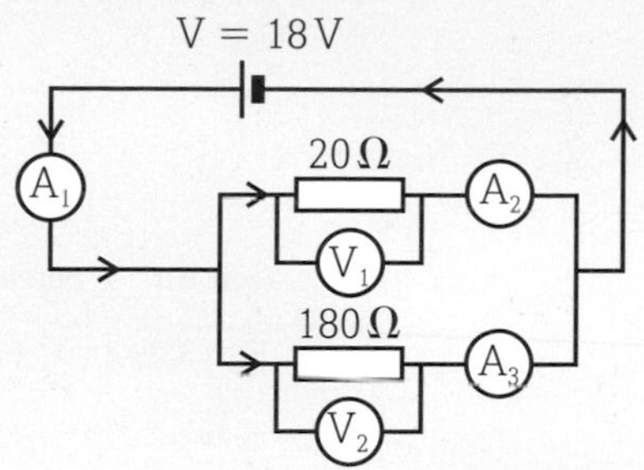

Figure 6

4 You are given three 3.0 Ω resistors and asked to make as many resistance values as possible using combinations of one, two or all three of the resistors. Write down all of the values you can make.

5 When a voltmeter V is added to a circuit to measure the p.d. it acts as a parallel resistor of resistance R. The circuit of Figure 7 consists of a 6.0 V supply of negligible internal resistance connected to two equal resistors and a 0–5 V voltmeter.

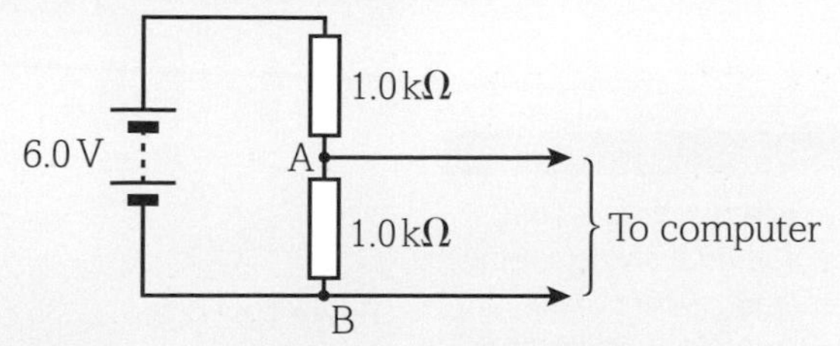

Figure 7

(a) Write down the total resistance in the circuit:

(i) before, and (ii) after the voltmeter is added.

(b) Show that the p.d. across one 1000 Ω resistor before the voltmeter is added is 3.0 V.

(c) The voltmeter reads 2.0 V when it is in the circuit. Calculate its resistance R.

(d) An accurate voltmeter should not alter the p.d. it is measuring. Suggest the best value of the resistance R to make the voltmeter more accurate.

The potential divider

By the end of this topic, you should be able to demonstrate and apply your knowledge and understanding of:

* potential divider circuit with components
* potential divider equations e.g. $V_{out} = \frac{R_2}{R_1 + R_2} \times V_n$ and $\frac{V_1}{V_2} = \frac{R_1}{R_2}$
* potential divider circuits with variable components e.g. LDR and thermistors
* techniques and procedures used to investigate potential divider circuits which may include a sensor such as a thermistor or LDR

Introduction

A typical **potential divider** circuit is shown in Figure 1. The circuit contains just two resistors, R_1 and R_2, connected in series. The input voltage is labelled V_{in} and the output voltage – the voltage across resistor R_2 – is labelled V_{out}. The resistance values of the two resistors are important because this determines how the potential difference will be shared across each.

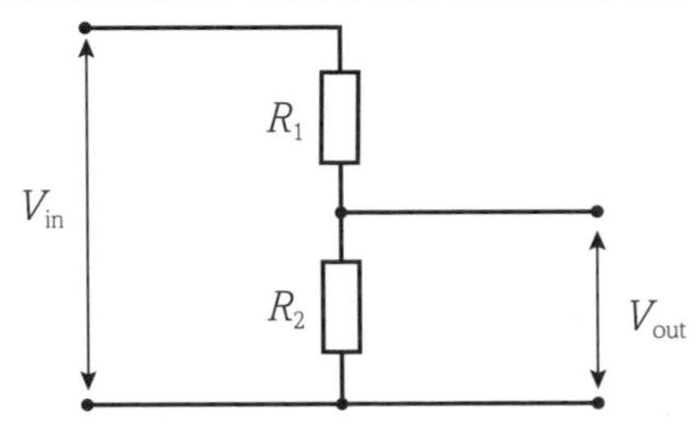

Figure 1 A potential divider circuit.

The size of the output voltage is calculated using:

$$V_{out} = V_{in} \times \frac{R_2}{(R_1 + R_2)}$$

You will find that the value of V_{out} is highest when V_{in} is large, R_2 is large and R_1 is small in comparison to R_2.

KEY DEFINITION

A **potential divider** circuit uses two resistors in series to split or divide the voltage of the supply in a chosen ratio so that a chosen voltage can be provided to another device or circuit.

WORKED EXAMPLE 1

In Figure 2, the supply voltage is 24 V and resistors R_1 and R_2 have resistances of 20 Ω and 80 Ω, respectively. What is the value of the output voltage, V_{out}?

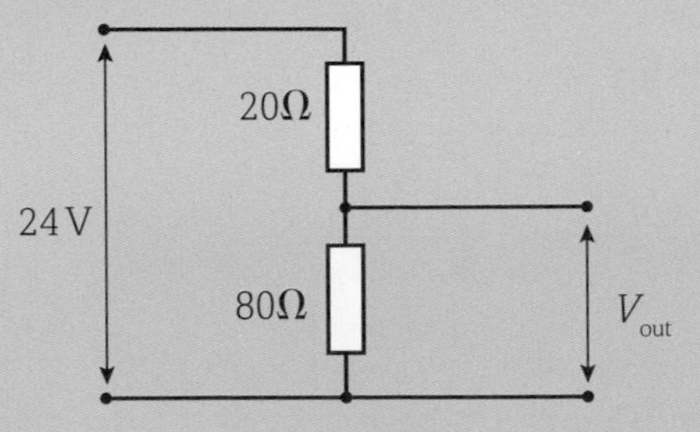

Figure 2

Answer

$$V_{out} = V_{in} \times \frac{R_2}{(R_1 + R_2)}$$

$$= 24\text{ V} \times \left(\frac{80\ \Omega}{(80 + 20)\ \Omega}\right)$$

$$= 19.2\text{ V}$$

There is an alternative way of working out V_{out} across R_2 using Ohm's law:

The total resistance of the resistors, $R_1 + R_2 = 100\ \Omega$

The current in the circuit is $I = \frac{V}{R}$, which is $\frac{24\text{ V}}{100\ \Omega} = 0.24\text{ A}$

So, the voltage across R_2, which is $V_{out} = I \times R_2$ is 0.24 V × 80 Ω, which is 19.2 V.

LEARNING TIP

It is possibly easier to work out values for potential dividers using the first method, rather than working out currents. However, it is always useful to have a second method, using Ohm's law and current values, to check your answer in examinations.

LEARNING TIP

It is always useful to know a second method – using Ohm's law and current values – that you can use to check your answer in examination conditions.

In Figure 1, we can find the p.d. across each resistor as:

p.d. across R_1, $V_1 = IR_1$

p.d. across R_2, $V_2 = IR_2$

Since $I = \dfrac{V_{in}}{(R_1 + R_2)}$

the ratio of the p.d.s across $R_1 + R_2$ is:

$$V_1/V_2 = \frac{V_{in}/(R_1 + R_2)\ R_1}{V_{in}/(R_1 + R_2)\ R_2}$$
$$= R_1/R_2$$

This only applies if the same current passes through each resistor. For example, if the supply p.d. is 12 V and the circuit contains two fixed resistors of 30 Ω and 10 Ω in series, the supply p.d. is split between the two resistors in the ratio of their resistances. So there will be 9 V across the 30 Ω resistor and 3 V across the 10 Ω resistor.

Potential divider circuits can have more than two resistors in series, but the method used is just the same. In the diagram shown in Figure 3 the output is across R_3, so the equation to calculate V_{out} is:

$$V_{out} = V_{in} \times \frac{R_3}{(R_1 + R_2 + R_3)}$$

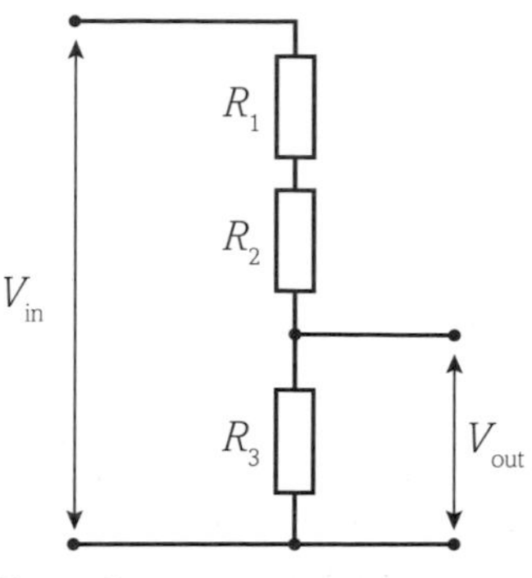

Figure 3

Variations on the potential divider circuit

Using a variable resistor

If one of the resistors is a variable resistor, the output p.d. can be easily adjusted.

The circuit shown in Figure 4 has a supply voltage, V_{in}, of 9.0 V and it is connected to a fixed value 2.0 kΩ resistor and a variable resistor that can have a range of resistance values from 0 Ω to 2.0 kΩ. The output voltage, V_{out}, will therefore range from 0 V when the variable resistor has a value of 0 Ω, to a value of 4.5 V when the variable resistor has a maximum value of 2.0 kΩ.

LEARNING TIP

When R_1 and R_2 are equal in size, the output voltage, V_{out}, is half the input voltage, V_{in}.

WORKED EXAMPLE 2

For the variable resistor circuit shown in Figure 4, what is the size of V_{out} when the input voltage is 9 V and the variable resistor is set to a resistance setting of one-quarter of its maximum value?

Answer

$V_{in} = 9$ V, $R_1 = 2000\ \Omega$, $R_2 = 500\ \Omega$, so:

$$V_{out} = 9 \times \frac{500}{2500} = 1.8\ \text{V}$$

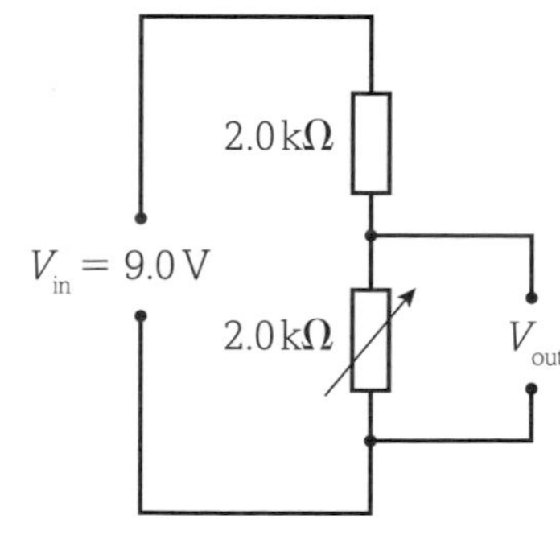

Figure 4 The output p.d. across the variable resistor will be 4.5 V at its maximum value. Both resistors will then have equal values, so each will have half the total p.d.

Using a thermistor in a temperature control circuit

A sensor can be used in a potential divider circuit to provide a variable output voltage, for example to control a light or alarm. The output may be connected to a suitable electromagnetic switch or relay that will be used to turn on a larger current in a more powerful device such as a heater.

At 0 °C the p.d. across the thermistor will be $9\ \text{V} \times \dfrac{12}{(12 + 4)} = 6.75\ \text{V}$

At 50 °C the p.d. across it will have fallen to $9\ \text{V} \times \dfrac{0.25}{(0.25 + 4)} = 0.53\ \text{V}$

The type of thermistor shown in Figure 5 could be used to control a heater. When the thermistor has an output of 6.75 V, the heater starts to generate heat because the voltage will be high enough to switch the heater on. When the temperature reaches 50 °C, the output p.d. falls to 0.53 V and the heater switches off.

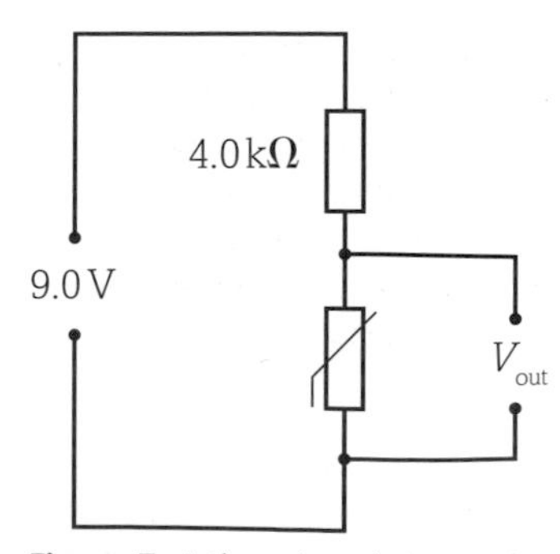

Figure 5 A thermistor has a resistance of 12 kΩ when the temperature is 0 °C and a resistance of 0.25 kΩ at 50 °C.

Using a light-dependent resistor in an illumination control circuit

A light-dependent resistor (LDR) has a high resistance when the incident light intensity is very low, such as in dark conditions. As the intensity of light falling on it increases, the resistance of the LDR falls. This means that an LDR can be used in a potential divider control circuit to switch lights on as it gets dark – street lights are controlled in this way. This is explained in worked example 3.

WORKED EXAMPLE 3

An LDR has a resistance of 1.0 MΩ in the dark, 100 kΩ in full moonlight, and 100 Ω in sunlight. At dusk, when we want street lights to come on, the resistance of the LDR is about 10 kΩ.

(a) Draw the potential divider circuit that you would need to use.

(b) if V_{in} is assumed to be 9.0 V, suggest values for R_1 such that the output voltage, V_{out}, is close to zero during the day and above 6.0 V at night.

Answers

(a) The potential divider circuit will look like the one shown in Figure 6, without the figures included at this stage.

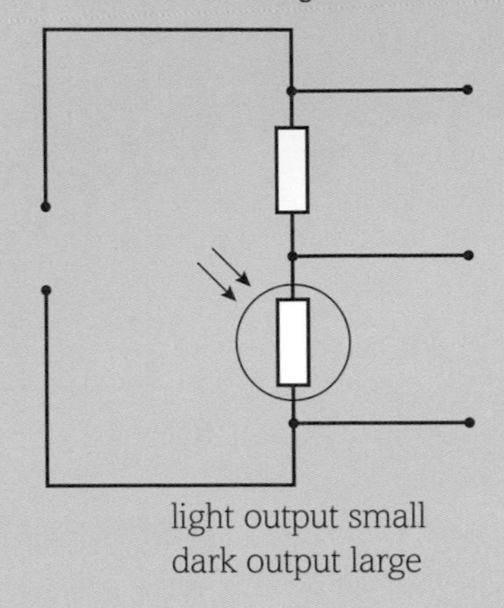

Figure 6

(b) Add a 9.0 V battery to the circuit shown in Figure 6. This means that 3.0 V will be across the fixed resistor when the LDR has your chosen value of 10 kΩ (at dusk) and has 6.0 V across it. The fixed resistor must therefore have a resistance of 5.0 kΩ (if there is 6 V across a 10 kΩ resistor there must be 3 V across a 5 kΩ resistor in series with it). You can check this using the potential divider equation. Remember: if R_2 is twice as big as R_1, then the voltage across it will also be twice as big – hence 6 V and 3 V respectively.

Now do a final check. When the LDR is in sunlight:

- its resistance will be low
- the output will be low
- the p.d. across the fixed resistor will be high.

In the dark, the LDR:

- will have high resistance
- will give an output higher than 6.0 V
- the light will stay on.

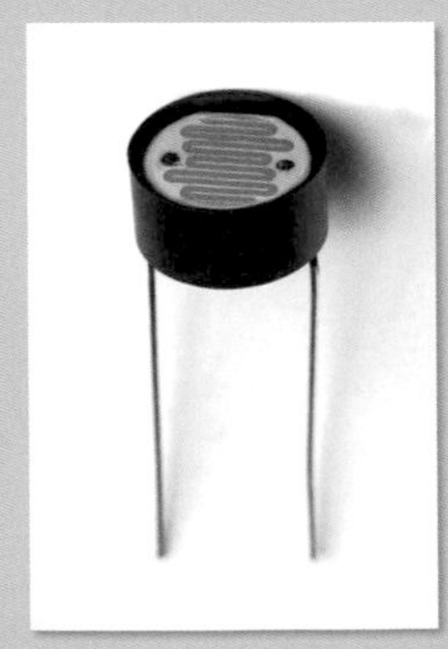

Figure 7

Question

1 This question is about the potential divider circuit shown in Figure 8.

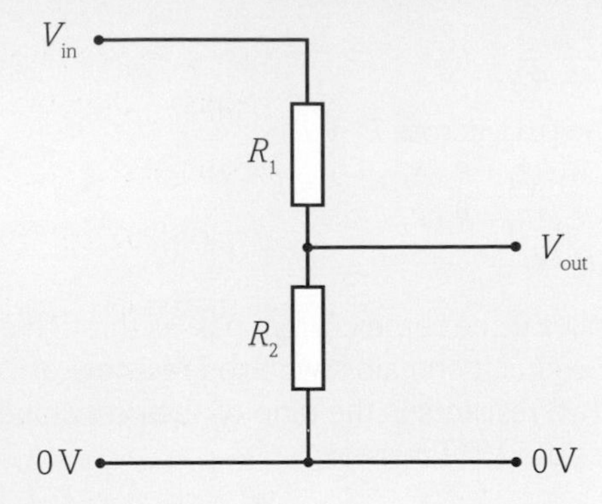

Figure 8

Copy and complete Table 1 by writing in the missing values:

V_{in}/V	R_1/Ω	R_2/Ω	V_{out}/V
12	24	48	
230		580	84
	1.6×10^3	3.5×10^3	40

Table 1

2 How could the circuit shown in Figure 6 be modified for a fridge? A fridge needs to be working when the temperature is high and off when the temperature is low.

3 A 12 V supply is connected across a variable resistor and a 100 Ω resistor. Describe what happens to the p.d. across the 100 Ω resistor as the variable resistor is adjusted to decrease its resistance.

4 Two fixed resistors are connected in series with a 6.0 V supply. What is the ratio of the resistances in the potential divider if the output voltage is 1.2 V?

5 Two fixed resistors $R_1 = 30\ \Omega$ and $R_2 = 20\ \Omega$ are connected in series across a supply of 5.0 V. What is the output voltage across the 20 Ω resistor?

5 Internal resistance

By the end of this topic, you should be able to demonstrate and apply your knowledge and understanding of:

* source of e.m.f.; internal resistance
* terminal p.d.; 'lost volts'
* the equations $\mathcal{E} = I(R + r)$ and $\mathcal{E} = V + Ir$
* techniques and procedures used to determine the internal resistance of a chemical cell; or other source of e.m.f.

Introduction

The size of the potential difference (p.d.) across a cell in a circuit is different when current is flowing compared with when current is not flowing.

If no current is flowing in the circuit, and the cell is effectively an open circuit (Figure 1), the value of the **terminal p.d.** registered will be the **electromotive force (e.m.f.)** of the cell – this is the maximum value of the potential difference across the cell.

KEY DEFINITIONS

The **terminal p.d.** is the potential difference recorded across the terminals of a cell. The difference between the e.m.f. and the terminal p.d. when charge flows in the cell is called the **'lost volts'**.

Figure 2 shows the same circuit with the switch closed and hence there is a current through the circuit. In this case the terminal p.d. recorded on the voltmeter will be lower than that in Figure 1.

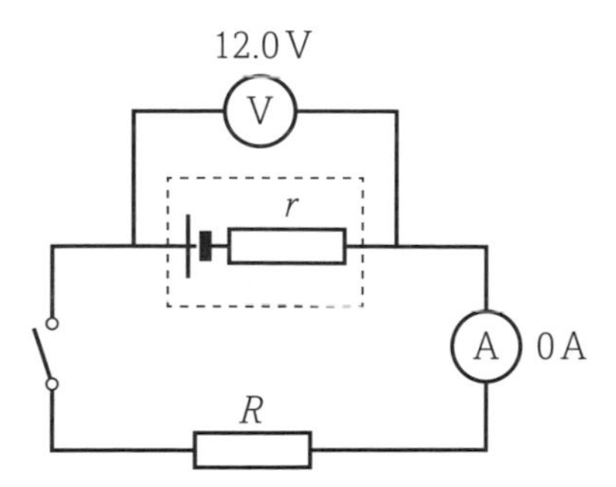

Figure 1 The e.m.f. is the maximum voltage across the cell when no current is flowing.

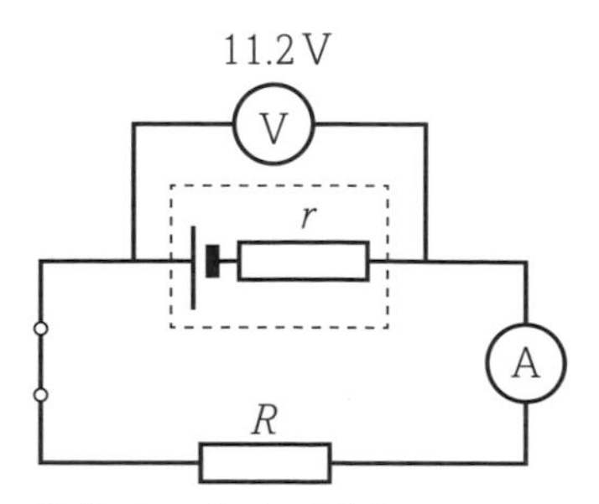

Figure 2 When a current flows in the circuit, the terminal p.d. is lower.

Why does this happen?

The answer lies in the idea of **internal resistance**. Every cell contains materials that offer a resistance to the flow of electric current. So when a current begins to flow in the cell a potential difference will develop across the internal resistance of the cell, reducing the energy per unit charge now available to the external circuit – the difference is sometimes called the '**lost volts**'.

KEY DEFINITION

The **internal resistance** of a source of e.m.f. is the resistance to electric current of the materials inside (chemicals, wires or components). When current flows, energy is transferred to these materials, resulting in the terminal p.d. dropping.

LEARNING TIP

There is a convention that uppercase letters (like 'R') are used for external quantities and lowercase letters (like 'r') are used for internal cell quantities.

The internal resistance is effectively connected in series with the cell and any other external components, such that the current flowing through the cell is the same as the current flowing through the internal resistance.

When a current flows in the series circuit (Figure 3) the same current flows through each component and the potential differences across the internal resistance and the load add up to give the cell's e.m.f.

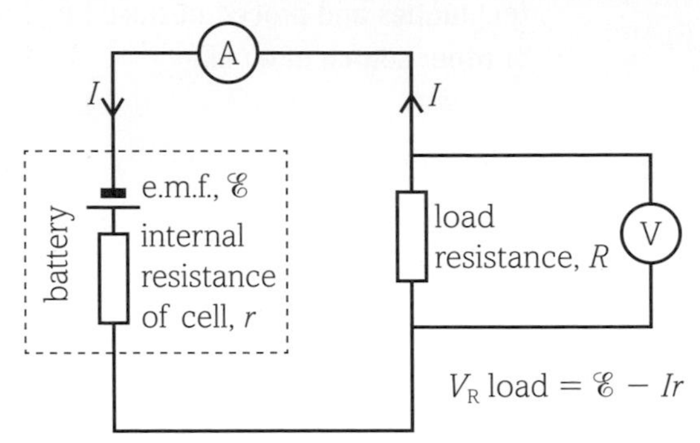

Figure 3

This means that:

$$\begin{aligned} \mathcal{E} &= IR + Ir \\ &= I(R + r) \\ &= V_R + Ir \end{aligned}$$

So when no current flows, $I = 0$ A and $\mathcal{E} = V_R$

WORKED EXAMPLE 1

A circuit is set up as shown in Figure 4.

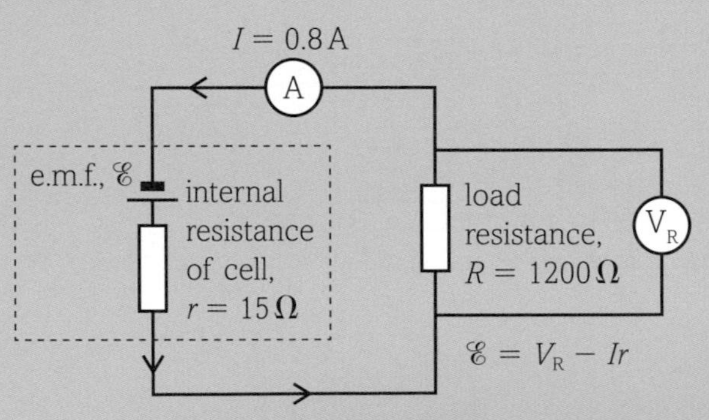

Figure 4

Work out:

(a) the value of V_R

(b) the value of the 'lost volts'

(c) the e.m.f. of the cell.

Answers

(a) $V_R = I \times R$

$= 0.8\text{ A} \times 1200\ \Omega = 960\text{ V}$

(b) The 'lost volts' is the potential difference developed across the internal resistance:

$I \times r = 0.8\text{ A} \times 15\ \Omega$

$= 12\text{ V}$

(c) $\mathcal{E} = V_R + Ir$

$= 960\text{ V} + 12\text{ V} = 972\text{ V}$

WORKED EXAMPLE 2

A cell with an e.m.f. of 12.0 V is connected in series with a load resistance of value $R = 8\,\Omega$ and left on open circuit so that no current flows in the circuit. When the switch is closed, a current of 1.2 A flows in the circuit and the voltage across the cell decreases to 9.6 V. Work out the internal resistance (r) of the cell.

Answer

Lost volts $= \mathcal{E} - V$

$= 12.0\text{ V} - 9.6\text{ V}$

$= 2.4\text{ V}$

So $I \times r = 2.4\text{ V}$

and $r = \frac{2.4\text{ V}}{1.2\text{ A}}$, so the internal resistance is $2.0\,\Omega$

Calculating an e.m.f. and internal resistance using simultaneous equations

The e.m.f. and internal resistance of a cell are constant values, so it is possible to have a situation in which different currents will flow for different load resistances. This means that exam questions can involve the use of simultaneous equations.

WORKED EXAMPLE 3

A current of 1.2 A flows in a circuit when the load resistance, R, is equal to $40\,\Omega$. When the resistance is lowered to $30\,\Omega$, the current increases to 1.5 A. Find the value of:

(a) the internal resistance, r, of the cell

(b) the e.m.f. of the cell.

Answers

(a) Forming simultaneous equations using $\mathcal{E} = IR + Ir$

$\mathcal{E} = 48 + 1.2r$ equation 1

$\mathcal{E} = 45 + 1.5r$ equation 2

Subtracting equation 2 from equation 1 gives:

$0 = 3 - 0.3r$

So, $0.3r = 3$ and r is $10\,\Omega$

(b) Substituting into either of the equations gives a value for the e.m.f. of 60 V.

INVESTIGATION

Finding the internal resistance of a cell

If we rearrange the equation $\mathcal{E} = I(R + r)$, and substitute V_R for IR:

$$V_R = -Ir + \mathcal{E}$$

This equation has the form $y = mx + c$, so if we plot a graph of V against I (Figure 6) we will get a straight line with a negative gradient and a positive y-intercept. The gradient of this graph will give us the value of r, the internal resistance, and the y-intercept will be equal to the e.m.f. of the cell.

To find the internal resistance of a cell, connect the cell in series with a variable resistor. Measure the current in the circuit, I, and the terminal p.d., V, as the variable resistor is adjusted. Plot a graph of V against I and draw a line of best fit.

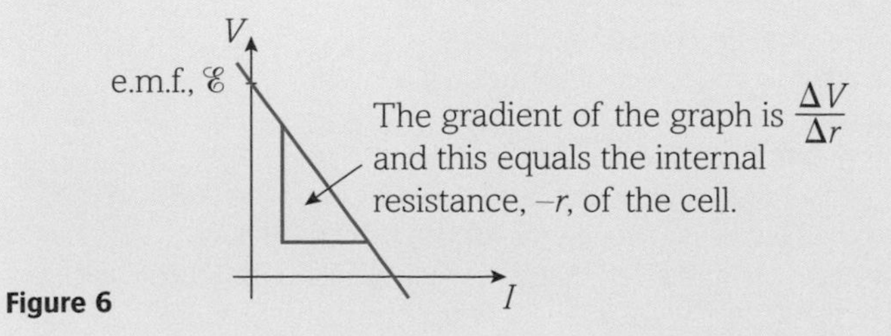

Figure 6

Questions

1. Use the equations in this spread to find:
   (a) $\mathcal{E}$ when $I = 3.0$ A, $R = 30\,\Omega$ and $r = 2\,\Omega$
   (b) R when $I = 1.2$ A, $\mathcal{E} = 48$ V and $r = 3.2\,\Omega$
   (c) V_R when $\mathcal{E} = 18$ V and the 'lost volts' = 0.8 V

2. A cell of e.m.f. 20 V is connected in series with a switch and a resistor. When the switch is closed, a current of 2.4 A flows through the circuit and the potential difference across the cell falls to 14.4 V. Use this information to calculate the value of the internal resistance, r, of the cell.

3. Use the following data to find the e.m.f., $\mathcal{E}$, and the internal resistance, r, of a cell:
   when $I = 1.8$ A, $R = 50\,\Omega$
   when $I = 1.0$ A, $R = 98\,\Omega$

4. Work out the e.m.f., $\mathcal{E}$, and the internal resistance, r, of the cell that gave the graph in Figure 7.

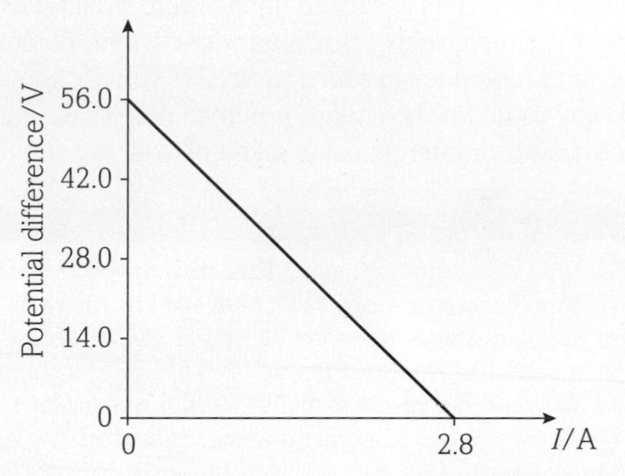

Figure 7

5. What is the name for the value recorded by the meter in Figure 8?

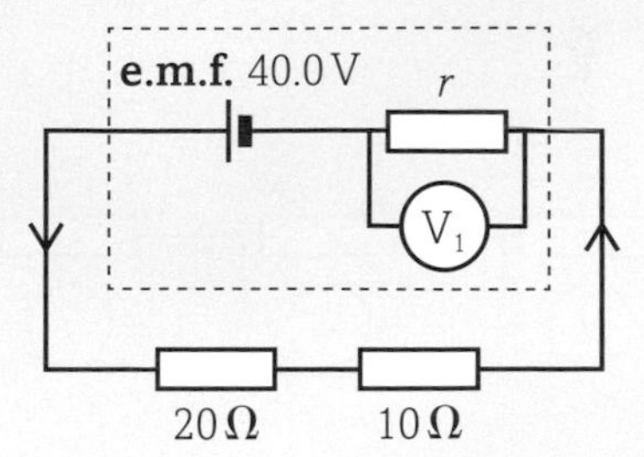

Figure 8

By the end of this topic, you should be able to demonstrate and apply your knowledge and understanding of:

* analysis of circuits with components, including both series and parallel

Circuit overview

Electric circuits vary from the simple battery–switch–bulb type to the very complex. The use of sophisticated electronic components or repetition of a simple circuit section can increase the complexity. However complex the circuit, the principle is to use electric current to get energy of the type required to the place where it is needed.

Analysis of series circuits

Analysing series circuits is very straightforward, so long as you remember the simple rules associated with current, potential difference and resistance in series circuits. If you do, you can calculate any value for the charge, potential difference, current, resistance, power or energy that is asked of you.

WORKED EXAMPLE 1

Figure 1 shows a circuit that contains a fixed resistor of size 50 Ω, a variable resistor that can have any value from 10 Ω to 100 Ω, and a thermistor with a resistance range from 200 Ω at 0 °C to 10 Ω when the temperature is 100 °C. Also in the circuit is a switch, an ammeter, voltmeters across each component and a cell with an e.m.f. of 12 V. The cell has negligible internal resistance and the variable resistor is initially set to a value of 40 Ω.

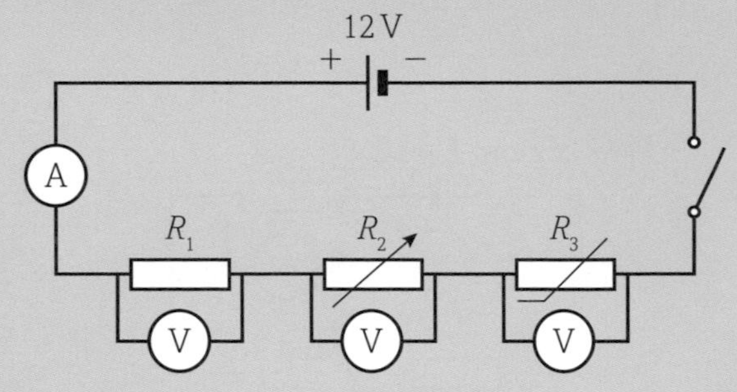

Figure 1 A typical circuit.

For the circuit shown, once the switch is closed, work out:

(a) the total resistance when the temperature is 0 °C
(b) the current flowing in the circuit when the temperature is 0 °C
(c) the potential difference across each of the components when the temperature is 0 °C
(d) the power dissipated as heat by the fixed resistor at 0 °C
(e) the quantity of charge that passes through the cell in a 10-minute period
(f) the number of electrons that pass through the cell in a 10-minute period
(g) the energy generated by the cell in an hour.

Answers

(a) The three components are connected in series, so:

$R_T = R_1 + R_2 + R_3$
$= 50\,\Omega + 40\,\Omega + 200\,\Omega = 290\,\Omega$

(b) $I = \frac{V}{R_T}$
$= \frac{12\text{ V}}{290\,\Omega} = 0.04\text{ A}$ (or 40 mA)

(c) Using $V = IR$ for each of the components:
for the fixed resistor, p.d. = $0.04\text{ A} \times 50\,\Omega = 2.0\text{ V}$
for the variable resistor, p.d. = $0.04\text{ A} \times 40\,\Omega = 1.6\text{ V}$
for the thermistor, p.d. = $0.04\text{ A} \times 200\,\Omega = 8.0\text{ V}$

(d) $P = I^2 \times R = 0.04^2 \times 50\,\Omega = 0.08\text{ W}$ (or 80 mW)

(e) $Q = I \times t = 0.04\text{ A} \times 10\text{ min} \times 60\text{ s} = 24\text{ C}$.

(f) The charge on one electron is 1.6×10^{-19} C. So, 1 coulomb of charge will have $\frac{1}{(1.6 \times 10^{-19})}$ electrons; that is 6.25×10^{18} electrons.
Because 24 C of charge pass through the cell in a 10-minute period, there will be $24 \times 6.25 \times 10^{18}$ electrons, or 1.5×10^{20}.

(g) $E = P \times t$ and $P = I \times V$. The energy generated in a hour can be calculated using E = p.d. × current × time = $12\text{ V} \times 0.04\text{ A} \times 3600\text{ s}$ = 1728 J of energy, which is 1700 J to 2 significant figures.

Analysis of parallel circuits

Like series circuits, analysis of parallel circuits is relatively simple provided that you know, understand and apply the rules of current, potential difference and resistance correctly.

WORKED EXAMPLE 2

A circuit contains three resistors that are connected in parallel. The resistors have resistances of 20 Ω, 40 Ω and 60 Ω respectively. The cell has an e.m.f. of 20 V and ammeters (of negligible resistance) placed as shown in Figure 2. Voltmeters are connected across each of the resistors and across the cell.

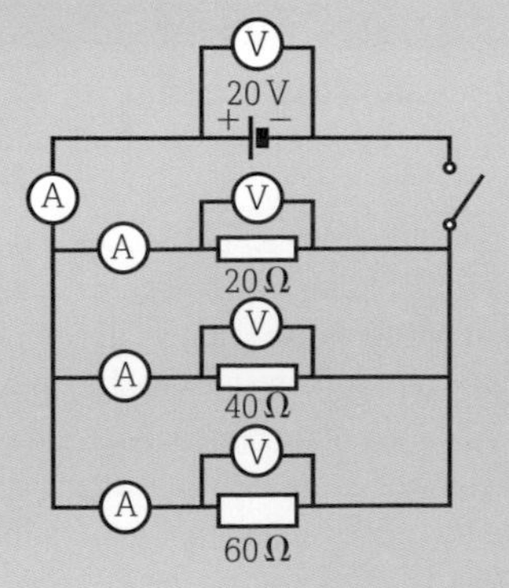

Figure 2

When the switch is closed, calculate:

(a) the effective resistance of the parallel network
(b) the current drawn from the cell
(c) the current through each of the resistors
(d) the power dissipated by the 20 Ω resistor
(e) the charge passing through the 40 Ω resistor each hour
(f) the energy from the cell that is made available to the circuit, each minute.

Answers

(a) The three resistors are connected in parallel, so:

$$\frac{1}{R_T} = \frac{1}{R_1} + \frac{1}{R_2} + \frac{1}{R_3}$$
$$= \frac{1}{20} + \frac{1}{40} + \frac{1}{60}$$
$$= 0.05 + 0.025 + 0.016 = 0.09\ \Omega^{-1}$$
$$R_T = \frac{1}{0.09} = 11\ \Omega \text{ to 2 significant figures.}$$

(b) $I = \frac{V}{R}$

$$= \frac{20\text{ V}}{11\ \Omega} = 1.8\text{ A to 2 significant figures.}$$

(c) All the three resistors experience a p.d. of 20 V. Using $I = \frac{V}{R}$ for each of these:

for the 20 Ω resistor, $I = 20\text{ V}/20\ \Omega = 1.0\text{ A}$; for the 40 Ω resistor, $I = \frac{20\text{ V}}{40\ \Omega} = 0.5\text{ A}$; for the 60 Ω resistor, $I = 20\text{ V}/60\ \Omega = 0.3\text{ A}$

(all answers stated to 1 decimal place).

(d) $P = I^2 \times R$

$= 1.0^2 \times 20 = 20\text{ W}$

(e) $Q = I \times t$

$= 0.5\text{ A} \times 3600\text{ s} = 1800\text{ C}$

(f) $E = V \times I \times t$

$= 20\text{ V} \times 1.8\text{ A} \times 60\text{ s} = 2160\text{ J}$

Analysis of series and parallel hybrid circuits

Many circuits have a parallel branch that is connected in series with other components. Analysis of these circuits is fairly easy provided that you take logical steps and apply the circuit rules. It is always a good idea to calculate the resistance of the parallel network first, before adding it to the resistance of any other components that are connected in series with it. From this you can work out the current leaving the cell, and subsequently you can determine the values of any other current, potential difference, charge, power, etc.

WORKED EXAMPLE 3

For the circuit in Figure 3, work out:

(a) the effective resistance
(b) the current leaving the cell
(c) the current through the 20 Ω resistor
(d) the potential difference across the 20 Ω resistor
(e) the potential difference across the 30 Ω resistor
(f) the power dissipated as heat in the 60 Ω resistor.

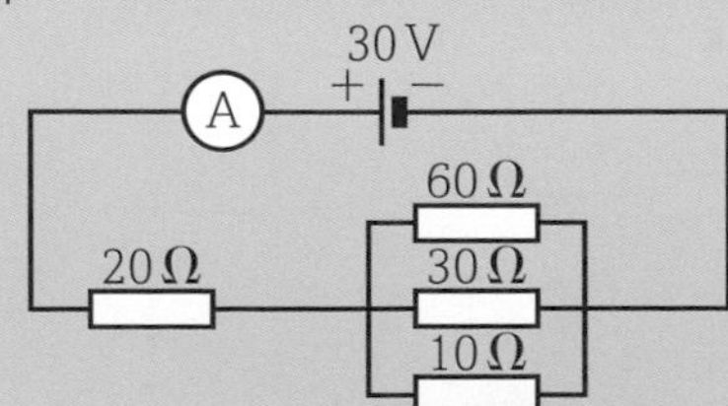

Figure 3

Answers

(a) The effective resistance is worked out by calculating the resistance of the parallel network, and then adding the 20 Ω series resistor.

$$\frac{1}{R_T} = \frac{1}{60} + \frac{1}{30} + \frac{1}{10}$$
$$= \frac{1}{60} + \frac{2}{60} + \frac{6}{60} = \frac{9}{60}$$

So $R_T = \frac{60}{9}$

= or 6.7 Ω to 1 decimal place.

Adding the 20 Ω resistor in series gives us a total circuit resistance of 26.7 Ω.

(b) $I = \frac{V}{R}$

$$= \frac{30\text{ V}}{26.7\ \Omega} = 1.1\text{ A}$$

(c) Using Kirchhoff's first law, the current leaving the cell must be equal to the current passing through the 20 Ω resistor because the current has not yet reached a junction where it can split.

(d) $V = I \times R$

$= 1.1\text{ A} \times 20\ \Omega = 22\text{ V}$

(e) The p.d. across the 30 Ω resistor will be equal to the cell voltage minus the p.d. across the 20 Ω resistor. This is equal to 30 V − 22 V = 8 V.
To check this, you can work out the current through the 30 Ω resistor and multiply it by its resistance.

(f) The current through the 60 Ω resistance is given by $\frac{V}{R}$. We know that there is a p.d. of 8 V across the parallel network, the current through the 60 Ω resistor is 8 V/60 Ω = 0.13 A.

$P = I^2 \times R$

$= 0.13^2 \times 60 = 1.0\text{ W}$ to 2 significant figures.

Questions

1 What are the resistances of the networks in Figure 4?

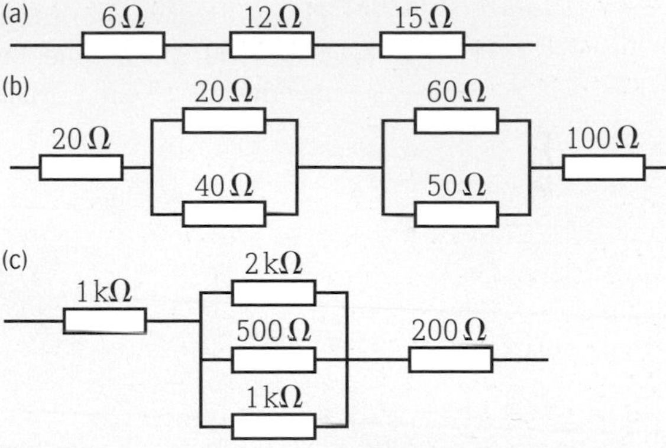

Figure 4

2 Each of the networks in Figure 4 is connected to an e.m.f. of 10 V. For each network calculate:

(a) the current through each resistor
(b) the potential difference across each resistor
(c) the energy leaving each cell each minute.

3 For the circuit shown in Figure 5, calculate the change in current when:

(a) switch 1 is closed
(b) switch 2 is closed
(c) switches 1 and 2 are closed together
(d) the 18 Ω resistor is removed from the circuit (with both switches open).

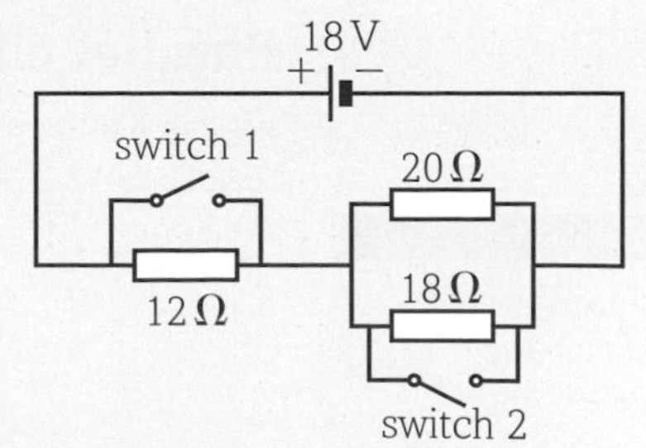

Figure 5

7 Circuit analysis 2

By the end of this topic, you should be able to demonstrate and apply your knowledge and understanding of:

* analysis of circuits with more than one source of e.m.f.

Introduction

The nature of how current, potential difference, resistance, energy and power behave in series circuits and parallel circuits has now been covered. Initially we considered relatively simple series circuits before moving on to more complex parallel circuits with one source of e.m.f..

In this topic, we are going to analyse further how to deal with circuits that have more than one source of e.m.f., starting with simple circuits and progressing to more complex ones. It would be advisable to re-familiarise yourself with the laws of how circuits work, with a special review of Kirchhoff's first and second laws.

Series circuits involving two sources of e.m.f.

Figure 1 shows a series circuit with two sources of e.m.f. – one of 9 V and one of 6 V. These are in series with one another but they have opposite polarities, and in accordance with Kirchhoff's second law the overall e.m.f. available is 3 V (from 9 V − 6 V).

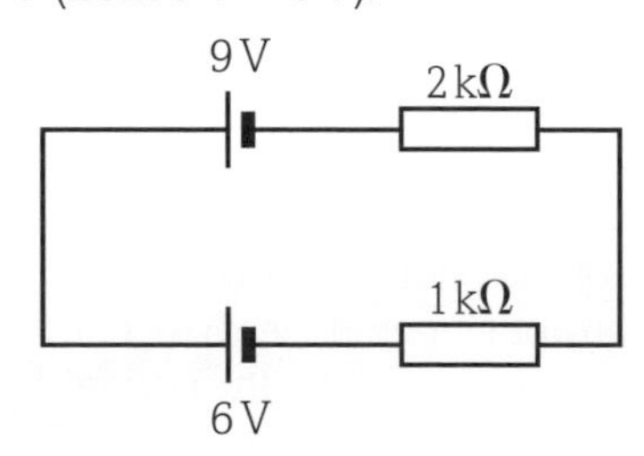

Figure 1

The total resistance in the circuit is 3 kΩ because the resistors are connected in series and there are no internal resistances shown for the cells. The current flowing in the circuit can be found using Ohm's law, $I = \frac{V}{R}$. The current is the same at any point, in accordance with Kirchhoff's first law, because there is only one loop.

The circuit shown in Figure 1 can be simplified to that shown in Figure 2. Using $I = \frac{V}{R}$ shows that a current of 0.001 A (1 mA) flows in the circuit.

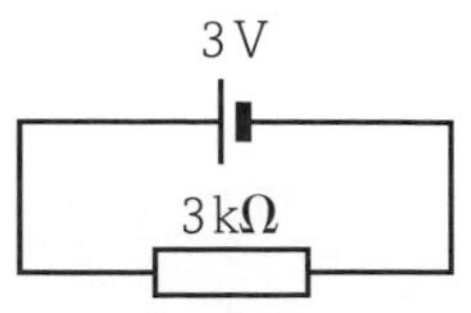

Figure 2 Simplified version of the circuit in Figure 1 with values.

Parallel circuits involving two sources of e.m.f.

Figure 3 shows two sources of e.m.f. arranged with opposing polarities. They are connected to a resistor and a lamp, which are connected in parallel with one another.

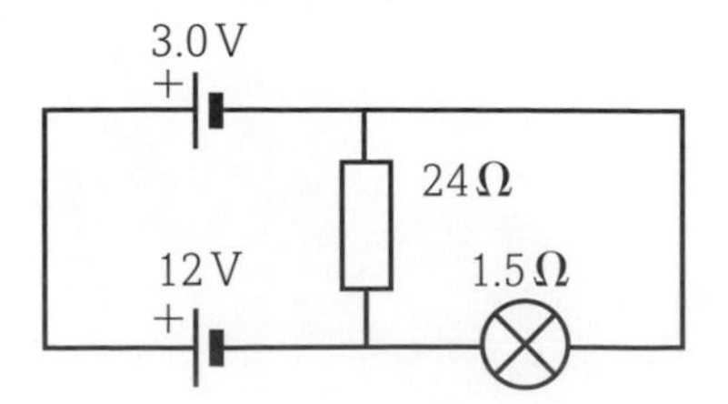

Figure 3

DID YOU KNOW?

If any number of resistors are connected in parallel, the effective resistance of the network is always less than the value of the lowest resistance in the network.

The overall e.m.f. available to the circuit, in accordance with Kirchhoff's second law is 9V (from 12 V − 3 V).

To calculate the combined effective resistance of the resistor and lamp network, arranged in parallel, we use the equation for resistors in parallel:

$$\frac{1}{R} = \frac{1}{R_1} + \frac{1}{R_2}$$

$$= \frac{1}{1.5} + \frac{1}{24}$$

$$= \frac{16}{24} + \frac{1}{24} = \frac{17}{24}$$

$$\text{So } R = \frac{24}{17} = 1.4\,\Omega$$

The current available to the resistor and the lamp from the net e.m.f. provided by the two cells is worked out using:

$$I = \frac{V}{R}$$

$$= \frac{9\text{ V}}{1.4\,\Omega} = 6.4\text{ A}$$

To work out the current through the lamp we use:

$$\frac{9\text{ V}}{1.5\,\Omega} = 6.0\text{ A}$$

To work out the current through the resistor we use:

$$\frac{9\text{ V}}{24\,\Omega} = 0.4\text{ A}$$

Checking this with Kirchhoff's first law, we should find that:

$$I_1 = I_2 + I_3$$

and this is indeed true.

WORKED EXAMPLE

Work out the values of I_1, I_2 and I_3 in the circuit shown in Figure 4.

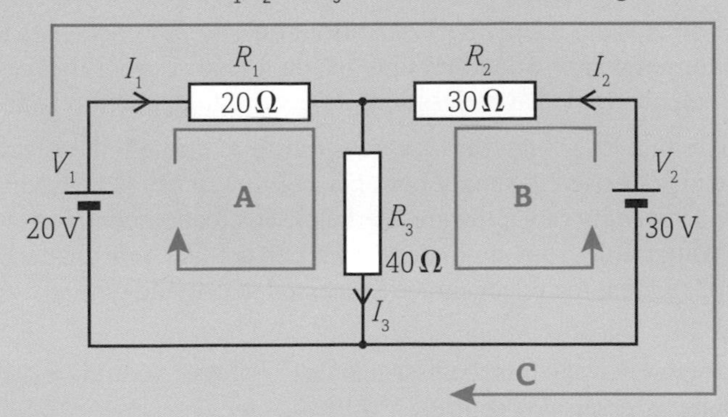

Figure 4

Answers

From Kirchhoff's first law, we know that $I_3 = I_1 + I_2$

From Kirchhoff's second law, we can formulate three equations for the three loops shown. Applying $\Sigma\varepsilon = \Sigma IR$ to each loop:

loop A gives: $20 = (R_1 \times I_1) + (R_3 \times I_3) = (20 \times I_1) + (40 \times I_3)$

loop B gives: $30 = (R_2 \times I_2) + (R_3 \times I_3) = (30 \times I_2) + (40 \times I_3)$

loop C gives: $20 - 30 = (20 \times I_1) - (30 \times I_2)$

Because I_3 is actually equal to $I_1 + I_2$ we can rewrite the equations:

$20 = (20 \times I_1) + 40 \times (I_1 + I_2) = (60 \times I_1) + (40 \times I_2)$ equation 1

$30 = 30 \times I_2 + 40 \times (I_1 + I_2) = (40 \times I_1) + (70 \times I_2)$ equation 2

These two simultaneous equations can be solved to give the values of I_1 and I_2.

Multiplying equation 1 by 2, and then equation 2 by 3 enables us to cancel the I_1 terms.

Substitution of I_1 in terms of I_2 gives us the value of $I_1 = -0.14$ A

Substitution of I_2 in terms of I_1 gives us the value of $I_2 = 0.43$ A

But $I_3 = I_1 + I_2$ so the current flowing in resistor R_3 is:
$-0.14\text{ A} + 0.43\text{ A} = 0.29\text{ A}$

Questions

1. For the circuit shown in Figure 5, work out values for:
   (a) the current flowing in the circuit
   (b) the charge passing through the 5 Ω resistor in 1 hour
   (c) the power rating of the 1.6 Ω resistor
   (d) the energy supplied to the circuit each minute.

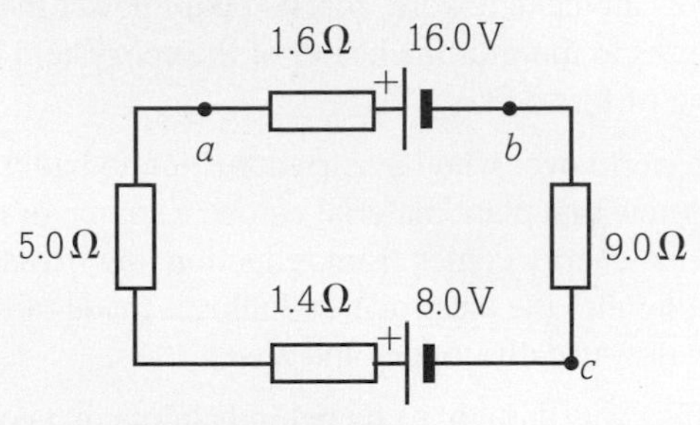

Figure 5

2. A cell of e.m.f. 24 V and internal resistance 2 Ω is arranged in parallel with another cell of e.m.f. 27 V and internal resistance 6 Ω. Across this network, a resistor of 4 Ω is then connected in parallel.
   (a) Draw the circuit.
   (b) Work out the currents through the 24 V cell, the 27 V cell and the 4 Ω resistor.
   (c) Calculate the energy dissipated as heat by the 4 Ω resistor in a 25 minute period.

THINKING BIGGER

TREE-MENDOUS ELECTRICITY

It is possible to obtain a small quantity of electrical energy from fruits such as lemons, as well as from potatoes. In the extract below, we will consider the use of trees to power electric circuits.

TREES COULD BE THE ULTIMATE IN GREEN POWER

Shoving electrodes into tree trunks to harvest electricity may sound like the stuff of dreams, but the idea is increasingly attracting interest. If we can make it work, forests could power their own sensor networks to monitor the health of the ecosystem or provide early warning of forest fires.

Children the world over who have tried the potato battery experiment know that plant material can be a source of electricity. In this case, the energy comes from reduction and oxidation reactions eating into the electrodes, which are made of two different metals – usually copper and zinc.

The same effect was thought to lie behind claims that connecting electrodes driven into a tree trunk and the ground nearby can provide a current. But last year Andreas Mershin's team at MIT showed that using electrodes made of the same metal also gives a current, meaning another effect must be at work. Mershin thinks the electricity derives from a difference in pH between the tree and the soil, a chemical imbalance maintained by the tree's metabolic processes.

Practical power

While proving that trees can provide a source of power is a significant step, a key question remains: can the tiny voltage produced by a tree be harnessed for anything useful?

Trees seem capable of providing a constant voltage of anywhere between 20 and a few hundred millivolts – way below the 1.5 volts from a standard AA battery and close to the level of background electrical noise in circuits, says Babak Parviz, an electrical engineer at the University of Washington in Seattle. "Normal circuits don't run from very small voltages, so we need ways to convert the small voltages to something that is usable," he says.

His team has managed to obtain a usable voltage from big-leaf maple trees by adding a device called a voltage boost converter. The converter spends most of its time in a kind of stand-by mode as it stores electrical energy from the tree, periodically releasing it at 1.1 volts.

To provide that periodic wake-up call, Parviz's team developed a clock, also powered by the tree, which keeps time by tracking the quantum tunnelling of electrons through thin layers of insulating material. It operates at 350 millivolts and uses just a nanowatt of power.

Parviz thinks trees could power gadgets to monitor their own physiology or their immediate surroundings, for ecological research. And, he adds, as electronic components continue to shrink and require less power, it is possible tree electricity could one day have a wide range of uses.

Green power race

Parviz's team isn't the only one trying to harness the tiny voltages trees can provide. Voltree Power, a company based in Canton, Massachusetts, patented a tree-powered circuit in 2005, says the company's CEO, Stella Karavaz.

Her firm is using energy harvested from trees to power sensors that monitor temperature and humidity inside forests. Earlier this year the company trialed awireless sensor network to detect forest fires.

Devices that lose water the way trees transpire through their leaves could also be used to supply power, according to Michel Maharbiz at the University of California, Berkeley. His team recently showed that evaporation from simulated leaves can act like a mechanical pump, and that the effect can be harnessed to provide power.

Source

http://www.newscientist.com/article/dn17767-trees-could-be-the-ultimate-in-green-power.html

Figure 1 Electrical devices can plug directly into trees for power.

Where else will I encounter these themes?

1.1 | 2.1 | 2.2 | 3.1 | 3.2 | 3.3

Let us start by considering the nature of the writing in the article.

1. The article above was taken from New Scientist magazine and is entitled 'Trees could be the ultimate in green power'. The research reported by the article was funded in part by the National Science Foundation. Consider the article and comment on the type of writing that is being used. Who is the audience? Is it for the typical citizen or is it for a scientific community? Does the report try to explain, persuade, or describe? Is there bias in the article? Are the findings open to interpretation by others? How might you change the article to make it more suitable for a younger or less-informed audience?

Let us now look at the physics that is in the article. Do not worry if the physics content or the mathematics is challenging at this stage. You can always return to the article later in your course, once some of the related topics have been studied in more depth. Use the timeline at the bottom of the page to help place this work in context with what you have already learned and what is ahead in your course.

2. What would a typical value be for the resistance of a tree?
3. Which factors may affect the size of the voltage measured in a study of this nature?
4. Explain, in terms of potential difference, how a potato can replace a battery in a circuit. How is this different from what is happening here with trees?
5. Suggest why the boost converter needs to be kept in stand-by mode.
6. How could the electrical output from trees be used to 'monitor their own physiology' as stated in the article?
7. How viable is the 'power from trees' technology as a future renewable energy resource?

For question 2, you could revisit ideas about orders of magnitude in Topic 2.1.2.

Activity

You are going to prepare a 5 minute presentation offering a compelling argument for using trees as a viable electrical power source. Your presentation must contain the following:

- A simple explanation of how typical circuits work.
- Electrical values involved in a circuit that is powered by trees.
- A comparison between trees and other sources of electrical energy of their efficiencies and reliabilities.
- The specifics of how you would use this tree-based electrical circuit technology.
- Typical costs compared with other forms of electrical energy generation.
- Advantages and disadvantages of this type of electrical technology.

4.3 Practice questions

1. Two resistors of resistance R and $2R$ are connected to a cell. Which of the following statements are true? [1]
   (i) The effective resistance of the resistors in series is $3R$.
   (ii) The effective resistance of the resistors in parallel is less than R.
   (iii) The power dissipated by R will be the same as that by $2R$.

   A (i), (ii) and (iii)
   B only (i) and (ii)
   C only (ii) and (iii)
   D only (i)

2. Three resistors of 5 Ω, 8 Ω and 12 Ω are connected in parallel and then connected to a cell of e.m.f. 12 V. Which of the following statements is correct? [1]

   A The effective resistance is the mean of the three values.
   B The current through the 8 Ω resistor is 1.5 times greater than the current through the 12 Ω resistor.
   C The effective resistance is 25 Ω.
   D The 5 Ω resistor has the greatest potential difference across it.

3. The following circuit is a potential divider arrangement.

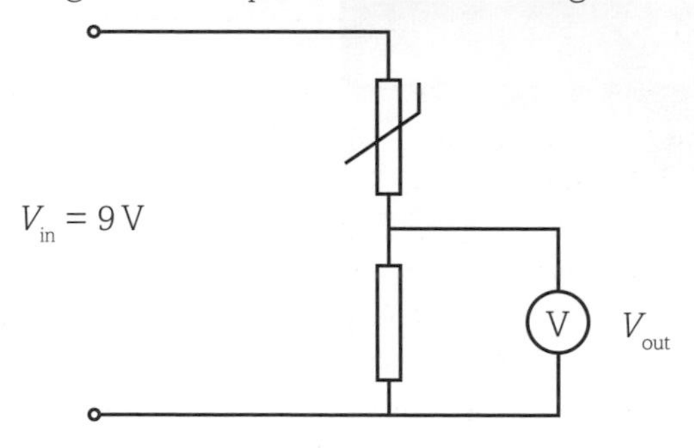

Figure 1

Which of the following statements is true for the circuit? [1]

   A The output will be 12 V.
   B The circuit can be used to switch a lamp on in dark conditions.
   C The circuit could be used to control a refrigerator.
   D The output will increase as the light intensity decreases.

4. Which statement regarding electrical circuits is not true? [1]

   A Kirchhoff's second law is the conservation of energy in relation to electrical circuits.
   B E.m.f. is the energy converted to electrical energy when unit charge passes through the cell.
   C The e.m.f. of a cell equals the terminal potential difference when no current is flowing in the circuit.
   D Kirchhoff's second law states that the sum of currents entering a junction is zero.

[Total: 4]

5. Figure 2 shows three resistors that are part of a circuit.

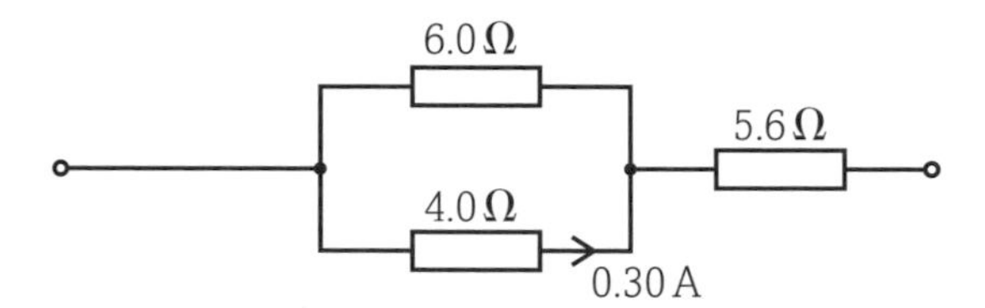

Figure 2

Calculate:

(a) the current through the 6.0 Ω resistor [2]
(b) the current through the 4.0 Ω resistor [1]
(c) the potential difference across the three resistors [3]
(d) the effective resistance of the three resistors. [3]

[Total: 9]

6. Four resistors have a resistance of 2 Ω, 4 Ω, 8 Ω and 20 Ω respectively. They are arranged in series and connected to a 16 V cell. They are then arranged in parallel and connected to the same 16 V cell.

   Calculate the ratio of the current through the 2 Ω resistor in the parallel arrangement to the current through the 2 Ω resistor in the series arrangement.

[Total: 5]

7. A student conducts an experiment using two identical filament lamps and a variable power supply of negligible internal resistance. The lamps are connected in series to the supply. The current in the circuit is 0.030 A and the lamps are dimly lit.

   The e.m.f. of the power supply is then doubled and the experiment repeated.

   The student expected the current to double, but the current only increased to 0.040 A. The lamps are brightly lit.

   Use your knowledge of physics to explain these observations. [3]

[Total: 3]

8. A battery of negligible internal resistance is connected across two resistors of resistance values R and $2R$ as shown in Figure 3.

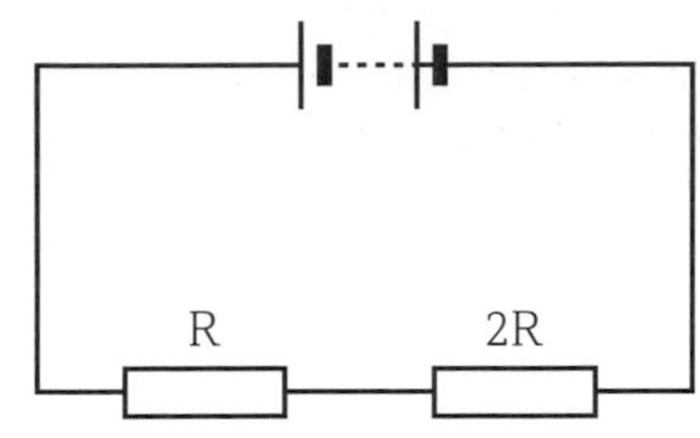

Figure 3

The same battery is now connected to the same resistors as shown in Figure 4.

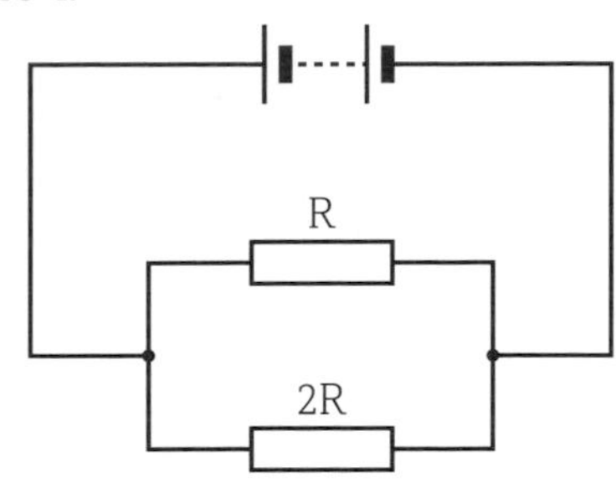

Figure 4

Calculate the following ratio:

$$\frac{\text{current from battery in circuit of Figure 3}}{\text{current from battery in circuit of Figure 4}}$$

[Total: 3]

[Q24(a), H156/01 sample paper 2014]

9. Figure 5 shows how the resistance of a thermistor varies with temperature.

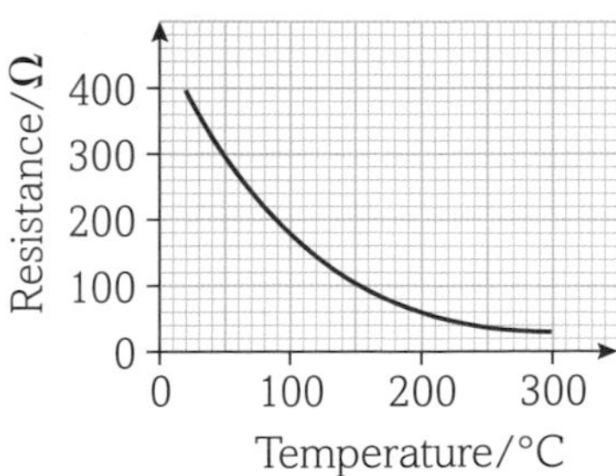

Figure 5

The thermistor is used in the potential divider circuit of Figure 6 to monitor the temperature of an oven. The 6.0 V DC supply has zero internal resistance and the voltmeter has infinite resistance.

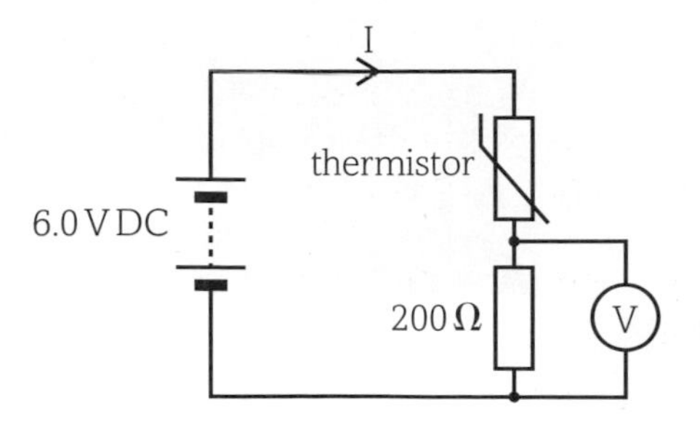

Figure 6

State and explain how the current, I, changes in the circuit as the thermistor is heated.

[Total: 3]

[Q3(a), G482 May 2014]

10. (a) The circuit in Figure 7 consists of a DC supply of e.m.f. 45 V and negligible internal resistance and three resistors.

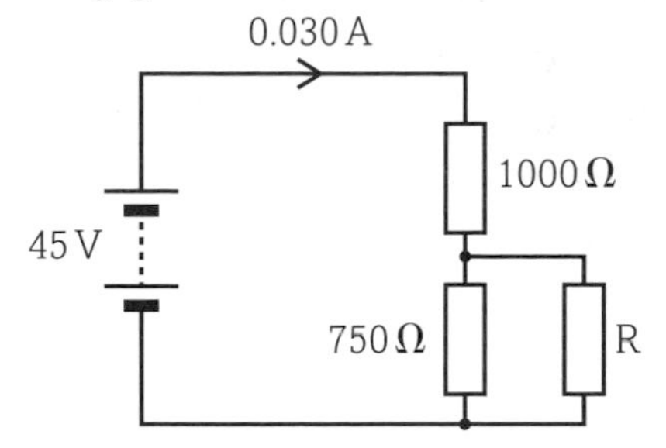

Figure 7

Two of the resistors have resistances of 1000 Ω and 750 Ω as shown. The current drawn from the supply is 0.030 A. Calculate the resistance of R. [4]

(b) Students are given a light dependent resistor (LDR) and asked to design a circuit for a light meter to monitor changes in light intensity. The meter reading must rise when the light intensity increases.

The incident light may cause the resistance of the LDR to vary between 1500 Ω and 250 Ω.

The students are asked to use the DC supply and one of the resistors from (a) above and either a voltmeter or ammeter.

(i) Draw a suitable circuit.

(ii) Explain why the reading on the meter increases with increasing light intensity.

(iii) Explain which of the three fixed resistors gives the largest scale change on the meter for the change in light intensity. [6]

[Total: 10]

[Q5, H156/02 sample paper 2014]

MODULE 4 Electrons, waves and photons

CHAPTER 4.4 WAVES

Introduction

Energy can be transferred from one place to another in a number of ways. When we eat food we convert the chemical energy stored in sugars and carbohydrates to kinetic energy for movement, sound energy for speech and heat energy to keep us warm. Waves transfer energy without any overall movement of matter. The light that reaches us from the Sun travels 150 million kilometres through empty space as an electromagnetic wave. It travels to us at the speed of light, an incredible 300 million metres per second, reaching us in just over 8 minutes. It provides us with infrared to keep us warm, visible light to help us see and UV for healthy skin and a feeling of wellbeing. Sound energy reaches us as a longitudinal wave, carrying our favourite music and the words and meaning of language. The information reaches us through a series of vibrations, but no air molecules accompany it. Seismic waves can tell us about the structure of the Earth but they can also cause mass devastation when they shake and vibrate buildings in the immediate aftermath of an earthquake or tsunami. In this chapter we will cover the fascinating topic of waves – how they behave, how they travel, how they can be described and explained and how they can be used.

All the maths you need

To unlock the puzzles of this chapter you need the following maths:

- Multiplication of values to determine wave speed
- Use of standard form
- Rearrangement of formulae, e.g. when trying to determine λ from the diffraction grating experiment or Young's double slit experiment
- Vector addition to show interference
- Use of trigonometry to determine angles of refraction, the critical angle and diffracted angles from a diffraction grating
- Use of the square function to relate the intensity of a wave to its amplitude

What have I studied before?

- Waves transfer energy from one place to another without any overall transfer of matter
- Waves can be longitudinal in nature (e.g. sound waves) or transverse (e.g. light)
- Waves can be described in terms of their wavelength, frequency, amplitude or speed and the Wave speed = frequency × wavelength
- The electromagnetic spectrum is composed of radio, microwaves, infrared, visible, ultraviolet, X-rays and gamma rays which all travel at the speed of light in a vacuum
- Sounds that are reflected are heard as echoes; light that is reflected off objects enables us to see things
- Refraction occurs when sound or light travels from one medium into another. Refraction involves a change in speed and (usually) a change in direction of the wave
- Diffraction is the spreading out of a wave when it passes through a gap or around an object. It is most pronounced when the wavelength of the wave is similar in size to the gap or width of the obstacle
- Waves that overlap at the same point in space may interfere, leading to constructive or destructive interference
- Seismic waves are produced during earthquakes and they can be analysed to tell us about the structure of the Earth

What will I study later?

- The period and frequency of an object in circular motion and how its motion can be modelled as a wave (AL)
- Simple harmonic motion and its relevance to wave motion (AL)
- Free and forced oscillations, resonance and natural frequency (AL)
- The condition for maxima when using a diffraction grating and the equation $n\lambda = d\sin\theta$ (AL)
- Wien's displacement law and the calculation of λ_{max} to estimate the peak surface temperature of a star (AL)
- The Doppler effect and the relationship between relative velocities and the change in frequency and wavelength of a moving object (AL)
- The production of X-rays from an X-ray tube (AL)
- The production, nature and behaviour of gamma rays in diagnostic medicine using medical tracers such as technetium-99m and fluorine-18 (AL)
- The gamma camera and its use in medicine (AL)
- Ultrasound techniques in a medical and non-medical setting (AL)

What will I study in this chapter?

- Progressive waves transfer energy through space and may be longitudinal or transverse in nature
- The difference in motion of particles in a wave can be expressed in terms of their phase difference
- Frequency and time period of a wave are related via the equation $f = \frac{1}{T}$
- Electromagnetic waves may display polarisation, and be plane-polarised. This can be demonstrated in the laboratory using light and microwaves
- The intensity of a progressive wave is given by intensity $= \frac{\text{power}}{\text{area}}$ or $I = \frac{P}{A}$ and the power of a wave is proportional to the square of its amplitude, via $I \propto A^2$
- Light will undergo refraction when travelling from one optically dense medium into another. This is explained mathematically by Snell's law, $n_1 \sin\theta_1 = n_2 \sin\theta_2$
- Light will undergo total internal reflection when travelling from a more to a less optically dense material if the angle of incidence is greater than the critical angle, C. The critical angle is calculated using the formula $\sin C = \frac{1}{n}$
- When two or more coherent waves overlap or superpose the resultant waveform will have an amplitude equal to the vector sum of their individual amplitudes. This is known as the principle of superposition. Constructive and destructive interference will occur, based on the phase difference and path difference of the overlapping waves
- The Young's double slit arrangement, and diffraction gratings, can be used to obtain a value for the wavelength of light
- Stationary or standing waves can be set up on strings or in air columns. They are formed when waves, travelling in opposite directions, interfere and superpose and can be described in terms of nodes, antinodes, modes of vibration and harmonics
- Techniques and procedures used to determine an accurate value for the speed of sound in air by using a tuning fork and a resonance tube to produce standing waves

4.4 ① Wave motion

By the end of this topic, you should be able to demonstrate and apply your knowledge and understanding of:

* progressive waves; longitudinal and transverse waves

Introduction

Waves transfer energy from one place to another without any net transfer of matter. There are many examples of waves in our lives, from the big water waves that crash against the sea wall to the tiny microwaves that cook our food and allow us to make mobile phone calls and send text messages.

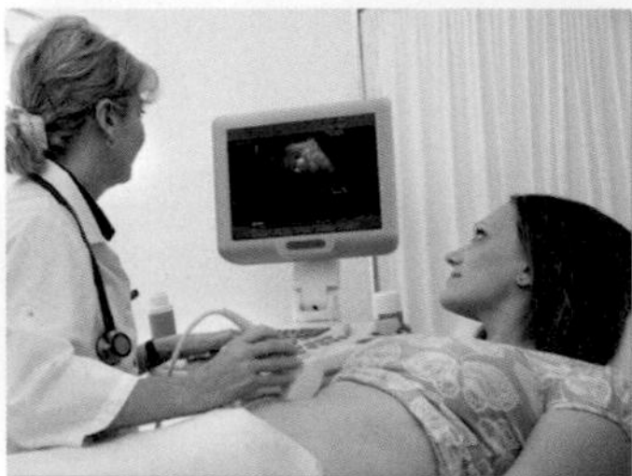

Figure 1 Examples of the uses and consequences of waves.

Waves can be **longitudinal** or **transverse**. Sound waves and ultrasound waves are examples of longitudinal waves, as are certain types of seismic waves called P waves. In longitudinal waves, the vibrations that cause them are parallel to the direction of energy transfer, as shown in Figure 2.

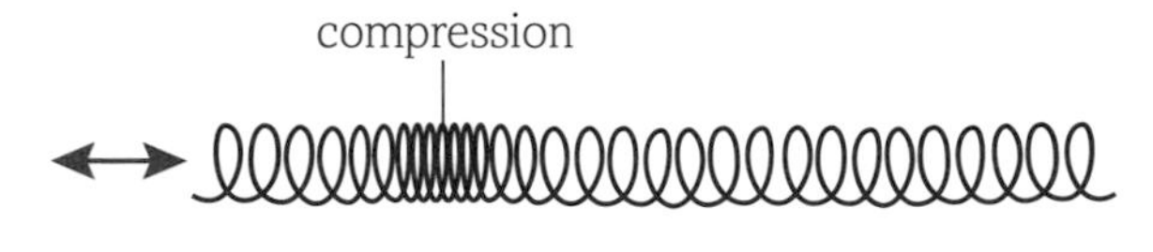

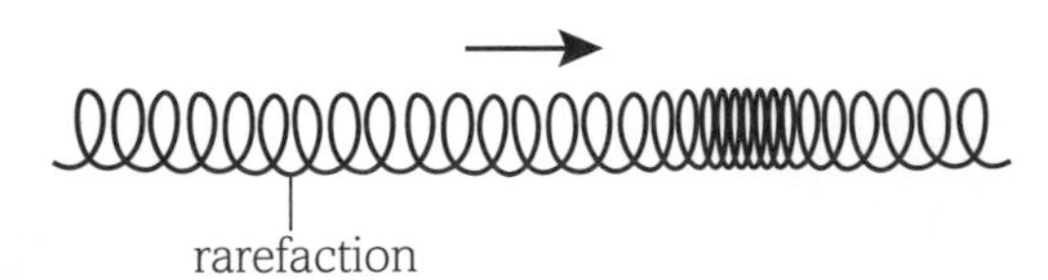

Figure 2 In a longitudinal wave, the vibrations are parallel to the direction of energy transfer, resulting in the formation of compressions and rarefactions.

Other types of waves, including the family of electromagnetic waves, are examples of transverse waves. Electromagnetic waves all travel at the same speed – the speed of light in a vacuum – and they are produced when the vibration causing them is at right angles to the direction of energy transfer.

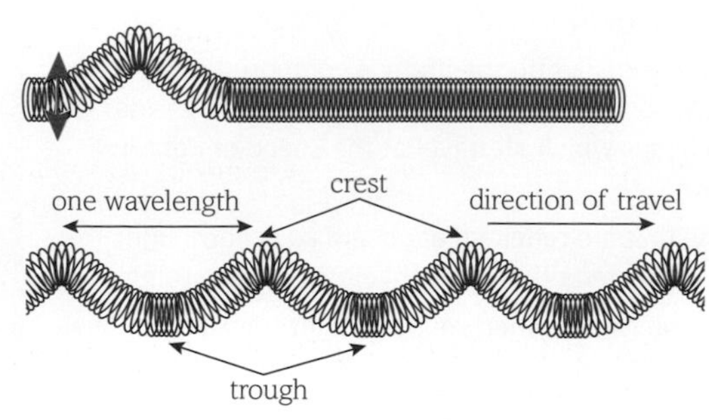

Figure 3 Transverse waves transfer energy from one place to another, but here the vibrations are at right angles to the direction of energy transfer.

The propagation of waves

Waves that move away from a source are called **progressive waves**. Figure 4 shows a wave moving forward, with two 'snapshots' – one taken quickly after the other. You can see that particle A is moving upwards while particle B is moving downwards to create the wave movement. All the particles oscillate vertically but they do not move forwards or backwards, even though the wave moves forward.

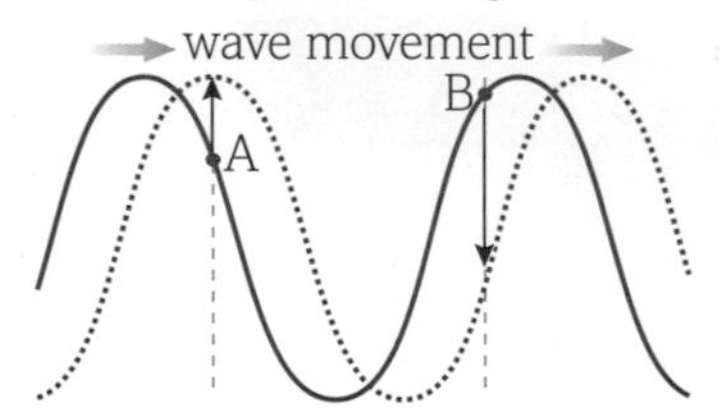

Figure 4

KEY DEFINITION

Waves that transfer energy away from a source are called **progressive waves**.

Questions

1. Copy and fill in the table to show which waves are longitudinal and which are transverse.
   Put the following kinds of waves in the correct columns: sound waves, X-rays, gamma rays, light rays, ultrasound waves, slinky waves.

Transverse waves	Longitudinal waves

2. How are longitudinal waves and transverse waves:
   (a) similar? (b) different?
3. How is an oscillation different from a wave?

Wave terminology

By the end of this topic, you should be able to demonstrate and apply your knowledge and understanding of:

* displacement, amplitude, wavelength, period, phase difference, and frequency of a wave
* the equation $f = \frac{1}{T}$
* graphical representations of transverse and longitudinal waves
* techniques and procedures used to use an oscilloscope to determine frequency

Introduction

One of the major difficulties of describing waves in a book or an examination paper is that they move – so representing waves can be awkward. For a sketch graph to make sense, you have to look at what is plotted against what.

When a wave, such as a water wave, travels from one point to another it is because a disturbance has caused particles in a material to be displaced. The particles closest to the initial disturbance get displaced first and this causes adjacent particles to do the same, slightly later. This is repeated until the disturbance reaches its final point.

We can look at a wave in two ways:

- We can look at a 'snapshot' of the wave profile as the disturbance travels outwards. Here we are looking at the *displacement* of the wave on the x-axis against the *distance* it has travelled outwards from the source.
- We can plot the oscillation of an individual particle in the medium against time. Unlike a progressive wave, which spreads out at right angles to the disturbance, the particles here will be vibrating up and down. This time we are plotting the *displacement* of the particle against the *time*.

In both cases, the *amplitude* (or maximum displacement) of the wave and the amplitude of oscillating particles will be the same, provided that energy is constantly being transferred so that the wave amplitude does not decrease over time. In reality, such as when a stone is dropped into a pond, the amplitude will decrease because energy is not being supplied in a repetitive way.

Compare Figure 1 with Figure 3 – these graphs represent the same water wave. The problem is partially overcome by drawing diagrams or sketch graphs of the waves at a specific instant in time and then stating what happens at a later time.

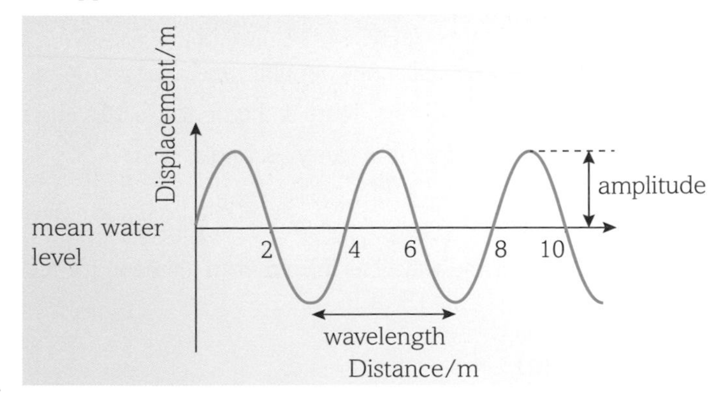

Figure 1 Displacement of wave plotted against distance.

Figure 2 This image shows what the graph may look like in reality – the initial disturbance at the centre has led to the formation of waves travelling outwards from the source in all directions.

The graph in Figure 1 has distance plotted on the x-axis. It is effectively a snapshot at a particular moment and tells you that the distance between successive identical points on the wave is 4.0 m. The vertical displacement of the wave has been plotted against the horizontal distance that the wave has moved, so we can show the maximum height of the wave (its amplitude) and its *wavelength*.

Figure 3 shows how the height of the wave changes. However, this time the *x*-axis shows the time for which the wave has been travelling, not the distance. So we can get values for the amplitude from the graph, but not the wavelength. The distance between successive identical points on the horizontal axis tells us the *time period* of the wave and also the oscillating particles.

These two sketch graphs look similar, and indeed give the same information about the height of the wave. However, the differences between them can be important and you must always check exactly which property is plotted on which axis.

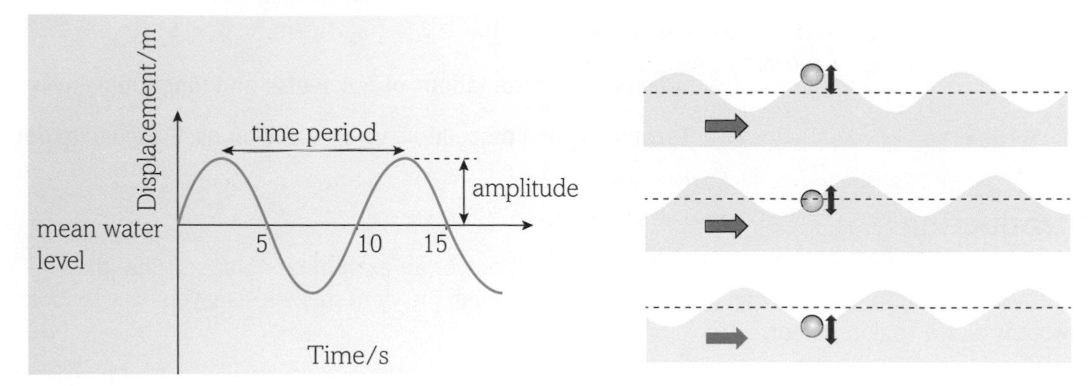

Figure 3 Displacement of an oscillating water particle in the wave against time.

Although the progressive wave is moving from left to right, the individual particles in the wave are oscillating up and down. This is shown by the ball in Figure 3 which bobs up and down vertically on the water's surface. The time taken for the wave to move a distance of one wavelength to the right is the time taken for the ball to complete one full vertical oscillation.

Terms and symbols used to describe waves

Wavelength λ, unit metre (m)

The *wavelength* of a wave is the distance between two successive identical points that have the same pattern of oscillation. It is also the distance the wave travels before the pattern repeats itself. In Figure 1 you can see that here the wavelength is 4.0 m, i.e. the wave has travelled 4.0 m before repeating its oscillation pattern.

Period T, unit second (s)

The *period* of a wave is the time it takes for one complete pattern of oscillation to take place at any point. Figure 3 shows this period to be 10.0 s – it takes 10.0 s for the wave to complete one oscillation pattern.

Frequency f, unit hertz (Hz)

The *frequency* of a wave is the number of oscillations per unit time at any point and is related to the time period T by the equation $f = \frac{1}{T}$. Using Figure 3, the period is 10 s and so the frequency $= \frac{1}{10} = 0.10$ Hz, that is there are 0.10 oscillations in every second.

Displacement x, unit metre (m)

Displacement is the distance any part of the wave has moved from its mean (or rest) position – it can be positive or negative.

Amplitude x_0, unit metre (m)

Amplitude is the maximum displacement – the distance from a peak or trough to the mean (rest) position.

Phase difference φ, unit radian or rad

Phase difference concerns the relationship between the pattern of vibration at two points. Two points that have exactly the same pattern of oscillation are said to be **in phase** – there is zero phase difference between them.

Wavelength can also be defined as the shortest distance between two points that are in phase. If the patterns of movement at the two points are exactly opposite then the waves are said to be in antiphase, as shown in Figure 4. They are half a cycle different from one another.

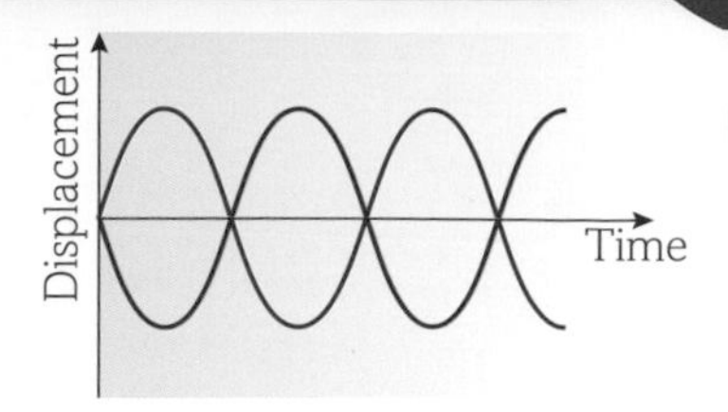

Figure 4 Waves in antiphase.

There is a strong relationship between circular motion and wave motion. Imagine the vector arrow shown in Figure 5 rotating at a constant rate. As it rotates, the tip of the arrow plots the graph of a circle – this is shown to the right of the circle in Figure 5. The displacement–time graph in Figure 5 has the form of a sine curve.

If the rotating vector makes an angle θ with the horizontal, the graph in Figure 5 represents $\sin\theta$ for one complete rotation of the vector arrow, 0–360°.

In radians, the angle for one complete rotation of the vector arrow is 2π radians, so one complete cycle of a wave is given as 2π rad. Waves that are in *antiphase* are π rad out of phase (Figure 4). Figure 6 shows two sine waves that are $\frac{\pi}{2}$ out of phase. If Figure 6 represented the displacement–time graph for two progressive transverse waves from $t = 0$, the blue wave would be a quarter of a cycle ahead of the red wave – there is a phase difference of 90° or $\frac{\pi}{2}$ rad.

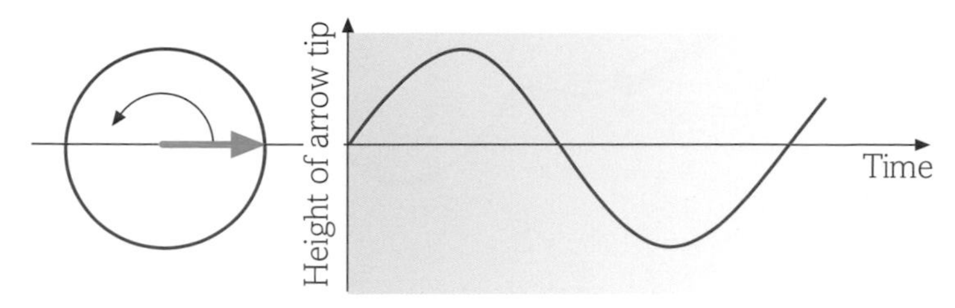

Figure 5 The vertical displacement of the tip of the arrow plots out the wave.

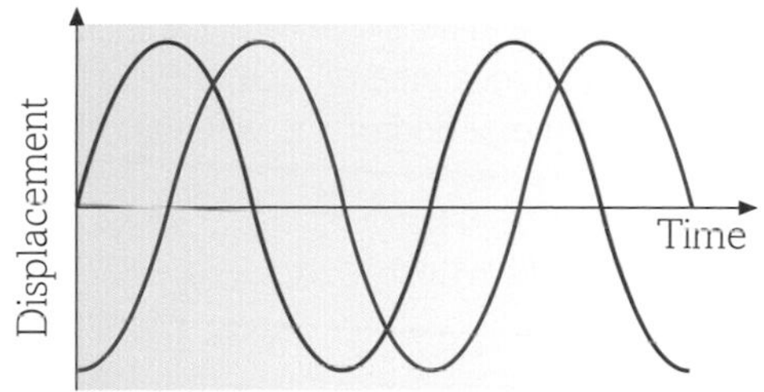

Figure 6 Two waves $\frac{\pi}{2}$ out of phase.

LEARNING TIP

You need to understand the relationship between degrees and radians and be able to translate from one to the other.

360° = 2π rad
180° = π rad
90° = $\pi/2$ rad

To convert an angle in degrees to an angle in radians, divide the angle by 180 and multiply by π. To convert an angle in radians to an angle in degrees, divide the angle by π and multiply by 180.

Representing longitudinal waves graphically

Up until now, the displacement–time graphs and displacement–distance graphs have been described for transverse waves, where it is easy to visualise the displacement of a wave. In a longitudinal wave, the displacement is in the direction of energy transfer and not at right angles to it. This makes plotting displacement–distance graphs tricky to plot so we usually represent a longitudinal wave with the displacement (measured along the *x*-axis) plotted on the *y*-axis. We can then plot displacement against position, as shown in Figure 7(a).

The shape is that of a sine wave.

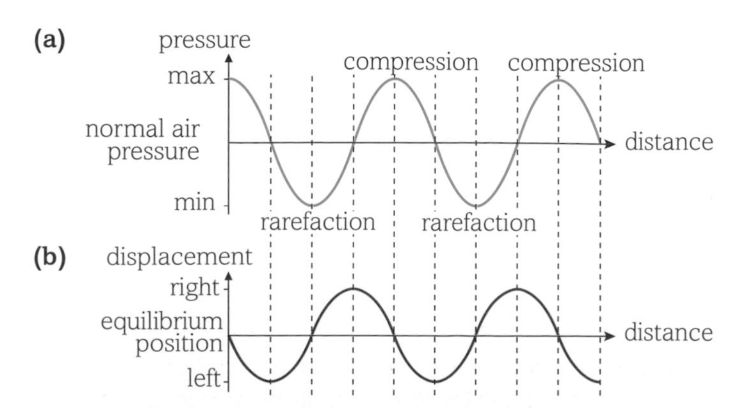

Figure 7 (a) Pressure–position and (b) displacement–position graphs for a longitudinal wave.

LEARNING TIP

Correct axis labelling is particularly important with longitudinal waves such as sound. Many students draw sound waves as transverse waves because of their appearance on an oscilloscope. This is acceptable provided the axes are labelled correctly. On an oscilloscope it is time on the *x*-axis not displacement; the *y*-axis shows the output from a microphone, which will usually be the pressure in the sound wave.

In Figure 7(a), different points along the wave have different phases, and the wavelength is the distance between two points where the oscillations are in phase.

Another way of representing a longitudinal wave is to plot pressure against position. You can see from Figure 7(b) that the displacement wave is 90° out of phase with the pressure wave. A position of maximum pressure (a compression) or minimum pressure (rarefaction) coincides with positions of zero displacement.

Using an oscilloscope to find the frequency of a wave

An oscilloscope (c.r.o.) displays a voltage–time signal and can be used as a voltmeter to display and measure the output from a microphone or signal generator (Figure 8). The time-varying voltage trace represents displacement against time for the longitudinal sound wave.

LEARNING TIP

The distance between peaks does *not* represent the wavelength. The horizontal axis is time so the distance between peaks is the time period.

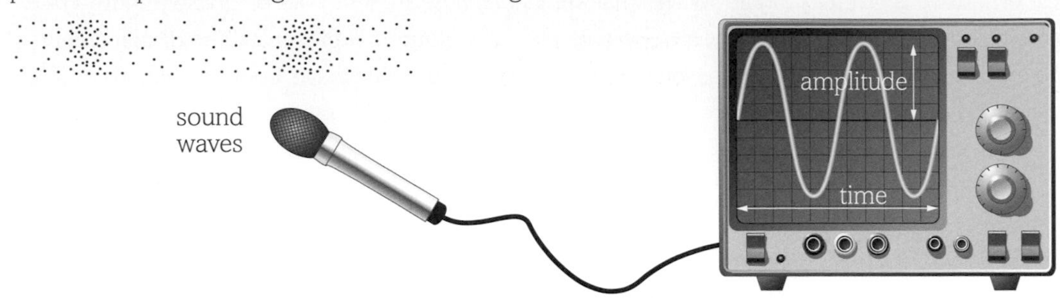

Figure 8 Using an oscilloscope.

The frequency of a wave that is displayed on an oscilloscope can be determined if you know the setting of the *time base* on the oscilloscope. This is the time taken for the luminous dot produced by the cathode ray tube to move a horizontal distance of 1 cm across the oscilloscope screen. Once you know this value you can work out the time period of the wave and from that you can find the frequency by using $f = \frac{1}{T}$.

period = distance between peaks × time base setting

Oscilloscope controls

Each horizontal division on the oscilloscope screen represents a unit of time. The time base control varies the seconds or milliseconds per division, e.g. 0.002 s/div. You can reduce the uncertainty in the frequency measurement by altering the time base such that one full wave has the widest possible range in the *x* direction.

If the time base is turned off, the spot no longer moves across the screen. This is useful if you are just looking at the intensity of the sound waves.

Each vertical division on the oscilloscope screen represents a unit of voltage. The *sensitivity* control (Y-gain) varies the volts per division, e.g. from 20 V/div (less sensitive) to 5 mV/div (more sensitive).

WORKED EXAMPLE

The time base setting on an oscilloscope is set at 5 ms cm^{-1}. Three full wave oscillations are displayed over a distance of 12 cm on the screen. Find:

(a) the time period, T

(b) the frequency, f.

Answers

(a) It takes 12 cm × 5 ms cm^{-1} = 60 ms for 3 complete oscillations. This means that 1 complete oscillation will be 20 ms, so T = 20 ms.

(b) $f = \frac{1}{T}$

$= \frac{1}{(20 \times 10^{-3})}$

$= 50$ Hz

It is important to convert the time from ms to s so that the frequency is in Hz.

A note of constant frequency (such as from a signal generator) gives a regular, sinusoidal waveform (Figure 9(a), 9(b)). The waveform for vibrations produced by musical instruments or a singer's vocal cords are more complex (Figure 9(c), 9(d)).

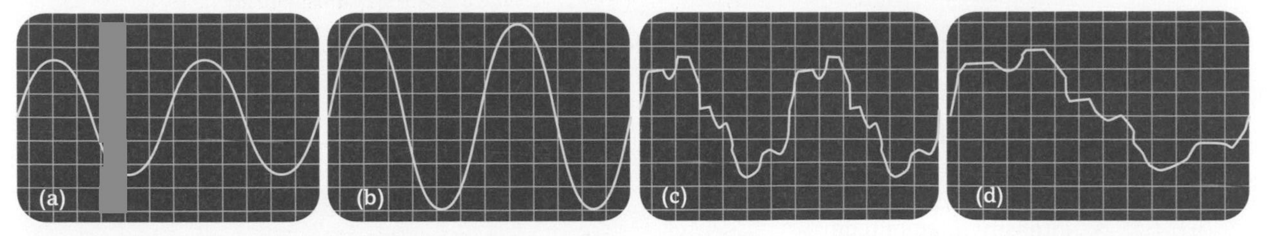

Figure 9 Waveforms for different types of sound on an oscilloscope.

Questions

1 Express the following angles in radians:

(a) 45°

(b) 135°

(c) 270°

(d) 90°

(e) 180°

2 A sound wave has a period of 0.002 s. Calculate the frequency of the wave.

3 Calculate:

(a) the frequency of a wave with period 1.5×10^{-5} s

(b) the period of a wave with frequency 4 MHz.

4 For each of the waveforms in Figure 10 work out the amplitude, the wavelength, the time period and the frequency where possible.

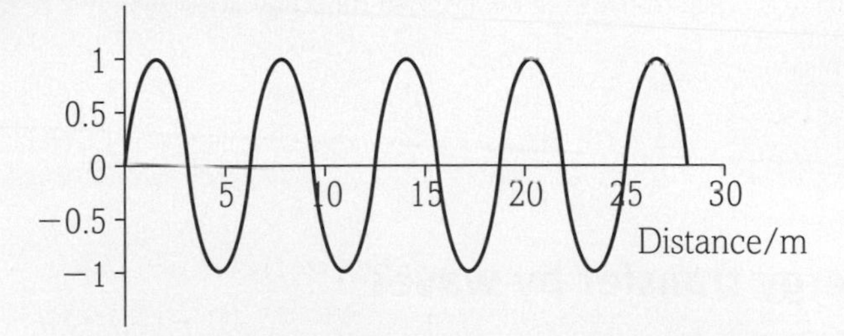

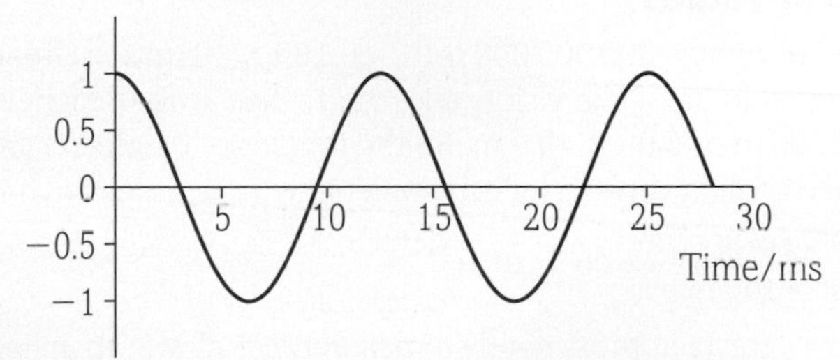

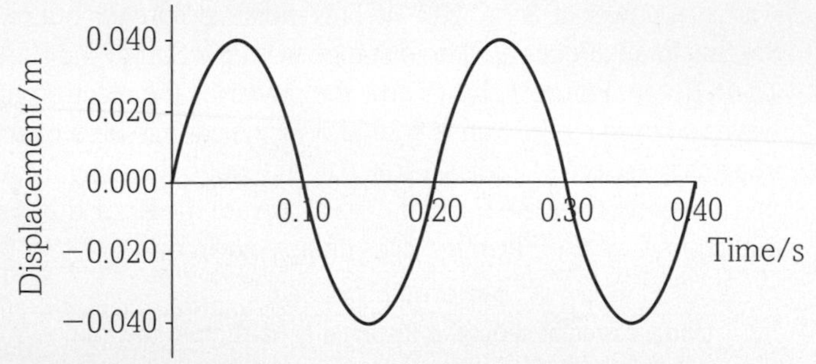

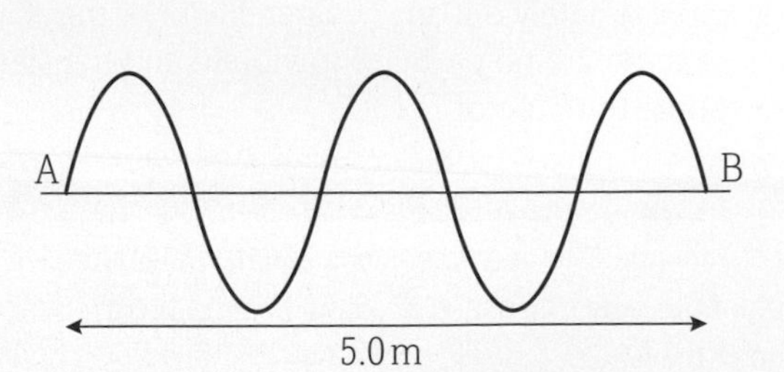

Figure 10

5 Sketch what you would observe on an oscilloscope screen for a sound wave of frequency 1600 Hz if the time base setting is 0.1 ms cm^{-1}. The screen is 10 cm wide.

6 Find the percentage uncertainty in the value for frequency, if a waveform displayed on the oscilloscope screen in question 5 has a wavelength of 7.1 cm.

4.4 ③ Wave speed and the wave equation

By the end of this topic, you should be able to demonstrate and apply your knowledge and understanding of:

* the wave equation $v = f\lambda$
* intensity of a progressive wave; $I = \frac{P}{A}$
* intensity α (amplitude)^2

The wave equation

The equation speed $= \frac{\text{distance}}{\text{time}}$ can also be applied to wave movement. In a time equal to one period, T, a wave travels one wavelength λ, therefore the speed v for any wave is:

$$v = \frac{\lambda}{T}$$

We know that frequency $f = \frac{1}{T}$, so substituting f for $\frac{1}{T}$ we get:

$$v = f \times \lambda$$

or velocity = frequency × wavelength. The S.I. units for this equation are velocity in m s^{-1}, frequency in Hz and wavelength in m.

Numerical values

The speed of light is 300 000 000 m s^{-1} or $3.00 \times 10^8\ \text{m s}^{-1}$. Green light – in the middle of the visible spectrum – has a wavelength of 0.000 000 500 m or 5.00×10^{-7} m. So the frequency of green light is the speed of light divided by its wavelength:

$$\frac{3.00 \times 10^8\ \text{m s}^{-1}}{5.00 \times 10^{-7}\ \text{m}} = 6.00 \times 10^{14}\ \text{Hz}$$

Sound waves travel approximately a million times slower than light waves. Their speed depends on what they are travelling through and is also temperature-dependent. In air at room temperature the speed of sound is approximately 340 m s^{-1}. Sound cannot travel in a vacuum because there are no particles to vibrate and transfer their energy from source to detector.

WORKED EXAMPLE 1

The frequency for the middle C note on a piano is 256 Hz. Calculate:

(a) The wavelength of the note (the speed of sound in air is 340 m s^{-1})

(b) The time period of the wave.

Answers

(a) $\lambda = \frac{v}{f}$

$= \frac{340\ \text{m s}^{-1}}{256\ \text{Hz}} = 1.33\ \text{m}$

(b) $T = \frac{1}{f}$

$= \frac{1}{256}\ \text{Hz} = 3.9 \times 10^{-3}\ \text{s}$

WORKED EXAMPLE 2

An X-ray machine is used to take photographs of broken bones. The X-rays travel at the speed of light ($3 \times 10^8\ \text{m s}^{-1}$) and have a wavelength of 1.2×10^{-10} m. Calculate:

(a) the number of X-ray pulses produced in 3 s

(b) the time between successive X-ray crests passing through a person's body.

Answers

(a) The number of X-ray pulses produced in 1 second is equal to the frequency. This is calculated by $f = \frac{v}{\lambda}$:

$\frac{3 \times 10^8}{1.2 \times 10^{-10}} = 2.5 \times 10^{18}\ \text{Hz}$

So in 3 seconds, there will be 7.5×10^{18} pulses produced.

(b) The time between one X-ray crest and the next passing through a body is equal to the time period, T, of the X-ray wave, which is equal to $\frac{1}{f}$:

$T = \frac{1}{f} = \frac{1}{2.5 \times 10^{18}\ \text{Hz}} = 4 \times 10^{-19}\ \text{s}$

Energy transfer by waves

A progressive wave transfers energy from one place to another. The Sun radiates electromagnetic energy at a rate of $3.7 \times 10^{26}\ \text{J s}^{-1}$ – it has a power of 3.7×10^{26} W. This radiation spreads out from the Sun in all directions. The distance from the Sun to the Earth is 1.5×10^{11} m. Figure 1 shows how the power of the electromagnetic waves emitted by the Sun is spread over the surface area of an imaginary sphere of radius 1.5×10^{11} m – but only a tiny fraction of this power reaches the Earth. So, each square metre of the upper atmosphere of the Earth receives energy given by:

$$\frac{\text{power output of Sun}}{\text{surface area of sphere at distance of Earth from the Sun}}$$

$$= \frac{3.7 \times 10^{26}}{4\pi \times (1.5 \times 10^{11})^2} = 1.3\ \text{kW m}^{-2}$$

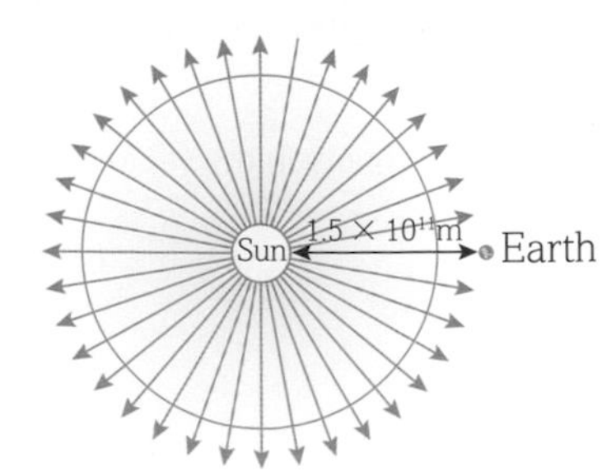

Figure 1 Electromagnetic waves spreading out from the Sun.

The power transmitted per unit area at perpendicular to the direction of wave propagation is called the wave **intensity**:

$$\text{intensity} = \frac{\text{energy/time}}{\text{area}}$$
$$= \frac{\text{power}}{\text{area}}$$

If the power spreads out equally in all directions:

$$\text{intensity} = \frac{\text{power}}{4\pi \times r^2}$$

KEY DEFINITION

The **intensity** of a progressive wave is defined as the rate at which energy is transferred from one location to another as the wave travels through space, perpendicular to the direction of wave travel. Intensity I is given by $I = P/A$, where I is the intensity in W m^{-2}, P is the power output of the source in W or J s^{-1} and A is the area over which the radiation falls in m^2.

WORKED EXAMPLE 3

A car stereo has two front speakers, each rated at 60 W.
Find the intensity of the sound waves produced by one 60 W speaker at a distance of 1.0 m from the speaker, at its maximum power.

Find the intensity of the sound waves produced by this speaker at a distance of 1.5 m, at its maximum power.

Answer

$$\text{intensity} = \frac{\text{power}}{\text{area}}$$

If the speaker emits sound waves uniformly in all directions,

$$\text{intensity} = \frac{\text{power}}{4\pi \times r^2}$$
$$= \frac{60}{(4 \times \pi \times 1^2)} = 4.8\ \text{W m}^{-2}$$

Since $I \propto \frac{1}{r^2}$, intensity × distance² is a constant.

$$\text{intensity} \times 1.5^2 = 4.8 \times 1^2$$
$$\text{intensity} = \frac{4.8}{1.5^2}$$
$$= 2.1\ \text{W m}^{-2}$$

Intensity and amplitude

The amplitude of a wave decreases as the wave spreads out from a source. The energy of a wave is proportional to the square of its amplitude. Hence the intensity of a progressive wave is also proportional to the square of its amplitude:

$$I \propto A^2$$

For example, if the amplitude of a wave decreases by a factor of 2 its intensity will decrease by a factor of 4.

Questions

1 All electromagnetic waves travel at $3.0 \times 10^8\ \text{m s}^{-1}$ in space and more slowly in transparent substances. For example, before entering glass, visible light has a wavelength of 600 nm. On entering the glass, its speed reduces to $2 \times 10^8\ \text{m s}^{-1}$. Calculate:

(a) the frequency of the light waves in air

(b) the frequency of the light waves in glass

(c) the wavelength of the light waves in glass.

2 (a) The receiving aerial for a UHF television is about 25 cm long. This is one half of the wavelength of the transmission signal. Calculate the frequency of the transmission.

(b) Sound waves travel in air at $340\ \text{m s}^{-1}$ on a warm day. The range of human hearing is 20 Hz to 20k Hz for a young person. Calculate the corresponding range of wavelengths.

(c) The speed of sound varies with the formula $331 + 0.6C$, where C is the temperature in degree Celsius. What will the wavelength of a 3000 Hz sound wave be when the air temperature is −12 °C?

3 Two fishing floats, a distance of 4.5 m apart, bob up and down 20 times in a minute. The floats always move in antiphase. There is always at least one wave crest between the floats, but never more than two.

(a) Show that the wavelength of the ripples on the river is 3.0 m. Hence, find the speed of the ripples on the surface.

(b) Near the bank, the depth of the river halves. The speed, v, of water waves in shallow water of depth d is given by $v = \sqrt{gd}$, where g is $9.8\ \text{m s}^{-2}$.

(i) What is the new frequency and wavelength of the waves near the bank?

(ii) What is the depth of the river near the bank?

4 What is the intensity at a given point on a circle of radius 3.5 m if radiation of energy 2.4×10^2 J is shone directly downward onto the circle each second?

5 How would the intensity change if:

(a) the area over which light was shone was reduced to 25% of its original area?

(b) the amplitude of the incident radiation was doubled?

6 Calculate the intensity of laser radiation from a laser pointer which has a power of 0.8 mW and produces a beam of diameter 1 mm. Give your answer in standard form with the correct S.I. units.

4.4  Common properties of waves

By the end of this topic, you should be able to demonstrate and apply your knowledge and understanding of:

* reflection, refraction and diffraction of all waves
* techniques and procedures used to demonstrate wave effects using a ripple tank

All waves can be reflected, refracted and diffracted. As we shall see in Topic 4.4.6, all transverse waves show polarisation. Longitudinal waves cannot be polarised.

Reflection

We usually associate **reflection** with light, but all wave types can be reflected. Figure 1 illustrates the reflection of a television signal using a dish aerial.

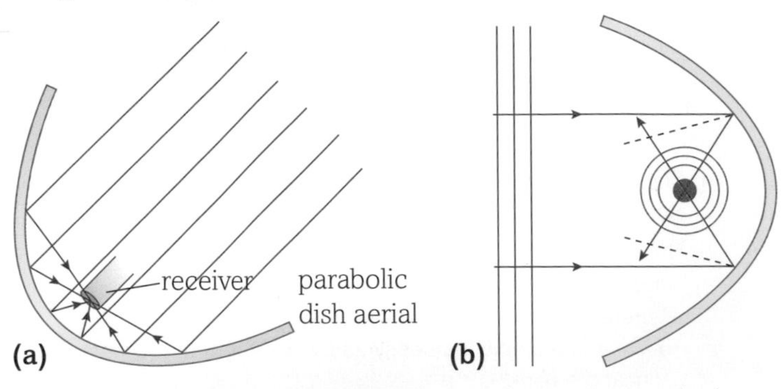

Figure 1 The reflection of a television signal: (a) showing the rays; (b) also showing the wavefronts.

Figure 2 shows an echo (which involves the reflection of sound waves). Note how the sound waves spread in all directions from the source, and how those that hit the wall continue to spread as they travel back towards the source. An echo is heard when returning waves reach the original source of the sound.

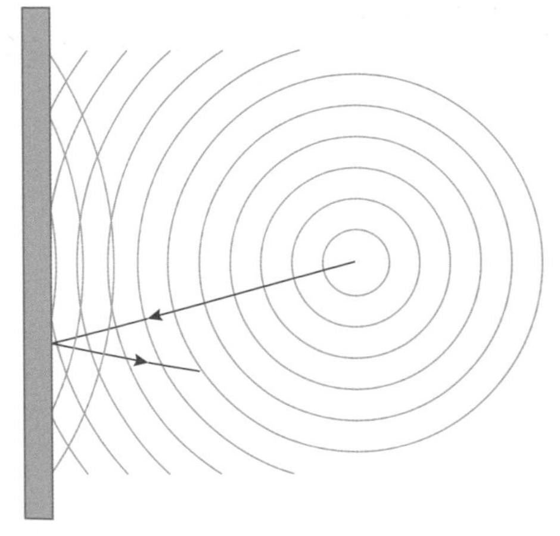

Figure 2 The reflection of sound waves – an echo.

A single line – known as a **ray** – has been drawn in Figure 2 to illustrate the direction in which the waves travel. Rays are often used to simplify diagrams in which the waves themselves are not shown. Rays are always drawn at right angles to the wavefronts. In illustrations such as this, rays are used to indicate the direction of the wave. **Wavefronts** are lines of constant phase, such as crests. They are drawn for each successive wave, with the distance between wavefronts representing wavelength. Notice how the wavelength does not change after the wave has been reflected.

Refraction

Refraction occurs when a wave moves from one material into another. In terms of the refraction of light, we say that the two materials have different optical densities. For example, water and glass are both transparent and will allow light to pass through them, but glass has the higher optical density.

Two things can be observed when refraction occurs:

- the wave will change its speed
- there may be a change in direction.

Refraction is probably best illustrated by light, as shown in Figure 3. Sound waves can also be refracted, as shown by the refraction of seismic waves in the Earth's crust.

Figure 3 The pencil appears bent due to the refraction of light as it moves from air into water.

Refraction occurs in the lenses that are prescribed by opticians. The optician knows exactly how much to refract light so that it is focussed on the retina (Figure 4) and causes a person who is short-sighted to be able to see clearly.

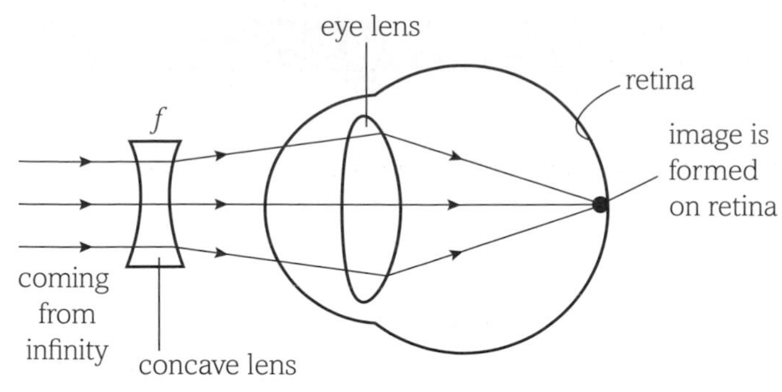

Figure 4 A diverging lens bending light so that it is focused on the retina of the eye.

Refraction and reflection, together, are responsible for the production of rainbows. When white light enters a raindrop, the different wavelengths (colours) of light are refracted by different amounts – this is called dispersion. The reflected rays reach the back of the raindrop, but instead of being refracted out, they are **totally internally reflected** (see Topic 4.4.8) before emerging again through the front of the raindrop. This only happens when the conditions for reflection, refraction and total internal reflection are correct.

INVESTIGATION

Investigating refraction and reflection in a ripple tank

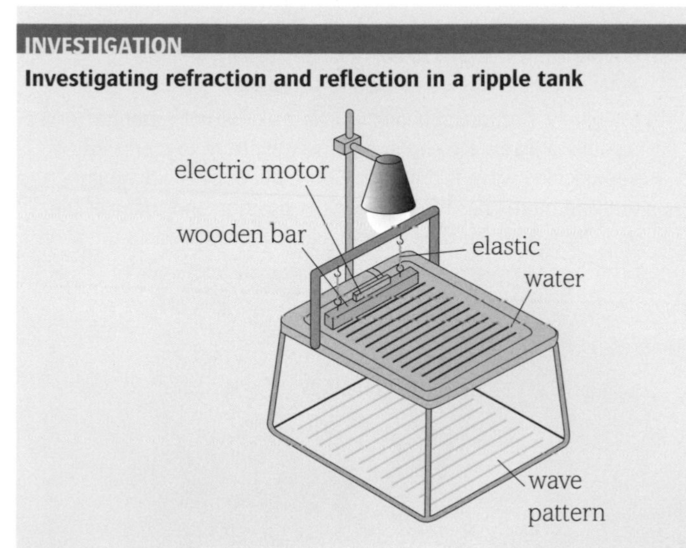

Figure 5 Ripple tank apparatus.

In a ripple tank (Figure 5), a motorised straight-edged bar produces plane (straight) waves while a small dipper produces circular waves.

When light is shone from above through the waves produced, the bright bands or curves of light seen on the screen below the tank show the wave crests. This makes it possible to measure the wavelength of the water waves, and investigate the angles of reflection and refraction.

Reflection at plane and curved surfaces can be investigated, and the angles of incidence and reflection measured with respect to the normal (the line drawn at right angles to the barrier surface at the point of incidence).

A glass sheet is used to decrease the water depth and so produce a region with a different wave speed. The water level can also be adjusted. If the separation of the wavefronts decreases this shows they are travelling more slowly. If the wavefronts are at a non-zero angle when they cross the barrier, the waves also changes direction.

The ripple tank can also be used to study interference and diffraction (Figures 6 and 7).

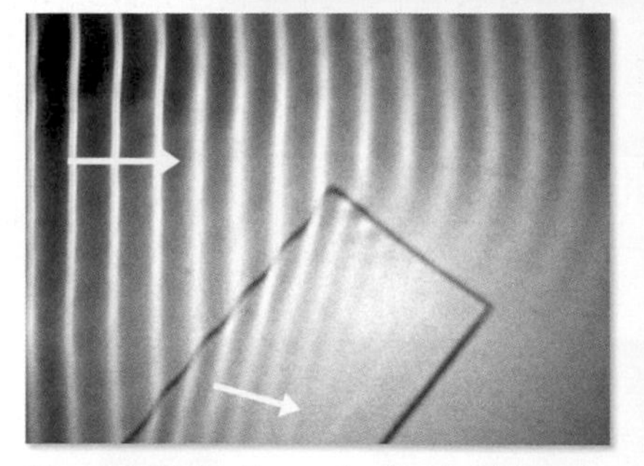

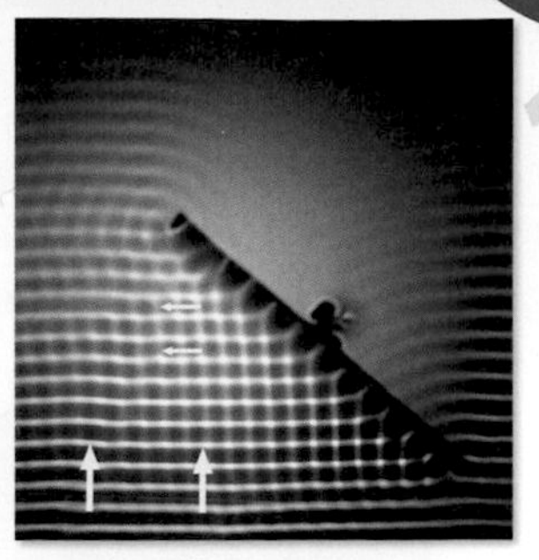

Figure 6 Refraction and reflection being demonstrated with a ripple tank.

Diffraction

Changes in the directions of waves can also occur when they meet an obstacle, or as they pass through an aperture. This process is called **diffraction** and it can be shown very clearly using water waves in a ripple tank, as shown in Figure 7.

Figure 7 There is not much diffraction of water waves at a wide gap.

KEY DEFINITION

Diffraction is the spreading out of a wave after passing around an obstacle or through a gap.

Diffraction is most pronounced when the wavelength of the wave being diffracted is the same size as the gap that they are travelling through. In Figure 7, you can see that there is a slight spreading out of the water waves as they pass through the gap. However, the spreading is very pronounced in Figure 8, where circular wavefronts are passing through the gap and the gap width is exactly the same size as the wavelength of the water waves.

Figure 8 Extensive diffraction happens with circular wavefronts at a narrow gap.

Interference

If there are two or more gaps, the waves spreading out from the gaps will overlap and interfere with each other – a process called **interference** (Figure 9). Diffraction and interference can occur for a large gap and a large wavelength or for a very small gap and a small wavelength.

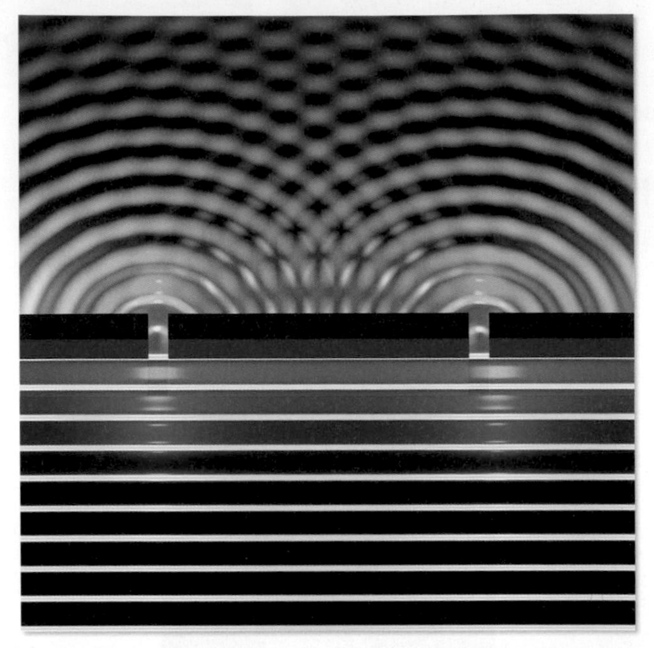

Figure 9 When diffracted waves overlap, interference occurs. The photo shows interference in a ripple tank.

KEY DEFINITION

Interference is the addition of two or more waves (superposition) that results in a new wave pattern.

Interference occurs when two or more waves come together, although we usually consider interference of just two sources, of the same frequency. For example, interference of radio wave signals can occur when waves from the transmitter interfere with waves reflected from a mountain or tall building.

In your course you may carry out experiments that demonstrate two-source interference using sound, light and microwaves (see Topics 4.4.9 and 4.4.10). As you will see, interference can cause variation with distance of the amplitude of a wave. In the radio interference example, when driving along a road between the two stationary sources there will be points where the sound is loud and points where the sound is quiet.

Questions

1. Give examples of the:
   (a) reflection of light
   (b) reflection of sound
   (c) refraction of light
   (d) diffraction and interference of long wavelength waves
   (e) diffraction and interference of short wavelength waves.

2. Diffraction is a property of all waves, but is only a significant effect when the wavelength of the diffracted wave is about the same as the aperture. Explain why the diffraction of sound is easily observed in everyday life but the diffraction of light is not.

3. Describe how you could carry out an experiment to demonstrate that diffraction of sound takes place.

4. Draw diagrams to illustrate how plane water waves are diffracted when they pass through a gap about (a) 2 wavelengths wide and (b) 10 wavelengths wide.

5. The direction of travel of some sea waves 100 m from the coast is almost parallel to the shore. However, as the waves move into shallower water they slow and turn towards the shore.
   (a) Explain why this happens and why the effect is most noticeable on flat sandy beaches (steep beaches usually require breakwaters to reduce erosion of the shore).
   (b) Imagine a theoretical beach where the depth of water decreases steadily towards the shore. A plane wavefront to represent an approaching wave is drawn in Figure 10. Copy the diagram and add more lines to represent the position and shape of the wavefront at equal time intervals as it moves towards the shore.

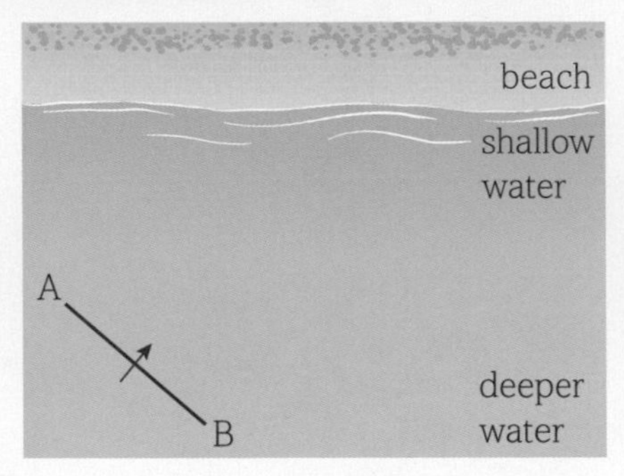

Figure 10

6. Road traffic noise barriers are designed to reduce noise levels experienced by houses alongside major roads.
   (a) Use the ideas of reflection and diffraction to explain how a solid barrier reduces the intensity of road noise behind the barrier.
   (b) Explain why the barriers reduce high-frequency sounds more effectively than low-frequency sounds.

7. Co-channel interference occurs for radio or TV signals when two or more signals of the same frequency are received from different transmitters. Usually, the distance between transmitters means that the signal from the distant transmitter out of range of the closer transmitter. However in certain weather conditions the upper parts of the atmosphere temporarily reflect radio waves back. Explain how this reflection can cause a temporary loss of signal quality.

Electromagnetic waves

By the end of this topic, you should be able to demonstrate and apply your knowledge and understanding of:

* electromagnetic spectrum; properties of electromagnetic waves
* orders of magnitude of wavelength of the principal radiations from radio waves to gamma rays

Common properties of electromagnetic waves

Like all waves, electromagnetic waves transfer energy from one place to another without any net movement of matter. The **electromagnetic spectrum** has a range of values for wavelength – values range from 10^{-16} m for gamma rays at one extreme to 10^{4} m for radio waves at the other.

Visible light represents a tiny part of the electromagnetic spectrum. Wavelengths for visible light range from approximately 3.70×10^{-7} m (370 nm) for violet to 7.40×10^{-7} m (740 nm) for red. All electromagnetic waves share these properties:

- They can all travel through a vacuum.
- All possess both a magnetic wave and an electrical wave interlocked and at right angles to each other.
- In free space, they all travel at the speed of light, c, or $2.98 \times 10^{8}\,\mathrm{m\,s^{-1}}$.
- They are all transverse waves.
- They can all be reflected, refracted and diffracted.
- They can all demonstrate interference.
- They can all be polarised.

increasing frequency/increasing danger (radio → gamma rays); increasing wavelength (gamma rays → radio)

	Wavelength/m	Frequency/Hz	Method of production	Method of detection	Uses
radio	$10^{-1} - 10^{4}$	$3\times10^{9} - 3\times10^{4}$	Electrons oscillated by electric fields in aerials	Resonance in electronic circuits	Television, radio, telecommunications
microwave	$10^{-4} - 10^{1}$	$3\times10^{12} - 3\times10^{9}$	Magnetron, klystron oscillators, using electrons to set up oscillations in a cavity	Heating effect, electronic circuits	Radar, mobile phones, microwave ovens, satellite navigation
infra-red	$7.4\times10^{-7} - 10^{-3}$	$4\times10^{14} - 3\times10^{11}$	Oscillation of molecules, from all objects at any temperature above absolute zero	Photographic film, thermopile, heating of skin	Heaters, night vision equipment, remote controls
visible light	$3.7\times10^{-7} - 7.4\times10^{-7}$	$8\times10^{14} - 4\times10^{14}$	From high-temperature solids and gases, lasers	Photographic film, retina of eye	Sight, communication
ultra violet	$10^{-9} - 3.7\times10^{-7}$	$3\times10^{17} - 8\times10^{14}$	From high-temperature solids and gases, lasers	Photographic film, phosphors, sunburn	Disco lights, tanning studios, counterfeit detection, by detergents
X-rays	$10^{-12} - 10^{-7}$	$3\times10^{20} - 3\times10^{15}$	Bombarding metals with high-energy electrons	Photographic film, fluorescence	Computer-aided tomography (CT) scans, X-ray photography, crystal structure analysis
gamma rays	$10^{-16} - 10^{-9}$	$3\times10^{24} - 3\times10^{17}$	Nuclear decay or in a nuclear accelerator	Photographic film, Geiger tube	Diagnosis and cancer treatment (radiotherapy)

Table 1 The electromagnetic spectrum. The wavelengths range over many orders of magnitude.

Categories of electromagnetic waves

Electromagnetic waves are grouped into seven major and distinct types, even though there is a continuous change in their energies – it is not possible to say when a radio wave becomes a microwave, for example.

DID YOU KNOW?

X-rays and gamma rays are often identical in terms of their energies, wavelengths and frequencies. The only difference between them is how they are produced. X-rays are made by accelerated electrons outside the nucleus, but gamma rays are always emitted from the nucleus of an atom.

Uses of electromagnetic radiation

Many of the uses of electromagnetic radiation mentioned in Table 1 are obvious: for example, the use of radio waves in radio and television; the use of X-ray photographs to examine broken bones; the use of visible light for sight. Electromagnetic radiation is so much part of everyday life that it is difficult to imagine life without it. Yet until the middle of the nineteenth century, with the exception of light and heat radiation (infrared), little was known about the electromagnetic spectrum. It was James Clerk Maxwell (Figure 1) who formulated a set of equations, called the Maxwell equations, that relate electric and magnetic fields and theoretically show that electromagnetic waves are possible.

Heinrich Hertz (Figure 2) confirmed Maxwell's equations experimentally by becoming the first person to produce radio waves. This early work paved the way for the discovery of the remainder of the electromagnetic spectrum – X-rays in 1895, gamma rays in 1896, microwaves (leading to radar) with the magnetron in the 1930s.

Figure 1 James Clerk Maxwell.

Figure 2 Heinrich Hertz.

Non-ionising electromagnetic radiation

Radio waves, microwaves, infrared and visible light do not have enough photon energy to remove electrons from the shells of atoms. They cannot produce ions and so are called non-ionising radiations. Ionisation leads to human cell mutations that can cause diseases such as cancer. For this reason, these types of electromagnetic radiations are deemed to be relatively safe.

Ionising electromagnetic radiation

Ultraviolet rays, X-rays and gamma radiation all have high photon energies. This means that they can cause ionisation by knocking electrons from the shells of atoms. This can lead to human cell mutations and so they are dangerous if exposure to them is sustained for long periods.

Ultraviolet radiation emitted by the Sun is often divided into three regions:

- UV-A: wavelength 3.15×10^{-7} to 4.00×10^{-7} m (315–400 nm); causes tanning when skin is exposed to sunlight (accounts for 99% of UV light).
- UV-B: wavelength 2.8×10^{-7} to 3.15×10^{-7} m (280–315 nm); causes damage such as sunburn and skin cancer.
- UV-C: wavelength 1.00×10^{-7} to 2.8×10^{-7} m (100–280 nm); filtered out by the atmosphere and does not reach the surface of the Earth.

Sunscreen creams contain chemicals designed to filter out UV-B, preventing sunburn and skin damage. Glass is an efficient absorber of ultraviolet – which is why you do not get sunburnt indoors, even if you sit for long periods in the sunlight.

DID YOU KNOW?

Substances known as phosphors glow when subjected to ultraviolet radiation, making UV radiation visible to the naked eye. Manufacturers of washing powders incorporate phosphors in their products, which is why white clothing glows a bright blue-white under the UV lights used in nightclubs.

X-ray photography has been used for over a century now, but many technical advances in the field are relatively recent. Once X-rays were essentially shadows, and therefore useful only for confirming bone fractures. The use of computers has enabled vast improvements in image contrasting. By linking X-ray machines to computers, we are now able to construct three-dimensional images from a series of cross-sectional planes. Known as computer-aided tomography (CT) scans, these allow much more accurate and extensive diagnoses. For example, X-ray photographs of the alimentary canal can now be taken and every twist and turn of the large intestine can be seen very clearly. X-rays are produced by firing high-energy electrons at a copper anode, as shown in Figure 3.

Gamma radiation is released when the nuclei of unstable atoms give out high-energy photons. Gamma radiation has many uses, such as the preservation of fruit and vegetables, the sterilisation of surgical equipment and the treatment of cancer cells. However, outside the body it is very harmful and can cause cancer, as

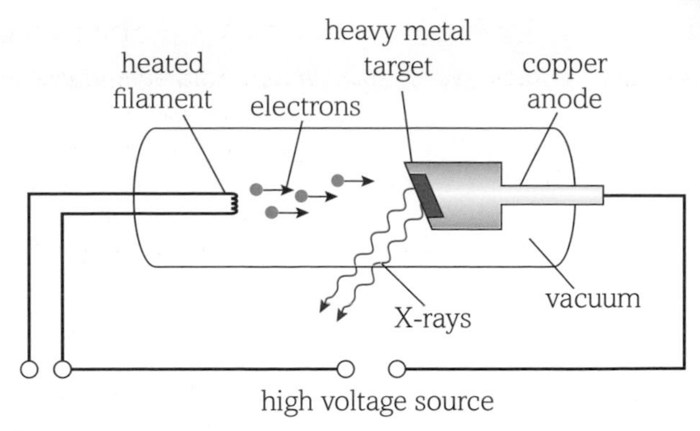

Figure 3 The production of X-rays.

happened to many people who lived in the town of Pripyat after the Chernobyl disaster of 1986. The radiation levels are considerably lower today than they were in April 1986, but many people were affected through contamination rather than irradiation.

Figure 4 Pripyat – the ghost town after the Chernobyl disaster. Today, levels of gamma radiation are thought to be safe.

Questions

1. Copy the Venn diagram and complete it by writing in the names of the members of the electromagnetic spectrum and indicating whether they are used in the home, the hospital or both.

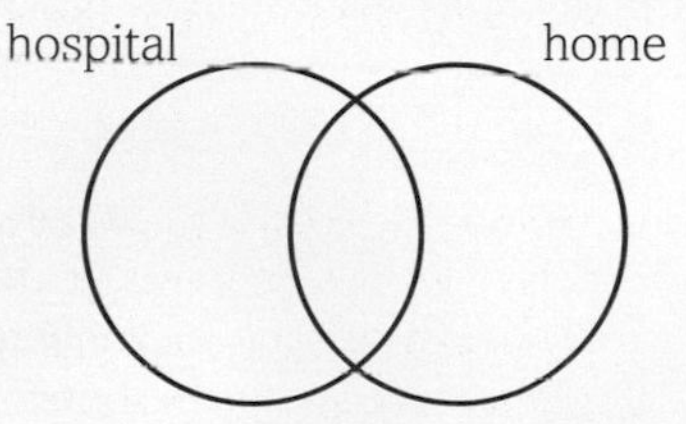

2. Can any type of electromagnetic radiation be stated to be completely safe? Explain your answer.

3. Name the type(s) of electromagnetic radiation which:
   (a) carries information
   (b) cooks food
   (c) could cause harm or injury to humans
   (d) can be reflected, refracted or diffracted
   (e) causes fluorescence
   (f) has a wavelength of 35 km
   (g) has frequency of 30 GHz
   (h) is man-made.

4. State two features, other than the speed of light, common to all types of electromagnetic waves.

5. Match the following approximate wavelengths to the name of a region of the electromagnetic spectrum:
   10^3, 10^{-2}, 10^{-5}, 5×10^{-6}, 10^{-8}, 10^{-10}, 10^{-12}

6. The receiving aerial for a UHF television is about 28 cm long. This is one half of the wavelength of the transmission. Calculate the frequency of the transmission.

4.4

6 Polarisation

By the end of this topic, you should be able to demonstrate and apply your knowledge and understanding of:

* polarisation of all waves
* plane polarised waves; polarisation of electromagnetic waves
* techniques and procedures used to observe polarising effects using microwaves and light

Polarisation is good evidence for the wave nature of light. All transverse waves (including all electromagnetic waves) can be polarised.

Plane-polarised waves

All electromagnetic waves are made of a magnetic field and an electric field oscillating at right angles to one another and the direction of travel, as shown in Figure 1.

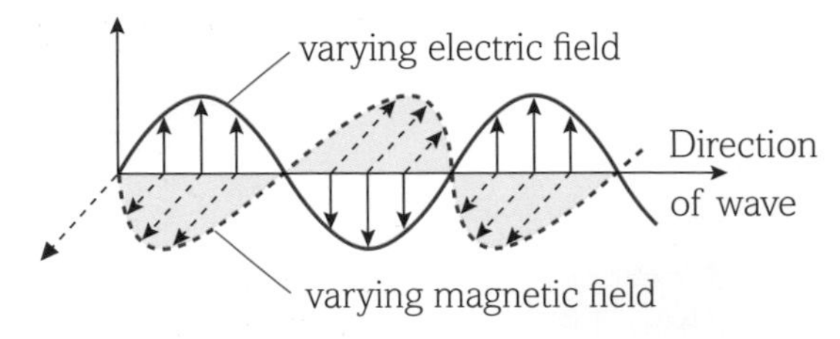

Figure 1 An electromagnetic wave.

The displacements of the oscillating field and the direction of travel define a plane, such as the blue or yellow planes in Figure 1. For electromagnetic waves we define the plane of the wave as the plane of oscillation of the *electric* field. Light from the Sun or a light bulb is **unpolarised** – the electric field can be in any number of planes. If you are facing a light wave coming towards you, there is a mixture of waves with many different planes, with all the waves oscillating perpendicular to the direction of energy travel (Figure 2).

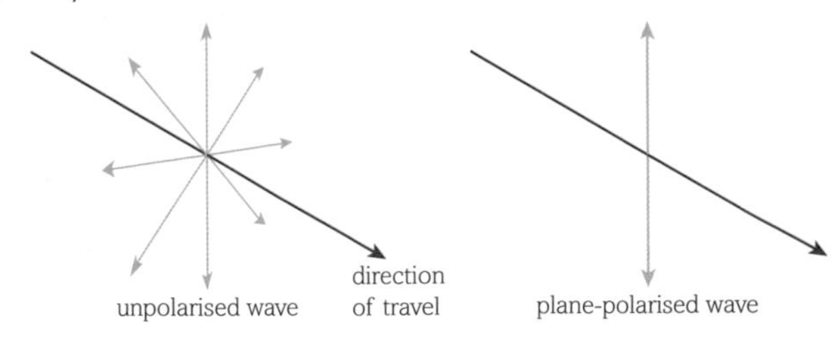

Figure 2 Polarised wave and unpolarised wave.

Some crystalline materials can cause the oscillating fields to happen in one plane only. These are known as **polarising filters** – for example 'Polaroid' filters. A wave that has fields only in one plane is called a **plane-polarised wave**.

Note that only transverse waves can be polarised (Figure 3) because longitudinal waves do not have oscillations at right angles to their direction of travel. The oscillations of these waves are in line with the wave's direction of travel.

INVESTIGATION

Observing polarisation of microwaves

The microwave transmitter in Figure 3 emits naturally polarised waves with a wavelength of 3 cm. You can easily show that the microwaves are polarised when they are emitted by placing a microwave receiver in front of the transmitter, and then rotating either around the line between them. The signal reception (detected either using an ammeter or an audio amplifier and loudspeaker so that the microwave signal can be 'heard') rises and falls in intensity as rotation occurs, and drops to zero when they are 'crossed' (at right angles).

When the receiver is at the maximum signal position, a metal grille is then placed between the transmitter and receiver. Figure 3 shows that as the grille is rotated (for example in 10° increments) the signal reception varies and is zero when the metal rods are aligned with the electric field vector of the emitted microwaves.

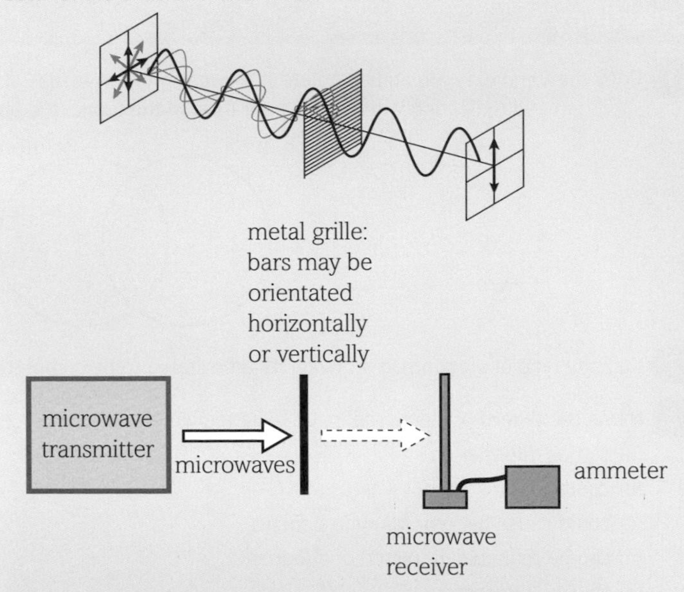

Figure 3 Demonstrating that microwaves can be plane-polarised.

The microwave transmitter is producing vertically plane-polarised radiation.

- If the bars of the metal grille are horizontally orientated, very few of the microwaves will be absorbed and the ammeter will show a high output.
- If the bars of the metal grille are vertically orientated, all of the microwaves will be absorbed – because they are vertically plane-polarised and the ammeter will show no output.

Rotating the plane of polarisation

An ordinary light wave from a domestic bulb is shown passing through sheets of Polaroid in Figure 4. The first Polaroid sheet is labelled a polariser. This produces plane-polarised light – polarised in the vertical direction in this example. The light wave then carries on to the second sheet of Polaroid, called the analyser, which is rotated through an angle θ relative to the polariser sheet. The analyser sheet polarises the light in a direction parallel to its long edge – the plane of polarisation will have been rotated through an angle θ.

All this is shown as a front view in Figure 5. If the amplitude of the light wave approaching the analyser sheet is A, then after it has had its plane of polarisation rotated by angle θ the amplitude will be $A \cos \theta$. Because the intensity of a wave is proportional to its amplitude squared, the intensity after the analyser is proportional to $\cos^2 \theta$. The intensity of the wave is reduced as it passes through the second filter.

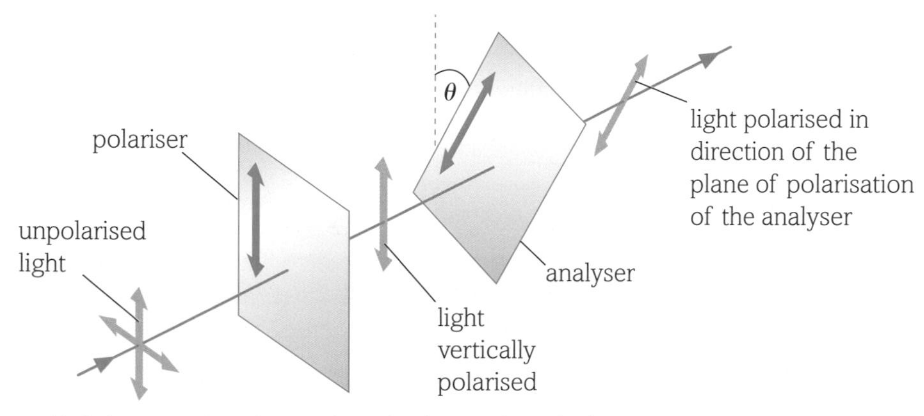

Figure 4 Visible light passes through an analyser that has its plane of polarisation at an angle θ to that of the polariser.

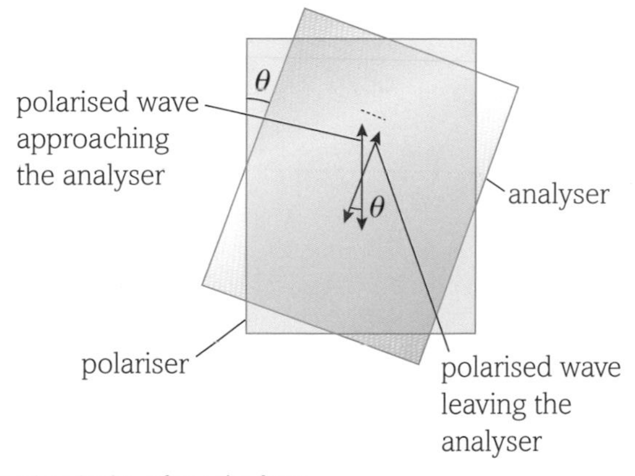

Figure 5 The view of Figure 4 when looking from the front.

This is known as Malus' law (after Etienne-Louis Malus). It states that when a perfect polariser is put in a beam of polarised light, the intensity, I, of the light that passes through it is given by $I = I_{max} \cos^2 \theta$, where I is the intensity transmitted at angle θ. Also, I_{max} is the maximum intensity transmitted (at $\theta = 0°$). This law also shows that if the analyser is at right angles to the polariser, then $\theta = 90°$ and no light will pass through. This situation is known as 'crossed Polaroids'.

Uses of polarisation

Reflected light and scattered light

Although sunlight and light from most sources are not polarised, some naturally occurring light can be *partially* polarised. This means that there is more light with the direction of oscillation in one direction than there is in any other direction. This mainly occurs in light that has been reflected. The light reflected from the surface of a lake, for example, is partially polarised, and so is the blue light from the sky. Blue light is more easily scattered by the dust particles and water vapour in the atmosphere than red light. It is the scattered blue light that is partially polarised. (It is because of blue light being scattered, creating blue skies, that sunsets are red.)

KEY DEFINITIONS

In a **plane-polarised wave** the oscillations of the field and the direction of travel are confined to a single plane.

A **polarising filter** produces plane-polarised light by selective absorption of one component of the incident oscillations – the filter transmits only the component of light polarised perpendicular to that direction.

INVESTIGATION

Investigating Malus's law

Using the two polarising filters as shown in Figures 4 and 5, you could use a calibrated light meter to measure intensity of the transmitted light as a function of polariser angle.

INVESTIGATION

Observing polarisation of light

1. Look through a polarising filter at different light sources (such as the Sun, a light bulb, a laptop or calculator screen, a reflection in glass). Try one filter, then with two filters aligned at different angles to each other. As filter B is rotated through different angles with respect to filter A, the intensity of transmitted light rises and falls.
2. Look through a single polarising filter at light reflected at a shallow angle from water, such as a pond, or from shiny surfaces in the lab. Rotate the polarising filter until you notice that the reflected image gets dimmer.

Figure 6 Polaroid filters enhance the colour of the sky in photographs.

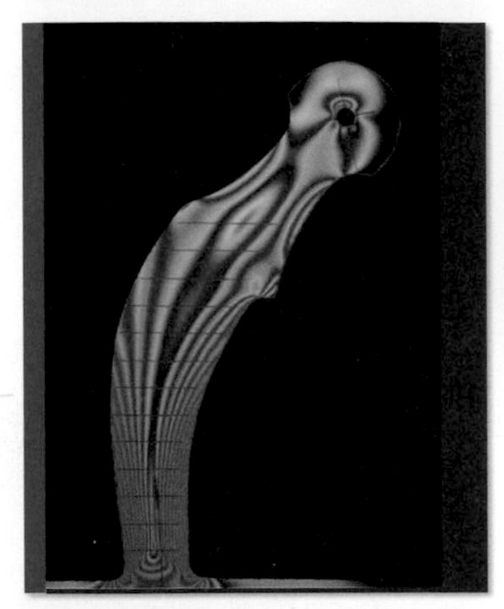

Figure 7 Using polarisation for strain analysis of a model femur.

DID YOU KNOW?

Animals such as bees, octopus and squid can detect the polarisation of light without wearing polarised sunglasses. Squid can even produce patterns on their body that reflect polarised light, giving their skin changing colour patterns which may be used for communication – even though most squid are colour-blind. The advantages of detecting differences in polarised light are not yet fully understood.

DID YOU KNOW?

Television transmission

Television transmitters send out horizontally polarised signals. In order to cover the whole country, low-power infill transmitters are used to provide signals in valleys. Because so many of these infill transmitters are needed, signals of the same frequency could interfere with those from the main transmitters. To overcome this, many of the infill transmitters are polarised vertically. The vertically polarised signals cannot interfere with the horizontally polarised ones from the main transmitters.

Anglers cannot usually see fish under the water surface because most of the light is reflected from the surface giving a 'glare'. However, Polaroid glasses cut out the horizontally polarised light reflected from the water surface, but allow the vertically polarised light reflected from objects below the surface to pass. This means that any fish in the water become visible.

Similarly, photographers often use Polaroid filters to enhance the colour of the sky. The filters remove some of the polarised light from a blue sky, so that the sky colour seems more intense, as shown in Figure 6.

Strain analysis

Another technique that makes use of polarisation is strain analysis. Certain plastics, such as those used for making rulers, protractors and even Sellotape, contain long chains of molecules, which become aligned during manufacture. These materials are able to rotate the plane of polarisation so that the transmitted light is polarised. When these plastics are placed between crossed Polaroids, coloured images are produced that change as the plastics are stretched or squashed. This is because the rotation of the plane of polarisation is different for different wavelengths.

This effect is shown in Figure 7 for a plastic model of a femur. Such models can also be used to analyse stresses in models of structures.

Detailed analysis of crystal shapes can also be carried out using this technique.

Questions

1 Suppose the intensity of a beam of unpolarised light incident on a linear polariser is I. Explain why the maximum possible intensity of the transmitted light, which is plane-polarised, is $\frac{1}{2}I$. We call this a perfect polariser. (In reality, when an unpolarised beam of light is shone onto a typical sheet of Polaroid, the transmitted beam is only about 30 to 35% of the incident intensity.)

2 Figure 8 shows a laboratory microwave transmitter T placed in front of a microwave detector D, which is connected to a microammeter.

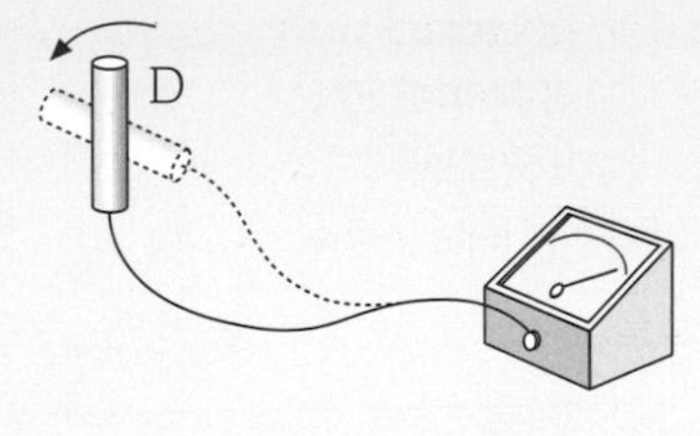

Figure 8

Initially the meter shows a maximum reading with D vertical. The detector is rotated through 360° in a vertical plane as shown. Figure 9 shows how the ammeter reading varies with angle.

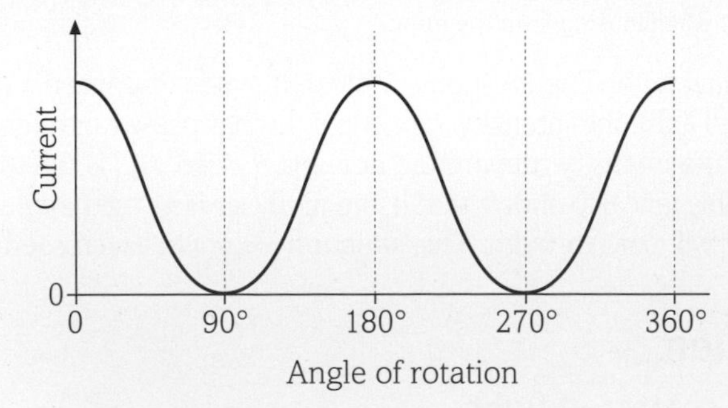

Figure 9

(a) Explain why there is a series of maxima and minima.

(b) What do the observations tell you about the nature of microwaves?

(c) What is the orientation of the transmitting aerial inside the microwave transmitter?

Refraction of light

By the end of this topic, you should be able to demonstrate and apply your knowledge and understanding of:

* refraction of light
* refractive index; $n = \frac{c}{v}$
* $n \sin \theta$ = constant at a boundary where θ is the angle to the normal
* techniques and procedures used to investigate refraction of light using ray boxes, including transparent rectangular and semi-circular blocks

Introduction

One of the main phenomena experienced by electromagnetic radiation is that it may undergo **refraction**. This phenomenon happens when a wave passes from one transparent (or translucent) medium into another.

Two things tend to happen when refraction occurs:

- the wave will change its speed
- the wave will change direction (unless it is travelling along the normal, in which case it will not).

For all types of electromagnetic radiation, the speed of the wave will decrease as it moves from a material of **lower refractive index** into a material of **higher refractive index**.

The speed of an electromagnetic wave in a material is related to the refractive index of the material by:

$$\text{refractive index, } n = \frac{\text{speed of light in a vacuum}}{\text{speed of light in the material}}$$

$$n = \frac{c}{v}$$

WORKED EXAMPLE 1

The speed of light in a vacuum is $3.0 \times 10^8\ \text{m s}^{-1}$. In a certain type of glass, the speed of light is found to be $1.8 \times 10^8\ \text{m s}^{-1}$. What is the refractive index of the glass?

Answer

$$n = \frac{c}{v}$$

$$= \frac{3.0 \times 10^8}{1.8 \times 10^8}$$

$= 1.7$ to 2 significant figures

The value 1.7 for the refractive index of the glass means that any electromagnetic radiation travelling through it travels 1.7 times slower than it would travel through a vacuum.

Refractive indices

Different materials have different refractive indices. The higher the refractive index, the slower electromagnetic radiation travels through this material compared with its speed in a vacuum. The refractive indices of a variety of materials are listed in Table 1.

Material	Refractive index
vacuum	1
air	1.00028
water	1.33300
glass	1.52300
diamond	2.41903

Table 1 Any transparent material will allow electromagnetic waves to travel through it and will have its own refractive index.

Figure 1 Diamond has an unusually high refractive index. Combined with the shape of the jewel, this maximises the amount of light that is totally internally reflected, giving diamonds their sparkle.

Snell's law

There is a link between the refractive indices of two materials and the directions at which the incident and refracted rays will travel with respect to the **normal**. The normal is a line drawn at 90° to the boundary between the two materials (Figures 2 and 3). An electromagnetic wave travelling at an angle, θ_1, to the normal in 'material 1' will travel at an angle, θ_2, to the normal in 'material 2'. This is explained by **Snell's law**, which is given by:

$$n_1 \times \sin \theta_1 = n_2 \times \sin \theta_2$$

or $n \sin \theta$ = constant

for a boundary between two materials of different reactive index. Note that the angle of incidence and angle of refraction are always measured to the normal line, and not to the boundary.

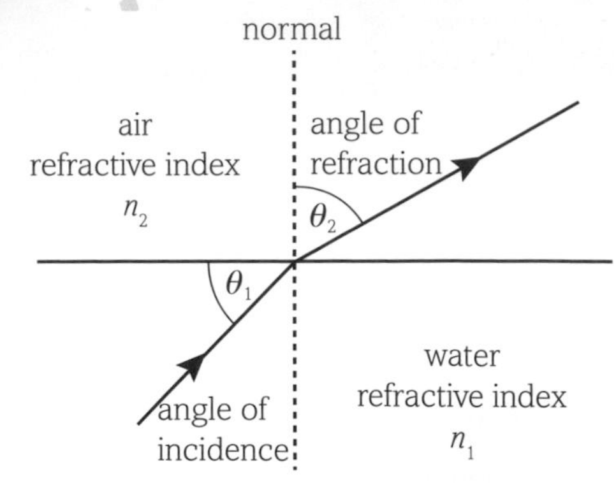

Figure 2 When a ray of light travels from water into air, its speed will increase and its direction will change causing it to refract away from the normal – so $\theta_2 > \theta_1$.

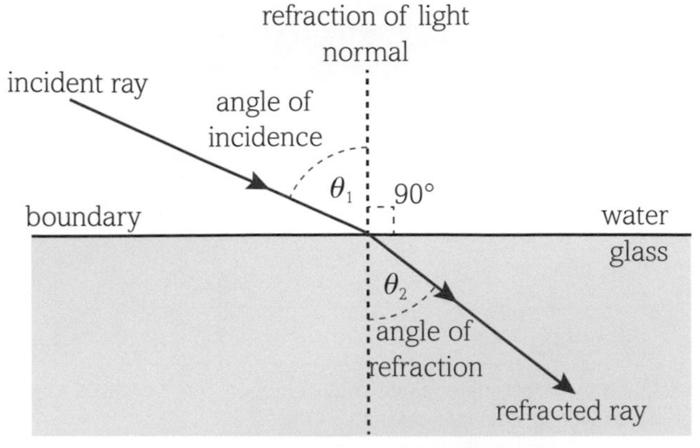

Figure 3 When light travels from water into glass, it is travelling from a material of lower refractive index into a material of higher refractive index – this means it will slow down and refract towards the normal.

WORKED EXAMPLE 2

Light travels from water of refractive index 1.33 into some glass of refractive index 1.51. The angle of incidence, θ_1, is 38°. Calculate the angle of refraction, θ_2, inside the glass.

Answer

From Snell's law:

$$n_1 \times \sin\theta_1 = n_2 \times \sin\theta_2$$

Substituting the values into the equation:

$$1.33 \times \sin 38° = 1.51 \times \sin\theta_2$$

$$\sin\theta_2 = \frac{(1.33 \times 0.616)}{1.51} = 0.543$$

$$\theta_2 = \sin^{-1} 0.543 = 32.9°$$

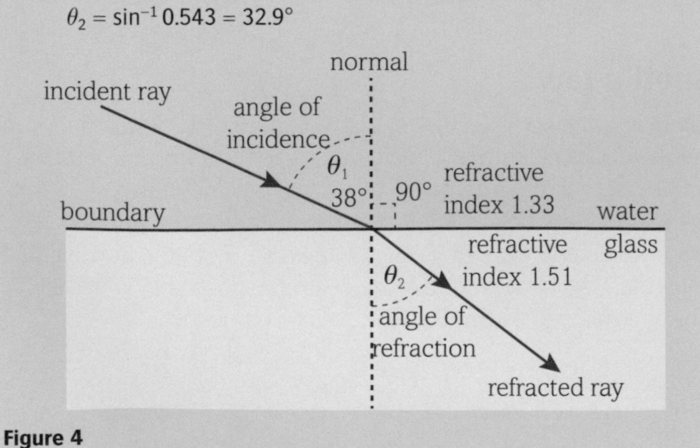

Figure 4

INVESTIGATION

Determining the refractive index of a transparent semi-circular block

The equipment is set up as shown in Figure 5. Measure the angle of the incidence, θ_1, and angle of refraction, θ_2, using a protractor. Since light rays exit the glass at 90° to the boundary of the semi-circular block, you do not have to deal with a refraction at this second interface. Use readings for θ_1 over a range of values between 0° and 80°. This will give sufficient data to plot a graph of $\sin\theta_1$ against $\sin\theta_2$. To measure values of θ_2 you will need to remove the glass block each time and join the points from where the emergent refracted ray left the block to the point at which it entered the block, and draw a normal line.

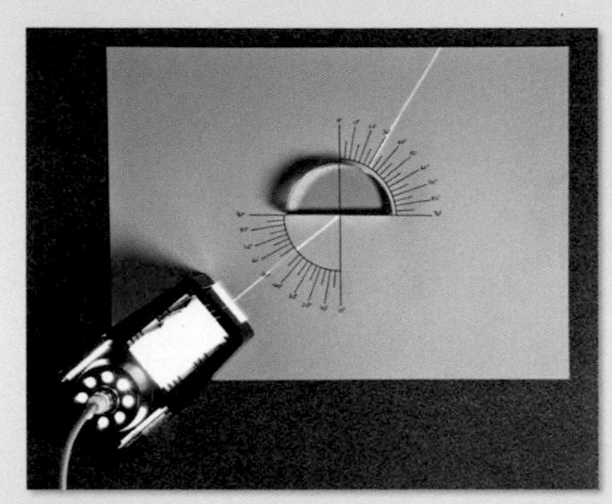

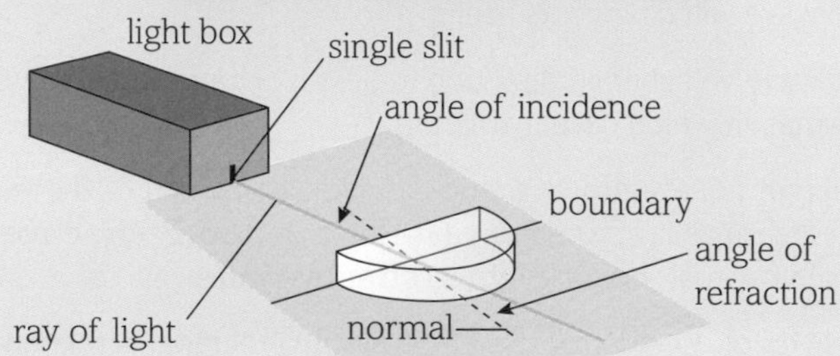

Figure 5 Measuring a refractive index.

The gradient of the graph $= \frac{\sin\theta_1}{\sin\theta_2} = \frac{n_2}{n_1}$

Since the refractive index of air is very close to 1, the equation simplifies to $n = \frac{\sin\theta_1}{\sin\theta_2}$

In Figure 6 the gradient $= \frac{1.0}{0.65}$ so the refractive index of glass is $\frac{1.0}{0.65} = 1.54$.

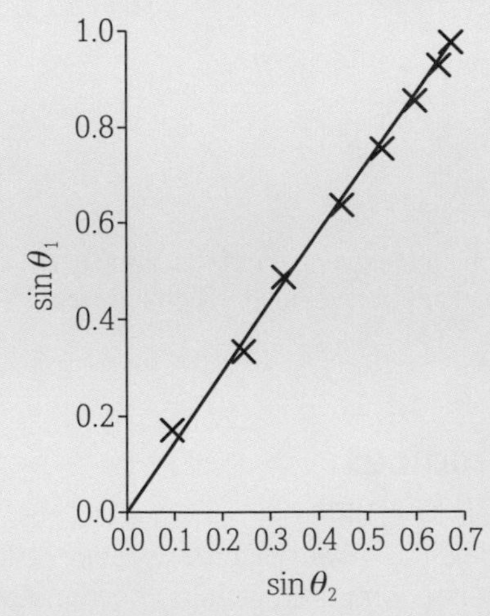

Figure 6 The gradient of this graph enables you to find the refractive index of the material that the semicircular block is made from.

Questions

1 A material allows light to travel through it at 2×10^8 m s^{-1}.

(a) What is the refractive index of the material?

(b) What might the material be?

2 Copy the table and fill in the missing values by using the equation for Snell's law:

$n_1 \sin \theta_1 = n_2 \sin \theta_2$

n_1	n_2	θ_1	θ_2
1.50	1.82	24°	
1.33	1.21		46°
	1.67	28°	16°
2.42		67°	82°

3 The diagram shows light entering both glass and water from the air. Use the values given to calculate the refractive indices of these materials and determine the speed that light is travelling through each material. Assume that the refractive index of air is 1.00.

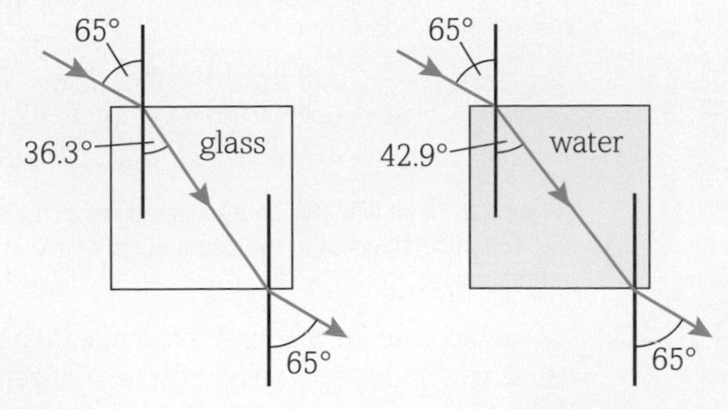

Figure 7

4 In question 3, explain why the emergent ray is parallel to the incident ray.

5 The lens in a human eye has a range of refractive indices that it can use. Based on what the human eye has to do, explain why this is so. Support your answer with values and diagrams.

6 Light travels at a speed of 3.00×10^8 m s^{-1} in air and 2.25×10^8 m s^{-1} in water.

(a) Calculate the refractive index for light travelling from air into water.

(b) A ray of light travelling from air into water enters the water at an angle of 50°. Calculate the angle of refraction.

4.4 8 Total internal reflection

By the end of this topic, you should be able to demonstrate and apply your knowledge and understanding of:

* total internal reflection for light
* critical angle; $\sin C = \frac{1}{n}$
* techniques and procedures used to investigate total internal reflection of light

Introduction

When light travels from a transparent material of refractive index n_1 into another transparent material of refractive index n_2 then refraction will occur as shown in Figure 1.

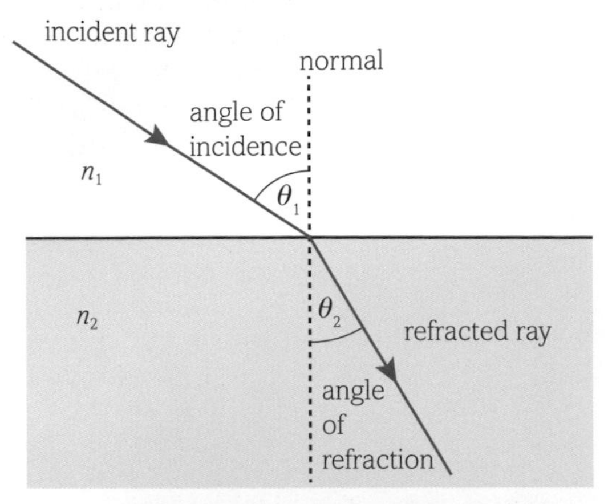

Figure 1a Because $n_1 < n_2$, light slows down on entry to the water and refracts towards the normal.

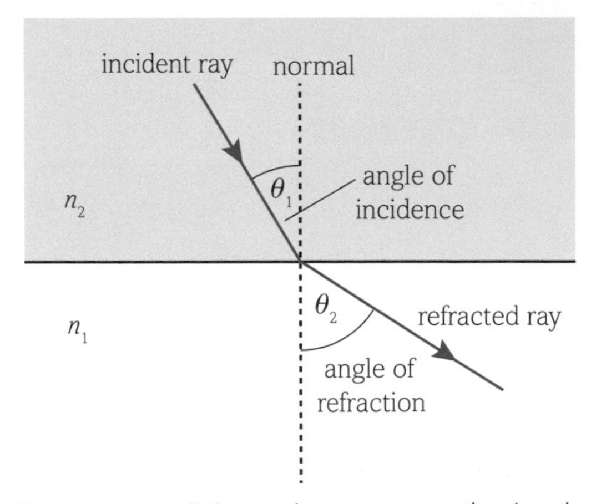

Figure 1b Because $n_1 > n_2$, light speeds up on entry to the air and refracts away from the normal.

Some of the wave's energy is also reflected, as shown in Figure 2.

Total internal reflection

Total internal reflection is a result of both reflection and refraction. We observe it when light, or other electromagnetic radiation, travels from a material of a higher refractive index to one of a lower refractive index. Initially, at an angle lower than the **critical angle**, refraction will occur but there will also be a weak reflected ray (Figure 2).

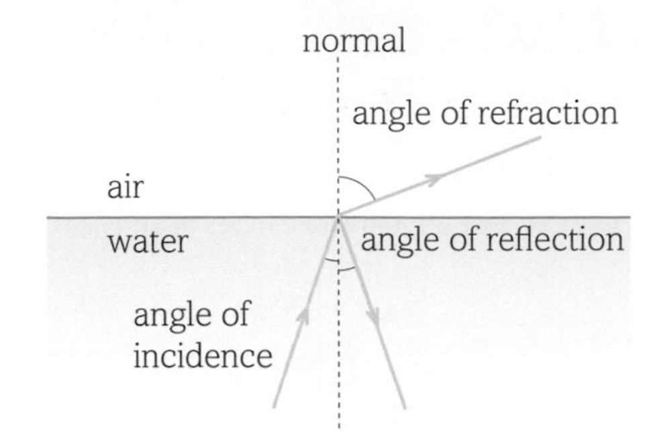

Figure 2 There will also be a reflected ray at the surface, with the angle of reflection being equal to the angle of incidence in accordance with the laws of reflection.

If we increase the angle of incidence then the angle of refraction increases and the partially reflected ray gets stronger. Eventually, when the angle of incidence is equal to the critical angle, the angle of refraction will be 90° and the refracted ray will travel along the boundary between the two materials as shown in Figure 3a.

If we increase the angle of incidence above the critical angle, no refraction will take place and all the light will be reflected – totally internally reflected (Figure 3b).

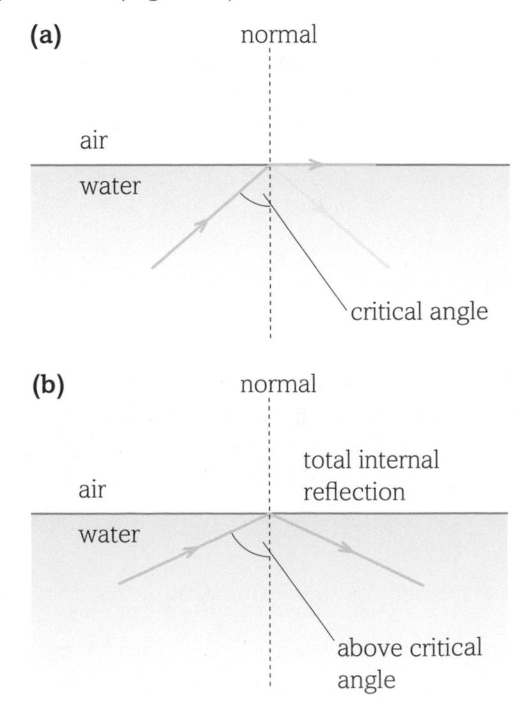

Figure 3 Increasing the angle of incidence eventually results in total internal reflection.

The critical angle

To find an equation that relates the critical angle, C, to the refractive indices involved we need to use Snell's law. Referring back to Topic 4.4.7, you know that this law can be written:

$$n_1 \sin \theta_1 = n_2 \sin \theta_2$$

In the specific case of total internal reflection between a material and air, we know that:

n_1 = the refractive index of the material of higher refractive index, for example water or glass.

$\sin \theta_1$ = the sine of the critical angle, which we shall call $\sin C$.

n_2 = the refractive index of air, which we shall call 1.

$\sin \theta_2$ = the sine of the angle of refraction – in this case $\sin 90°$ is 1.

Substituting in Snell's law:

$$n_1 \times \sin C = 1 \times 1$$

which leads to:

$$\sin C = \frac{1}{n_1}$$

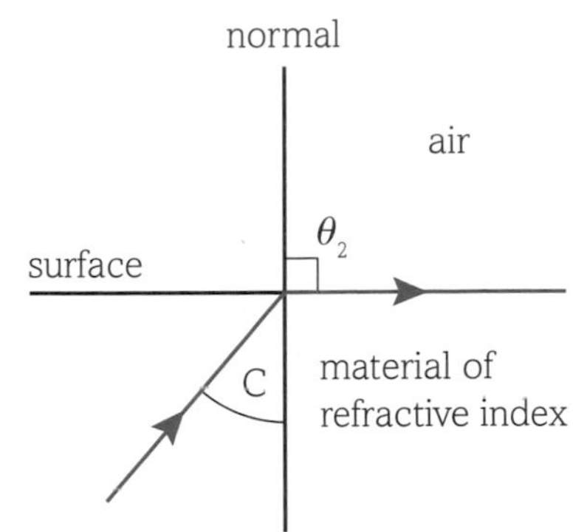

Figure 4 The critical angle.

WORKED EXAMPLE 1

A ray of light passes from water into air, such that the refracted ray travels along the surface of the water–air boundary. The refractive index of air is 1 and the refractive index of water is 1.33. Calculate the critical angle in this case.

Answer

$$\sin C = \frac{1}{n}$$

$$= \frac{1}{1.33} = 0.75$$

$$C = \sin^{-1} 0.75 = 48.6°$$

For examples in which the two materials in contact do not involve air, we can use Snell's law to find an equation that relates the critical angle to the refractive indices of the two materials in contact, n_1 and n_2. This equation is:

$$\sin C = \frac{n_2}{n_1}$$

LEARNING TIP

The ratio $\frac{n_2}{n_1}$ must always be less than 1, so $n_2 < n_1$. Make sure that the smaller refractive index goes on the top of the fraction. The refractive index of material 1 (the material that the ray is travelling from) must be higher than the refractive index of material 2.

WORKED EXAMPLE 2

A ray of light travels from glass of refractive index 1.66 into water of refractive index 1.33. What will be the critical angle for this arrangement?

Answer

$$\sin C = \frac{n_2}{n_1}$$

$$= \frac{1.33}{1.66} = 0.801$$

$$C = \sin^{-1} 0.801 = 53.2°$$

INVESTIGATION

Determining the critical angle for a transport material

To determine the critical angle of a material, use a semicircular shaped block of the material. Draw round the shape on a piece of paper, make a mark at the centre of the straight edge and draw a normal at that point. Shine a ray of light from a ray box towards the curved edge of the block in an arc until the ray emerges along the surface of the boundary between the material and air. The angle the ray makes with the normal is the critical angle. The critical angle can then be used to find the refractive index of the material in question using $\sin C = \frac{1}{n}$.

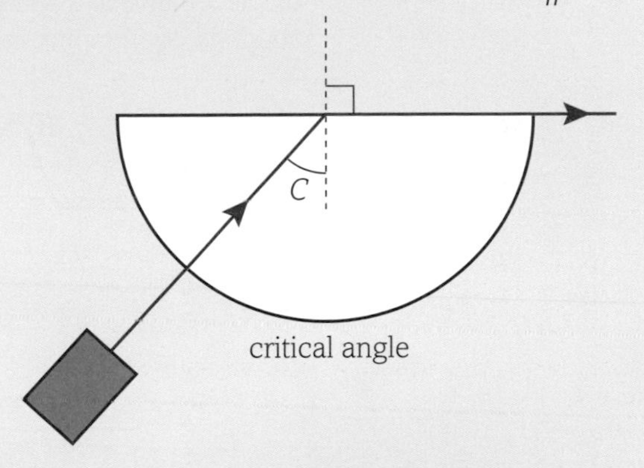

Figure 5 The critical angle of transparent blocks can be found by experiment.

Questions

1. Find the critical angle for a glass block of refractive index 1.58 in contact with air.
2. Diamond has a refractive index of 2.42. A gem is in contact with water of refractive index 1.33. What is the critical angle for this arrangement?
3. How do the following make use of total internal reflection:
   (a) binoculars
   (b) optical fibres
   (c) transmission of radio waves in the upper atmosphere
   (d) endoscopes?
   Draw a diagram for each and explain why total internal reflection is needed in each case.
4. What conditions are necessary for light to be totally internally reflected?

9 Interference

By the end of this topic, you should be able to demonstrate and apply your knowledge and understanding of:

* the principle of superposition of waves
* graphical methods to illustrate the principle of superposition
* two-source interference with sound and microwaves
* interference, coherence, path difference and phase difference
* constructive interference and destructive interference in terms of path difference and phase difference

The principle of superposition

The **principle of superposition** states that when two or more waves of the same type meet, the resultant wave can be found by adding the displacements of the individual waves.

The principle, illustrated in Figure 1, can be applied to calculate the resultant wave (i.e. the net displacement) at any time. The diagrams show two wave pulses moving towards each other at $t = 0$ seconds, overlapping at $t = 1$ second, before moving away from each other.

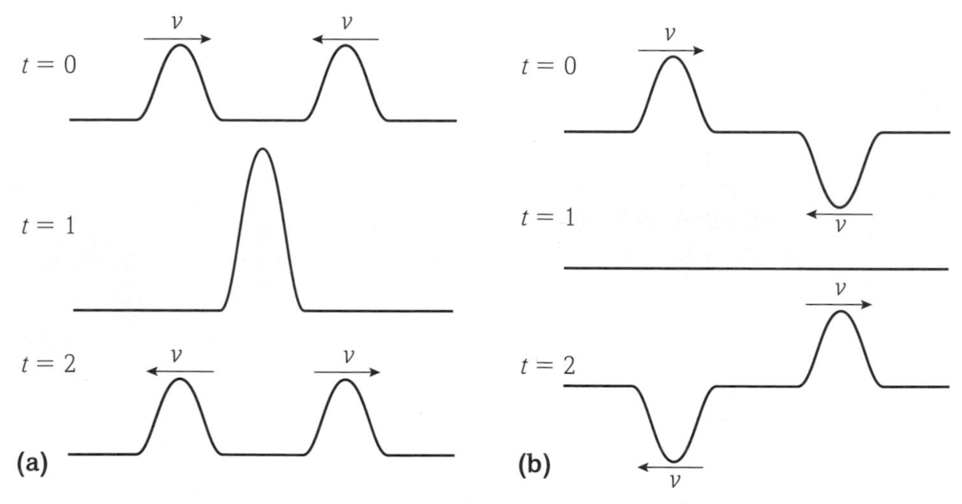

Figure 1 Superposition – two waves moving towards one another will overlap and interfere. The resulting amplitude of the wave is the vector sum of their individual amplitudes at any point where they overlap.

The waves in Figure 1(a) are moving towards each other and overlap constructively. The waves in Figure 1(b) are moving towards each other and overlap destructively.

Remember that displacement is a vector, and take care when adding negative and positive displacements.

For example, if a displacement of -3 is added to a displacement of $+5$, the resultant is $+2$.

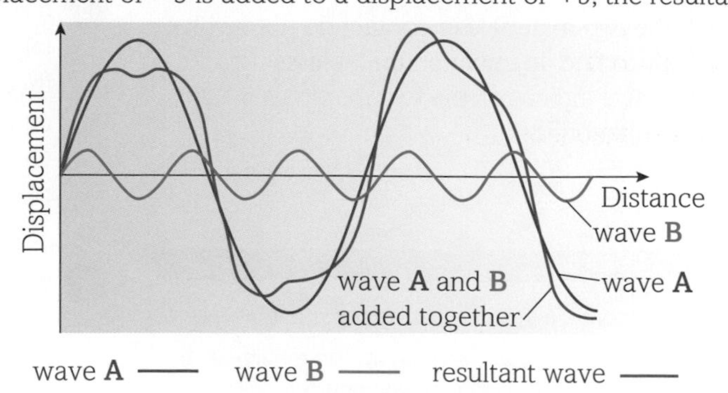

Figure 2 A displacement–distance graph for two waves travelling along the same line. When two waves are superposed, the resultant wave is the sum of the two individual waves.

Interference

If two waves, A and B, with the same amplitude exist at the same point and are travelling in phase, the amplitude of the resultant wave will be twice that of the individual waves (see Figure 3a). This is known as **constructive** interference. If the two waves are in antiphase, they will cancel each other out, and the resultant wave will have an amplitude of zero (see Figure 3b). This is known as **destructive** interference.

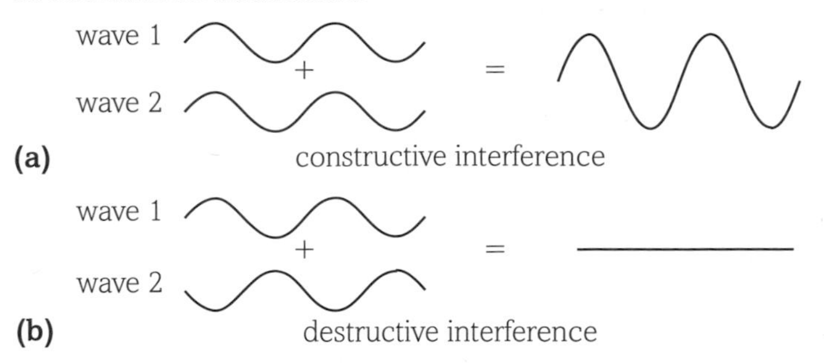

Figure 3 (a) Constructive interference occurs when waves meet in phase. (b) Destructive interference occurs when waves arrive at a point completely out of phase.

Coherence

In reality, things are rather more complicated than these two situations just described. The two waves that overlap and interfere with each other may have different amplitudes and frequencies, or may change their phase relationship with one another. In order to calculate a meaningful resultant using the principle of superposition, the two waves must have a constant **phase difference** – they must be **coherent**. This is true of the waves in Figures 4 and 5, but notice that the green wave in Figure 6 does not follow a regular pattern, so these waves are not coherent.

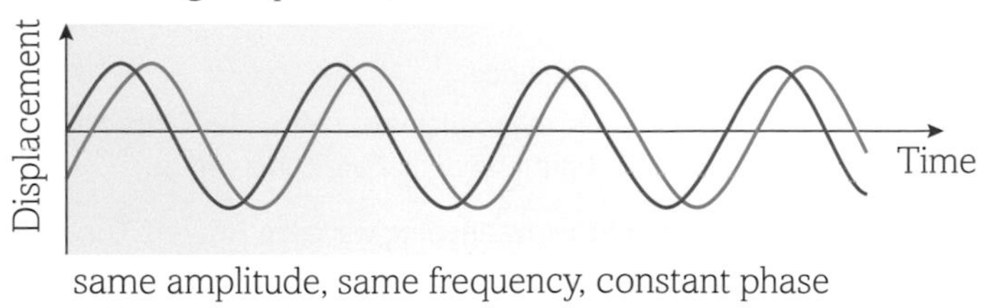

Figure 4 Coherent waves with identical amplitudes.

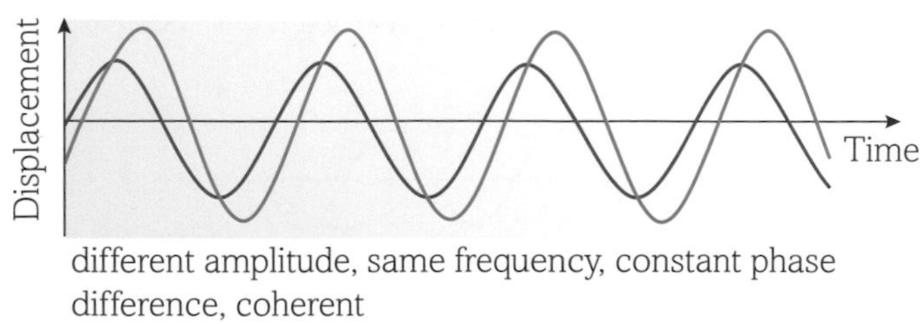

Figure 5 Coherent waves with different amplitudes.

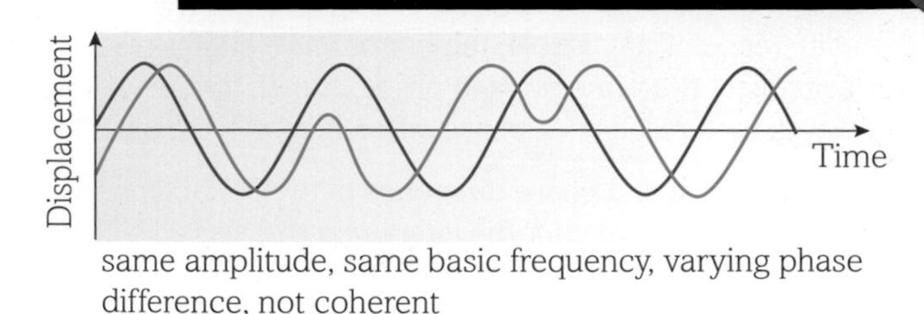

Figure 6 Non-coherent waves.

The light waves emitted by a filament bulb have random phase differences so the light is not coherent. Laser light on the other hand is coherent and all the waves emitted are in phase.

Path difference and phase difference

Path difference and phase difference can be used to determine whether interference is constructive or destructive.

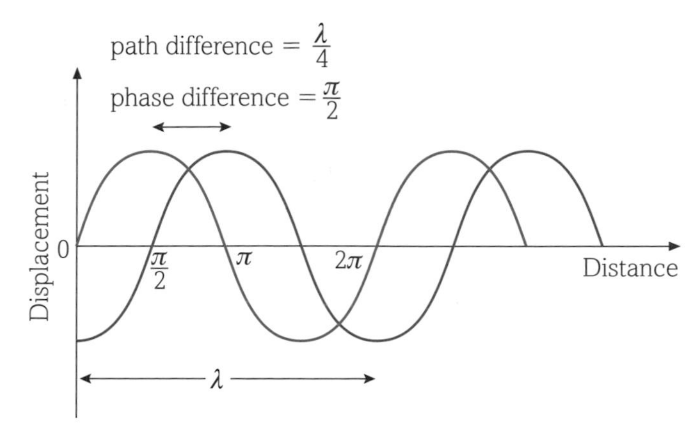

Figure 7 Two coherent waves with their path difference and phase difference shown.

Path difference, which we refer to in relation to the wavelength, is the difference (in metres) between the distances travelled by two waves arriving at the same point. In Figure 7, the path difference between the red wave and the blue wave is $\frac{\lambda}{4}$ or one-quarter of a wavelength.

Phase difference (measured in radians) is the difference in the phases of two waves of the same frequency. In Figure 7, the blue wave leads the red wave by a phase difference of $\frac{\pi}{2}$ (or 90°).

A path difference of any whole number of wavelengths results in waves arriving at the same point in phase. A path difference of half a wavelength results in a phase difference of π radians.

The path difference and phase difference between two coherent waves will determine whether the interference observed will be constructive or destructive, as outlined in Table 1.

	Path difference		Phase difference	
			In degrees	In radians
Constructive interference	whole number of wavelengths	e.g. 0, λ, 2λ, 3λ...	0, 360, 720...	0, 2π, 4π...
Destructive interference	odd number of half wavelengths	e.g. ½λ, 1½λ, 2½λ...	180, 540, 900...	π, 3π, 5π...

Table 1

So far, we have discussed two-source interference in quite general terms. We also saw in Topic 4.4.4 that circular waves from two coherent sources in a ripple tank produce an interference pattern. Two-source interference can also be shown with sound waves, with light waves and with microwaves.

Interference using sound waves

Interference with sound waves can be demonstrated using two loudspeakers connected to the same signal generator. As you walk along in front of the loudspeakers, you will hear a loud sound where the sound waves reinforce one another and a quiet sound where the waves partially cancel one another out. This variation is clearer if you cover one ear. The distance between the loud and quiet regions is longer for low frequencies.

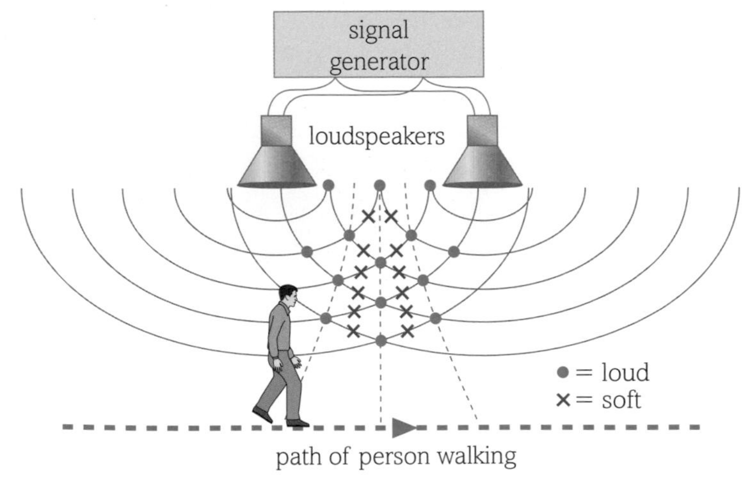

Figure 8 Interference of two sound sources.

Interference using microwaves

In radar systems, microwaves of wavelength 5.0 cm travel in tubes called waveguides. Figure 9 shows a system in which the microwave energy is split along two different paths, A and B, before joining again. By adjusting the lengths of the two paths, it is possible to create either constructive interference or destructive interference when the waves rejoin.

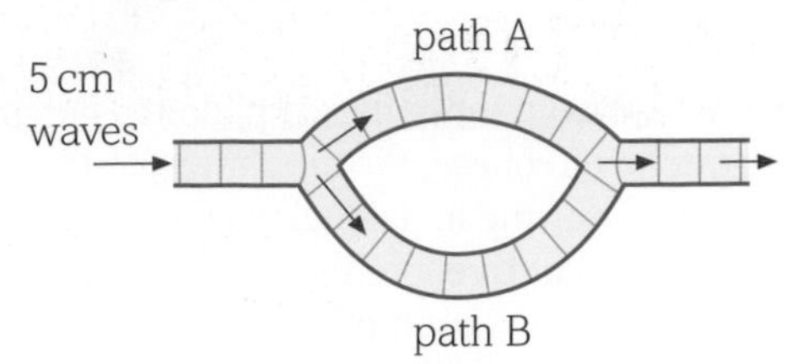

Figure 9 Waveguides allow microwaves two routes before rejoining.

For example, if path A is 35 m long and path B is 45 cm long, waves in A will travel 7 wavelengths and waves in B will travel 9 wavelengths before the two sets rejoin. They have a path difference of 10 cm or 2 wavelengths, and will therefore rejoin in phase resulting in constructive interference.

By altering the dimensions of the waveguides, it is possible to create destructive interference. Imagine, for example, that path A is 37.5 cm long and path B is 45 cm long. In this case, waves that follow path A will travel 7.5 wavelengths before they rejoin waves that follow path B, which have travelled 9 wavelengths. They have a path difference of 7.5 cm, or 1.5 wavelengths, and so will be in antiphase cancelling each other out.

While the actual procedure employed in radar systems is more complex than this, the principle is the same. This method is used to prevent outgoing waves from swamping weak returning echoes.

DID YOU KNOW?

New 4G mobile services at 800 MHz use similar frequencies to Freeview digital TV. Homes located close to a 4G transmitter may experience problems with Freeview picture quality due to interference.

Questions

1. Explain:
   (a) what is meant by coherence
   (b) why it is much easier to produce two radio waves that are coherent than two light waves that are coherent.

2. Look at the images of the overlapping waves in Figure 10. In each case draw the resulting wave and state whether the waves are coherent or not.

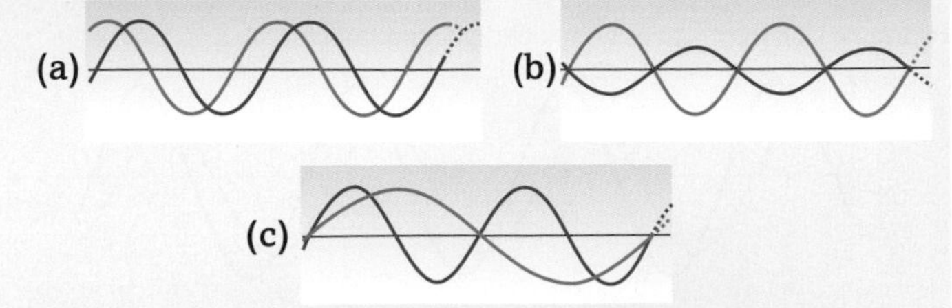

Figure 10

3. State whether the following examples will lead to constructive interference, destructive interference or something in between:
   (a) two waves with a path difference of 8λ
   (b) two waves with a phase difference of 1080°
   (c) two waves with a path difference of 9.5λ
   (d) two waves out of phase by $\frac{3\pi}{2}$ radians.

10 The Young double-slit experiment

By the end of this topic, you should be able to demonstrate and apply your knowledge and understanding of:

* Young double-slit experiment using visible light
* $\lambda = \frac{aX}{D}$ for all waves, where $a \ll D$
* techniques and procedures used to determine the wavelength of light using a double slit

Introduction

The coloured patterns you see in a soap bubble (Figure 1) or in an oil spill on water are a result of the interference of light waves. In order to make measurements of the wavelength of light, two conditions have to be satisfied:

- The light must be monochromatic – that is all the light waves have the same wavelength.
- There must be an accurate method of producing very small path differences, and of measuring these. When we discussed interference of microwaves in Topic 4.4.9, we used radiation of wavelength 5.0 cm, but light waves have a much smaller wavelength.

Figure 1 A soap bubble.

The Young double-slit experiment

Thomas Young was the first person to measure the wavelength of light successfully in 1801 – and in doing so established the wave theory of light. Until then many people, including Newton, thought of light as a stream of tiny particles called corpuscles. The method that Young used is illustrated in Figure 3.

Young used a monochromatic red light source, which he placed behind a single slit in a piece of black card, X. Light passes through the slit and spreads out by diffraction, until it reaches another obstacle, Y, in which there are two parallel narrow slits. The light from these two slits is coherent because it comes from a single slit.

Waves from both slits that are in phase will interfere constructively. Alternate bright and dark vertical bands ('fringes') are seen on the other screen.

Figure 2 Thomas Young.

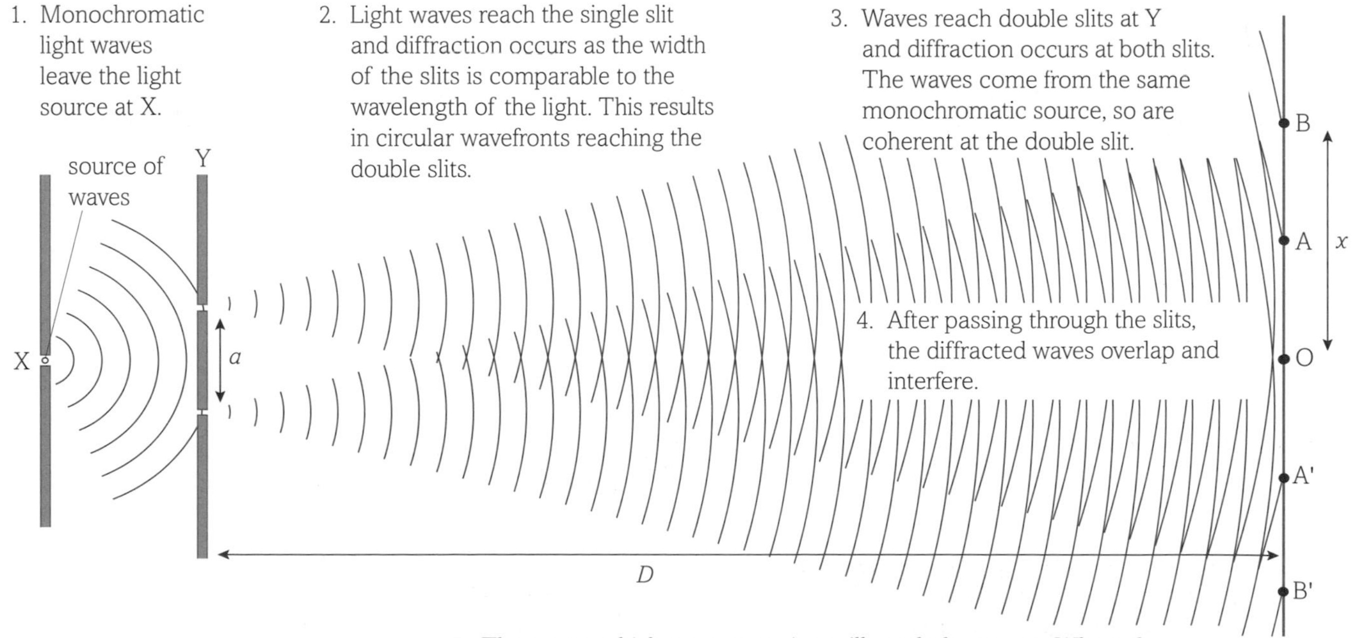

Figure 3 Young's double-slit experiment but shown with a wavelength of 4 mm.

The values shown in Figure 3 are not those for light waves. For example, the gap between the double slits would be about 0.1 mm, not 17 mm as drawn in the text. The wave crests, shown as red arcs, would be much closer together because the wavelength of light is very much smaller than 4 mm, and the 39 mm distance between the fringe and the central position would also be much smaller. Typical values for light be $a = 0.1$ mm, $D = 2$ m and $x = 12$ mm.

INVESTIGATION

Determining the wavelength of light using the double-slit apparatus

The experiment described in Figure 3 must be done in a darkened room as the light intensity is low and the fringes are hard to see.

Using white light produces blurred fringes as each colour produces its own set of fringes, which overlap slightly. A coloured filter is used to produce fringes with sharp edges so a more accurate value of λ can be found. The experiment can also be done with a laser as a source of monochromatic light.

To reduce the percentage error in the fringe separation x it is best to measure across all the fringes and then divide by the number of fringes. Increasing the slit-to-screen distance increases the fringe separation allowing a more precise measurement, but reduces the light intensity at the screen.

The equation for working out the wavelength of the light source

For constructive interference to take place, the path difference must be a whole number of wavelengths. In Figure 3, light from the bottom slit must travel one wavelength further than light from the top slit to reach the first maxima at point B. For destructive interference to take place, the waves must arrive out of phase with a path difference of half a wavelength.

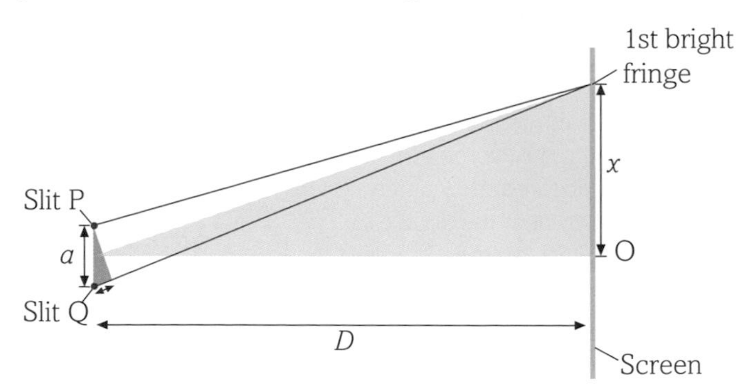

Figure 4 Geometry of the double slit experiment.

In Figure 4 the two shaded triangles are a similar shape so their angles are equal. If θ is the angle away from the central fringe, $\sin\theta = \frac{\lambda}{a}$ and $\tan\theta = \frac{x}{D}$. If the angle θ is small enough, $\sin\theta \approx \tan\theta$ so:

$$\frac{\lambda}{a} = \frac{x}{D} \text{ giving } \lambda = \frac{ax}{D}$$

The equation only applies if $a \ll D$, or the angle away from the central fringe is less than 10°.

WORKED EXAMPLE 1

Monochromatic green light of wavelength 5.5×10^{-7} m (550 nm) is used in a double-slit experiment. The slit separation is 0.2 mm and the distance from the double-slit to the screen is 2.0 m. What is the fringe separation for this set up?

Answer

$$\lambda = \frac{ax}{D}$$

Rrearranging this equation:

$$x = \frac{\lambda D}{a} = \frac{(5.5 \times 10^{-7}\text{ m} \times 2\text{ m})}{2 \times 10^{-4}\text{ m}}$$

$= 0.0055$ m or 5.5 mm

All waves will experience superposition and interference, regardless of whether they are longitudinal sound waves or transverse electromagnetic waves. This means that we should be able to apply the double slit equation not only to coherent light waves, but to all waves, including sound waves.

WORKED EXAMPLE 2

Two loudspeakers are connected to a signal generator. The loudspeakers are separated by a distance of 2.0 m and produce coherent sound waves of the same audible frequency, 200 Hz. An observer at a distance of 20.0 m from the loudspeakers notices that adjacent points of maximum loudness occur every 16.0 m as they walk along a line parallel to the line connecting the speakers.

Find the speed of sound in air from this experiment.

Answer

The loudspeakers are connected to the *same* signal generator, so the two sound sources are coherent.

From Young's double slit experiment, we know that $\lambda = \frac{ax}{D}$ when $a \ll D$.

In this case, we can say that:
a = separation of the speakers, 2.0 m
x = the horizontal distance between adjacent maxima of loudness, 16.0 m
D = the distance from the loudspeakers to the observer, 20.0 m.

Using $\lambda = \frac{ax}{D}$ we obtain $\lambda = (2.0 \times 16.0)/20.0$, giving a value for the wavelength of the sound waves as $\lambda = 1.6$ m.

We know that the frequency of the sound waves produced by the speakers is 200 Hz, so using $v = f\lambda$, we obtain a value for the speed of sound in air equal to 320 m s^{-1}.

INVESTIGATION

Interference using microwaves

The double slit experiment can also be performed with microwaves of about 3 cm. A microwave transmitter is placed 20 cm or so in front of a metal sheet with two gaps about 5 cm apart. The microwaves are diffracted at the two gaps and interfere in the region behind the metal sheet. A microwave detector can be moved in a straight path across the interference pattern (always the same distance from the gaps) to find the positions of maximum intensity (constructive interference) and minimum intensity (destructive interference).

It is possible to connect an amplifier and loudspeaker to the receiver so that the microwave signal can be 'heard'.

Questions

1. When red monochromatic light is used in the double slit arrangement, the separation of the fringes is found to be 16 mm when the double slit width is 1.125×10^{-4} m and the distance from the double slits to the screen is 2.4 m. Use this information to find a value for the wavelength of the red light source.

2. Look at the double-slit experiment arrangement. Use the equation $\lambda = \frac{ax}{D}$ to decide what will happen to the fringe width when you:
   (a) increase the value of *a*
   (b) increase the value of *D*
   (c) double the values of *a* and *D*.

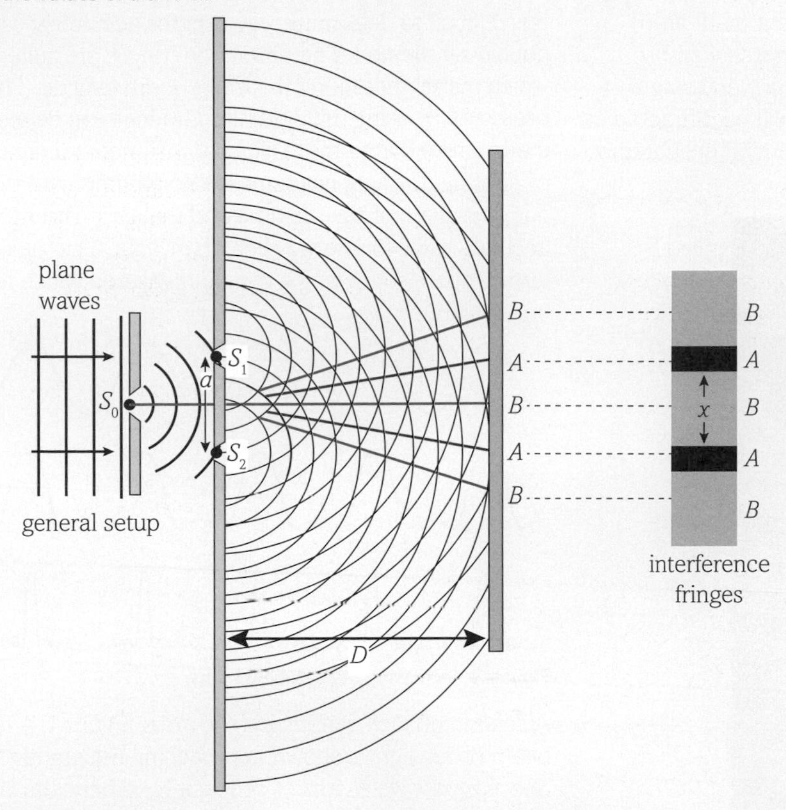

Figure 5

3. In the worked example for interference of sound waves, explain why the distance between loud and soft regions would be longer for low frequencies.

4. The microwave transmitter described above has a wavelength of 2.8 cm. The separation between the slits is 5.0 cm and the receiver is a distance of 80 cm from the slits.
   (a) Calculate the separation between adjacent maxima.
   (b) If the screen was moved to a distance of 60 cm from the slits, what would the fringe separation become?

5. Laser light of wavelength 6.5×10^{-7} m passes through two narrow parallel slits, the centres of which are at a distance of 0.25 mm apart. A screen is placed 1.5 m in front of the slits.
   (a) Calculate the distance between adjacent dark fringes seen on the screen.
   (b) How will the pattern change when either (i) the separation of the slits or (ii) the distance between the slits and the screen is doubled?
   (c) Suppose that one slit is half the width of the other but the centres are still the same distance apart. How will this alter the pattern?

6. Explain why the fringes are closer together for green light than for red light.

7. Show that for a slit separation $x = 1.4 \times 10^{-3}$ m and slit-to-screen distance D = 1.20 m, $\sin\theta = \tan\theta = \frac{x}{D}$ is a very good approximation.

11 The diffraction grating

By the end of this topic, you should be able to demonstrate and apply your knowledge and understanding of:

* techniques and procedures used to determine the wavelength of light using a diffraction grating

Introduction

A *diffraction grating* is a piece of optical equipment made from glass, onto which many thousands of very thin, parallel and equally spaced grooves have been accurately engraved using a diamond. Light that passes through the grating will be diffracted at different angles based on the wavelength of the incident light and the separation of the grooves (Figure 1).

Figure 1 A diffraction grating with 600 lines per millimetre.

Figure 2 shows how the apparatus is used.

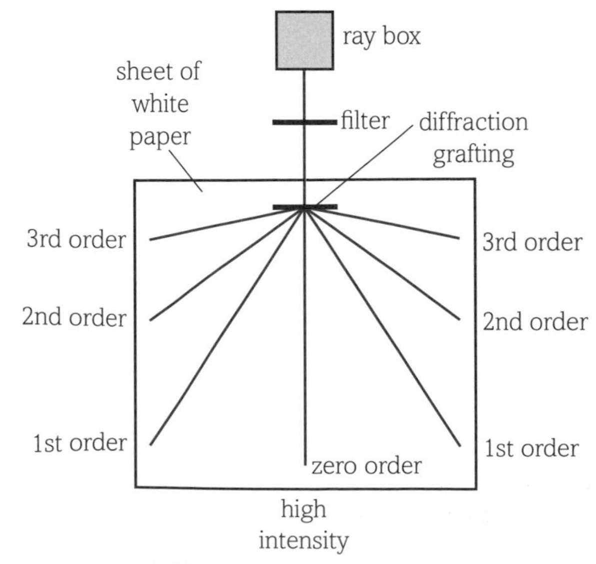

Figure 2 The experimental setup.

Multiple slits

It is difficult to determine wavelengths accurately using the Young double-slit method. The diffraction fringes are quite blurred, which makes measuring the fringe width difficult. This problem is overcome by using multiple slits. Gratings can be made in which there can be 600 or more slits per millimetre. Increasing the number of slits that the light has to pass through improves the brightness and sharpness of the maxima and makes it easier to measure an accurate value for the wavelength of light. The maxima are also further apart, so the angle can be measured with a lower percentage uncertainty.

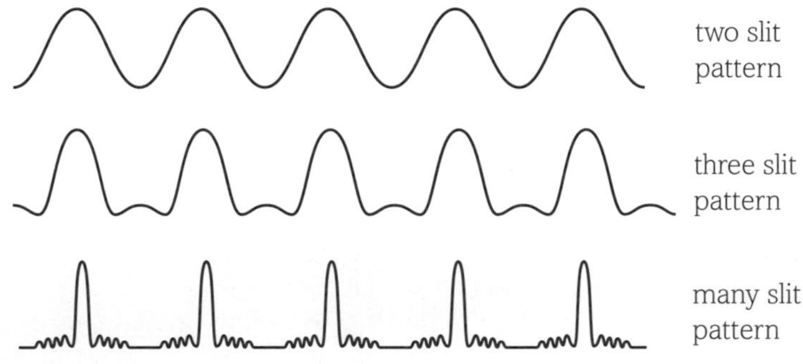

Figure 3 Increasing the number of slits.

A section of such a grating is shown in Figure 4. A fine laser beam of red light is shown approaching the grating from the left. Successive wavelengths of red light have been drawn on the beam after they have passed through each of the slits. You will notice that in order for constructive interference to occur, the waves must all be in phase with each other at points A to F. When this is the case, the waves will all move off at an angle θ and form a constructive interference bright fringe on the screen.

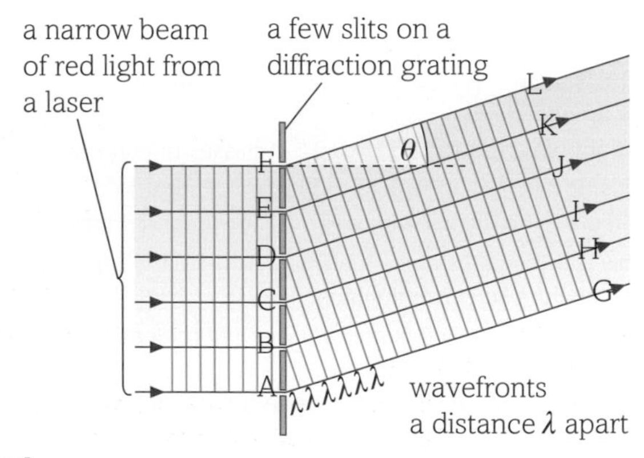

Figure 4

If this is to work then the successive distances travelled by neighbouring waves must be a whole number of wavelengths different. So lengths of AG, BH, CI etc. in Figure 4 must be different in length by a whole number of wavelengths.

This being the case, for constructive interference to occur and for a maximum to be produced on the screen, there must be a whole number of wavelengths in each of lengths AG, BH CI etc. and they must occupy a distance that is equal to $d \sin \theta$. This is shown in the close-up image in Figure 5.

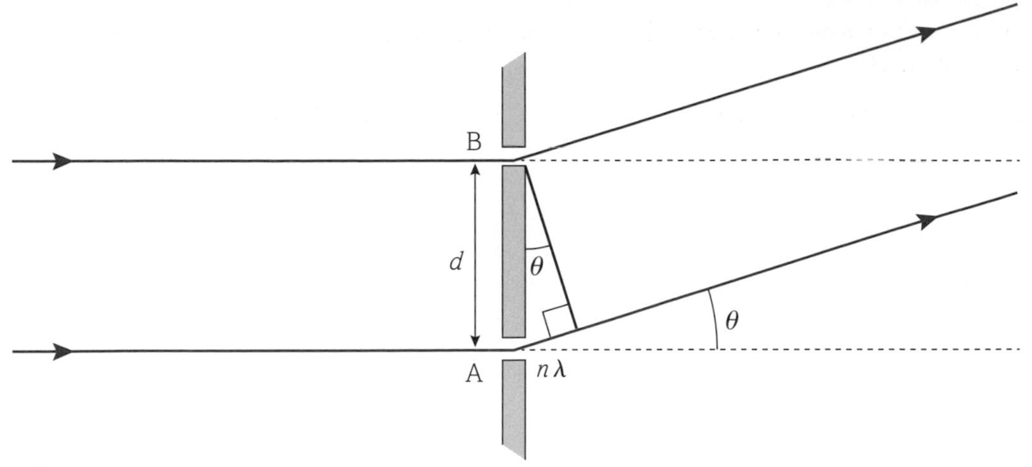

Figure 5

Because the length of n wavelengths, $n\lambda$, must be equal to $d \sin \theta$, we get the equation for a diffraction equation

$$n\lambda = d \sin \theta$$

where n is called the **order** of the maximum (1, 2, 3 ...), λ is the wavelength of the incident monochromatic light, d is the separation of the slits in the grating and θ is the angle that the beam makes with the grating.

KEY DEFINITION

The number of the pattern, n, on either side of the central maximum is known as the **order**.

INVESTIGATION

Measuring the wavelength of light using a diffraction grating

Look at the experiment illustrated in Figure 6. A diffraction grating is put in a parallel beam of monochromatic light from a sodium lamp. The angle between the light that passes straight through the grating and the diffracted beam is measured. It is found that there are a number of angles at which light is seen – these are called maxima. They occur when the path difference between light from adjacent slits is 0, λ, 2λ or 3λ or 4λ or .... (rather than just λ) and reinforcement (constructive interference) occurs.

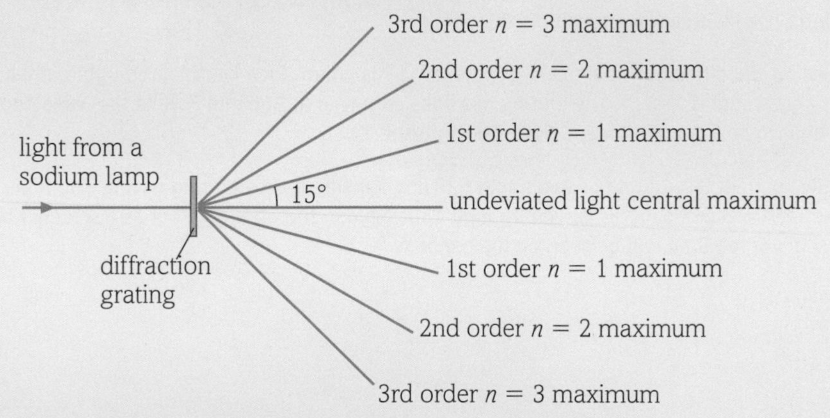

Figure 6 The spectral pattern seen from a sodium lamp.

The experiment can also be done with a laser as a source of monochromatic light.

The wavelength is found using $n\lambda = d \sin \theta$. If the grating has X slits per metre, the slit spacing, $d = \frac{1}{X}$. Remember to convert all lengths to metres.

The angle θ is found with a protractor. The percentage uncertainty in θ is decreased if θ is as large as possible (maxima further apart). This occurs for smaller slit separations and also for higher order fringes. However the intensity of the higher order fringes is lower, making them difficult to locate.

You could also find θ by geometry, using $\tan \theta = \frac{x}{D}$ where x is the distance from the central maxima and D is the distance of the grating from the screen. If the screen is far away compared to the distance between the slits ($D >> d$), $\tan \theta \approx \sin \theta \approx \theta$ (in rad).

WORKED EXAMPLE

The wavelength of sodium light is 589 nm. The first order spectrum ($n = 1$) in Figure 6 occurs at an angle of 15.0°. Calculate:

(a) the grating spacing, d

(b) the number of lines per mm on the grating

(c) the angle for the second and third order spectra.

Answers

(a) $n\lambda = d\sin\theta$:

$1 \times 589 \times 10^{-9} = d \times \sin 15.0$

$d = \dfrac{589 \times 10^{-9}}{0.26} = 2.28 \times 10^{-6}\,\text{m}$

(b) Number of lines $= \dfrac{1}{2.28 \times 10^{-6}\,\text{m}} = 440\,000\,\text{m}^{-1}$

So, the number of lines per mm is 440.

(c) For the second order:

$2 \times (589 \times 10^{-9}) = 2.28 \times 10^{-6} \times \sin\theta_2$

$\sin\theta_2 = \dfrac{2 \times 589 \times 10^{-9}}{2.28 \times 10^{-6}} = 0.517$

So $\theta_2 = 31.1°$

For the third order:

$3 \times (589 \times 10^{-9}) = 2.28 \times 10^{-6} \times \sin\theta_3$

$\sin\theta_3 = \dfrac{3 \times 589 \times 10^{-9}}{2.28 \times 10^{-6}}$

$= 0.775$

So $\theta_3 = 50.8°$

There is no fourth order – can you explain why?

LEARNING TIP

It is not possible to have a value for $\sin\theta$ that exceeds 1, so when $\dfrac{n\lambda}{d} > 1$ no further maxima are observed. This is actually very helpful when answering questions. If asked how many maxima are possible from a particular arrangement, simply divide d by λ. The answer is whatever the whole number is – ignore what comes after the decimal point. So, if $\lambda = 600\,\text{nm}$ and $d = 1.3 \times 10^{-3}\,\text{mm}$, then $\dfrac{d}{\lambda} = \dfrac{1.3 \times 10^{-6}\,\text{m}}{6.00 \times 10^{-7}\,\text{m}} = 2.167$, so $n = 2$ will be the highest order maximum observed.

Questions

1. What wavelength of light would produce a second order maximum at an angle of 28° for a diffraction grating with 580 lines per millimetre?

2. At what angle would the $n = 3$ maximum be observed for a diffraction grating of 550 lines per millimetre when the wavelength of light was 5.7×10^{-7} m?

3. What is the highest order maximum that would be observable for a diffraction grating with 600 lines per millimetre used with a wavelength of 700 nm?

4. What would the difference in angle be for the $n = 1$ maximum for a beam of red light compared with a beam of green light? Assume the grating has lines engraved at 560 mm^{-1}. Take the wavelength of red and green light to be 7.0×10^{-7} m and 5.3×10^{-7} m respectively.

5. A narrow beam of laser light of wavelength 630 nm is incident on a grating having 300 lines per millimetre. A piece of paper is curved through 180° beyond the grating to act as a screen. Calculate how many spots of red light will be seen on the paper.

Stationary waves

By the end of this topic, you should be able to demonstrate and apply your knowledge and understanding of:

* explain how stationary waves are formed
* graphical representations of a stationary wave similarities and differences between stationary and progressive waves
* nodes and antinodes
* the idea that the separation between adjacent nodes (or antinodes) is equal to $\frac{\lambda}{2}$, where λ is the wavelength of the progressive wave

Introduction

We often think of waves being produced at a source and transferring energy outwards from that point. Ripples on a pond, light from the Sun and seismic waves from earthquakes are all examples of this. These types of waves are called *progressive* waves, because energy is transferred and the positions of the peaks and troughs change as the wave moves outwards.

However, it is also possible to confine a wave, and its energy, to a fixed position. These waves are called *standing waves* or *stationary waves* – examples include waves on strings, including stringed instruments, and air columns in tubes or pipes.

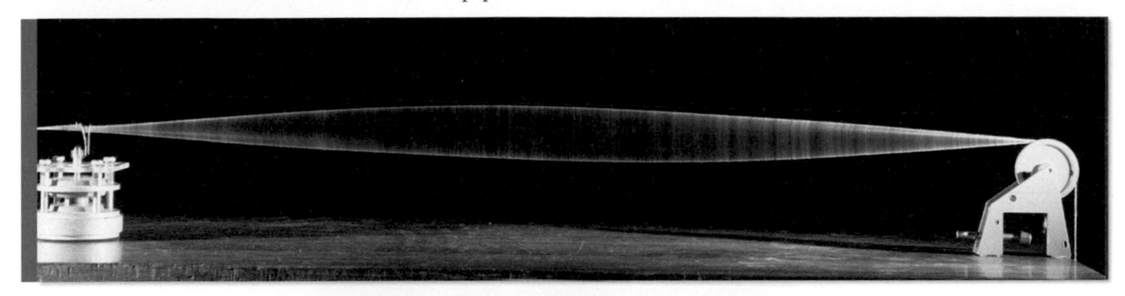

Figure 1 A stationary wave is confined between the two fixed ends.

Figure 2 Stringed instruments use stationary waves to produce sounds.

The formation of stationary waves

Stationary waves are produced by interference in accordance with the *principle of superposition*. This states that the resultant displacement of two or more waves is equal to the vector sum of the waves' displacements at that point (see Topic 4.4.9).

In order for a stationary (standing) wave to be produced on a string or in a pipe, the two waves that overlap must be travelling in opposite directions, have the same frequency and have approximately equal amplitudes.

The diagrams in Figure 3 help to illustrate this.

(a) Initially, there are two progressive waves moving in opposite directions – the green wave is moving to the left and the red wave is moving to the right. The two waves are completely out of phase – they have a phase difference of 180° (π radians) and we say that they are in antiphase. This leads to a resultant wave of zero amplitude because the waves cancel each other out.

(b) A fraction of a second later, the red wave has moved a quarter of a cycle to the right, and the green wave has moved a quarter of a cycle to the left. The waves have moved so that their crests are in the same place, so they are now completely in phase with a phase difference of zero. Adding these two waves together gives the 'vector sum' – in this case, a wave of twice the amplitude of either of the single waves.

(c) This shows the wave pattern half a cycle after the initial disturbance of the wave. The waves are in antiphase again and so the displacement along the resulting wave is zero.

(d) Three-quarters of a cycle after the start, the waves will have shifted a further quarter of a cycle in opposite directions, resulting in the two waves being in phase again and with a resulting wave of twice their respective amplitudes. However, note how a crest in Figure 3(b) has now become a trough. This accounts for why the standing wave appears to move up and down but not sideways.

(e) The wave pattern for a whole cycle, $t = T$, will take us back to the diagram shown in Figure 3(a).

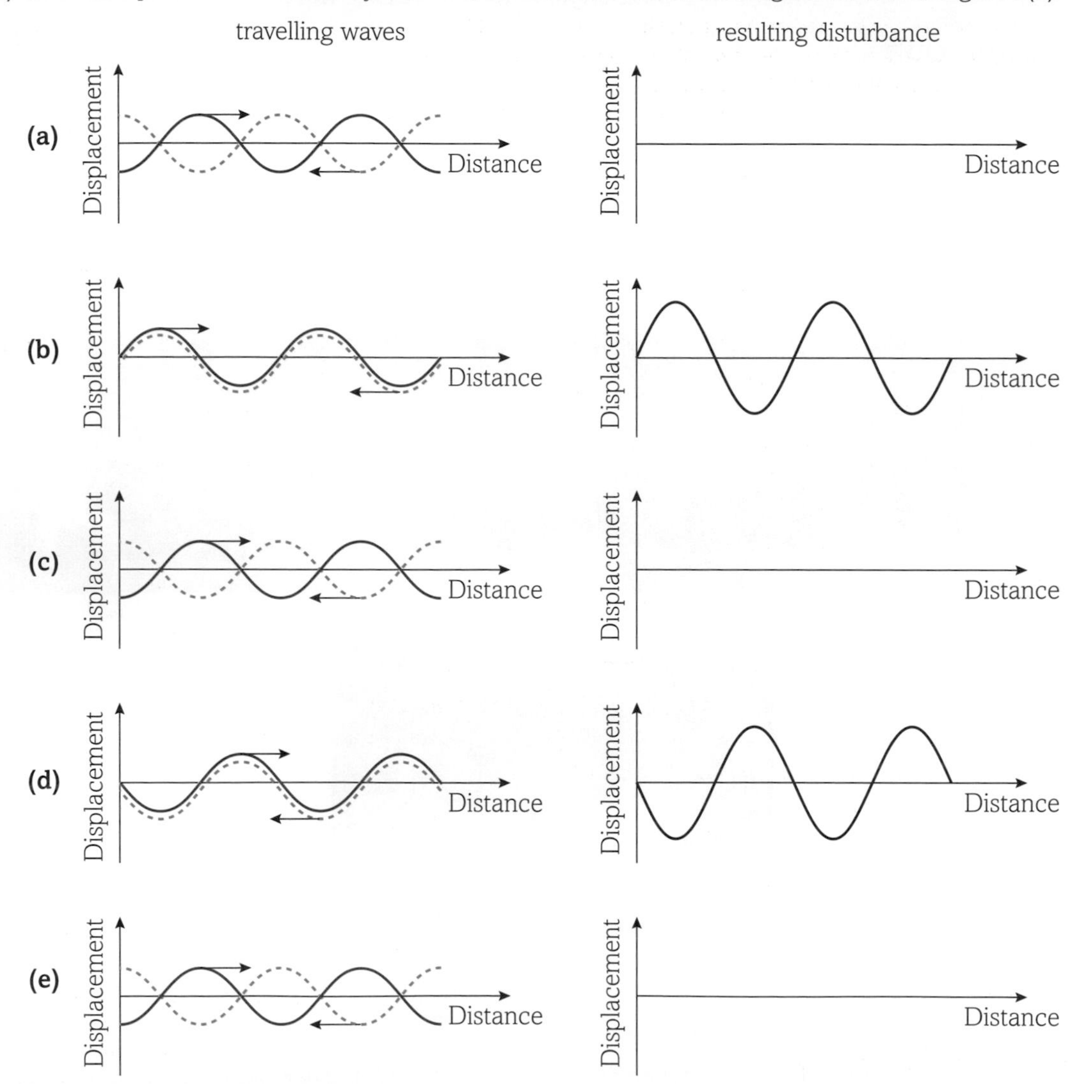

Figure 3 The changes as stationary waves are formed.

Nodes and antinodes

At some points of the resultant wave in Figure 3 the displacement is always zero – these points are called **nodes**. At other points the displacement varies from maximum positive to zero to maximum negative and back – these points are called **antinodes**.

The distance between a node and an antinode is a quarter of a wavelength; and the distance between two adjacent nodes or antinodes is a half a wavelength. The wave pattern for a stationary wave is shown in Figure 4.

KEY DEFINITIONS

Nodes are points in a stationary wave at which there is no displacement of the particles at any time. At **antinodes**, the displacement of the particles in a stationary wave varies by the maximum amount.

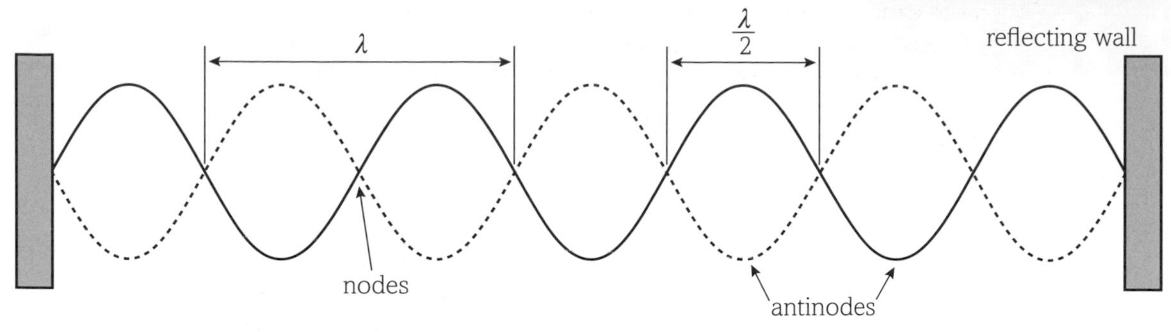

Figure 4 Nodes and antinodes.

DID YOU KNOW?

Stationary waves can happen on bridges when the wind speed has a particular value that matches the 'natural frequency' of the bridge. This causes nodes and antinodes to be clearly visible and can lead to bridges collapsing. One example of this is the Volgograd bridge in Russia. When a bridge oscillates at its maximum amplitude due to a driving force such as the wind it is said to be resonating.

Figure 5 The first Tacoma Narrows Bridge – stationary waves can be damaging.

Questions

1 (a) State the conditions needed for a stationary wave to be produced.

(b) How are stationary waves different from progressive waves?

(c) How are stationary waves similar to progressive waves?

(d) Give two examples of stationary waves and two examples of progressive waves.

2 What is the distance between:

(a) successive nodes

(b) successive antinodes

(c) a node and an adjacent antinode?

3 Two wave pulses move at 1.0 m s^{-1} in opposite directions along a string as shown in Figure 6. They are drawn at time $t = 0$. Copy this diagram and do the following:

(a) Draw the profile of the string at 1.0 s intervals for times up to 3.0 s.

(b) Sketch the profile of the string between 2.0 m and 5.0 m, at 1.5 s and 2.5 s.

(c) Mark the position(s) of the point(s) on the string that never move.

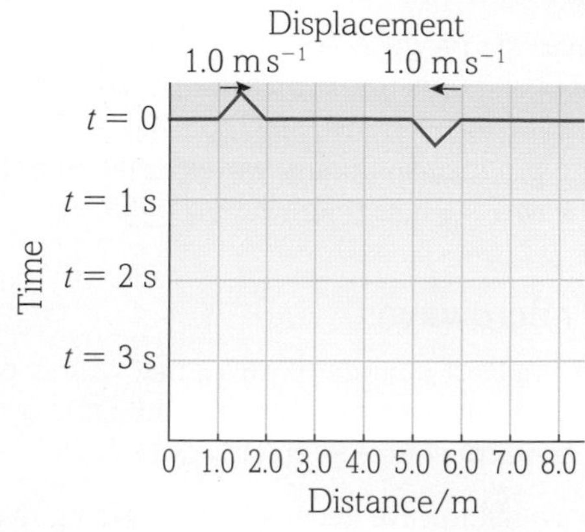

Figure 6

4 To repeat the exercise shown in Figure 3, draw Figure 7 on graph paper.

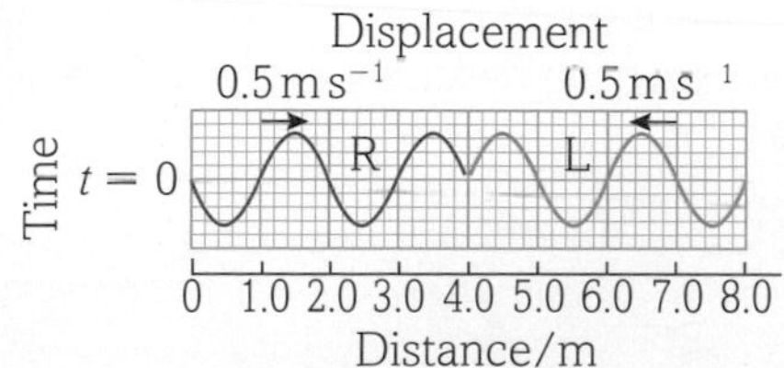

Figure 7

Two sine wave pulses, each of two wavelengths, travel along a string towards each other at 0.5 m s^{-1}. The wave labelled R stretches from 0 to 4.0 m travelling to the right along the string, and the one labelled L, from 4.0 m to 8.0 m, travels to the left.

(a) Using the same arrangement as used in question 3, draw R and L in their new positions at 1.0 s and the resultant shape of the string.

(b) Repeat the exercise at 2.0 s, 3.0 s and 4.0 s.

(c) In a single diagram, draw the resulting shapes of the string between 3.0 m and 5.0 m, at 1.0 s, 2.0 s, 3.0 s and 4.0 s.

(i) Which points on the string appear to have no displacement while the waves overlap?

(ii) What is the separation of these points?

(iii) What are these points called?

(d) Both waves carry energy. How is this energy stored at 1.0 s and 2.0 s?

13 Stationary wave experiments

By the end of this topic, you should be able to demonstrate and apply your knowledge and understanding of:

* stationary (standing) waves using microwaves and stretched strings
* standing wave patterns for a stretched string
* fundamental mode of vibration (1st harmonic); harmonics

Introduction

Stationary waves are produced on strings when progressive waves, travelling in opposite directions, are superposed. A description and explanation of this was given in Topic 4.4.12. However, it is also possible to produce stationary waves using other longitudinal or transverse waves, such as microwaves.

Stationary microwaves

A common example of stationary waves that we can produce in school laboratories involves microwaves, and a microwave oven is a convenient, portable source of microwaves.

Figure 1 shows a microwave generator transmitting microwaves towards a metal sheet. The microwaves are reflected back from the sheet along their initial path, resulting in the formation of a stationary wave. When you move a microwave detector along the stationary wave, it registers the microwaves as signals. These signals are particularly strong every half wavelength along the wave.

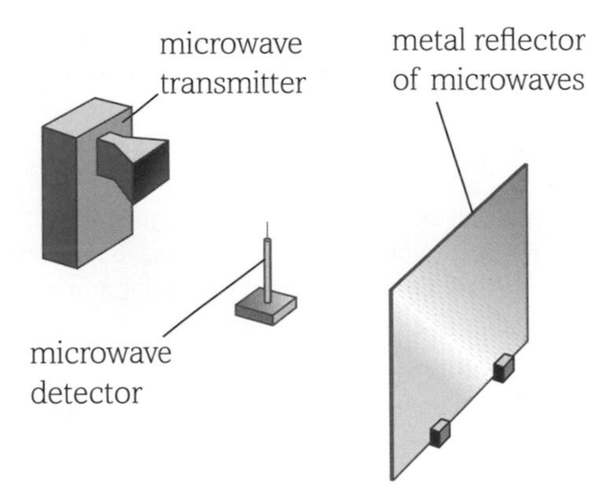

Figure 1 Measuring the wavelength of microwaves can be done in a laboratory.

If you multiply the frequency of the microwaves by the measured wavelength values from the standing wave, you can work out the speed of the electromagnetic wave. For greatest accuracy, do not measure the distance between one maximum and its adjacent one (as shown in Figure 2), but measure from one maximum to the one furthest from it. For example, you may measure a distance of 0.328 m from the first maximum to the fifteenth maximum. This will give you 14 half-wavelengths or 7 full wavelengths, giving:

$$\lambda = \frac{0.328}{7}$$

$$= 0.0469 \text{ m, or about } 5 \text{ cm}$$

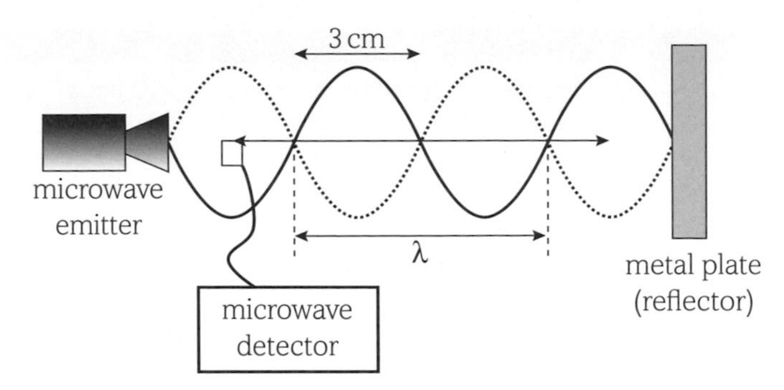

Figure 2 If you know the frequency of the microwaves you can determine the speed of the microwaves.

Stationary waves on strings

Many musical instruments make use of vibrating strings in order to create a note of a particular frequency. Obviously the string section of an orchestra – the violins, violas, cellos and double basses – use strings and so do banjos, pianos and guitars. The vibrating stationary wave on a guitar string is shown in Figure 3.

Figure 3 Vibrating stationary wave on a guitar string.

For stringed instruments, the string itself may be bowed or plucked or, as in a piano, it can be struck with a hammer. Sounding boards or amplifiers are used to increase the note volume. The frequency of the vibrations of a string is governed by:

- its mass per unit length
- its tension – this is adjusted when tuning the instrument
- the length of the string, which is increased or decreased by the musician's fingers.

The effect of bowing a violin string is to create progressive transverse waves that travel in opposite directions along the string away from the bow. When these waves reach the points where the string is fixed to the wood, they are reflected back along the string. They then interfere with each other producing a stationary wave.

WORKED EXAMPLE

A stationary wave in a guitar string obeys the equation:

$$f = \frac{1}{2}L\sqrt{\frac{T}{\mu}}$$

where f is the frequency of vibration in Hz, L is the length of the string in m, T is the tension on the string in N and μ is the mass per unit length of the string in kg m^{-1}.

What frequency would be produced if the length of the string was 80 cm, the tension was 70 N, and the mass per unit length was 0.91 g m^{-1}?

Answer

Converting into S.I. units and substituting into the equation gives:

$$f = \frac{1}{2} \times 0.8 \times \sqrt{\frac{70}{9.1 \times 10^{-4}}}$$
$$= 110.9 \text{ Hz}$$

Fundamental mode of vibration and harmonics

The sound you hear from a musical instrument depends on the production of stationary waves. A specific type of stationary (sound) wave is produced when you pluck a stretched string. The simplest stationary wave that can be set up on a violin string is shown in Figure 4.

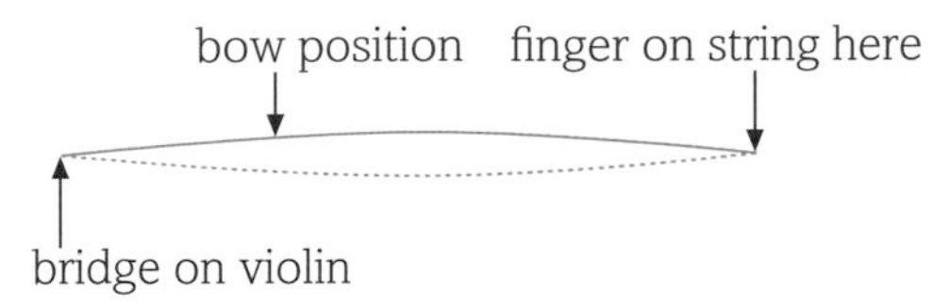

Figure 4 The fundamental oscillation of a violin string.

This type of oscillation is called the **fundamental mode of vibration**, where the length of the string is half the wavelength of the note. The fundamental frequency is the lowest frequency, highest wavelength that can be produced on a string. However, other notes of higher frequency can also be produced and these are called **harmonics**. Common examples of these are shown in Figure 5. The frequency of a harmonic is always a whole number multiple of the fundamental frequency.

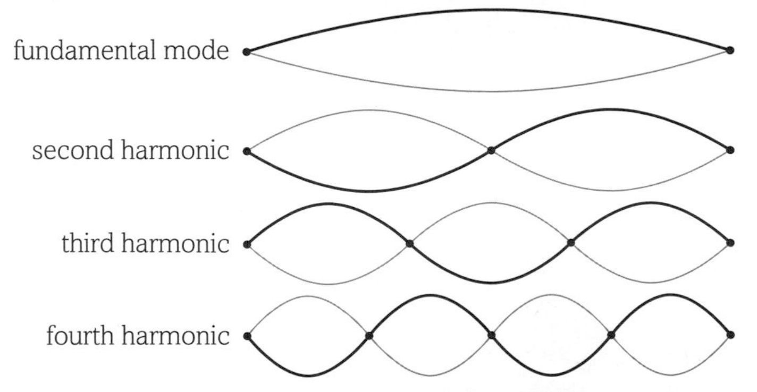

Figure 5 The fundamental has the lowest frequency and longest wavelength.

KEY DEFINITIONS

In the **fundamental mode of vibration**, the length of the string is half the wavelength. This produces the lowest possible frequency called the **first harmonic**. Other **harmonics** are whole number multiples of this frequency.

Figure 6 shows the oscilloscope trace for the sound of a particular note from a violin. It is a complex pattern because many different modes of vibration of the string are possible and occur simultaneously.

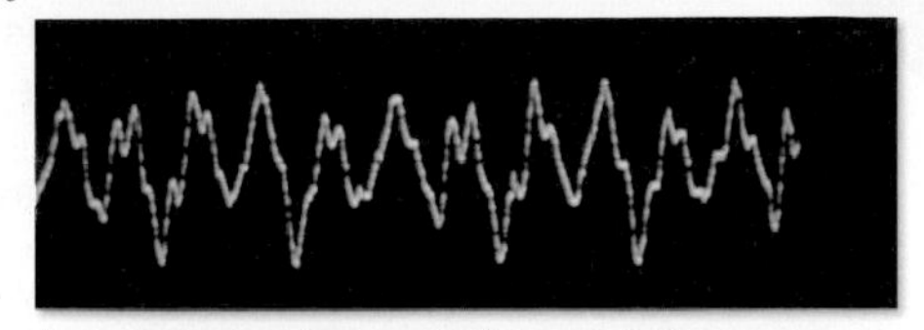

Figure 6 The oscilloscope trace for a note from a violin.

Questions

1. Figure 7 shows, at a given instant, the shape of the stretched string on which a stationary wave has been produced.
   (a) State the phase difference between the oscillations of the particles at:
   (i) W and Y; (ii) Y and Z in the stationary wave.
   (b) State which of the particles labelled in Figure 7 are at:
   (i) a node; (ii) an antinode.

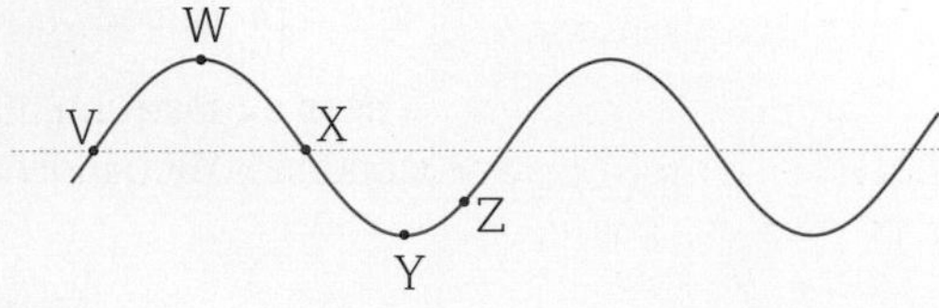

Figure 7

2. One end of a string is tied to a fixed support at X and the other to a weight. The string passes over a pulley and hangs vertically. The string is set in motion so that it vibrates at a frequency of 180 Hz, as shown in Figure 8.
   (a) What is the lowest possible (fundamental) frequency of vibration of the string?
   (b) The length of the vibrating string is reduced to two-thirds by moving the pulley towards X. What is the new fundamental frequency of vibration?
   Give your reasoning.

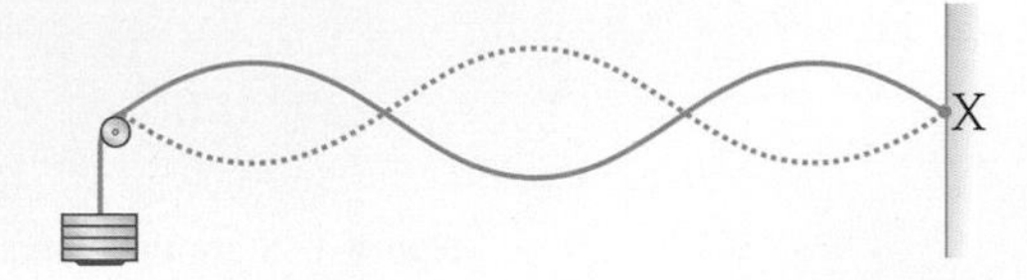

Figure 8

3. The fundamental frequency, f, of a standing wave on a stretched string is given by:
   $f = \frac{1}{2}L\sqrt{\frac{T}{\mu}}$, where L is the length of the string between the supports, T is the tension, and μ is the mass per unit length of the string.
   (a) The weight on the end of the string in Figure 8 is doubled. Calculate the frequency of the new fundamental note, taking the fundamental frequency before the extra weight was added to be 60 Hz.
   (b) In practice, the mass per unit length decreases because the string becomes thinner when it is stretched. State and explain how this will change the fundamental frequency.

4.4 (14) Stationary longitudinal waves

By the end of this topic, you should be able to demonstrate and apply your knowledge and understanding of:

* stationary (standing) waves using air columns
* stationary wave patterns for air columns in closed and open tubes
* techniques and procedures used to determine the speed of sound in air by formation of stationary waves in a resonance tube

Introduction

Many musical instruments that do not produce sound using strings do so by means of stationary waves in air columns. These air columns may be tubes or pipes of different shapes and sizes. One of the earliest known musical instruments was the reed pipe. In a modern orchestra, wind instruments vary in size from the small piccolo to the large bassoon in the woodwind section, and from a cornet to a euphonium in the brass section. When blown, the air inside these instruments vibrates. The sound produced varies depending on the length of the pipe and the structure of the instrument.

These air instruments can be simple like a flute, or complex like a church organ, some of which can contain 30 000 individual pipes.

Figure 1 Stationary longitudinal waves can be set up in the pipes of a church organ.

Figure 2 Stationary longitudinal waves can also be produced in simple woodwind instruments such as the flute.

Stationary waves in closed tubes

A stationary longitudinal wave in a tube is a specific type of stationary wave, created by blowing across one end of the tube. As with the stationary wave produced on a string, the stationary wave in the tube is produced when progressive waves travel in opposite directions through one another.

A progressive wave can be started at one end of a closed tube by blowing across the open end. This wave travels down the tube and is reflected at the closed end. This produces two progressive waves travelling in opposite directions, which then interfere to produce a stationary wave. However, to produce a stationary wave the length of the tube must be such that a node is formed at the closed end (where the air is stationary) and an antinode at the open end. This is illustrated in Figure 3, which shows how the amplitude of vibration of air particles varies along the tube (displacement against distance). It also shows the position of the nodes (N) and antinodes (A) for the fundamental mode of vibration, as well as for several harmonics.

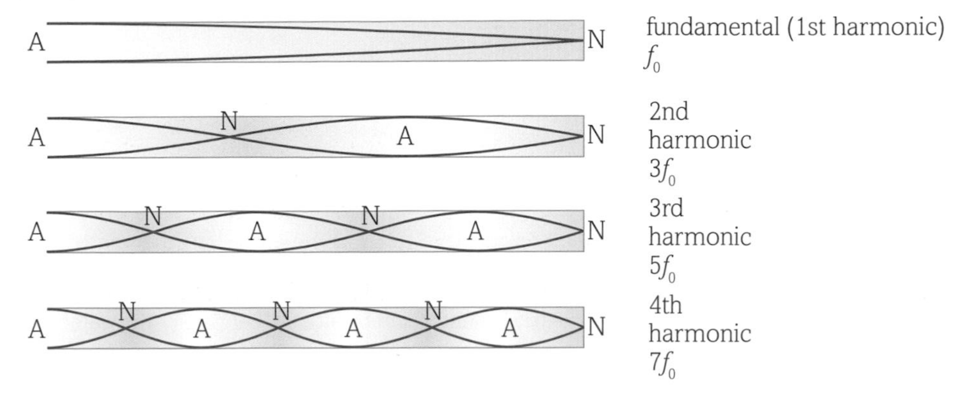

Figure 3 Stationary waves in closed pipes, ranging from the fundamental to the fourth harmonic. Notice how there is always a node at the closed end and an antinode at the open end. Although the wave looks transverse, it is actually longitudinal and the vibrations of the air molecules are highest at an antinode and zero at a node.

At a node the air displacement is zero but the air pressure varies. A node for displacement is always an antinode for pressure – the particles experience maximum squeezing toward that point. In contrast, there is little change in the spacing between the particles at a displacement antinode. Consequently, the pressure at displacement antinodes barely changes. If you push a small microphone along an air column in which a stationary sound wave has been set up, it will register a loud sound at the displacement nodes, but hardly any sound at the displacement antinodes.

Stationary waves in open tubes

Many musical instruments have tubes where the two ends are open, yet stationary waves are still formed. The reason for this is that the sound wave will also be reflected at the open ends of the pipe. Since the tube is open at both ends, an antinode is present at both ends. An example is a simple recorder. The displacement of air particles against distance along an open tube is shown in Figure 4. Also shown are the positions of the nodes and antinodes for the fundamental mode of vibration, as well as for several harmonics. As in Figure 3, the wave shown is not transverse, it is longitudinal in nature, so the greater the amplitude of the wave shown, the greater the horizontal oscillation of the air particles.

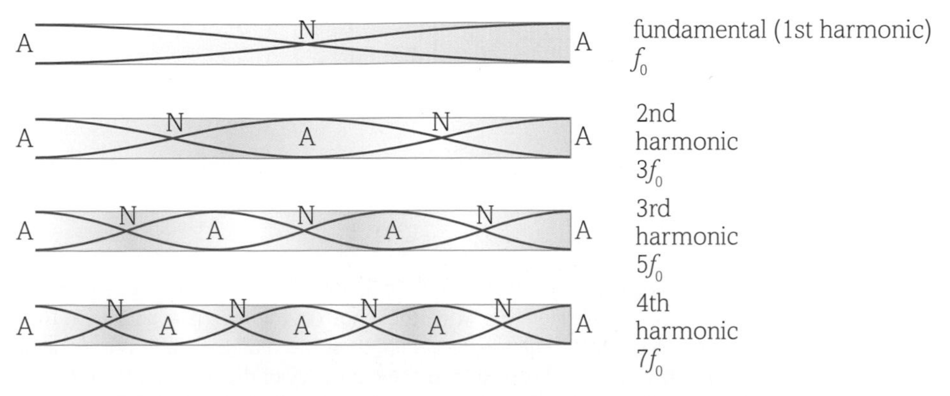

Figure 4 Stationary longitudinal waves can also be produced in pipes that are open at both ends.

INVESTIGATION

Measuring the speed of sound in air

A simple but effective way of determining the speed of sound in air is to use a tuning fork and a tube of water, as shown in Figure 5.

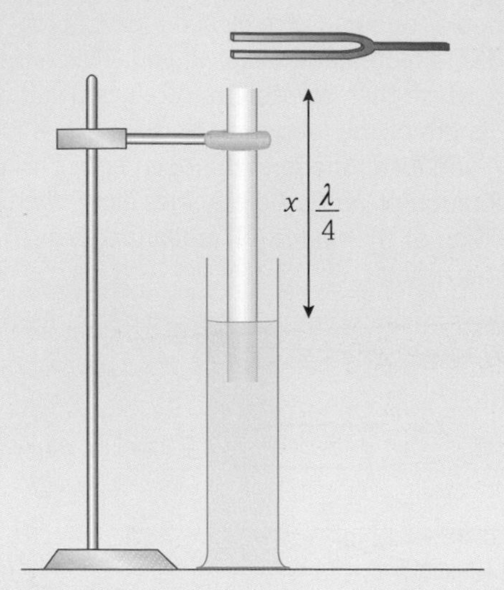

Figure 5 This apparatus can be used for setting up a stationary wave in a closed tube.

The tube is held by a clamp and is moveable so that its length can be altered. Because of the water in the measuring cylinder, the tube is effectively closed at one end.

When a tuning fork of known frequency is struck and held at the open end, the air molecules in the tube will vibrate and a stationary wave will be set up in the tube. By listening carefully, the fundamental frequency can be obtained, when the sound is loudest for at the minimum length. This is achieved when the length of the tube is equal to $\frac{1}{4}\lambda$, as shown in Figure 6.

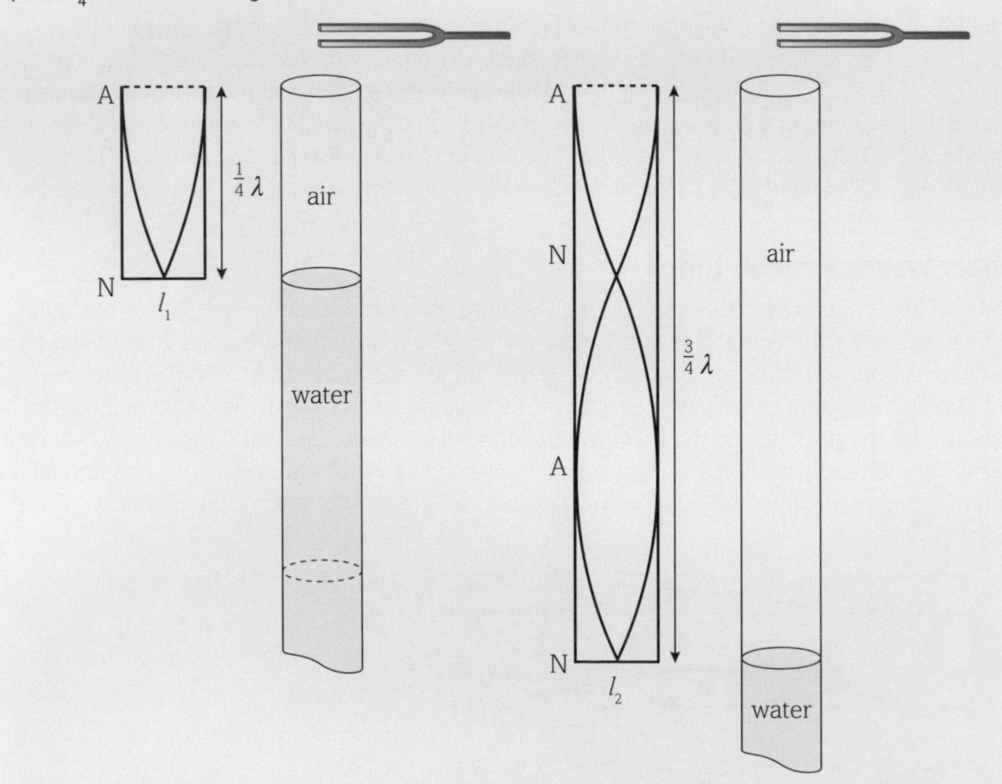

Figure 6 Setting up stationary waves in an air column.

The tube can then be lengthened by loosening the clamp, and the loudness of the sound will reduce initially before increasing again to a second maximum loudness when the length of the tube is equal to $\frac{3}{4}\lambda$. The difference between these two lengths is equal to half the wavelength of the sound, $\frac{1}{2}\lambda$, and the speed of sound can be determined by multiplying this value by 2 and then by the frequency of the tuning fork in Hz.

WORKED EXAMPLE

A tuning fork of 440 Hz is struck and held over a tube that is closed at one end. When the tube is lengthened to 18 cm, a maximum loudness is heard. This occurs again when the tube is lengthened to 54 cm. Calculate:

(a) the wavelength of the sound wave

(b) the speed of sound in air.

Answers

(a) We can use $\frac{1}{4}\lambda = 18$ cm, $\frac{1}{2}\lambda = 36$ cm or $\frac{3}{4}\lambda = 54$ cm to get a value for the wavelength as 72 cm, or 0.72 m.

(b) We know that $\lambda = 0.72$ m and that $f = 440$ Hz.

$v = f\lambda$

$= 0.72 \times 440 = 316.8\ \text{m s}^{-1}$, which is relatively close to the true value of $340\ \text{m s}^{-1}$.

Questions

1. For sound waves produced in a closed pipe, the relationship between the number of nodes (N) and the number of antinodes (A) is $N = A$. What is the relationship for the number of nodes and antinodes for waves produced in an open pipe?

2. A vibrating tuning fork is held above the glass tube shown in Figure 5. The length of the air column is adjusted by raising or lowering the tube in the water, until a loud sound is heard. The standing wave formed in the air column is the fundamental.
   (a) Sketch the tube and label the position of a node (N) and an antinode (A).
   (b) Add arrowed lines at the three points: N, A and half-way between N and A, to show the direction of movement and relative amplitudes of the air at these positions.
   (c) The length of the air column when the fundamental frequency is heard is 0.32 m. Calculate the wavelength (in m) of the standing wave in the tube.
   (d) The speed of sound in air is $330\ \text{m s}^{-1}$. Calculate the frequency (in Hz) of the tuning fork.
   (e) Determine the value of the lowest frequency of the note produced in an open tube of length 0.32 m. Show your reasoning.

3. The air in the open pipe shown in Figure 7 is made to vibrate at the lowest resonance frequency (fundamental frequency) of the pipe.
   Describe the similarities and differences in the displacement of the air molecules at points:
   (a) P and S
   (b) P and Q
   (c) P and R.

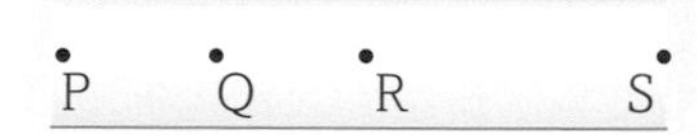

Figure 7

GROUND-SHAKING DISCOVERIES

Seismic waves provide geophysicists with a large amount of information about the internal structure of the Earth. In this activity you will explore the types of seismic waves and how we can use our knowledge of waves to determine the location of an earthquake's epicentre.

EARTHQUAKES

Seismology is the study of seismic waves that move through and around the Earth. These seismic waves are caused by sudden movements of rock or plates in the Earth's crust or the release of energy that may have come from nuclear tests, landslides, volcanic activity or mine blasts.

There are different kinds of seismic waves; the two main types are **body waves** and **surface waves**. Earthquakes radiate seismic energy from the source as both body waves and surface waves.

Body waves

These include P waves and S waves. P waves are the faster of the two waves and can move through solid rock and fluids. P waves are also known as compressional or push–pull waves. An S wave is slower than a P wave and so is the second type of body wave to arrive at the surface after an earthquake. S waves can only move through solid rock. S waves move particles up and down, at right angles to their direction of motion.

Surface waves

Travelling only through the crust, surface waves are of a lower frequency than body waves and can easily be distinguished as such on a seismogram as a result. Though they arrive after body waves, surface waves are almost entirely responsible for the damage and destruction associated with earthquakes. The two surface waves are Love waves and Rayleigh waves.

The seismometer and the lag time for P waves and S waves

A seismometer is used to detect and measure seismic waves from an earthquake. P waves will arrive at the seismometer first, followed some time later by S waves. The difference in time between the first of each of those types of wave arriving is called the lag time.

In order to interpret the lag time, we need to use a graph which tells us the travel time of P and S waves and the distance that the waves have travelled from the epicentre. This relationship is shown on the graph:

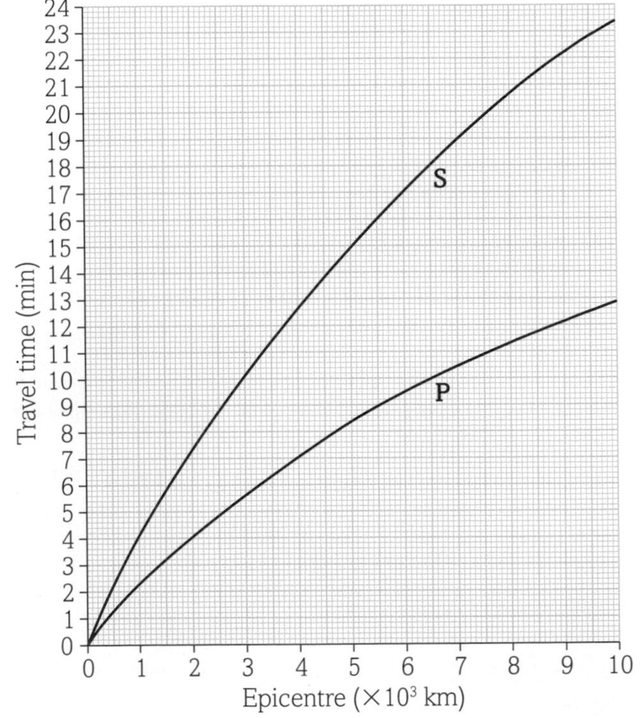

Figure 1 Travel time of P and S waves against their distance from the epicentre.

If we know the lag time between the arrival, at the location of the seismometer, of the P waves and the S waves, we find this time gap on the vertical axis of the graph. Reading down onto the x-axis then tells us the distance of this seismometer from the epicentre. For example, if I know that the lag time is 3 minutes and 20 seconds, then I must be 2000 km from the epicentre.

Source

Michigan Technological University, 2007, UPSeis, an educational site for budding seismologists, Houghton, MI, USA, viewed 20 January 2015, http://www.geo.mtu.edu/UPSeis/waves.html

Where else will I encounter these themes?

1.1 2.1 2.2 3.1 3.2 3.3

Let us start by considering the nature of the writing in the article. The text above is the voiceover from a short YouTube video posted by an anonymous user.

1. Consider the extract and comment on the type of writing that is being used. Who is the audience? Does the video try to explain, persuade or describe? Are the findings open to interpretation by others? How might you change the video script to make it more suitable for a scientifically informed audience? What diagrams, images and equations would you expect to be used in a video for this audience?

Is there a difference between the credibility of a video source compared with a text source found online? What would be the characteristics of a high-quality source of information?

We will now look at the physics that is in the article. Do not worry if the physics content or the mathematics is challenging at this stage. You can always return to the article later in your course, once some of the related topics have been studied in more depth. Use the timeline at the bottom of the page to help you put this work in context with what you have already learned and what is ahead in your course.

2. How can seismic waves be produced?
3. How are body waves and surface waves:
   a. similar to each other
   b. different from each other?
4. What evidence is there in the text to suggest that:
   a. P waves are longitudinal waves
   b. S waves are transverse waves?
5. How far away will the epicentre be if the lag time is 5 minutes?
6. What will the lag time be if the epicentre is 3400 km away?
7. Waves can be explained in terms of their 'phase velocity' and their 'group velocity'. Find out what these are, how they are different and how they are important in the study of earthquakes.
8. How can stationary waves be set up during an earthquake?
9. What do seismologists need know in order to model the travel time for seismic waves between two points on the Earth's surface?
10. How is a graph such as that shown in Figure 1 produced? Will it be valid for every earthquake or do certain assumptions need to be made?

DID YOU KNOW?

An earthquake happens somewhere in the world once every 30 seconds. Each year, there are about a million earthquakes around the world, but only about 100 of these cause serious damage. Steel, reinforced concrete and wood are good building materials for an earthquake-resistant house because they flex somewhat without breaking. Homes built completely out of brick are not as safe because they can break apart easily.

Activity

As part of their Voluntary Service Overseas (VSO) work, a set of students are designing an education programme to help volunteers who are going to remote parts of the world know what to do in the event of an earthquake. The education programme will consist of models, graphs, video and textual material explaining how earthquakes occur and how the level of devastation caused depends on certain factors such as location, proximity to the focus of the earthquake and the substructure of the ground in that point. Write a script and storyboard for a short video that will provide scientific information for the VSO volunteers in South America on:

- what seismic activity is and why it occurs
- the dangers of seismic activity
- the factors that affect how much damage will be caused by the earthquake.

Practice questions

1. Which of the following statements is true, based on the wave shown below? [1]

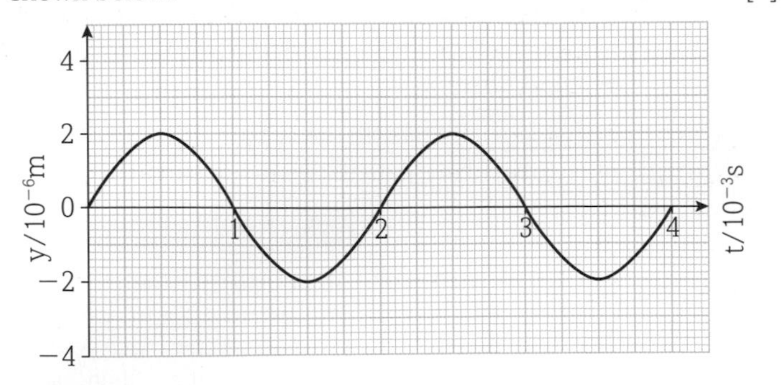

Figure 1

A The amplitude of the wave is 2 cm.

B The wavelength of the wave is 2 m

C The time period of the wave is 2 s.

D The frequency of the wave is 500 Hz.

2. Which of the following is true in relation to electromagnetic waves? [1]
   (i) They all travel at the speed of light in a vacuum.
   (ii) The energy carried by the waves increases with their wavelengths.
   (iii) They are all highly ionising.

A (i), (ii) and (iii)

B only (i) and (ii)

C only (ii) and (iii)

D only (i)

3. Which of the following wave behaviours is displayed by longitudinal waves? [1]
   (i) diffraction
   (ii) refraction
   (iii) polarisation

A (i), (ii) and (iii)

B only (i) and (ii)

C only (ii) and (iii)

D only (i)

4. Monochromatic light of wavelength 5.8×10^{-7} m is directed onto a diffraction grating of 600 lines per mm. At what angle will the second maxima be observed? [1]

A 44 degrees

B 54 degrees

C 64 degrees

D 68 degrees

[Total: 4]

5. (a) Define the following terms:
   (i) amplitude [1]
   (ii) frequency [1]
   (iii) coherent [1]
   (iv) period [1]
   (v) wavelength [1]

   (b) A wave was found to have a frequency of 12 kHz and be longitudinal in nature.
   (i) Explain whether the wave was likely to be a sound wave or a light wave. [1]
   (ii) Calculate the wavelength of the wave based on this explanation. [2]
   (iii) How would the wave appear different if its frequency was changed? [1]
   (iv) How would the wave appear different if its amplitude was changed? [1]

[Total: 10]

6. Figure 2 shows an incomplete illustration of the electromagnetic spectrum.

Radio	**A**	**B**	Visible	UV	**C**	Gamma

Figure 2

(a) Identify the areas labelled A, B and C. [2]

(b) State one example of how the types of radiation A and C are:
   (i) similar [1]
   (ii) different. [1]

[Total: 4]

7. A monochromatic source of light is shone through a diffraction grating that has 600 lines per millimetre. The second order maximum is observed at an angle of 38 degrees from the straight through position, as shown in Figure 3.

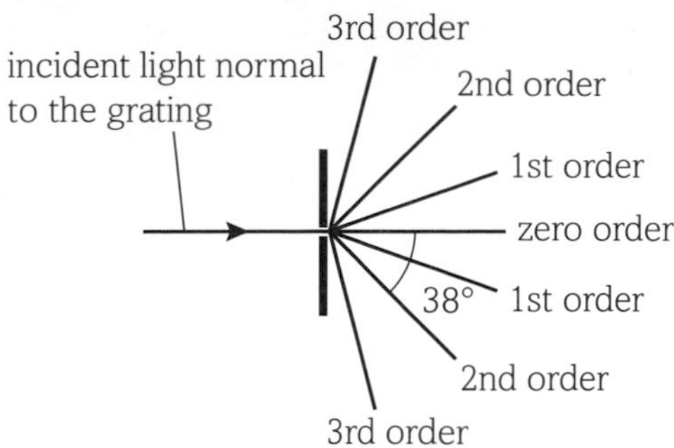

Figure 3

(a) Calculate:
 (i) the value for d in the equation $n\lambda = d\sin\delta$ [2]
 (ii) the wavelength of the light being used in the investigation [2]
 (iii) the highest order of maxima observable for this arrangement. [2]

(b) Explain how the angle of the second order maxima that was observed would change if:
 (i) the monochromatic source had a greater wavelength [2]
 (ii) the diffraction grating had 450 lines per mm instead of 600 [2]
 (iii) the laser was moved further away from the diffraction grating. [2]

[Total: 12]

8. The apparatus shown in Figure 4 was set up to determine the value for the wavelength of light using Young's double slits.

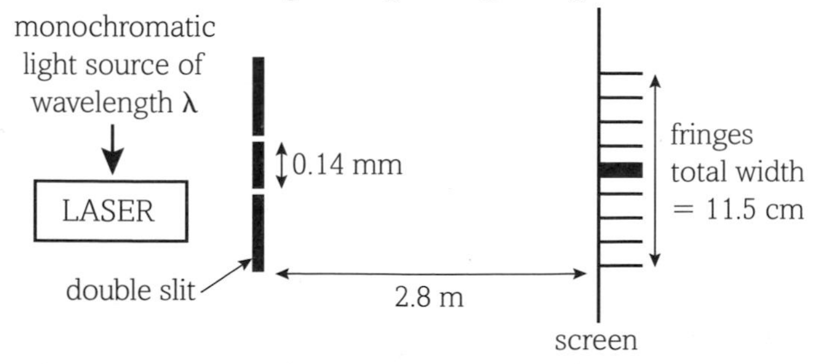

Figure 4

- D = 2.8 m
- W = 0.14 mm

Use the information provided to obtain a value for the wavelength of the light. [3]

[Total: 3]

9. In an investigation of standing waves, sound waves are sent down a long pipe, with its lower end immersed in water. The waves are reflected by the water surface. The pipe is lowered until a standing wave is set up in the air in the pipe. A loud note is then heard. See Figure 5.

Length l_1 is measured. The pipe is then lowered further until a loud sound is again obtained from the air in the pipe. Length l_2 is measured.

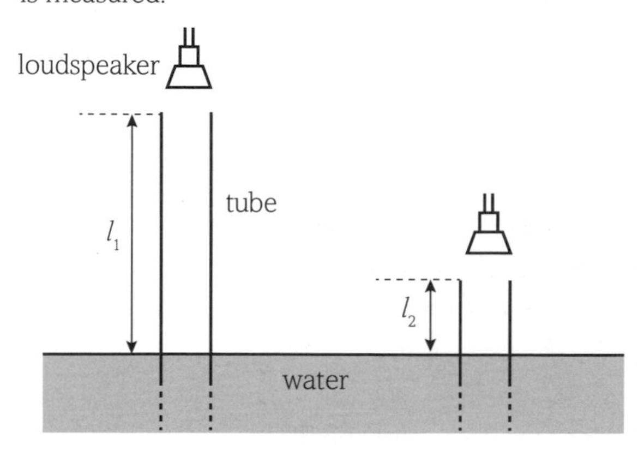

Figure 5

(a) A student obtained following results in the experiment.

Frequency of sound/Hz	Length l_1/m	Length l_2/m
500	0.506	0.170

Use data from the table to calculate the speed of sound in the pipe. Show your reasoning. [4]

(b) The student repeats the experiment, but sets the frequency of the sound from the speaker at 5000 Hz.
Suggest and explain whether these results are likely to give a more or less accurate value for the speed of sound than those obtained in the first experiment. [2]

[Total: 6]

[Q6(a) and (b), H156/02 sample paper 2014]

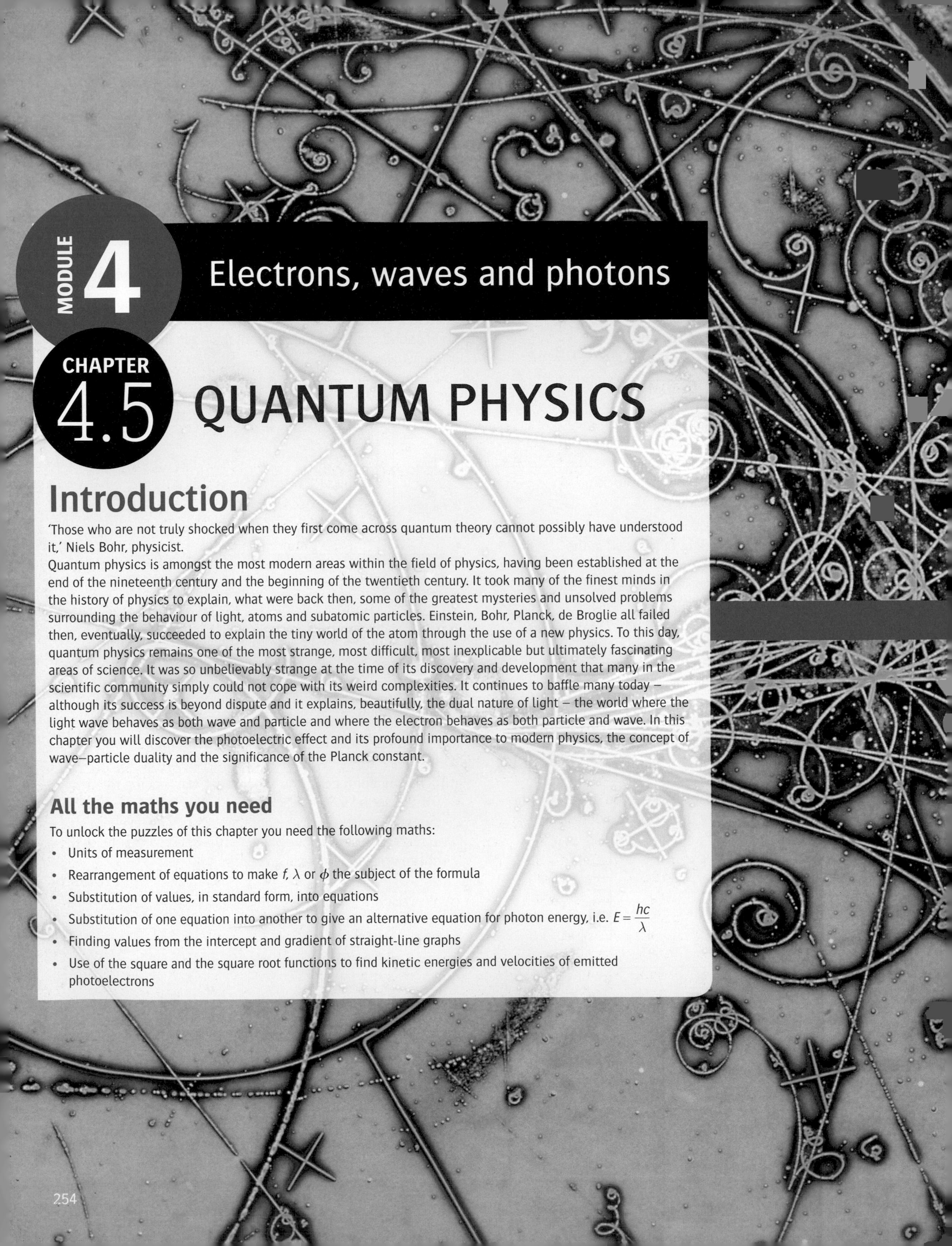

MODULE 4 Electrons, waves and photons

CHAPTER 4.5

QUANTUM PHYSICS

Introduction

'Those who are not truly shocked when they first come across quantum theory cannot possibly have understood it,' Niels Bohr, physicist.

Quantum physics is amongst the most modern areas within the field of physics, having been established at the end of the nineteenth century and the beginning of the twentieth century. It took many of the finest minds in the history of physics to explain, what were back then, some of the greatest mysteries and unsolved problems surrounding the behaviour of light, atoms and subatomic particles. Einstein, Bohr, Planck, de Broglie all failed then, eventually, succeeded to explain the tiny world of the atom through the use of a new physics. To this day, quantum physics remains one of the most strange, most difficult, most inexplicable but ultimately fascinating areas of science. It was so unbelievably strange at the time of its discovery and development that many in the scientific community simply could not cope with its weird complexities. It continues to baffle many today – although its success is beyond dispute and it explains, beautifully, the dual nature of light – the world where the light wave behaves as both wave and particle and where the electron behaves as both particle and wave. In this chapter you will discover the photoelectric effect and its profound importance to modern physics, the concept of wave–particle duality and the significance of the Planck constant.

All the maths you need

To unlock the puzzles of this chapter you need the following maths:

- Units of measurement
- Rearrangement of equations to make f, λ or ϕ the subject of the formula
- Substitution of values, in standard form, into equations
- Substitution of one equation into another to give an alternative equation for photon energy, i.e. $E = \frac{hc}{\lambda}$
- Finding values from the intercept and gradient of straight-line graphs
- Use of the square and the square root functions to find kinetic energies and velocities of emitted photoelectrons

What will I study later?

- The concept of negative energy levels of electrons in isolated gas atoms and quantum explanations of emission and absorption spectra (AL)
- The notion of quantised energy levels in atoms and the application of the equations $\Delta E = hf$ and $\Delta E = \frac{hc}{\lambda}$ to determine the frequencies and wavelengths of emitted photons (AL)
- The concept of the nuclear atom, the behaviour of subatomic particles and the strong nuclear force (AL)
- Alpha particle scattering and evidence for the small, positively charged nucleus (AL)
- The femtometre as a unit of subatomic distance (AL)
- Fundamental particles, their classification, behaviour and the laws by which they interact with one another (AL)
- The quark model of the proton and the classification of particles as hadrons and leptons (AL)
- Radioactive decay, half-life and carbon dating, nuclear fission and(AL) fusion (AL)
- The use of X-rays in medical imaging, via X-rays machines and CAT scanning (AL)
- The quantum world of medical diagnostics – medical tracers, gamma cameras and positron emission tomography (AL)

What have I studied before?

- Electromagnetic waves carry energy through space at the speed of light
- Electromagnetic waves carry energy in packets called photons
- The energy of electromagnetic waves is related to the frequency of the radiation
- Gamma rays are the most energetic electromagnetic waves and radio waves are the least energetic
- UV, X-rays and gamma rays are examples of ionising electromagnetic radiation

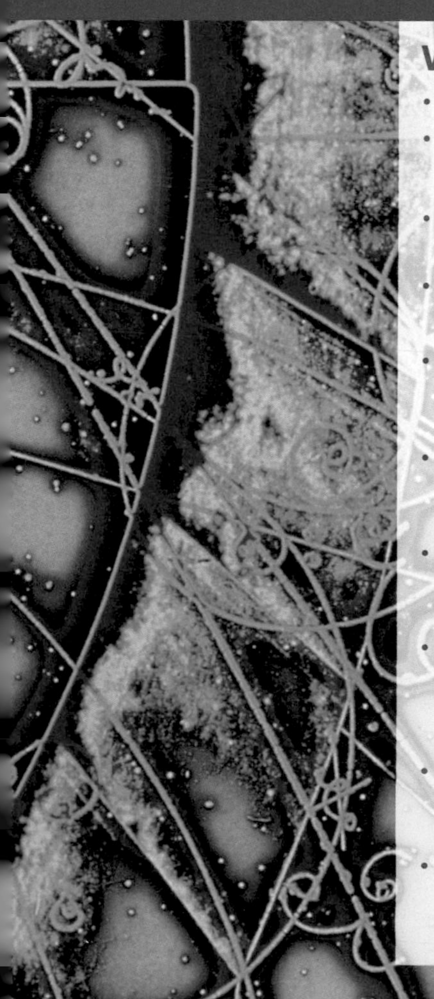

What will I study in this chapter?

- The photon is a quantum of electromagnetic energy
- The energy of a photon is calculated using the equations $E = hf$ or $E = \frac{hc}{\lambda}$
- The electronvolt is a unit of energy used to describe systems at atomic level. An energy of 1 eV is equal to 1.6×10^{-19} J
- Light emitting diodes, or LEDs, can be used to estimate a value for the Planck constant using the equation $eV = \frac{hc}{\lambda}$
- The photoelectric effect is an example of the how light behaves as a particle. It can be shown via simple laboratory equipment using a zinc plate, a UV source and an electroscope
- One photon will release one electron from the surface of a metal if the photon has sufficient energy and has a frequency above the threshold frequency of the metal
- The photoelectric effect connects the work function of the metal, the maximum kinetic energy of the emitted photoelectron and the incident photon energy via the equation $hf = \phi + \frac{1}{2}mv^2$
- The rate of emission of photoelectrons from the surface of a metal is directly proportional to the intensity of the incident radiation above the threshold frequency. The maximum kinetic of an emitted photoelectron is independent of the intensity of the incident radiation
- Electron diffraction is most pronounced when the wavelength of the electrons is similar to the interatomic spacing of the material they are being fired at, e.g. graphite
- Wave–particle duality is described, mathematically, by the de Broglie equation, $\lambda = \frac{h}{p}$, where p is the momentum of the particles, e.g. the electrons

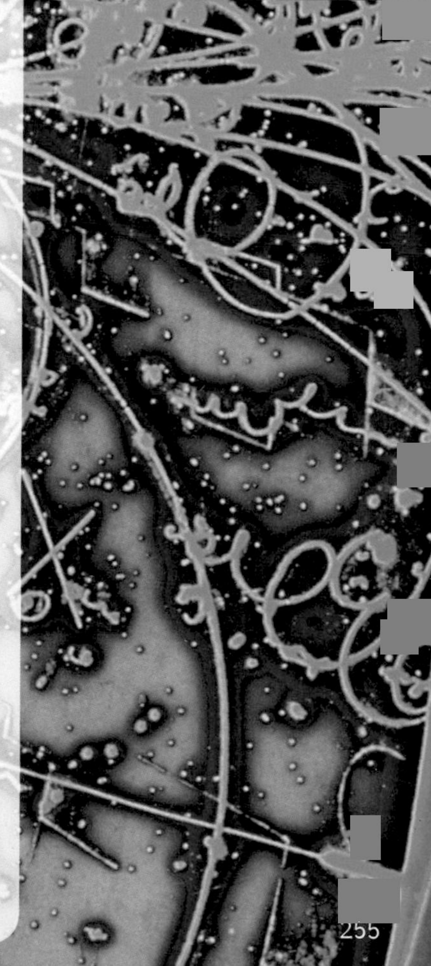

4.5 (1) The photon

By the end of this topic, you should be able to demonstrate and apply your knowledge and understanding of:

* the particulate nature (photon model) of electromagnetic radiation
* photon as a quantum of energy of electromagnetic radiation
* energy of a photon; $E = hf$ and $E = \frac{hc}{\lambda}$
* using LEDs and the equation $eV = \frac{hc}{\lambda}$ to estimate the value of Planck constant h
* determine the Planck constant using different coloured LEDs

A brief history of ideas about light

Figure 1 Isaac Newton.

Figure 2 Christiaan Huygens.

Figure 3 Thomas Young.

In 1672, the English physicist Isaac Newton suggested that light was composed of a stream of tiny 'corpuscles' (particles), moving in straight lines from a light source. He supported this theory with his own laws of motion and showed that both reflection and refraction could be explained in terms of these particles either bouncing off a surface or travelling more quickly as they move from a less dense to a more dense medium.

Conversely, the Dutch physicist Christiaan Huygens suggested in 1678 that light was a wave, like sound. His wave theory could also explain refraction and reflection. Unlike Newton's corpuscular theory, Huygens' theory predicted that a light wave would *slow down* when it passed from a less dense to a more dense medium, such as from air into water. This was later found to be correct. Huygens' theory was also used by others to predict that light would diffract if it passed through a small gap or around very small objects, and that interference would then occur.

Although most scientists of that time dismissed Huygens' theory, in 1801 Thomas Young's double slit experiment showed that light did indeed behave like a wave. By passing a coherent source of light through a surface with two small slits, Young showed that the light diffracted, overlapped and interfered to produce bright and dark 'fringes' of light – something that could only be explained by a wave theory of light (see Topic 4.1.10).

The quantum model

In 1900, Max Planck's work on the electromagnetic radiation emitted by hot objects, such as stars or a red-hot glowing lump of metal, made him realise that the radiated energy could not have continuous values. Instead Planck hypothesised that the total radiated energy had to be emitted in packets, with each packet or **quantum** having a definite, fixed amount of energy.

The photon model

At the time, Planck and other physicists did not realise the full implications of the quantum idea. However in 1905 Einstein used this idea to show that atoms absorb and emit light in individual packets of energy. As we shall see later, Einstein's explanation opened up the paradox that light sometimes behaves in ways that can only be explained by thinking of it as consisting of particles. The French aristocrat and physicist Louis de Broglie showed that light could behave as *both* a particle and a wave. Initially he was ridiculed and his PhD thesis proposing this was rejected, but later this idea was shown to be valid. Quanta of light or other electromagnetic energy became known as **photons**.

The energy of a single photon is given by the equation

$$E = hf$$

where E is the photon energy in J, h is the **Planck constant** and f is the frequency of the radiation in Hz. The Planck constant, h, has a value of 6.626×10^{-34} J s. The energy of a photon is directly proportional to its frequency, and photon energies are always emitted in whole number multiples:

$$E = nhf \; (n = 0, 1, 2, 3 \ldots)$$

KEY DEFINITIONS

A **quantum** (plural **quanta**) is a small discrete unit of energy.
A **photon** is a quantum associated with electromagnetic radiation.
The **Planck constant**, h, has a value of 6.626×10^{-34} J s – photon energies are always emitted in multiples of this.

The equation relating the speed of light, c, the frequency of a wave, f, and its wavelength, λ, is

wave speed = frequency × wavelength or $c = f\lambda$

Rearranging this equation and substituting into $E = hf$ gives an alternative equation for calculating the energy of a photon:

$$E = \frac{hc}{\lambda}$$

To calculate the photon energy, we use:

- $E = hf$ if we know the frequency of the radiation
- $E = hc/\lambda$ if we know the wavelength of the radiation.

LEARNING TIP

The energy of an individual photon depends on a single frequency. However, the electromagnetic spectrum is the range of all possible frequencies of electromagnetic radiation. Not all photons of visible light have the same energy. The total energy transferred by an electromagnetic wave is equal to the sum of the energies of the photons emitted by the light source.

WORKED EXAMPLE 1

Calculate the energy of a photon with:
(a) a frequency of 4.6×10^{18} Hz
(b) a wavelength of 4.2×10^{-8} m
The speed of light is 3×10^{8} m s^{-1}.

Answers

(a) $E = hf$

$= 6.626 \times 10^{-34} \times 4.6 \times 10^{18} = 3.0 \times 10^{-15}$ J

(b) We have the value for the wavelength so we use:

$$E = \frac{hc}{\lambda}$$

$$= \frac{(6.626 \times 10^{-34} \times 3 \times 10^{8})}{4.2 \times 10^{-8}} = 4.7 \times 10^{-18}\ \text{J}$$

WORKED EXAMPLE 2

A monochromatic source of red light has a power of 12 W and a wavelength of 720 nm. Calculate:
(a) the frequency of light with this wavelength
(b) the energy of light with this wavelength
(c) the number of photons emitted per second.

Answers

(a) To calculate the frequency of the red light we use $c = f\lambda$ after converting nm to m and rearranging:

$$f = \frac{3 \times 10^{8}\ \text{m s}^{-1}}{7.2 \times 10^{-7}\ \text{m}}$$

$= 4.2 \times 10^{14}$ Hz

(b) To calculate the energy of the light we us either $E = \frac{hc}{\lambda}$ or $E = hf$.
Substituting the values into either gives a value of 2.8×10^{-19} J.

(c) The energy of one photon is 2.8×10^{-19} J. A source of light with a power of 12 W provides 12 J of light energy each second. So the number of photons emitted each second can be calculated by dividing the total energy by the energy of one photon:

$$\frac{12}{2.8 \times 10^{-22}} = 4.3 \times 10^{19}$$

INVESTIGATION

Using light-emitting diodes to determine a value for the Planck constant

The light emitted by LEDs comes in a range of different colours. Because the colour or wavelength of the light being emitted is related to photon energy, we can use different LEDs to determine a value for the Planck constant. We need the following equipment:

- four LEDs of different colours – red, orange, green and blue, of known wavelengths
- a 6 V DC power supply
- an ammeter and a voltmeter
- a 1 kΩ variable resistor.

We set up the apparatus as shown in Figure 4.

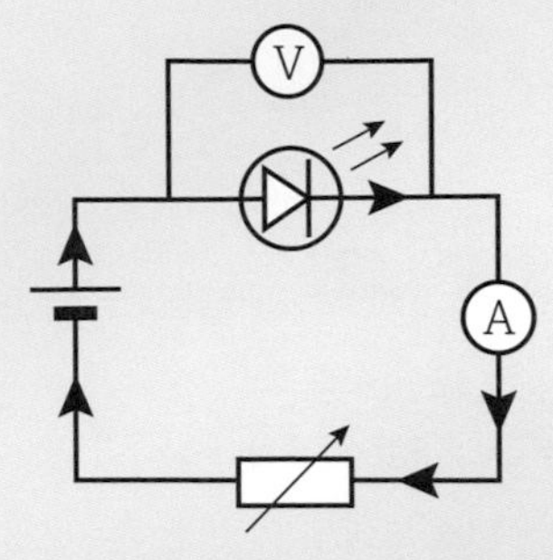

Figure 4 Determining the Planck constant.

We then measure the current flowing as we increase the p.d. across the LED by adjusting the variable resistor. Be aware that some current may be registered by the microammeter even before the LED has emitted light. Obtain at least six values for V and I, in steps of 0.05 V over the range where the LED starts to emit light.

For each LED, we plot a graph of current against voltage,

We then determine the activation voltage from each graph – this is the maximum voltage at which the current begins to increase linearly and the LED starts to emit photons. Monitoring current is a more accurate method of determining when the LED starts to emit light than visually looking for first emission from the LED. The activation voltage can be read from the graph by extrapolating the straight line until it touches the x-axis. Repeat this measurement several times for each LED to get a repeatable result. Typical values are shown in the table:

LED colour	Typical wavelength/nm	Activation voltage/V
red	625	1.77
orange	586	1.90
green	568	2.00
blue	468	2.44

Table 1

Plotting a graph of activation voltage against $\frac{1}{\text{wavelength}}$ gives a graph like that shown in Figure 5.

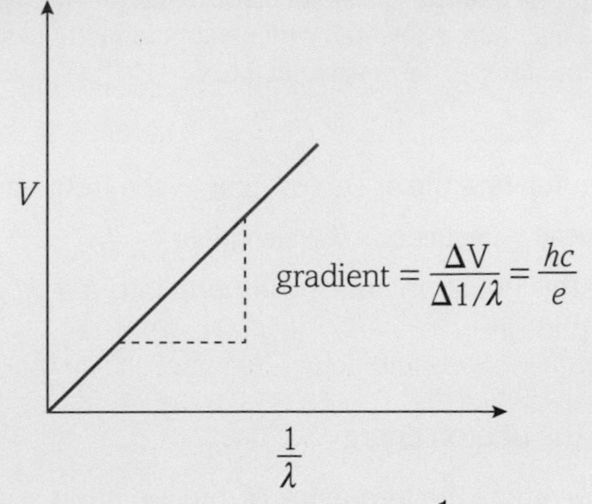

Figure 5 Graph of activation voltage against $\frac{1}{\text{wavelength}}$.

The energy of a photon is equal to the energy of the electrons that are excited in the semiconductor material of the LED:

$$eV = \frac{hc}{\lambda}$$

which can be rearranged:

$$V = \frac{hc}{e}\frac{1}{\lambda}$$

Comparing this to the form of a straight line graph ($y = mx + c$), we can see that h is the gradient of the graph multiplied by a constant, $\frac{e}{c}$.

If we measure the gradient of the graph of activation voltage against $\frac{1}{\lambda}$, and then multiply it by the charge on an electron and divide it by the speed of light then we will get a value for h.

Questions

1. How have the ideas about the behaviour of light changed over the last 400 years?
2. What equations are used to find the energy of a photon?
3. Orange light has a wavelength of 600 nm. Calculate:
   (a) the frequency of orange light
   (b) the energy of a photon of orange light.
4. Calculate the photon energy of:
   (a) radio waves of frequency 5×10^5 Hz
   (b) X-rays of frequency 2.8×10^{18} Hz
   (c) microwaves of wavelength 5 m
   (d) infrared light of wavelength 4.8×10^{-6} m
5. How many photons are emitted per second from a UV source with a power of 24 W and a frequency of 840 terahertz?
6. Use Table 1 and the graph in Figure 6 to obtain a value for the Planck constant.

2 The electronvolt

By the end of this topic, you should be able to demonstrate and apply your knowledge and understanding of:

- the electronvolt (eV) as a unit of energy

Introduction

The energies associated with individual photons are very low, as can be seen in the worked examples in the previous topic. When expressed in joules, these very small numbers require an answer expressed in standard form with a large negative power. A more convenient unit often used by scientists is called the **electronvolt**.

KEY DEFINITION

The **electronvolt** is defined as the kinetic energy gained by an electron when it is accelerated through a potential difference of 1 volt.

(a)

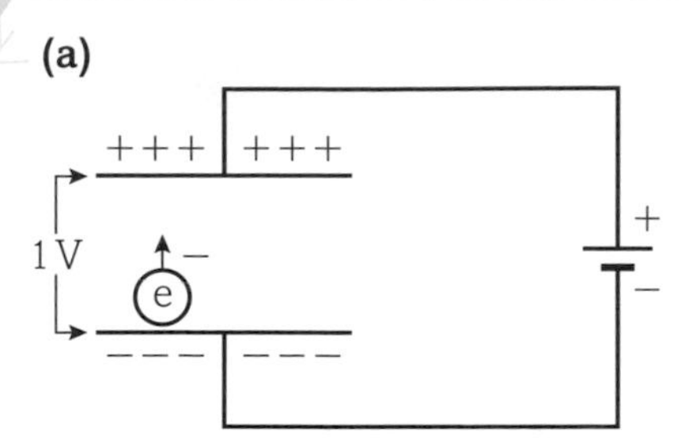

(b)

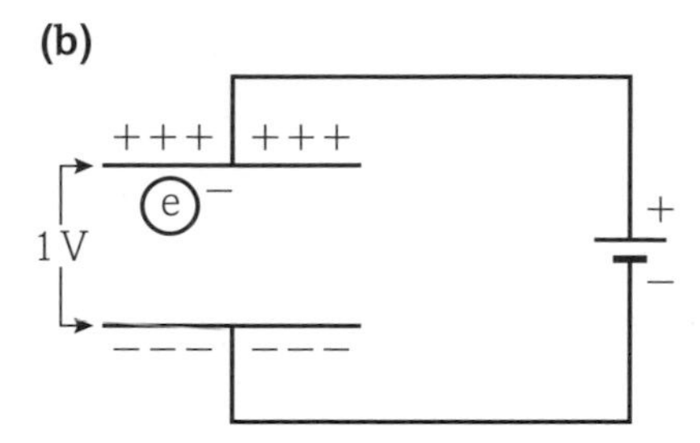

Figure 1 The kinetic energy of an electron will increase by 1 eV when it is accelerated across a potential difference of 1 V. (a) Shows an electron leaving the negative plate. (b) Shows the electron arriving at the positive plate having gained 1 eV of kinetic energy.

We calculate the energy gained by an electron (in Figure 1) using:

energy gained = charge × potential difference, or $E = QV$

The charge on an electron is 1.602×10^{-19} C and the potential difference is 1 V, so:

$E = 1.602 \times 10^{-19}\,\text{C} \times 1\,\text{V} = 1.602 \times 10^{-19}\,\text{J}$

This means that 1 eV is equal to 1.602×10^{-19} J

If we increase the particle's charge or the potential difference between the plates, then the kinetic energy acquired by the particle will increase in direct proportion.

So, if an electron of charge *e* is accelerated through a potential difference of 100 V it will gain 100 eV of kinetic energy.

(a)

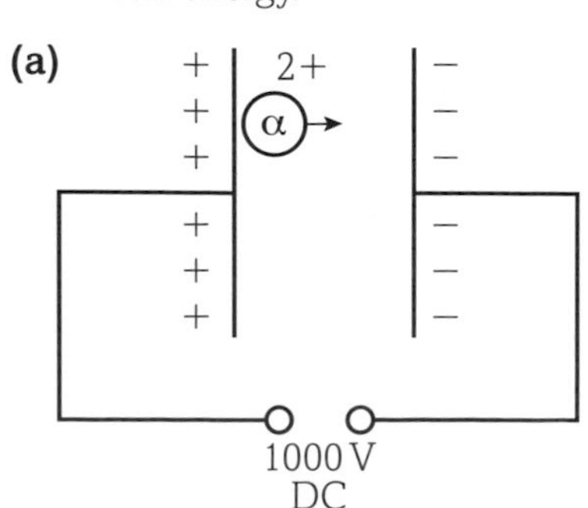

(b)

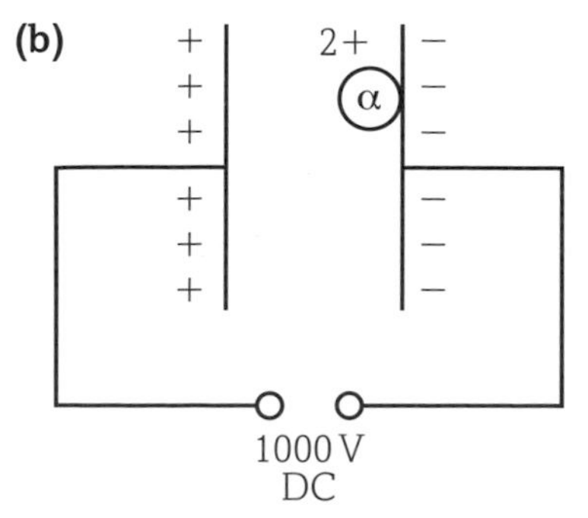

Figure 2 When an alpha particle, of charge +2*e*, is accelerated by a potential difference of 1000 V, it will increase its kinetic energy by 2000 eV.

- To convert from electronvolts (eV) to joules (J) multiply by 1.602×10^{-19}
- To convert from joules (J) to electronvolts (eV) divide by 1.602×10^{-19}

LEARNING TIP

The unit of the electronvolt is a clue to how it is calculated – eV is the unit, and $e \times V$ is the calculation to perform. For example, if a question asks how many electronvolts an electron gains when accelerated through 200 V then the answer is $e \times 200\ \text{V} = 200\ \text{eV}$. To obtain an answer in joules, simply use the absolute value of the charge in coulomb, instead of e, which for an electron is 1.602×10^{-19} C. The answer in joules would then be $1.602 \times 10^{-19}\ \text{C} \times 200\ \text{V} = 3.204 \times 10^{-17}$ J.

LEARNING TIP

Always remember to look at the prefixes and to interpret them correctly before substituting into calculations, for example a value of 5 cm is 0.05 m, a value of 500 MeV is 500 million eV or 5×10^8 eV.

WORKED EXAMPLE

(a) Convert 200 eV to joules.

(b) Convert 4.808×10^{-13} J to eV.

Answers

(a) $200\ \text{eV} = 200 \times 1.602 \times 10^{-19} = 3.204 \times 10^{-17}\ \text{J}$

(b) $\dfrac{4.808 \times 10^{-13}\ \text{J}}{1.602 \times 10^{-19}\ \text{J}} = 3 \times 10^6\ \text{eV} = 3\ \text{MeV}$

Questions

1. Convert the following from electronvolts (eV) into joules (J).
   (a) 5 eV (b) 8 MeV (c) 20 GeV

2. Convert the following from joules (J) into electronvolts (eV).
   (a) 3.204×10^{-13} J (b) 48 nJ (c) 1.5 J (d) 100 J

3. In each of the following cases, explain whether electronvolts or joules would be the most appropriate unit to use:
   (a) the energy required to boil a kettle of water
   (b) the energy of an electron orbiting an atomic nucleus
   (c) the energy released by a light bulb over the course of an evening
   (d) the energy absorbed by the skin when a photon of UV light hits it.

4. An electron (mass $= 9.1 \times 10^{-31}$ kg) and an alpha particle (mass $= 6.6 \times 10^{-27}$ kg) are both accelerated through a potential difference of 450 V. Find:
   (a) the kinetic energy gained by each of the particles in eV
   (b) the kinetic energy gained by each of the particles in J
   (c) the increase in velocity of each particle in m s^{-1}.

3 The photoelectric effect 1

By the end of this topic, you should be able to demonstrate and apply your knowledge and understanding of:

* photoelectric effect, including a simple experiment to demonstrate this effect
* demonstration of the photoelectric effect using, e.g. gold-leaf electroscope and zinc plate
* a one-to-one interaction between a photon and a surface electron
* work function; threshold frequency
* Einstein's photoelectric equation $hf = \varphi + KE_{max}$

Introduction

When electromagnetic radiation of a particular frequency is shone on the surface of a metal, electrons are emitted from its surface. This phenomenon is known as the *photoelectric effect* and the electrons that are released are called *photoelectrons*.

Demonstrating the photoelectric effect with a gold-leaf electroscope

The gold-leaf electroscope is composed of a brass stem to which a thin gold leaf is attached. There is a metal cap attached to the top of the stem and the metal to be irradiated with electromagnetic radiation is laid on the metal cap (Figure 1).

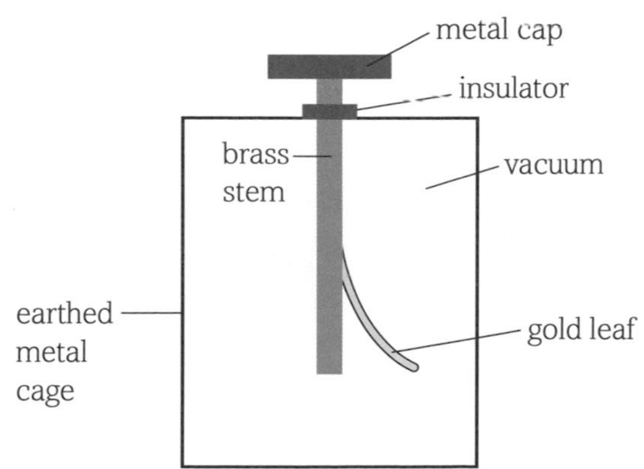

Figure 1 The photoelectric effect can be demonstrated using a gold-leaf electroscope.

A metal plate (usually zinc) is placed on the metal and is then charged negatively by touching it with a negatively charged polythene rod, or by electrostatic induction. When this is done, the metal stem and gold leaf will also become negatively charged, meaning that the stem and the leaf will repel each other.

It is also possible to make the zinc plate, metal stem and the gold leaf positively charged.

Observations of the photoelectric effect

Figure 2 shows how the photoelectric effect is investigated for two different sources of radiation – visible light from a standard desktop lamp and UV light from a UV light source.

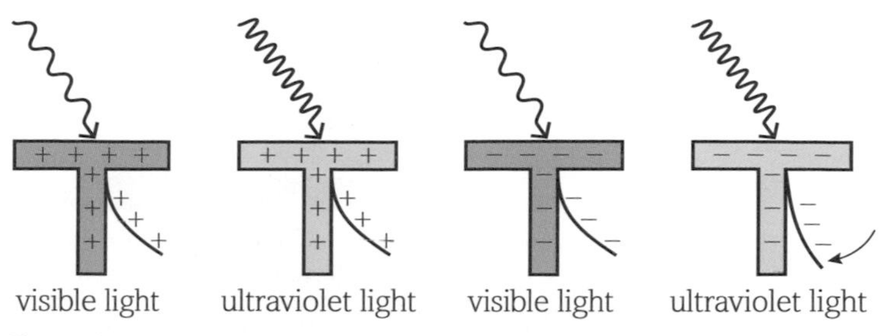

Figure 2

When visible light is incident on a positively charged metal plate, there is no movement of the gold leaf. The same is also true when UV light is shone on the positively charged plate.

Shining visible light on the negatively charged zinc plate also causes no movement of the gold leaf. No matter how bright or intense the visible light beam is, no movement of the gold leaf is seen.

However, when UV light is shone on the negatively charged zinc plate, the gold leaf falls.

This shows that the metal plate loses its negative charge through the emission of electrons, which are repelled by the negative charge on the electroscope. The discharge of the electroscope cannot be caused by ions in the air, because the electroscope is in a sealed vacuum.

Further experiments show that:

- Electrons will be emitted from the surface of a metal only if the incident radiation is above a minimum frequency called the **threshold frequency**. This is determined by the metal itself and is different for different metals. Below the threshold frequency, no electrons are emitted.
- Emission of electrons starts the instant the surface starts to be irradiated, provided that the incident radiation exceeds the threshold frequency.

KEY DEFINITION

The **threshold frequency** is the lowest frequency of radiation that will result in the emission of electrons from a particular metal surface. For most metals, this frequency occurs in the ultraviolet region of the electromagnetic spectrum.

Explaining the photoelectric effect

From around 1887, scientists noticed that photoelectrons could be released from the surface of metals by UV light but they could not explain why visible light would not cause electrons to be emitted, even for high light intensities. It wasn't until 1905 that Albert Einstein suggested that the incident radiation was not behaving as a wave but as packets of energy that arrive with fixed energy values depending on the frequency of the light. These packets of energy are called *photons.*

The work function

Einstein's explanation for the photoelectric effect was that in order for electrons to be released from a metal, the frequency of the incident radiation must exceed the threshold frequency for that metal. This is needed to provide at least the minimum energy required to release an electron from the surface. We call this minimum energy the **work function** of the metal.

> **KEY DEFINITION**
>
> The **work function** of a metal is the minimum energy required to release an electron from its surface, overcoming the electrostatic attraction between the electron and the positive metal ions.

Figure 3 can be interpreted like this:

1. The incident radiation is below the threshold frequency – this means the photons do not have enough energy to overcome the work function of the metal.
2. The incident radiation has a frequency equal to the threshold frequency – this will cause electrons to be omitted from the metal surface, but with zero kinetic energy.
3. The incident radiation is above the threshold frequency – this will cause electrons to be ejected with some kinetic energy. Increasing the frequency further increases the kinetic energy of the photoelectrons.
4. Increasing the incident wave intensity (of a single frequency that is above the threshold frequency) ejects more electrons from the metal surface per second, but it will not affect their kinetic energy.

Einstein's photoelectric equation

Einstein also suggested that each single photon could only eject one electron from the metal surface – either the photon energy was larger than the work function of the metal and the electron would be released with some kinetic energy; or the photon would not have enough energy and the emitted electron would stay on the metal surface. Applying the principle of conservation of energy, Einstein suggested an equation to explain the photoelectric effect:

$$hf = \varphi + KE_{max}$$

where:

hf = the photon energy of an incident photon and h = Planck's constant

φ = the work function of the metal

KE_{max} = the maximum kinetic energy of an electron once it has been ejected from the surface of the metal.

This equation explains why there is a threshold frequency for emission of photoelectrons, and shows that light can behave like a stream of particles. The photoelectric effect cannot be explained using a wave model since in the wave model, the energy of a wave depends on its amplitude (intensity) not its frequency.

Einstein's detailed explanation of the photoelectric effect led to him winning the Nobel Prize for physics in 1921.

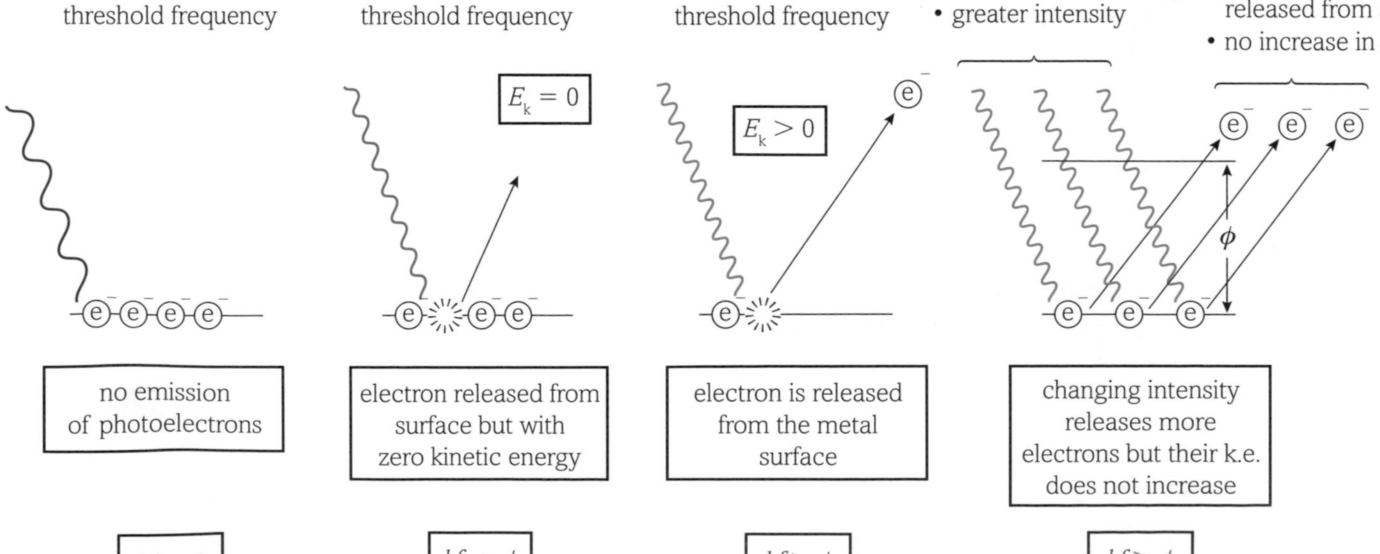

Figure 3 Changing the conditions for photoelectron emission.

WORKED EXAMPLE 1

A monochromatic source of UV light has a frequency of 5×10^{16} Hz. The work function of zinc is 3.6 eV. Calculate:

(a) the photon energy of the incident UV source, in J

(b) the value of the work function of zinc, in J

(c) the maximum kinetic energy of the emitted photoelectrons.

Answers

(a) $E = hf$

$E = 6.626 \times 10^{-34}\ \text{J s} \times 5 \times 10^{16}\ \text{Hz}$

$= 3.3 \times 10^{-17}\ \text{J}$

(b) To convert from electronvolts to joules, you multiply the eV value by 1.602×10^{-19}.
So a work function of 3.6 eV is equal to $3.6 \times 1.602 \times 10^{-19} = 5.8 \times 10^{-19}\ \text{J}$

(c) The photoelectric equation is $hf = \varphi + \frac{1}{2}mv^2$

Rearranging this:

$\frac{1}{2}mv^2 = hf - \varphi$

$= 3.3 \times 10^{-17} - 5.8 \times 10^{-19}$

$= 3.2 \times 10^{-17}\ \text{J}$

WORKED EXAMPLE 2

Calculate the threshold frequency for copper.

Answer

Copper has a work function of 4.6 eV, which is equal to 7.4×10^{-19} J.
The photoelectric equation is $hf = \varphi + \frac{1}{2}mv^2$, and at the threshold frequency, the kinetic energy of the photoelectrons on the verge of being emitted is zero, leading to $h \times f_0 = \varphi$ and:

$$f_0 = \frac{\varphi}{h}$$

$$= \frac{(7.4 \times 10^{-19})}{(6.626 \times 10^{-34})} = 1.1 \times 10^{15}\ \text{Hz}$$

A frequency higher than this will release electrons from the surface of copper.

Questions

1 Explain what is meant by the following terms:

(a) photoelectrons

(b) threshold frequency

(c) work function

(d) photons.

2 Explain why neither visible light nor ultraviolet light discharges a positively charged electroscope.

3 UV light is shone on a negatively charged zinc plate fitted to the top of a gold leaf electroscope, but the leaf does not move. Explain why the leaf will move if the zinc surface is first cleaned with sandpaper.

4 What effect will the following have on the kinetic energy of an emitted photoelectron:

(a) increasing the frequency of the incident radiation

(b) increasing the intensity of the incident radiation

(c) using a metal with a lower work function

(d) using X-rays instead of UV light?

5 The work function of sodium is 2.4 eV.

(a) What is the threshold frequency for photoelectric emission to take place?

(b) Calculate the frequency of the electromagnetic radiation that will release an electron with a maximum kinetic energy of 2.6 eV.

6 When light with a wavelength of 170 nm irradiates a metallic surface, electrons are emitted with kinetic energies ranging up to 9.6×10^{-19} J.

(a) What is the work function of the metal?

(b) What is the threshold frequency?

The photoelectric effect 2

By the end of this topic, you should be able to demonstrate and apply your knowledge and understanding of:

* **the idea that the maximum kinetic energy of the photoelectrons is independent of the intensity of the incident radiation**
* **the idea that rate of emission of photoelectrons above the threshold frequency is directly proportional to the intensity of the incident radiation**

Introduction

By 1902, many things about the release of photoelectrons from the surface of metals by UV light could not be disputed. These included:

- Electrons are emitted from metals that are exposed to UV light.
- The number of photoelectrons emitted increases as the light intensity increases.
- The kinetic energy, and hence velocity, of the emitted photoelectrons does not depend on the intensity of the incident light ray but it does depend on the frequency of the radiation.
- Below a certain frequency of incident radiation, called the *threshold frequency*, no electrons are emitted from the surface of the metal.

One unusual observation was that the kinetic energy of the emitted photoelectrons did not decrease when the intensity of the ultraviolet light was reduced. This meant that the kinetic energy of the electrons did not depend on the intensity of the light.

Determining the maximum kinetic energy of photoelectrons

Figure 1 shows a photocell circuit used to investigate the energy of the emitted electrons. The frequency of the incident radiation is above the threshold frequency for this metal. The photocell is enclosed in a vacuum so that there are no collisions between the electrons and gas molecules in the air.

As shown in Figure 1, plate A is connected to the positive terminal of the power supply. The p.d. across the photocell is initially zero. The photoelectrons that are emitted from A due to the incident UV radiation cross the gap and this will register as a current on the microammeter.

If the potential divider is adjusted, then the potential difference between A and B is increased and electrons emitted from A are attracted back to A. The current decreases until, eventually, even the most energetic photoelectrons are prevented from leaving plate A and the microammeter reading is zero (Figure 2). The p.d. at which this occurs is called the *stopping potential*.

We can equate the work done to accelerate or decelerate charged particles with the kinetic energy transferred to the electrons, giving:

$$eV = \frac{1}{2}mv^2$$

At the stopping potential, all the emitted electrons have been brought to rest, so we obtain a value for the maximum kinetic energy of the photoelectrons by multiplying the stopping potential (V_0) by the charge on the electron (e)

$$eV_0 = KE_{max}$$

So, if the stopping potential is found to be 1.83 V, the maximum kinetic energy of the photoelectrons must be 1.83 eV or 2.93×10^{-19} J.

Figure 3 shows that the stopping potential does not depend on the intensity of the incident light.

Rate of emission of photoelectrons

The circuit in Figure 1 can also be used to investigate the rate of emission of photoelectrons. The current in the circuit depends on the number of photoelectrons emitted each second. For incident radiation of a single frequency above the threshold frequency, observations show that doubling the light intensity doubles the number of electrons emitted, but does not affect the energies of the emitted electrons.

- The number of emitted photoelectrons is directly proportional to the intensity of the incident radiation, provided that the radiation is of a single frequency and is above the threshold frequency.
- The kinetic energy of the emitted photoelectrons is not affected by the intensity of the incident radiation.
- The kinetic energy of the emitted photoelectrons is affected by the frequency of the incident photons. If the frequency of the incident radiation is increased, the kinetic energy of the photoelectrons also increases.

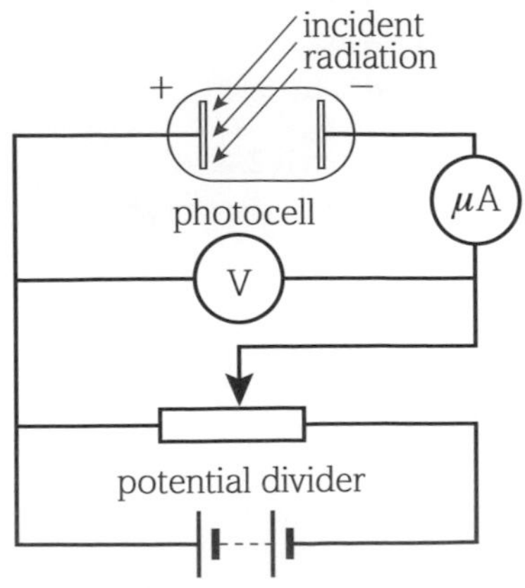

Figure 1 Measuring electron kinetic energy.

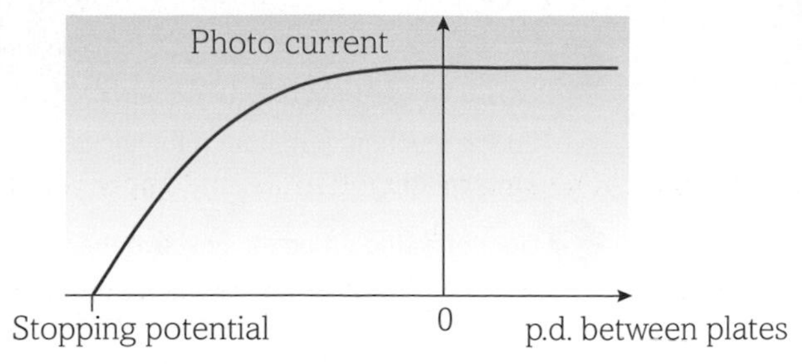

Figure 2 How the photoelectric current depends on the p.d. between the plates shown in Figure 1.

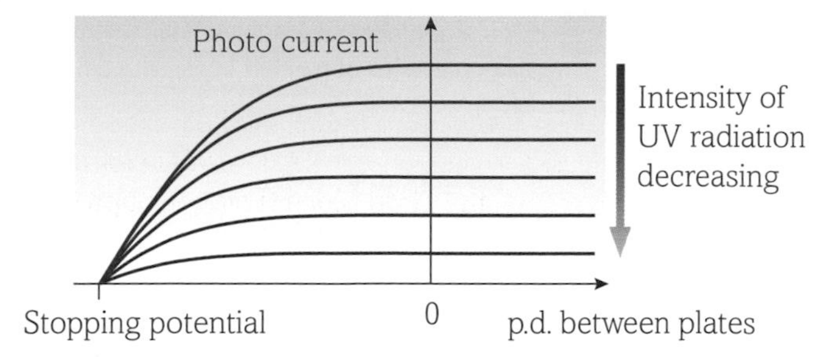

Figure 3

Einstein's quantum explanation can account for all these observations. A single photon interacts with a single surface electron so increasing the light's intensity (i.e. the number of incident photons/second) only increases the number of released electrons and not their kinetic energy. However, once a photon has enough energy to overcome the work function any excess energy is transferred to the electron as kinetic energy, so kinetic energy increases with the frequency of the incident photons.

Millikan's photoelectric experiment

The photoelectric effect was explained by Albert Einstein in 1905.

It was in 1916 that the American physicist Robert Millikan carried out a series of very detailed and accurate experiments that would completely verify Einstein's explanation.

Millikan irradiated the metals sodium, potassium and lithium with monochromatic light. By applying a positive potential to the target metal, Millikan could decelerate the electrons. He increased the size of the potential difference until the most energetic electrons were unable to reach the cathode, causing the current to fall to zero. This enabled him to determine the stopping potential, V_s, of each metal as described above, and show that this depends on the frequency of the radiation as predicted by Einstein.

Not only did Millikan's experimental work obtain accurate values for the work functions of the metals, he also managed to obtain an accurate value for Planck's constant.

We know that $eV_s = KE_{max}$ and that $hf = \varphi + KE_{max}$

Substitution leads to:

$$hf = \varphi + eV_s$$

If we divide both sides by the electron charge, e, and rearrange:

$$V_s = \left(\frac{h}{e}\right)f - \frac{\varphi}{e}$$

This equation is of the form $y = mx + c$, meaning that if we plot a graph of stopping potential (V_s) against frequency (f), we will get a graph of gradient $\frac{h}{e}$ and a y-intercept equal to $-\frac{\varphi}{e}$. If we measure the gradient and multiply it by e we get a value for Planck's constant; if we read the y-intercept and multiply it by $-e$ we will get a value for the work function of the metal.

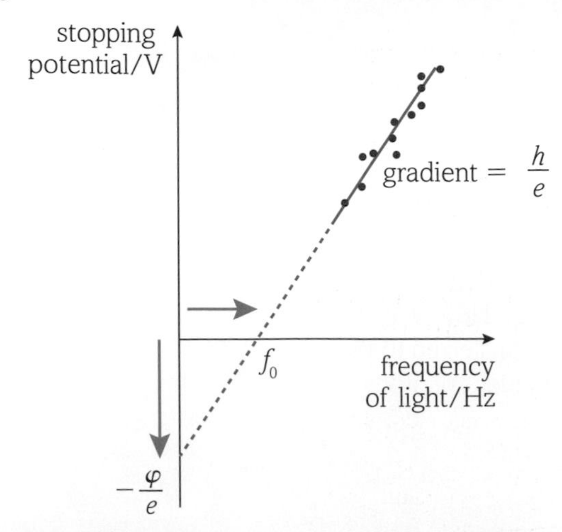

Figure 4 Millikan's graph.

Questions

1 Use Table 1 to find:
(a) the threshold frequency of aluminium
(b) the maximum kinetic energy of photoelectrons that will be emitted from the surface of calcium when exposed to monochromatic UV radiation of frequency 4×10^{15} Hz.

Metal	Work function/eV
sodium	2.4
potassium	2.3
caesium	1.9
barium	2.5
calcium	2.9
zirconium	4.1
magnesium	3.7
aluminium	4.2
copper	4.6
silver	4.6
zinc	3.6

Table 1 The values of work functions for a variety of metals.

2 Use Table 1 to find:
(a) the threshold frequency of zirconium
(b) the stopping potential if the incident light has a wavelength of 580 nm.

3 Light of a single frequency above the threshold frequency falls on a photocell and the potential difference is increased until the current just stops. Describe and explain how the stopping potential will change if the wavelength of the incident monochromatic light is decreased without changing its intensity.

4 If you plot graphs of V against f for any metal, as shown in Figure 5, you get a series of parallel lines, with different y-intercepts. Explain why.

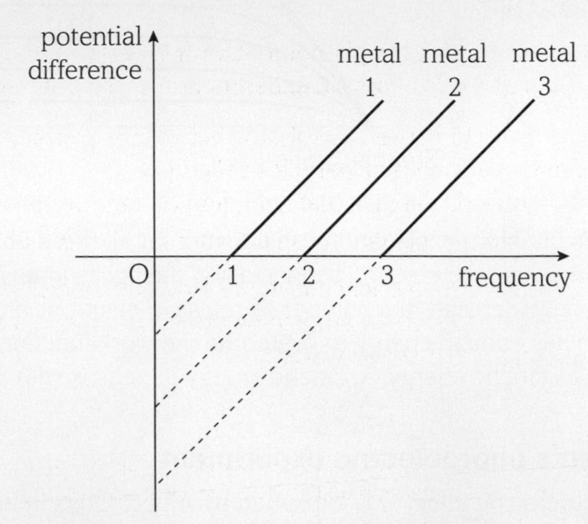

Figure 5 Graphs of V against f for these different metals.

5 Wave–particle duality

By the end of this topic, you should be able to demonstrate and apply your knowledge and understanding of:

- electron diffraction, including experimental evidence of this effect
- diffraction of electrons travelling through a thin slice of polycrystalline graphite by the atoms of graphite and the spacing between the atoms
- the de Broglie equation $\lambda = \frac{h}{p}$

Introduction

In Chapter 4.4 you met the phenomena of interference and diffraction. Interference and diffraction effects – such as the Young's slit experiment – could only be explained by the wave model of light.

However, in order to explain the photoelectric effect experiment, Einstein proposed that light waves could behave as a stream of particles called photons. The thinking in the scientific community was now shifting towards the idea that light, and other types of electromagnetic radiation, could behave as a wave but also as a stream of particles. This is called *wave–particle duality*.

The concept of wave–particle duality for light soon led other scientists to thinking of a wave–particle dual nature for matter. In the strange world of quantum physics, waves can behave like particles – can particles of matter behave as waves?

LEARNING TIP

There are two models that we can use to explain the behaviour of radiation – the **wave model** and the **particle model**. Sometimes the particle model provides the best description, such as in the case of the photoelectric effect. On other occasions, such as when we encounter diffraction and interference, the wave model is better. Reflection is an example of where both models work well.

What is the evidence to support wave–particle duality?

In 1923, the French physicist Louis de Broglie suggested something that, at the time, sounded totally preposterous. He claimed that all matter, regardless of its mass, could display wave-like properties. Evidence for this suggestion was provided by an experiment which showed that electrons showed wave-like behaviour as well as behaving like particles.

To prove that particles can also act as waves, you have to show the particles exhibiting a wave-like characteristic or property, such as diffraction or interference. For example, electrons can be diffracted just as light can be – a wave property. The apparatus used for such an experiment is shown in Figure 1. As we shall see below, the wavelength of electrons is much smaller than that of light. The slit spacing of a diffraction grating is very large in comparison with an electron's wavelength, so the spacing between atoms – which is very similar to the electron's wavelength – is used instead.

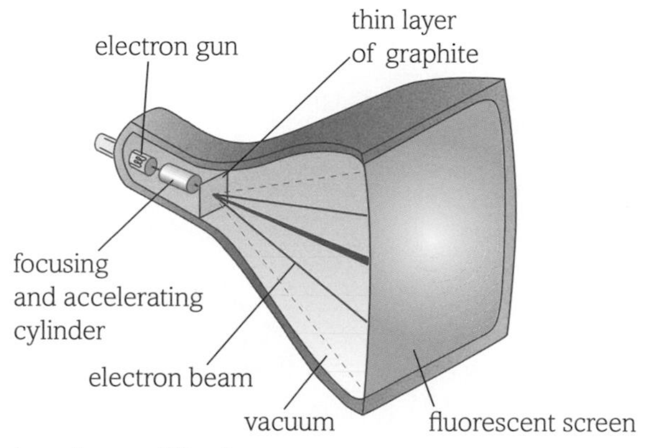

Figure 1 Apparatus used to show electron diffraction.

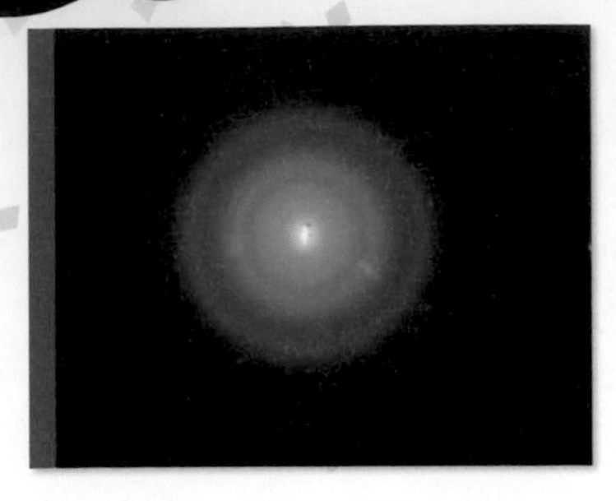

Figure 2 An electron diffraction pattern.

Electrons from an electron gun are accelerated through a vacuum towards a layer of polycrystalline graphite. A polycrystalline material is made up of many tiny crystals, each consisting of a large number of regularly arranged atoms. The electrons diffract as they emerge from the gaps between the atomic layers in the graphite film, and interfere constructively. Since the graphite atoms are not all lined up in the same direction as in a diffraction grating, this gives a circular pattern instead of the parallel lines seen when light diffracts. An electron diffraction pattern is shown in Figure 2. The image seen on the fluorescent screen is created when electrons striking the screen cause light to be emitted.

The wavelength of electrons

De Broglie proposed that if electrons and other particles travel through space as a wave, they have an associated wavelength. By combining the idea of an energy quantum with $E = mc^2$ he derived a formula for the wavelength, λ of a particle:

$$\lambda = \frac{h}{mv}$$

where m is the mass of the particle and h is the Planck constant. This is known as the *de Broglie equation* – mv is momentum which is often given the symbol p, so this equation is also expressed as

$$\lambda = \frac{h}{p}$$

In 1927, two American physicists called Davisson and Germer confirmed de Broglie's equation by observing the behaviour of electrons that had been diffracted from the surface of a nickel crystal. By accelerating electrons, of charge e, through a potential difference of V, they observed a pattern of electron diffraction (Figure 2), from which they could calculate the electron's wavelength.

The predicted wavelength is found by equating the work done to accelerate the electrons with the kinetic energy transferred to the electrons. Substituting

$$eV = \frac{1}{2}mv^2$$

into the de Broglie equation gives:

$$\lambda = \frac{h}{\sqrt{(2mVe)}}$$

Davisson and Germer found that their values for the wavelength of electrons from measurements of the diffraction pattern were similar to those predicted by de Broglie's equation. De Broglie's equation also predicts that the wavelength of the electron can be changed by varying its velocity. As Davisson and Germer increased the accelerating voltage, the rings in the diffraction pattern got bigger showing the wavelength must be getting smaller as the electrons move faster.

WORKED EXAMPLE

1. Calculate the wavelength of an electron moving at 3×10^7 m s^{-1}. (Mass of electron = 9×10^{-31} kg)

Answer

$\lambda = \frac{h}{p}$, so:

$$\frac{h}{mv} = \frac{6.626 \times 10^{-34}}{(9 \times 10^{-31} \times 3 \times 10^7)}$$

$$= 2.46 \times 10^{-11}\text{ m} = 0.0246\text{ nm}$$

2. Calculate the wavelength of an electron accelerated through a potential difference of:
   (a) 10 kV
   (b) 300 MV.

Answers

(a) $\lambda = \frac{h}{\sqrt{(2mVe)}} = 1.16 \times 10^{-11}$ m

(b) $\lambda = \frac{h}{\sqrt{(2mVe)}} = 7.09 \times 10^{-14}$ m

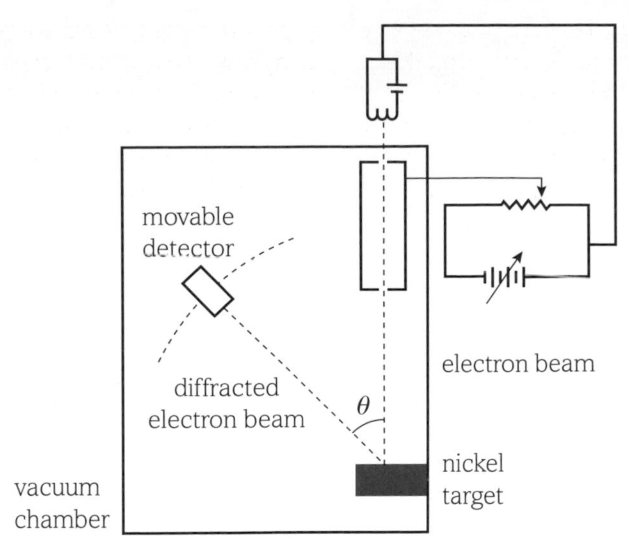

Figure 3 The arrangement used by Davisson and Germer.

Electron diffraction can be used to determine atomic spacing. It is also useful for determining other information about the structure of matter. Increasing the speed of the electrons results in a decrease in their wavelength, and so much smaller values of inter-atomic spacing can be measured. Today, high-speed electrons can be used to determine the arrangement of atoms in crystalline structures or to measure the diameter of a nucleus.

Wave–particle duality for matter

Davisson and Germer's experiment showed that matter could exhibit wave properties. According to de Broglie's equation, all moving objects have a wavelength, including solid objects such as a tennis ball or a human being. The greater the mass, the shorter the wavelength, and macroscopic objects have such tiny wavelengths that diffraction effects are negligible and they can be considered to be particles for all practical purposes. For example, a person does not diffract as they walk through a door.

In modern physics, the wave nature of a particle is considered a probability wave. It represents the probability of an observer measuring a particle at a given place and time. For example, in a diffraction pattern on a screen, the intensity is interpreted as the probability of the particle hitting the screen at a certain position. This is true for electrons as well as photons.

Questions

1. What is the difference between a photon and a wave?
2. What eveidence is there to show that light can behave as:
   (a) a particle
   (b) a wave?
3. Use the de Broglie equation, $\lambda = \frac{h}{mv}$, to find:
   (a) the wavelength of an electron travelling at 3% of the speed of light
   (b) the wavelength of a car of mass 800 g travelling at 18 m s^{-1}.
   (c) Why do we not notice the wave nature of objects such as cars?
4. Explain how the electrons in the second worked example question above can be used to investigate the structure of the atom and the nucleus.
5. For diffraction effects to be observed, the diffracting aperture must be of the order of the wavelength of the waves. Neutrons can also be made to behave like waves of wavelength close to the atomic spacing of atoms in crystals. This can give us information about atomic structure. Neutrons, uncharged particles of mass 1.67×10^{-27} kg, are generated in nuclear reactors and can be slowed down. At room temperature or lower, they are called thermal neutrons. At room temperature, thermal neutrons move at about 3000 m s^{-1}.
   Show that the wavelength associated with a thermal neutron is of the order of the spacing of atoms in a crystal.

SAILING BY THE SUN

There have been many space missions in the past century, fuelled by, and limited by, the power supplied by chemical rocket systems that these spacecraft had to carry for their propulsion. In this activity, you will examine how a vehicle on a space mission can be powered by light.

DOPPLER SHIFT AND ENERGY TRANSFER TO A SOLAR SAIL

The solar sail is now being tested as a practical space-craft drive. Lightweight Mylar, coated with thin aluminium, provides a useful mirror and does not present any friction problems in deep space, where few gas molecules are present. At planetary distances, a low pressure of about 10^{-5} to $10^{-4}\,N\,m^{-2}$ is provided by the momentum of photons from the Sun. It is not hard to see that after a sizable, but low mass, mirror is unfurled in space, a considerable velocity may be obtained in a few months.

If a single photon strikes a mirror at rest in space (assuming a fictitious perfect reflectivity, for the sake of argument) one would expect that it should be reflected with no frequency change. Since the energy of a photon, E, is the product of its frequency, f, and the Planck constant, h, the photon's energy should not change.

The photon's momentum is

$$p = hf/c$$

where c is the speed of light. The photon's momentum will change because of the reversal of its direction. If it strikes along a normal line, the mirror is imparted a momentum which should have a magnitude of twice that of the photon, by conservation of momentum. Therefore, the mirror, of mass m, should obtain a velocity, v, and consequently a kinetic energy of $\frac{1}{2} mv^2$. However, this violates conservation of energy if the photon suffers no energy loss.

The resolution of this paradox is that an observer measuring the reflected photon would notice that it is Doppler shifted slightly towards the red because the mirror, that is now its source, is receding very slowly away from it.

Conservation of energy predicts that the kinetic energy of the mirror is

$$\frac{1}{2} mv^2 = hf_0 - hf$$

where f_0 is the initial frequency and f is the Doppler shifted frequency of the photon. Conservation of momentum for the system is satisfied by

$$mv = hf_0/c + hf/c$$

[…]

For small changes in frequency and velocity […] therefore

$$\Delta f/f_0 = -2v_0/c$$

The resulting relative frequency change differs from the usual Doppler shift formula by a factor of two because there are two Doppler shifts – one for the photon initially absorbed and one for the re-emitted or reflected photon. It is evident that an incident photon imparts less momentum to a moving mirror than to a stationary one. […]

Source

Hirsch, W. and Kobrack, M., Doppler shift and energy transfer to a solar sail. *Physics Education* vol. 37, No. 5, pp. 422–433.

DID YOU KNOW?

It takes photons of light about 500 seconds to reach us from the surface of the Sun, but it takes them tens of thousands of years to reach the surface of the Sun from its centre. The photons are produced by fusion reactions in the core of the Sun and travel outwards through the convective zone and radiative zone of the Sun before being emitted from the Sun's surface.

Where else will I encounter these themes?

1.1 | 2.1 | 2.2 | 3.1 | 3.2 | 3.3

Let us start by considering the nature and context of the writing in the article. The article above was taken from the publication *Physics Education*, which is produced by the Institute of Physics (IOP).

1. **Consider the article and comment on the type of writing that is being used. Who is the audience? What are the words or terms in the article that would influence you when determining who the article is intended for? How might you change the article to make it more suitable for a younger or less-informed audience?**

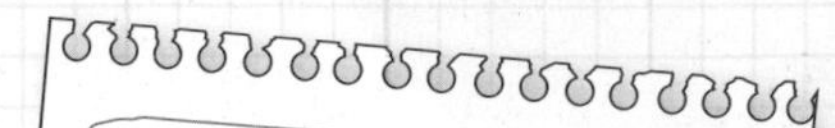

How does this article connect the theoretical and practical areas of science?

We will now look at the physics that is in the article. Do not worry if the physics content or the mathematics is challenging at this stage. You can always return to the article later in your course, once some of the related topics have been studied in more depth. Use the timeline at the bottom of the page to help you put this work in context with what you have already learned and what is ahead in your course.

2. What would be the benefits to mankind of launching such a solar sail? What are the ethical and economic issues involve with a project of this nature?
3. What would a typical momentum be of a photon of visible light from the Sun, using the equation $p = hf/c$?
4. Justify the statement 'At planetary distances, a low pressure of $10^{-5}\,\mathrm{N\,m^{-2}}$ is provided by photons from the Sun.'
5. Provide a calculation to show that, for a solar sail, that 'a considerable velocity may be obtained in a few months.'
6. Explain the significance of:
   a. 'The photon's momentum will change because of the reversal of its direction'
   b. 'This violates the conservation of energy if the photon suffers no energy loss.'
7. Explain how the second equation demonstrates the conservation of energy.
8. Explain why the third equation contains the term $+hf/c$? and not $-hf/c$.
9. What is the difference between Doppler shift and double-Doppler shift? Give examples of where each is observed in modern life.

Figure 1 A space craft powered by a solar sail.

Activity

Nearly 400 years ago, Johannes Kepler proposed the idea of exploring the galaxy using sails that would be propelled by a solar breeze. While the idea of such a breeze has been disproved, scientists believe that sunlight does indeed exert enough force to propel objects through space.

Students in your A level physics lessons are convinced that the content of this article is nonsense and that solar sail technology is impossible. How would you convince them otherwise?

You could find out about:

- the key components of a solar sail
- history of their usage and development
- how the craft LightSail-1 will get into space.

Write a summary of notes explaining how you could demonstrate that solar sail technology is in fact correct.

4.5 Practice questions

1. Which of the following statements is/are true about photons? [1]
   (i) The speed of a photon changes at the boundary between air and water.
   (ii) Photons possess a positive electrical charge.
   (iii) The energy of a photon is related to its frequency by $E = hf$.

   A (i), (ii) and (iii)
   B only (i) and (ii)
   C only (i) and (iii)
   D only (iii)

2. A photon travelling in a vacuum is found to have a wavelength of 580 nm. Which of the following are true? [1]
   (i) The photon will have a frequency of 5.2×10^{12} Hz.
   (ii) The photon will be an X-ray.
   (iii) The photon will be travelling at the speed of light.

   A (i), (ii) and (iii)
   B only (i) and (ii)
   C only (i) and (iii)
   D only (iii)

3. Electrons start to be released from the surface of a metal when UV light of wavelength 200 nm is used to illuminate it. Which of the following is true? [1]

   A The photoelectrons released from the metal surface will travel at the speed of light.
   B UV light of wavelength 210 nm will release electrons from the metal surface.
   C The work function of the metal is 5.3 eV.
   D The threshold frequency of the metal is 1.5×10^{15} Hz.

   [Total: 3]

4. A light-emitting diode (LED) emits photons of red light.
   (a) State what is meant by a *photon*. [1]
   (b) The intensity of the light emitted from the LED is doubled. Explain the effect this has on the energy of a photon. [2]

   [Total: 3]

5. Figure 1 shows part of the apparatus for an experiment in which electrons pass through a thin slice of graphite (carbon atoms) and emerge to produce concentric rings on a fluorescent screen.

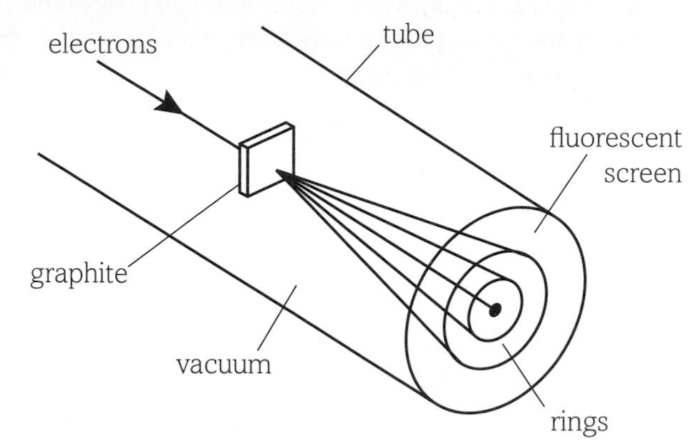

Figure 1

   (a) Explain how this experiment demonstrates the wave-nature of electrons. [3]
   (b) The beam of electrons in the apparatus shown in Figure 1 is produced by accelerating electrons through a potential difference of 1200 V.
   Show that the de Broglie wavelength of the electrons is 3.5×10^{-11} m. [3]

   [Total: 6]

   [Q26(b), H156/01 sample paper 2014]

6. Which of the following values is typical for the wavelength of visible light? [1]

   A 3×10^{-4} m
   B 5×10^{-7} m
   C 1×10^{3} m
   D 5×10^{-9} m
   E 5×10^{-10} m

   Use your answer to determine how many photons of visible light would be needed to produce an energy of 4.2 J. [4]

   [Total: 5]

7. Monochromatic light is shone onto a metal of work function 2.8 eV. Electrons just start to be released from the surface of the metal.
   (a) Explain the meaning of the following terms.
      (i) Monochromatic [1]
      (ii) Work function [1]
   (b) Calculate:
      (i) the work function of the metal in J [2]
      (ii) the threshold frequency of the metal [2]
      (iii) the maximum wavelength for which electrons can be released from the surface of the metal. [2]

[Total: 8]

8. (a) The Planck constant h can be measured in an experiment using light-emitting diodes (LEDs).

   Each LED used in the experiment emits monochromatic light. The wavelength λ of the emitted photons is determined during the manufacturing process and is provided by the manufacturer.

   When the p.d. across the LED reaches a specific minimum value V_{min}, the LED suddenly switches on emitting photons of light of wavelength λ.

   V_{min} and λ are related by the energy equation $eV_{min} = \frac{hc}{\lambda}$.

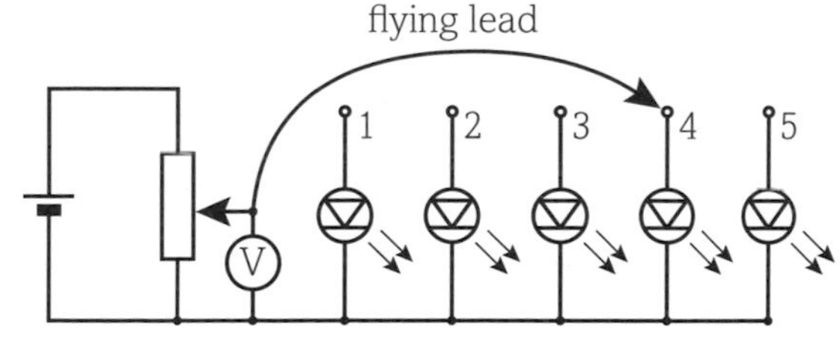

Figure 2

LED	Wavelength, λ/nm	V_{min}/V
1 red	627	1.98
2 yellow	590	2.10
3 green	546	2.27
4 blue	468	2.66
5 violet	411	3.02

Discuss how you could use the circuit of Figure 2 to determine accurate values for V_{min} and how data from the table can be used graphically to determine a value for the Planck constant. [6]

(b) A beam of ultraviolet light is incident on a clean metal surface. The graph Figure 3 shows how the maximum kinetic energy KE_{max} of the electrons ejected from the surface varies with the frequency f of the incident light.

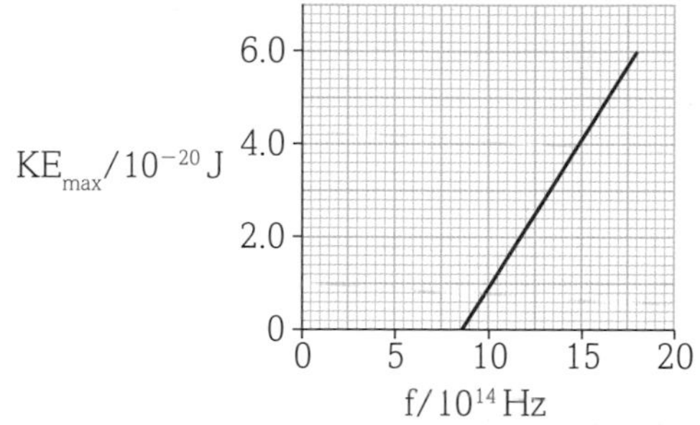

Figure 3

Explain how the graph shown in Figure 3 cannot be explained in terms of the wave-model for electromagnetic waves. [2]

Use data from Figure 3 to find a value of:

1. the Planck constant [2]
2. the threshold frequency of the metal [1]
3. the work function of the metal. [2]

[Total: 13]

[Q7, H156/02 sample paper 2014]

Maths skills

In order to be able to develop your skills, knowledge and understanding in Physics, you will need to have developed your mathematical skills in a number of key areas. This section gives more explanation and examples of some key mathematical concepts you need to understand. Further examples relevant to your AS/A level Physics studies are given throughout the book. In particular, Chapter 2.1 explores important ideas about units and estimation.

Arithmetic and numerical computation

Using standard form

Dealing with very large or small numbers can be difficult. To make them easier to handle, you can write them in the format $a \times 10^b$. This is called standard form.

To change a number from decimal form to standard form:

- Count the number of positions you need to move the decimal point by until it is directly to the right of the first number which is not zero.
- This number is the index number that tells you how many multiples of 10 you need. If the original number was a decimal, your index number must be negative.

Here are some examples:

Decimal notation	Standard form notation
0.000 000 012	1.2×10^{-8}
15	1.5×10^{1}
1000	1×10^{3}
3 700 000	3.7×10^{6}

EXAMPLE

Calculate the momentum of a bullet of mass 5.2 grams fired at a speed of 825 m s⁻¹ in kg m s⁻¹.

The momentum of this bullet would be:

$$p = m \times v$$
$$p = 5.2 \times 10^{-3} \times 825$$
$$p = 4290 \times 10^{-3}$$
$$p = 4.29\ \text{kg m s}^{-1}$$

Algebra

Changing the subject of an equation

It can be very helpful to rearrange an equation to express the variable that you are interested in in terms of the variables it is related to. Always remember that any operation that you apply to one side of the equation must also be applied to the other side.

EXAMPLE

Consider the following equation which relates v = velocity, u = initial velocity, a = acceleration and s = displacement:

$$v^2 = u^2 + 2as$$

If we wished to rearrange this equation to make u the subject, we would first subtract $2as$ from each side to obtain:

$$v^2 - 2as = u^2$$

Now to obtain the formula in terms of u, we have to 'undo' the square term by doing the opposite of squaring for each side. In essence, we have to take the square root of each side:

$$\sqrt{u^2} = \sqrt{v^2 - 2as}$$
$$u = \sqrt{v^2 - 2as}$$

Handling data

Using significant figures

Often when you do a calculation, your answer will have many more figures than you need. Using an appropriate number of significant figures will help you to interpret results in a meaningful way.

Remember the 'rules' for significant figures:

- The first significant figure is the first figure which is not zero.
- Digits 1–9 are always significant.
- Zeros which come after the first significant figure are significant unless the number has already been rounded.

Here are some examples:

Exact number	To one s.f.	To two s.f.s	To three s.f.s
45 678	50 000	46 000	45 700
45 000	50 000	45 000	45 000
0.002 755	0.003	0.002 8	0.002 76

Graphs

Understand that $y = mx + c$ represents a linear relationship

Two variables are in a linear relationship if they increase at a constant rate in relation to one another. If you plotted a graph with one variable on the x-axis and the other variable on the y-axis, you would get a straight line. Any linear relationship can be represented by the equation $y = mx + c$ where the gradient of the line is m and the value at which the line crosses the y-axis is c. An example of a linear relationship is the relationship between degrees Celsius and degrees Fahrenheit, which can be represented by the equation $F = \frac{9}{5}C + 32$ where C is temperature in degrees Celsius and F is temperature in degrees Fahrenheit.

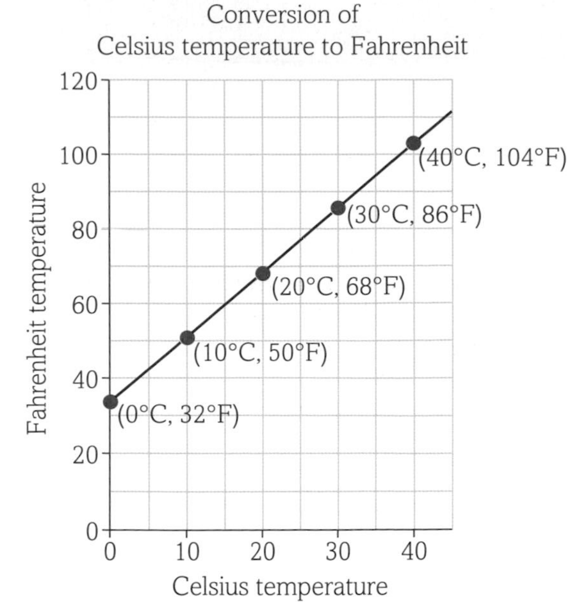

Draw and use the slope of a tangent to a curve as a measure of a rate of change

Sir Isaac Newton was fascinated by rates of change. He drew tangents to curves at various points to find the rates of change of graphs as part of his journey towards discovering the calculus – an amazing branch of mathematics. He argued that the gradient of a curve at a given point is exactly equal to the gradient of the tangent of a curve at that point.

Technique:

1. Use a ruler to draw a tangent to the curve.
2. Calculate the gradient of the tangent using the technique given for a linear relationship. This is equal to the gradient of the curve at the point of the tangent.
3. State the unit for your answer.

Geometry and trigonometry

Using sin, cos and tan with right-angled triangles

For a right-angled triangle, Pythagoras' Theorem can be used to find the length of a side given the length of two other sides. For problems that involve angles as well as side lengths, we can use the trigonometric ratios sine, cosine and tangent.

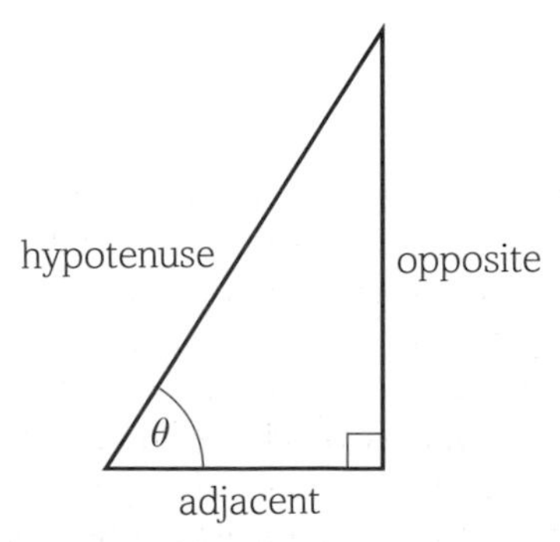

$$\sin\theta = \frac{\text{opposite}}{\text{hypotenuse}}$$

$$\cos\theta = \frac{\text{adjacent}}{\text{hypotenuse}}$$

$$\tan\theta = \frac{\text{opposite}}{\text{adjacent}}$$

To help you to remember these rules, you can learn a mnemonic for the letters 'SOH CAH TOA'. For example:

Some Old Horses

Can Always Hear

Their Owners Approaching

Finding the length of an unknown side

Follow this four-step procedure:

EXAMPLE

Calculate the length of side x.

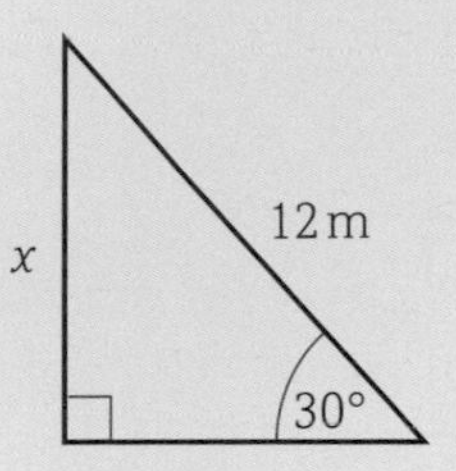

1. Summary of data

   First, write down all of the data involved:

   angle = 30°

   hypotenuse = 12 m

   opposite = x

2. Selection

   Next, select the appropriate trigonometric ratio using 'SOH CAH TOA'. It can be helpful to place a small tick above any variable involved. In this case we see:

   ✓✓ over SOH, ✓ over CAH, ✓ over TOA

   SOH CAH TOA

3. Substitution

   Write out the trigonometric ratio that you have chosen. You may need to rearrange it to express the value you are interested in.

   $$\sin A = \frac{\text{opposite}}{\text{hypotenuse}}$$

   opposite = sin A × hypotenuse

   opposite = sin 30 × 12 m

4. Calculate and check

   Calculate your answer using the required function on your scientific calculator. Then perform three checks: check that your answer is sensible, check whether it needs to be rounded and check whether you have used the correct units.

   opposite = 0.5 × 12 m = 6 m

Finding an unknown angle

Follow the same four-step procedure. The only difference is that you will need to use the inverse function on your calculator.

EXAMPLE

Calculate the length of side x.

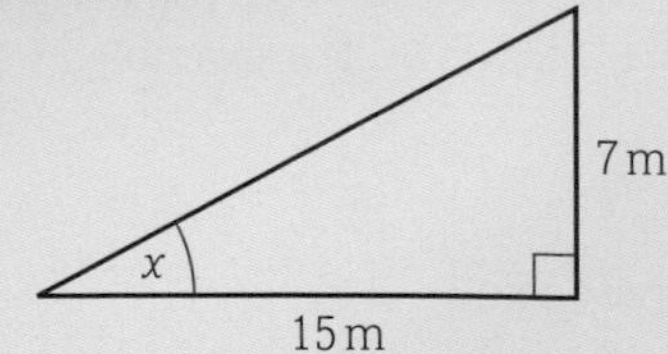

1. Summary of data

   opposite = 7 m

   adjacent = 15 m

   angle = x

2. Selection

   ✓ ✓ ✓✓
   SOH CAH TOA

   The correct ratio to use is tangent.

3. Substitution

   $$\tan\theta = \frac{\text{opposite}}{\text{adjacent}}$$

   $$\theta = \tan^{-1}\left(\frac{\text{opposite}}{\text{adjacent}}\right)$$

   $$\theta = \tan^{-1}\left(\frac{7}{5}\right)$$

4. Calculate and check

   $x = 25.016\,89$

   $x = 25.0°$ (1 decimal place)

Applying your skills

You will often find that you need to use more than one maths technique to answer a question. In this section, we will look at three example questions and consider which maths skills are required and how to apply them.

EXAMPLE

A tree stands in a field. The tree is 50 m north of the gate to the field and is 20 m east of the gate.

(a) Calculate the distance of the tree from the gate.

(b) Calculate the angle of the direction to the tree from north.

You recognise a right angle triangle. You need to use Pythagoras.

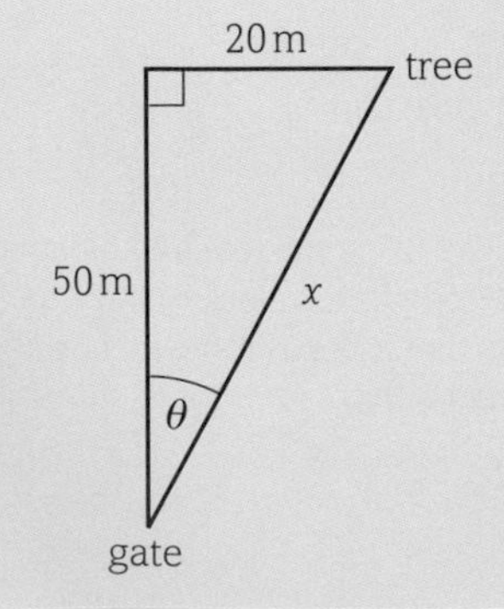

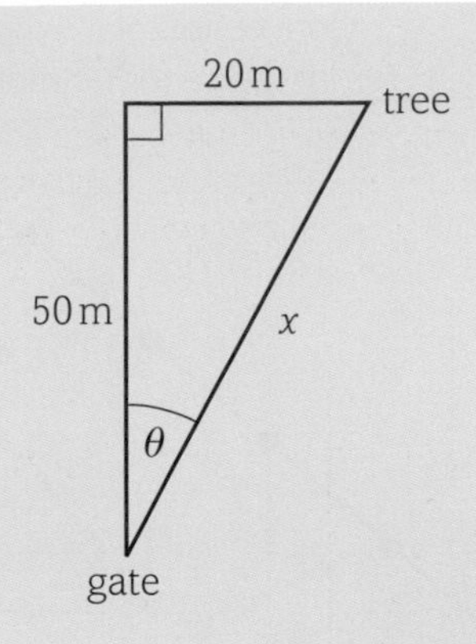

(a) $x^2 = 50^2 + 20^2$

$x^2 = 2500 + 400 = 2900$

$x = \sqrt{2900}$

$x = 54$ m

(b) $\tan\theta = \frac{20}{50}$

so $\theta = 22°$

EXAMPLE

A boat sails due east (090°) at 3 m s^{-1}. The tide comes from the south-east (135°) at 1 m s^{-1}.

Draw a vector diagram to show the movement of the boat over the ground and calculate the speed.

You need to draw a diagram that represents the information in the question. If you draw a scale diagram, choose a simple scale like 1 cm : 10 m.

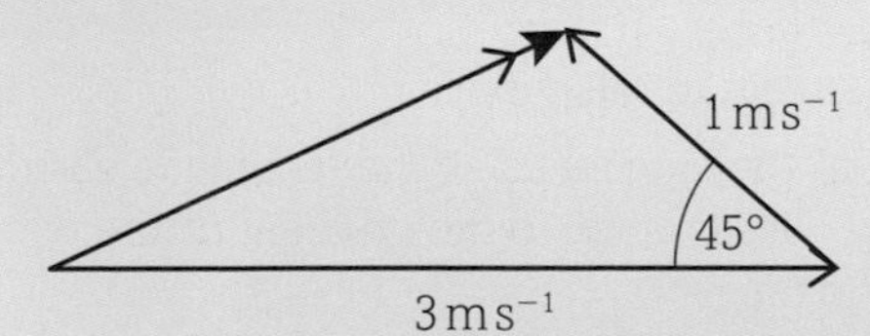

Now we must split both of the vectors up into two components, one going north and one going east.

North The boat speed vector 3 m s^{-1} has no component going north. The tide speed vector of 1 m s^{-1} has a component 1 m s^{-1} × sin 45 = **0.707 m s^{-1} going north.**

East The boat speed vector 3 m s^{-1} is going due east. The tide speed vector of 1 m s^{-1} has a component 1 m s^{-1} × cos 45 = −0.707 m s^{-1} going east; the minus sign means that the component is to the west and so we must subtract it since it reduces the boat speed to the east. So the **boat speed to the east** is (3 − 0.707) = 2.293 or **2.29 m s^{-1}** using 3 significant figures.

Now we must combine the north and east components to find the net speed. You recognise a right angle triangle. You need to use Pythagoras.

$(\text{speed})^2 = 0.707^2 + 2.29^2 = 0.500 + 5.24 = 5.74$

so speed $= \sqrt{5.74} = 2.40$ m s^{-1}.

EXAMPLE

A student measures the resistivity of some metal in the form of a wire. She starts by looking up the definition of resistivity which is $p = R \times A/l$ where R is the resistance of a length l of a wire of cross-sectional area A. She measures the resistance of different lengths of the wire and plots a graph of her readings.

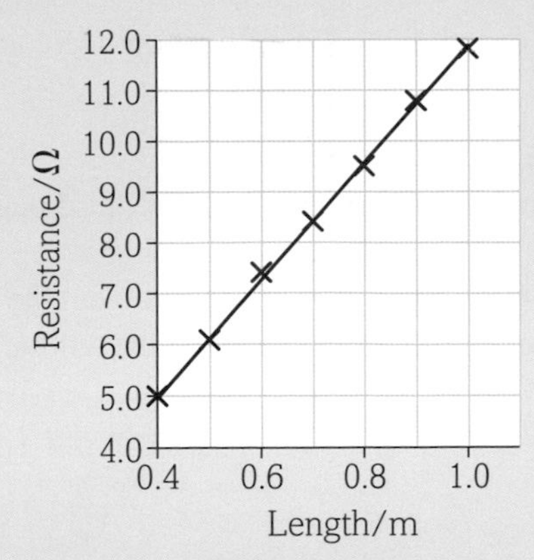

(a) Explain whether the readings should give a line of best fit that passes through the origin.
(b) She measures the diameter of the wire as 0.234 mm. Determine the cross-sectional area A.
(c) Calculate the resistivity.

(a) You need to recall the equation of a straight line $y = mx + c$ where the data is plotted on the x and y axes and m is the gradient and c is the intercept on the y-axis.
You must then compare this with the equation for resistivity where she has measured R and l.
Since $R = \frac{p \times l}{A}$ we can see that if R is plotted on the y-axis and l on the x-axis then the gradient of the line will be $\frac{p}{A}$ which is a constant so the line will be straight. There are no other terms in the resistivity equation so the value for c will be zero and the line should pass through the origin.
Note that the origin is not marked on this graph, it starts at (0.4, 4.0)

(b) She measures the diameter of the wire as 0.234 mm. Determine the cross-sectional area A.
You need to remember the formula for the area of a circle $A = \pi \times r^2$
Note you can only measure the diameter of a piece of wire so you must divide by 2 to get the radius r.

$$\text{So } A = \pi \times \left(\frac{(0.234 \times 10^{-3})}{2}\right)^2 = 4.30 \times 10^{-8}\,\text{m}^2$$

(c) Take measurements from the graph to show that the gradient m of the line of best fit is about $m = 11.7\,\Omega\,\text{m}^{-1}$ and therefore calculate a value for the resistivity of the metal of the wire.
Gradient = rise over run, so take measurements from the line of best fit – not the plots.

$$\text{gradient} = \frac{11.9 - 4.9}{1.0 - 0.4} = \frac{7}{0.6} = 11.7 = \frac{p}{A}$$

$$\text{so } p = 11.7\,\Omega\,\text{m}^{-1} \times 4.30 \times 10^{-8}\,\text{m}^2 = 5.03 \times 10^{-8}\,\Omega\,\text{m}$$

EXAMPLE

A ball is dropped from a height of 1.50 m above the ground.
(a) Show that it hits the ground at about 5.4 m s⁻¹.
(b) The ball is now thrown horizontally at a height of 1.5 m with a horizontal velocity of 20 m s⁻¹. Calculate the angle at which it hits the ground.

(a) You need a suvat equation. You know $u = 0\,\text{m s}^{-1}$, $g = 9.81\,\text{m s}^{-2}$ and $s = 1.50$ m. You need to find v.
The equation you need is $v^2 = u^2 + 2as$

$$\text{So } v^2 = 0 + 2 \times 9.81 \times 1.50 = 29.4 \text{ so } v = \sqrt{29.4} = 5.42\,\text{m s}^{-1}$$

(b) Remember for projectiles that the horizontal velocity remains constant. While the ball is falling it is travelling horizontally at a constant velocity of 20 m s^{-1}.
It hits the ground horizontally at 20 m s^{-1} and vertically at 5.42 m s^{-1}.

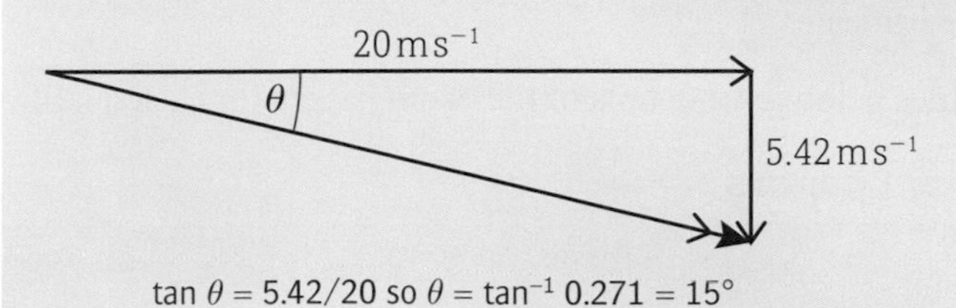

$$\tan\theta = 5.42/20 \text{ so } \theta = \tan^{-1} 0.271 = 15°$$

Preparing for your exams

Introduction

The way that you are assessed will depend on whether you are studying for the AS or the A level qualification. Here are some key differences:

- AS students will sit two exam papers, each covering 50% of the content of the AS specification.
- A level students will sit three exam papers, each covering content from both years of A level learning. The third paper will consist of synoptic questions that may draw on two or more different topics.
- A level students will also have their competency in key practical skills assessed by their teacher in order to gain the Practical endorsement in physics. The endorsement will not contribute to the overall grade but the result (pass or fail) will be recorded on the certificate.

The tables below give details of the exam papers for each qualification.

AS exam papers

Paper	Paper 1: Breadth in physics	Paper 2: Depth in physics
Topics covered	Modules 1–4	Modules 1–4
% of the AS qualification	50%	50%
Length of exam	1 hour 30 minutes	1 hour 30 minutes
Marks available	70 marks	70 marks
Question types	20 marks multiple-choice followed by 50 marks short answer structured response, problem solving and calculations	Short answer structured questions and extended writing including two level of response questions
Experimental methods?	Yes	Yes
Mathematics	A minimum of 40% of the marks across both papers will be awarded for mathematics at GCSE higher tier level or above	

A level exam papers

Paper	Paper 1: Modelling physics	Paper 2: Exploring physics	Paper 3: Unified physics	Practical endorsement in physics
Topics covered	Modules 1, 2, 3 and 5	Modules 1, 2, 4 and 6	Modules 1–6	Assessed by teacher throughout course. Does not count towards A level grade but result (pass or fail) will be reported on A level certificate. It is likely that you will need to maintain a separate record of practical activities carried out during the course
% of the A level qualification	37%	37%	26%	
Length of exam	2 hours 15 minutes	2 hours 15 minutes	1 hour 30 minutes	
Marks available	100 marks	100 marks	70 marks	
Question types	15 marks multiple choice followed by 85 marks short answer structured response and extended writing including two level of response questions	15 marks multiple choice followed by 85 marks short answer structured response and extended writing including two level of response questions	Short answer structured response and extended writing including two level of response questions	
Experimental methods?	Yes	Yes	Yes	
Mathematics	A minimum of 40% of the marks across all three papers will be awarded for mathematics at GCSE higher tier level or above			

Exam strategy

Arrive equipped

Make sure you have all of the correct equipment needed for your exam. As a minimum you should take:

- pen (black, ink or ball-point)
- pencil (HB)
- ruler (ideally 30 cm)
- rubber (make sure it's clean and doesn't smudge the pencil marks or rip the paper)
- calculator (scientific)

Ensure your answers can be read

Your handwriting does not have to be perfect but the examiner must be able to read it! When you're in a hurry it's easy to write key words that are difficult to decipher.

Plan your time

Note how many marks are available on the paper and how many minutes you have to complete it. This will give you an idea of how long to spend on each question. Be sure to leave some time at the end of the exam for checking answers. A rough guide of a minute a mark is a good start, but short answers and multiple choice questions may be quicker. Longer answers might require more time.

Understand the question

Always read the question carefully and spend a few moments working out what you are being asked to do. The command word used will give you an indication of what is required in your answer.

Be scientific and accurate, even when writing longer answers. Use the technical terms you've been taught.

Always show your working for any calculations. Marks may be available for individual steps, not just for the final answer. Also, even if you make a calculation error, you may be awarded marks for applying the correct technique.

Plan your answer

Questions marked with a * are level of response questions. The examiners will be looking for a line of reasoning in your response. Here, marks will be awarded for your ability to logically structure your answer showing how the points that you make are related or follow on from each other where appropriate. Read the question fully and carefully (at least twice!) before beginning your answer.

Make the most of graphs and diagrams

Diagrams and sketch graphs can earn marks – often more easily and quickly than written explanations – but they will only earn marks if they are carefully drawn and fully annotated. Spend time on a diagram, don't make it a quick sketch, use a ruler and make it large.

If you are asked to read a graph, pay attention to the labels and numbers on the x- and y-axes. Remember that each axis is a number line. You do not necessarily need the origin, your scale should spread out the plotted points across the paper.

If asked to draw or sketch a graph, always ensure you use a sensible scale and label both axes with quantities and units. If plotting a graph, use a pencil and draw small crosses (×) or dots with a circle around them (⊙) for the points.

Diagrams must always be neat, clear and fully labelled or annotated.

Check your answers

For open-response and extended writing questions, check the number of marks that are available. If three marks are available, have you made three distinct points? It can be helpful to number or bullet-point your response.

For calculations, read through each stage of your working. Substituting your final answer into the original question can be a simple way of checking that the final answer is correct. Another simple strategy is to consider whether the answer seems sensible. Pay particular attention to using the correct units.

Sample AS questions – multiple choice

To find the resistivity of a metal a student takes measurements of a piece of wire. She records the following values with the estimated uncertainty

$R = 10.2 \pm 0.2\,\Omega$

$d = 0.234 \pm 0.002$ mm

$l = 90.2 \pm 0.2$ cm

To calculate a value for the resistivity she uses the equation

$\rho = (\pi R d^2)/(4l)$.

What is the percentage uncertainty in her value?

A: 2% B: 3% C: 4% D: 5%

You are allowed to write on the exam paper or you can use scrap paper. Note that you do not need to calculate a value for resistivity. So you start by calculating the percentage uncertainty for each of the quantities.

for R %U $= 100 \times \frac{0.2}{10.2} = 2.0\%$

for d %U $= 100 \times \frac{0.002}{0.234} = 0.85\%$

for l %U $= 100 \times \frac{0.2}{90.2} = 0.22\%$

Since the %U values are so close to each other it is a good idea to use 2 significant figures in your calculations.

Question analysis

Multiple choice questions look easy until you try to answer them. Very often they require some working out and thinking.

In multiple choice questions you are given the correct answer along with three incorrect answers (called distractors). You need to select the correct answer and write the appropriate letter in the box provided.

If you change your mind, put a line through the box and write your new answer next to it.

Multiple choice questions always have one mark and the answer is given! For this reason students often make the mistake of thinking that they are the easiest questions on the paper. Unfortunately, this is not the case. These questions often require several answers to be worked out and error in one of them will lead to the wrong answer being selected. The three incorrect answers supplied (distractors) will feature the answers that students arrive at if they make typical or common errors. The trick is to answer the question before you look at any of the answers.

Sample student answer

B

This question actually tests your understanding of how uncertainties are combined. Answer B is achieved if you simply add together the percentages. The correct combination is to double the percentage uncertainty for d since it is squared in the equation for ρ.

This gives the sum $2\% + 2 \times 0.85\% + 0.22\% = 3.9\%$ so the correct response is C.

Even if the numbers don't work out exactly you should always select the answer closest to the one you calculated – if it is not very close to any of the answers given it is worth doing your calculation again.

Verdict

This is a weak answer because:

This candidate has probably made the mistake of adding together the percentage uncertainties without doubling the value for *d*.

If you have any time left at the end of the paper go back and check your answer to each part of a multiple choice question so that a slip like this does not cost you a mark.

Sample AS questions – short structured

(a) State what is meant by the *photoelectric effect.* [1]

(b) The photoelectric effect cannot be explained in terms of the wave model of electromagnetic waves. Discuss how the new knowledge of the photon model was used to explain the photoelectric effect and so to validate the photon model. [3]

You should learn definitions which are prompted by the command term 'state'.

The command term 'discuss' does not mean that you need to write a great deal, rather you should write bullet points to make an explanation that leads from one point to the next. In this case three bullet points is enough.

You do not need to give further description. This would only waste your limited time in the examination.

Question analysis

- Generally one piece of information is required for each mark given in the question. The first part is a definition for one mark and there are three marks for the second part of this question so make sure you make three distinct points.
- Clarity and brevity are the keys to success on short structured questions. For a definition you need to be able to use the right words in a short sentence. Making your points as bullet points makes your answer clearer.

Sample student answer

(a) Photoelectric effect is when electrons are emitted when light shines.

(b) The wave model cannot explain the threshold frequency but because $E = hf$ some photons have more energy than others and so some emit electrons and some don't.

Your answer for a 'state' question should be precise and use technical terms correctly. Here the answer is a bit vague.

There are only two valid points in (b) and although some linkage is indicated the second half does not do enough to explain the correct first part.

You should always start by describing briefly the phenomenon you are going to explain and then explain it clearly in terms of the correct physics.

Verdict

This is a weak answer because:

In (a) the student does not mention a surface nor the fact that it should be a metal, also the effect is caused by electromagnetic radiation of a wide range of frequencies, notably ultraviolet which is not just 'light'.

The student has identified the weakness in that the threshold frequency cannot be explained by the wave model but they then lose their way. The quantum equation is quoted but this is on the formula sheet so simply stating it gets no credit.

There is no mention of work function so why a more energetic photon is better is not explained. The key aspect of the photon model that one electron is emitted by one photon is also missing; this is an essential part of explaining the photoelectric effect and should be included.

Sample AS questions – extended writing

A star radiates energy produced from fusion reactions from within its core.

Explain what is meant by fusion and explain the conditions necessary for fusion to occur in the core of a star. [4]

You need to be careful with a question like this as there are two parts to it. You must be careful to answer both parts and not get distracted by your first answer and forget the second part.

Both the command words in this question are 'explain'. This means that you need to provide a reason for your statements, so here you must state the problem that the conditions you describe will overcome. Four marks are available so four distinct points should be made and it is a good idea to aim for two marks for each part.

Question analysis

- With any question worth four or more marks, think about your answer and the points that you need to make before you write anything down. It may be worth writing a few notes to help organise your thoughts – these could be written at the back of the examination paper.
- Keep your answer concise, and the information you write down relevant to the question. You will not gain marks for writing down physics that is not relevant to the question (even if correct) but it will cost you time.
- Remember that you can use bullet points or annotated diagrams in your answer. A well-placed diagram can be worth a lot of words.

Sample student answer

Fusion occurs when small particles hit each other and join together.

They must hit each other very hard because they repel each other because they have the same charge. They will only stick together (fuse) if they get very close. So it must be very hot.

At this level, your answers need technical terms and clarity in expression otherwise you will find yourself losing marks. It is also important to follow the rubric of the question. You must use A level words and ideas to explain yourself in an A level physics paper.

Verdict

This is a weak answer because:

- The student has not used A level ideas or terms – it is nuclei or protons that fuse, particles is not precise enough so although the first answer has the idea of joining it does not score a mark.
- The repulsion of like charges scores a mark but there is no indication why they will stick together.
- An A level physics student knows that the core of a star is hot, either an indication of how hot (~10^7 K) or why high temperatures are needed would be better.

To improve this answer

- The idea that the fusion products have less mass and that this is the source of the energy that is radiated would be worth a mark as it explains how it is that the star lose energy.
- The strong force is an idea that students should be familiar with and this should be used to explain why they fuse at all.
- A high temperature is needed so that the protons are travelling fast enough to get close enough

The density must be high too so that there is a good chance that the will collide at all

The answer is rather short but if it contained the key words and ideas it would score marks, the ideas are not really too difficult to understand but they must be used in a long answer.

Sample AS questions – extended writing level of response

The graph shows how the resistance of a thermistor varies with temperature.

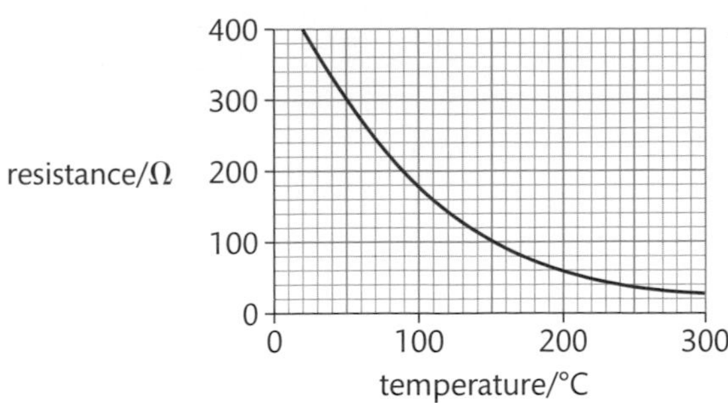

The circuit below shows a potential divider circuit which uses this thermistor. The circuit is designed to monitor the temperature of an oven in the range 200 °C to 300 °C.

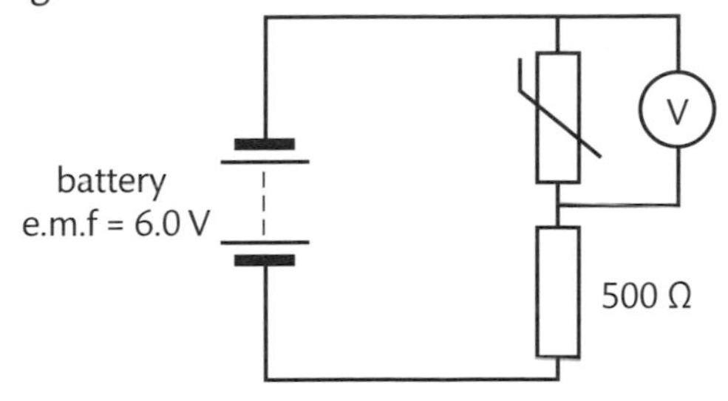

The voltmeter has very high resistance and has a full scale deflection (f.s.d.) of 6.0 V. The fixed resistor has a resistance of 500 Ω.

Explain how the circuit works and use calculations to discuss a significant limitation of this design. [6]

[Q17(c), H556/02 sample paper 2014]

Six marks are available so six points need to be made but there must be enough points for your answer to show clearly a line of reasoning. The command word is 'explain' so you must state what happens and then say how that is helpful in monitoring temperature.

You need to organise your answer so that it follows a logical line of reasoning and you can fit your calculations in to help the flow.

Question analysis

- There will usually be two questions in each examination which will test your ability to organise your response with a clear and logical structure and to follow a line of reasoning.
- These questions will be allocated 6 marks.
- Your response will be assessed for the level of organisation and whether the information presented is relevant.
  - For level 1 (1–2 marks) the information is basic and presented in an unstructured way.
  - For level 2 (3–4 marks) there is a line of reasoning and some structure is evident. The information is for the most part relevant and is supported by some evidence.
  - For level 3 (5–6 marks) there is a well-developed line of reasoning and a logical structure. The information is relevant and substantiated.
- It is vital to plan out your answer before you write it down. There is always space given on an exam paper to do this so just jot down the points that you want to make before you answer the question in the space provided. This will help to ensure that your answer is coherent and logical and that you don't end up contradicting yourself.

Sample student answer

As the temperature increases the resistance of the thermistor decreases and this changes the reading on the voltmeter.

The resistance of the thermistor is about 50 Ω so there will be 0.9 V reading on the voltmeter. This is quite a small voltage for this voltmeter.

The graph is quite flat at these temperatures so the voltmeter reading will not change very much and the resolution will be poor.

The answer is quite short but addresses the question. There is no linkage between the resistance varying and the change in voltmeter reading but in contrast there is a link between the shape of the graph and the resolution of the temperature scale. There is a calculation but it adds little to the line of reasoning since it is not specific to the temperatures mentioned and the 'small' reading on the voltmeter is not explained.

The answer is worth about 3 out of 6.

Verdict

This is an average answer because:

The student has kept to relevant detail but omitted some important linkage.

The student has organised the information so that a line of thought is fairly clear and so would score one 'reasoning' mark.

The answer should consider the effect of the changing resistance on the whole circuit and the consequent change in current. This links the temperature change to the voltmeter reading.

The question suggests that more than one calculation is needed and the graph has been used poorly. It is not clear how the value of 0.9 V is arrived at. It would be better if calculations had been done using the resistance readings from the graph at the two temperatures quoted.

The suggestion that the resolution is poor is a valid conclusion and explained from the shape of the graph, so two calculations would add strength to it.

Sample AS questions – practical

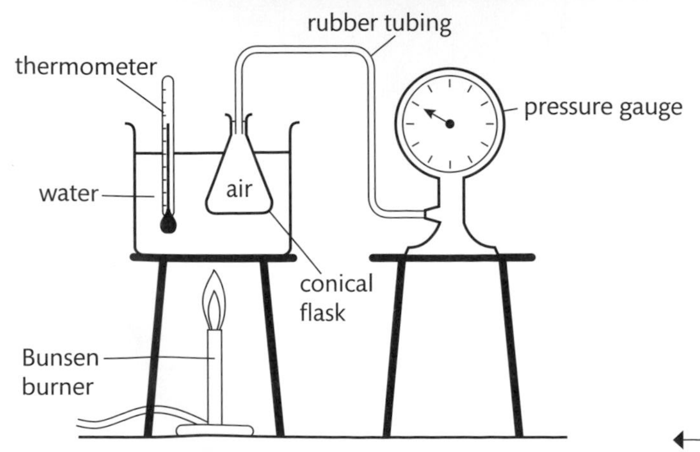

Describe how this apparatus can be set up and used to ensure accurate results. [4]

[Q20(a), H556/01 sample paper 2014]

Notice that the question says 'describe how...'. The command word 'describe' requires a suitable level of detail. Accurate means 'how close to the true value' so that must be the focus for the answer.

Question analysis

- There will be questions in your exams which assess your understanding of practical skills and draw on your experience of the core practicals. For these questions you might need to think about:
  - how the apparatus is set up
  - the method of how the apparatus is to be used
  - how readings are to be taken
  - how to control any variables
  - how you will get the answer required from the readings taken, using a graph
  - possible limitations and improvements.
- You will not be asked about every aspect, as in this question, but you should be able to answer questions as suggested by paragraph 1.1 in the specification. These can be set on any practical aspect of the work contained in the specification.
- Some questions on practical work will be worth 6 marks and you will be awarded marks for following a line of reasoning and linking the steps together, as was explained in the previous question.
- It is vital to plan out your answer before you write it down. There is always space given on an exam paper to do this so just jot down the points that you want to make before you answer the question in the space provided. This will help to ensure that your answer shows a line of reasoning and that you don't end up contradicting yourself.

Sample student answer

Heat the water steadily with the Bunsen burner and keep the thermometer close to the conical flask. Keep your eye level with the dial of the pressure gauge to avoid parallax and take a good number of readings.

Use ice to start at 0 °C and watch for it to boil at 100 °C.

A suitable level of detail means stating suitable techniques, notice that you are not being asked to explain these techniques so you should not write too much. Your detail should be specific and here there is not enough focus on accuracy.

Verdict

This is a weak answer because:

The student has given some detail that does not address accuracy. Keeping the thermometer close to the flask will mean that the temperature it reads is likely to be the same as the air but the temperature at the start and the finish are not so important.

A good number of readings is very important but at A level you are expected to state that there should be at least six for a graphical method and spaced over about 50 °C would be a sensible minimum.

The dial is fairly flat and parallax is often mentioned by candidates who can think of nothing else to say – it is usually not relevant, as here.

Heating the water steadily is not important because it is taking the readings that matters. When taking a reading the Bunsen should be removed and the water should be stirred. This helps to ensure the water all around the flask is at the same temperature and you have allowed enough time for the thermal energy to conduct through the glass – so that the temperature of the air is at the same temperature as the water.

Additionally, the whole flask should be submerged and the tubing should be thin since it contains air whose temperature is not changing.

Sample AS questions – calculation

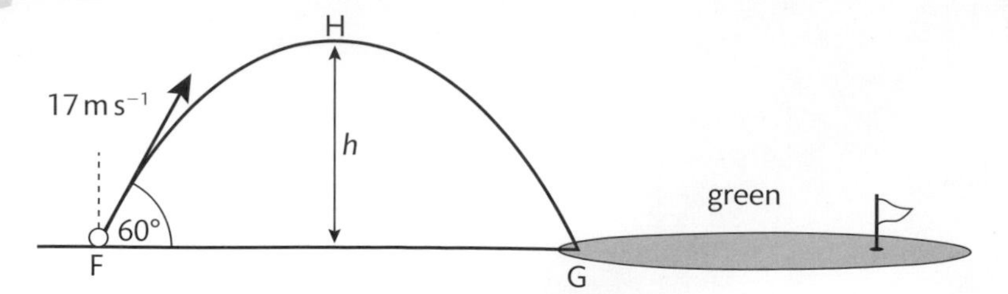

(a) Show that the speed of the ball at the highest point H of the trajectory is between 8 and 9 $m\,s^{-1}$.

speed = ... $m\,s^{-1}$ [2]

(b) At t = 1.5 s the ball reaches point H. Calculate

(i) the maximum height h of the ball

h = ... m [3]

(ii) the distance between F and G.

distance FG = ... m [2]

[Q2(a) and (b), H156/02 sample paper 2014]

The first command word here is 'show that' and in order to get both marks you must explain why your calculation is the correct one to perform. When the command term is 'calculate' you need to obtain a numerical answer to the question, showing relevant working. If the answer has a unit, this must be included.

Calculating the numerical answer requires you to use the appropriate numbers from the question, make sure that you get these right.

Question analysis

The important thing with calculations is that you must show your working clearly and fully, on the examination paper there will always be plenty of room to do that. The correct answer on the line will usually gain all the available marks. However, an incorrect answer can gain all but one of the available marks if your working is shown and is correct.

Show the calculation that you are performing at each stage and not just the result. When you have finished, look at your result and see if it is sensible.

Using 'quantity algebra' is a really good idea. To do this, every time you write down a number you write down the unit for that number.

Sample student answer

(a) $v = 17\,m\,s^{-1}\,(\cos 60°) = 8.5\,m\,s^{-1}$

(b) (i) The ball travels vertically for 1.5 s with an initial velocity upwards of $17\,m\,s^{-1}\,(\sin 60) = 14.7\,m\,s^{-1}$

So $h = ut + \frac{1}{2}at^2 = 14.7\,m\,s^{-1} \times 1.5\,s - \frac{1}{2} \times 9.81 \times (1.5)^2$
$= 22.05 - 11.04 = 11.0\,m$

(ii) $FG = 17\,m\,s^{-1}\,(\cos 60°) \times 1.5\,s = 8.5\,m\,s^{-1} \times 1.5\,s = 12.8\,m$

Verdict

This is a good answer because:

- (a) The speed at H is correct but they have ignored the vertical velocity; although it is zero here the candidate has not explicitly said so and loses a mark.
- (b) (i) NB easier is $s = (u + v)t/2 = (14.7\,\mathrm{m\,s^{-1}} + 0)1.5/2 = 11.0\,\mathrm{m}$ and $v^2 = u^2 + 2as$ is also valid
- If $g = 10\,\mathrm{m\,s^{-2}}$ is used then a mark is lost.
- The candidate makes the mistake of using 1.5 s for the time when this is the time to the top, they should have used $t = 3.0$ s, so they score the compensatory mark.
- Note how the use of 'quantity algebra' allows the correct unit to fall out.

Glossary

absolute uncertainty: a measurement showing how large the uncertainty actually is, and has the same units as the quantity being measured.
acceleration of free fall, *g*: the acceleration of a body falling under gravity. On Earth it has the value of 9.81 m s^{-2}.
acceleration, *a*: the rate of change of velocity, measured in metres per second squared (m s^{-2}); a vector quantity.
accuracy: the degree to which a value obtained by an experiment is close to the actual or true value.
ammeter: a device used to measure electric current, connected in series with the components.
ampere, A: S.I. unit for electric current, e.g. 4 A.
amplitude: the maximum displacement of a wave from its mean (or rest) position, measured in metres (m).
anomalous: values in a set of data that do not fit the overall trend and so are judged not to be part of the inherent variation.
antinode: the displacement of the particles in a stationary wave varies by the maximum amount.
Archimedes' principle: the upward buoyant force (upthrust) exerted on an object immersed in a fluid, whether fully or partially submerged, is equal to the weight of the fluid that the object displaces.
area, *A*: a physical quantity representing the size of part of a surface, measured in metres squared (m^2).
average speed: a measure of the total distance travelled in a unit time.
braking distance: the distance a vehicle travels while decelerating to a stop.
brittle: a material that breaks with little or no plastic deformation.
centre of gravity: the point at which the entire weight of an object can be considered to act.
centre of mass: the single point at which all of the mass of the object can be assumed to be situated. For a symmetrical body of constant density, this will be at the centre of the object.
charge: *see* **electric charge**.
closed system: any system in which all the energy transfers are accounted for; energy or matter cannot enter or leave.
coherence: two waves with a constant phase relationship.
components of a vector: the results from resolving a single vector into horizontal and vertical parts.
components: parts of electric circuits, including bulbs, LDRs, thermistors, etc.
compressive force: two or more forces that have the effect of reducing the volume of the object on which they are acting, or reducing the length of a spring.
conductor: a material with a high number density of conduction electrons and therefore a low resistance.
conservation of charge: physical law stating charge is conserved in all interactions; it cannot be created or destroyed.
conservation of energy: physical law stating that energy cannot be created or destroyed, just transformed from one form into another or transferred from one place to another. This is the situation in any closed system.
conventional current: a model used to describe the movement of charge in a circuit, from + to –.
coulomb, C: unit of electric charge, e.g. 1.6 × 10^{-19} C. 1 C = 1 A × 1 s.
couple: two forces that are equal and opposite to each other but not in the same straight line.
current: *see* **electric current**.
de Broglie equation: an equation expressing the wavelength of a particle as a ratio of Planck's constant and the particle's momentum, *mv*.
deformation: the change in shape or size of an object; if it returns to its original shape then the deformation is elastic.
degree Celsius: unit for temperature, e.g. 100 °C (not the S.I. unit; see kelvin).
density, *ρ*: defined as mass per unit volume (kg m^{-3}).
diffraction: when a wave spreads out after passing around an obstacle or through a gap.
displacement, *s* or *x*: the distance travelled in a particular direction, measured in metres (m), e.g. 3 m; a vector quantity.
displacement–time graph: a motion graph showing displacement against time for a given body.
distance, *d*: how far one position is from another, measured in metres (m), e.g. 12 m; a scalar quantity.
drag coefficient: a characteristic that determines the amount of drag that acts on an object.
drag: the resistive force that acts on a body when it moves through a fluid.
drift velocity: the average velocity of an electron as it travels through a wire due to a p.d.
ductile: can be drawn into wires and show plastic deformation under tensile stress before breaking.
efficiency: the ratio of useful output energy to total input energy.
elastic deformation: the object will return to its original shape when the deforming force is removed.
elastic limit: the point at which elastic deformation becomes plastic deformation.
elastic potential energy: the energy stored in a stretched or compressed object (for example a spring), measured in joules (J); a scalar quantity.
elasticity: the property of a body to resume its original shape or size once the deforming force or stress has been removed.
electric charge, *Q* or *q*: = current × time, measured in coulombs (C); a scalar quantity.
electric current, *I*: a flow of charge. An S.I. quantity, measure in amperes (A); a vector quantity.
electrolyte: a fluid that contains ions that are free to move and hence conduct electricity.
electromagnetic wave: a self-propagating transverse wave that does not require a medium to travel through.

electromotive force, e.m.f.: the energy gained per unit charge by charges passing through a supply, when a form of energy is transferred to electrical energy carried by the charges. It is measured in volts (V) or joules per coulomb ($J\,C^{-1}$).

electron diffraction: the process of diffracting an electron through a gap (usually between atoms in a crystal structure, for example graphite). An example of wave-particle duality.

electron flow: the movement of electrons (usually around a circuit), from – to +.

electron: negatively charged sub-atomic particle. Conduction electrons travel around circuits creating an electric current.

electronvolt: the kinetic energy gained by an electron when it is accelerated through a potential difference of 1 volt.

energy, *E*: the stored ability to do work, measured in joules (J); a scalar quantity.

equations of motion: the equations used to describe displacement, acceleration, initial velocity, final velocity and time when a body undergoes a constant acceleration.

equilibrium: when all the forces acting on an object in the same plane are balanced – there is zero net resultant force. In terms of motion, the object is either stationary or is travelling at constant velocity.

error bars: on a graph these represent the absolute uncertainty in measurements and can be plotted in the *x* and *y* directions.

extension, *x*: the change in length of an object when subjected to a tension, measured in metres (m).

fiducial mark: an object placed in the field of view for the observer to use as a point of reference.

first harmonic: in the fundamental mode of vibration, the length of the string is half the wavelength. This produces the lowest possible frequency called the first harmonic.

fluid: a material that can flow from one place to another (i.e. a liquid or a gas).

force constant, *k*: the constant of proportionality in Hooke's law, measured in newtons per metre ($N\,m^{-1}$).

force, *F*: a push or a pull on an object, measured in newtons (N); a vector quantity.

free fall: when an object is accelerating under gravity (i.e. at $9.81\,m\,s^{-2}$).

frequency, *f*: the number of oscillations per unit time, measured in hertz (Hz), e.g. 50 Hz.

fundamental frequency: the lowest frequency in a harmonic series where a stationary wave forms.

fundamental mode of vibration: where the length of a string is half the wavelength, producing the lowest possible frequency called the first harmonic. Other harmonics are whole number multiples of this frequency.

fuse: an electrical component designed to heat up, melt and break the circuit (hence stop the current) when a specified amount of electric current passes through it. Used as a safety device.

gamma rays: a form of electromagnetic wave with wavelengths between 10^{-16} m and 10^{-9} m. Used in cancer treatment.

gradient of a graph: the change in *y*-axis over the change in the *x*-axis (rise over step).

gravitational force: the force due a gravitational field acting on an object's mass.

gravitational potential energy, E_p: the energy stored in an object (the work an object can do) by virtue of the object being in a gravitational field, measured in joules (J); a scalar quantity.

hard: materials that resist plastic deformation by surface indentation or scratching.

harmonics: whole number multiples of the fundamental frequency of a stationary wave.

Hooke's law: the extension of an object is proportional to the force that causes it, provided that the elastic limit is not exceeded.

impulse: the product of a force *F* and the time Δt for which the force acts.

infrared: a form of electromagnetic wave with wavelengths between 7.4×10^{-7} and 10^{-3} m. Used in remote controls.

instantaneous speed: the speed of an object at a given moment in time.

insulator: a material with a small number density of conduction electrons and therefore a very high resistance.

intensity: (of a progressive wave) the rate at which energy is transferred from one location to another as the wave travels through space, perpendicular to the direction of wave travel. Intensity *I* is given by $I = P/A$, where *I* is the intensity in $W\,m^{-2}$, *P* is the power output of the source in W or $J\,s^{-1}$ and *A* is the area over which the radiation falls in m^2.

interference: the addition of two or more waves (superposition) that results in a new wave pattern.

internal resistance, *r*: (of a source of e.m.f.) the resistance to electric current of the materials inside (chemicals, wires or components). When current flows, energy is transferred to these materials, resulting in the terminal p.d. dropping.

***I–V* characteristic:** a graph to show how electric current through a component varies with potential difference across it.

joule: unit of energy (J), e.g. 1200 J. 1 J is the work done when a force of 1 N moves its point of application 1 m in the direction of the force.

kelvin: S.I. unit of temperature (K), e.g. 373 K.

kilowatt: unit of power (kW), e.g. 3.5 kW. 1 kW = 1000 W.

kilowatt-hour (kW h): 1000 watts for 3600 seconds. It is therefore 3 600 000 J.

kinetic energy: the work an object can do by virtue of its speed, measured in joules (J); A scalar quantity.

Kirchhoff's first law: the sum of the currents entering any junction is always equal to the sum of the currents leaving the junction (a form of conservation of charge).

Kirchhoff's second law: the sum of the e.m.f. is equal to the sum of the p.d. in a closed loop (a form of conservation of energy).

light dependent resistor, LDR: a component that changes its resistance with changes in the light intensity (dark = high resistance, light = low resistance).

light emitting diode, LED: a component that only allows electric current to pass through it in one direction and that emits light when a p.d. is applied across it.

longitudinal wave: a wave where the oscillations are parallel to the direction of wave propagation, e.g. sound.

lost volts: the difference between the e.m.f. and the terminal p.d. when charge flows in the cell.

malleable: can be hammered or beaten into flat sheets and show extensive plastic deformation when subjected to compressive forces.

Malus' law: a physical law describing the change in intensity of a transverse wave passing through a Polaroid analyser.

mass, *m*: S.I. quantity, measured in kilograms (kg), e.g. 70 kg; a scalar quantity.

microwaves: a form of electromagnetic wave with wavelengths between 10^{-4} and 10^{-1} m. Used in mobile phones.

moment of force: the product of a force and the perpendicular distance of its line of action from the point or axis. Also called **turning moment**.

monochromatic light: light waves with a single frequency (or wavelength).

newton: the force that causes a mass of one kilogram to have an acceleration of one metre per second every second.

nodes: points in a stationary wave at which there is no displacement of the particles at any time.

ohm: unit of resistance (Ω), e.g. 24 Ω. $1\,\Omega = 1\,V\,A^{-1}$.

Ohm's law: the current through a conductor is directly proportional to the potential difference across it, provided that physical conditions remain constant.

Order: the number of the pattern, *n*, on either side of the central maximum.

parallel circuit: a type of circuit where the components are connected in two or more branches and therefore provide more than one path for the electric current.

percentage difference: the difference between two values, divided by the average and shown as a percentage.

percentage uncertainty: the difference between a measured and a true value expressed as a percentage.

period, *T*: the time taken for one complete pattern of oscillation, measured in seconds (s).

phase difference, ϕ: the difference by which one wave leads or follows another. In-phase waves are in step with each other; in completely out-of-phase waves one wave is half a wavelength in front of the other. Measured in radians (rad).

photocell: a component that reduces its resistance when light shines on it due to photoelectric emission of electrons.

photoelectric effect: the emission of electrons from the surface of material when electromagnetic radiation is incident on the surface.

photon: a quantum associated with electromagnetic radiation.

Planck constant, *h*: has a value of 6.626×10^{-34} J s – photon energies are always emitted in multiples of this.

plane-polarised wave: a transverse wave oscillating in only one plane.

plastic deformation: the object will not return to its original shape when the deforming force is removed, it becomes permanently distorted.

polarisation: the process of turning an unpolarised wave into a plane-polarised wave (for example, light passing through a Polaroid filter).

polymeric: made of long chains of molecules called polymers.

polymeric material: a material made of many smaller molecules bonded together, often making tangled long chains. These materials often exhibit very large strains (e.g. 300%), for example rubber.

potential difference, p.d.: the measured across a component is the energy transferred per unit charge by the charges passing through the component. It is measured in volts (V) or joules per coulomb ($J\,C^{-1}$).

potential divider: a type of circuit that uses two resistors in series to split or divide the potential difference of the supply in a chosen ratio so that a chosen p.d. can be provided to another device or circuit.

potential divider: a type of circuit containing two components designed to divide up the p.d. in proportion to the resistances of the components.

potential energy: a form of stored energy (*see* gravitational potential energy, elastic potential energy and Topic 1.3.3).

power, *P*: the rate of doing work, measured in watts (W); a scalar quantity.

precision: the degree to which repeated values, collected under the same conditions in an experiment, show the same results.

pressure, *p*: force per unit area at right angles to (normal to) the area.

principle of conservation of energy: the total energy of a closed system remains constant. Energy can neither be created nor destroyed, only transferred from one form to another.

principle of conservation of momentum: the total momentum before a collision is always equal to the total momentum after the collision, provided that no external forces are involved.

principle of moments: for an object to be in rotational equilibrium, the sum of the clockwise moments must equal the sum of the anticlockwise moments.

progressive waves: waves that transfer energy away from a source.

quanta: the plural of quantum.

quantum: a small discrete unit of energy.

radian, rad: unit of angle or phase difference, e.g. 3π rad. One radian is the angle subtended at the centre of a circle by an arc of circumference that is equal in length to the radius of the circle. $2\pi = 360°$.

radio waves: a form of electromagnetic wave with wavelengths between 10^{-1} and 10^{4} m. Used in telecommunications.

random errors: give measurements that are scattered randomly above and below the true value when the measurement is repeated. A better result can be obtained by finding the mean value of the results of several readings.

reflection: when waves rebound from a barrier, changing direction but remaining in the same medium.

refraction: when waves change direction when they travel from one medium to another due to a difference in the wave speed in each medium.

resistance, *R*: a property of a component that regulates the electric current through it. Measured in ohms (Ω), e.g. 24 Ω.

resistivity, *ρ*: the ratio of the product of resistance and cross-sectional area of a component and its length (best defined by using the equation $\Omega = \frac{RA}{l}$).

resolution of vectors: splitting a vector into horizontal and vertical components (use to aid vector arithmetic).

resultant force: a single force which has the same effect as the sum of all the forces acting on a body.

resultant vector: the sum of the two vectors forms the third side of the triangle.
scalar: a quantity that has magnitude but not direction.
semiconductor: a material with a lower number of conduction electrons than a conductor and therefore a higher resistance.
series circuit: a type of circuit where the components are connected end to end and therefore provide only one path for the electric current.
significant figures: the number of digits in a measured or calculated quantity that have a meaning and about which we can be certain.
spectrum: a collection of waves with a range of frequencies, for example, visible spectrum and electromagnetic spectrum.
speed, *s*: the distance travelled per unit time, measured in metres per second ($m\,s^{-1}$), e.g. $12\,m\,s^{-1}$; a scalar quantity.
spring constant: force per unit extension.
standing wave: an alternative name for a stationary wave.
stationary wave: a wave formed by the interference of two waves travelling in opposite directions.
stiffness: the ability of a material to resist a tensile force.
stopping distance: the sum of the thinking distance and the braking distance (i.e. the total distance required to stop a vehicle from seeing the need to stop to vehicle becoming stationary).
strain: extension per unit length. It has no units and is therefore dimensionless. It is given the symbol ε.
stress: force per unit cross-sectional area. It has units of $N\,m^{-2}$ or Pa. It is given the symbol σ.
superposition: the principle that when two or more waves of the same type exist at the same place the resultant wave will be found by adding the displacements of the individual waves.
systematic error: an error that does not happen by chance but instead is introduced by an inaccuracy in the apparatus or its use by the person conducting the investigation.
tensile force: usually two equal and opposite forces acting on a wire in order to stretch it. When both forces have the value T, the tensile force is also T, not $2T$.
tensile stress: the tensile force per unit cross-sectional area.
terminal p.d.: the potential difference recorded across the terminals of a cell. The difference between the e.m.f. and the terminal p.d. when charge flows is called the 'lost volts'.
terminal velocity: the velocity at which an object's drag equals its accelerating force. Therefore there is no resultant force and zero acceleration.
thermistor: a component that changes its resistance depending on its temperature. An NTC thermistor's resistance reduces as the temperature increases.
thinking distance: the distance travelled from seeing the need to stop to applying the brakes.
threshold frequency: the lowest frequency of radiation that will result in the emission of electrons from a particular metal surface. For most metals, this frequency occurs in the ultraviolet region of the electromagnetic spectrum.
thrust: a type of force due to an engine.
time interval, *t*: S.I. quantity, measured in seconds (s), e.g. 60 s; a scalar quantity.
torque: the turning moment due to a couple is the product of one of the forces and the perpendicular distance between them. The units are N m.
transverse wave: a wave where the oscillations are perpendicular to the direction of wave propagation, e.g. water waves, electromagnetic waves, etc.
triangle of forces: if three forces acting at a point can be represented by the sides of a triangle, they are in equilibrium.
turning forces: forces that if unbalanced will cause a rotation.
ultimate tensile strength: the maximum stress a material can withstand while pulled or stretched before it fails or breaks.
ultimate tensile stress: the maximum stress that can be applied to an object before it breaks.
ultraviolet: a form of electromagnetic wave with wavelengths between 10^{-9} and 3.7×10^{-7} m. Causes sun tanning.
upthrust: a force on an object due to a difference in pressure when immersed in a fluid.
vector: a quantity has magnitude and direction.
vector triangle: a type of scale diagram with two vectors, drawn tiptotail, to show how they can be added together.
velocity, *v*: the displacement per unit time, measured in metres per second ($m\,s^{-1}$), e.g. $330\,m\,s^{-1}$; a vector quantity.
velocity–time graph: a motion graph showing velocity against time for a given body.
volt: unit of potential difference and e.m.f. (V), e.g. 230 V. $1\,V = 1\,J\,C^{-1}$.
voltmeter: device used to measure the p.d. across a component. It is connected in parallel across a component.
volume, *V*: a physical quantity representing how much 3D space an object occupies, measured in metres cubed (m^3).
watt: unit of power (W), e.g. 60 W. $1\,W = 1\,J\,s^{-1}$.
wave: a series of vibrations that transfer energy from one place to another.
wavelength, λ: the smallest distance between one point on a wave and the identical point on the next wave (e.g. the distance from one peak to the next peak), measured in metres (m).
wave-particle duality: the theory that all objects can exhibit both wave and particle properties.
weight, *w*: the gravitational force on a body, measured in newtons.
work done: or energy transferred is the product of the force and the distance moved by the force and the force in the direction of movement.
work function: the minimum energy required to release an electron from a metal's surface, overcoming the electrostatic attraction between the electron and the positive metal ions.
worst fit: the worst acceptable line, still passing through all the error bars. This will be either the steepest possible line of fit, or the least steep line of fit.
x-rays: a form of electromagnetic wave with wavelengths between 10^{-12} and 10^{-7} m. Used in X-ray photography.
Young modulus, *Y*: the ratio between stress and strain, measured in pascals (Pa).
Young's double slit experiment: an experiment demonstrating the wave nature of light via superposition and interference.
zero error: a type of systematic error caused when an instrument is not properly calibrated or adjusted, and so gives a non-zero value when the true value is zero.

Index